Edited by
Bruno Pignataro

New Strategies in Chemical Synthesis and Catalysis

Related Titles

Christmann, Mathias, Bräse, Stefan (eds.)

Asymmetric Synthesis

More Methods and Applications

2012

ISBN: 978-3-527-32921-2 (Hardcover)
ISBN: 978-3-527-32900-7 (Softcover)

Pignataro, Bruno (ed.)

Molecules at Work

Selfassembly, Nanomaterials, Molecular Machinery

2012

ISBN: 978-3-527-33093-5

Andersson, Pher (ed.)

Innovative Catalysis in Organic Synthesis

Oxidation, Hydrogenation,and C-X Bond Forming Reactions

2012

ISBN: 978-3-527-33097-3

Garcia-Martinez, Javier,
Serrano-Torregrosa, Elena (eds.)

The Chemical Element

Chemistry's Contribution to Our Global Future

2011

ISBN: 978-3-527-32880-2

Nicolaou, K. C. , Chen, Jason S.

Classics in Total Synthesis III

Further Targets, Strategies, Methods

2011

ISBN: 978-3-527-32958-8 (Hardcover)
ISBN: 978-3-527-32957-1 (Softcover)

Carreira, E. M., Kvaerno, L.

Classics in Stereoselective Synthesis

2009

ISBN: 978-3-527-29966-9

Edited by Bruno Pignataro

New Strategies in Chemical Synthesis and Catalysis

WILEY-VCH Verlag GmbH & Co. KGaA

The Editor

Prof. Bruno Pignataro
Università di Palermo
Dipartimento di Chimica
"S. Cannizzaro"
Viale delle Scienze ed. 17
90128 Palermo
Italy

Library of Congress Card No.: applied for

British Library Cataloguing-in-Publication Data
A catalogue record for this book is available from the British Library.

Bibliographic information published by the Deutsche Nationalbibliothek
The Deutsche Nationalbibliothek lists this publication in the Deutsche Nationalbibliografie; detailed bibliographic data are available on the Internet at <http://dnb.d-nb.de>.

Cover Design Adam-Design, Weinheim
Composition Laserwords Private Limited, Chennai, India
Printing and Binding Markono Print Media Pte Ltd, Singapore

Print ISBN: 978-3-527-33090-4
ePDF ISBN: 978-3-527-64585-5
oBook ISBN: 978-3-527-64582-4
ePub ISBN: 978-3-527-64584-8
Mobi ISBN: 978-3-527-64583-1

Contents

Preface

The international scenario of chemical research, alongside the development of the traditional areas of chemistry, sees a growing development of those areas of chemistry that are multi- and cross-disciplinary dealing more and more with recent challenges and opportunities in chemistry. The focus of interest is always our global society and its "healthy" and sustainable future.

The aim of the most advanced meetings in chemistry is often to have a better quality of life for all people and to showcase knowledge, advanced products, and services which improve the efficiency of chemical professionals, the local and global environment and our well-being. As shown in the last conference of the European Association for Chemical and Molecular Sciences – 3rd EuCheMs Chemistry Congress in Nürnberg (29 August to 2 September 2010) – chemistry is considered "a creative force" and scientists are convinced that it will shape the future.

In the road map for the development of chemistry you can find different trends (see, for instance, the document "Chemistry: Developing Solutions in a Changing World" produced by EuCheMs). One of these is related to the fact that the advancement in molecular design and its control becomes more complete. Chemists engineer their synthetic products more and more on the molecular scale exploiting and guiding in an increasingly controlled manner not only the strong bond but also the weak bonds ($\pi-\pi$ interaction, metal ligand coordination, hydrogen bonds, hydrophobic interactions, van der Waals interactions, etc.). Taking into account that Nature still has a lot of things to teach us for preparing useful chemical systems, in this trend one can find the effort to close the gap between synthetic and natural products. This even if we must here stress the fact that those systems chemists can create may have characteristics or properties that are present or not in nature!

The present synthetic efforts in this area of chemistry are directed at overcoming self-assembling and obtaining control over the kinetic instability of the covalent architectures going from self-assembling to the far-from-equilibrium self-organization. This is in order to have molecular superstructures with particular well-defined conformations and therefore functions.

In connection with the advancements in understanding the phenomena and behaviors at molecular level, another trend in the development of chemical science sees an increasing tendency to look, in a more and more different manner, to the properties and reactions of the chemical systems also in order to throw further bridges between Chemistry and other disciplines such as Molecular Biology, Electronics, or Material Science. The power of synthetic methods directed at obtaining new functional nanosystems is now well documented and the products of such synthetic efforts embrace a large spectrum of sophisticated applications such as gene transfection, catalysis, lithium storage or sensors, and, in general, materials science and technology for a variety of applications.

In addition, learning also from the behavior of green plants, a research line is developing molecular photovoltaic devices having power conversion efficiencies of the order of 10%. This brings us closer to identifying "environmentally friendly" solutions for the world energy problem.

Another very important international trend follows from the fact that new discoveries and technological advancements improve our capacity to obtain better and better spatial, temporal, and energy resolutions. This is for one of these quantities alone or for these quantities in combination. In various fields, we are close to achieving physical limits. One astonishing recent achievement, which exploits these improved capacities, is, for instance, that reported by Paul Corkum, who launched at Ottawa the attosecond science. These researches showed that we can measure electronic orbitals and we might film the orbital modification during a photochemical reaction. This area of research then passes from femtochemistry led by Ahamed Zewail, allowing for the production of movies of the rupture and formation of chemical bonds, to this type of measures where theory and experiment are more and more interwoven. In addition, today we have the ability to measure smaller and smaller weights or other physical quantities such as picojoules, piconewtons, fractions of nanometer, femtograms, femtoamperes, kilodaltons, and so on with always increasing facility and reliability.

Jumping from the nanoscopic to the macroscopic world and to complex systems, the number of data that computers are able to manage continues to increase in a dramatic way. The impact of computational methods has become extraordinarily important in the development of science and technology. Simulations that were unthinkable a few years ago are now possible and allow us to start thinking about extremely complex predictions.

Just to finish this survey, in the frontier area with Life Sciences the challenges that chemists had sought a few years ago seem less and less ambitious and it appears more and more clear that chemistry plays an essential role in understanding life itself.

This book is placed in this international scenario. In particular, it represents one of the two books comprising contributions of selected scientists from the last edition of the European Young Chemist Award (EYCA 2010) presented during the 3rd EuCheMS Chemistry Congress. It is aimed to cover the generic area of chemical synthesis and catalysis while the other book encloses contributions from the area of material chemistry and is entitled "Molecules at work: SelfAssembly, NanoMaterials, and Molecular Machinery."

As for the EYCA 2010, it was the third time this Award has been given. The aim of EYCA is to showcase and recognize the excellent research being carried out by young scientists working in the chemical sciences. In particular, it is intended to honor and encourage younger chemists whose current research displays a high level of excellence and distinction. It seeks to recognize and reward younger chemists (less than 35 years old) of exceptional ability who show promise for substantial future achievements in chemistry-related research fields.

The applications presented by the best candidates during the two previous editions of the Award were so stimulating that together with Wiley, EuCheMs, SCI, RSC, and GDCh, I decided to collect them into books. Thus, from the first edition of the Award was published the book *Tomorrow's Chemistry today: Concept in Nanoscience, Organic Materials, and Environmental Chemistry* (Wiley 2009) and from the second edition of the Award the three books (Wiley 2010) entitled *Ideas in Chemistry and Molecular Sciences: Advances in Synthetic Chemistry*, *Ideas in Chemistry and Molecular Sciences: Where Chemistry meets Life*, and *Ideas in Chemistry and Molecular Sciences: Advances in Nanotechnology, Materials, and Devices.*

The work from this third edition of the Award was once more very stimulating and again pushed by Wiley, EuCheMs, SCI, RSC, and GDCh, I planned to collect the best contributions into two books.

The scientific standing of the award applicants was undoubtedly very high and their research achievements are remarkable, especially in relation to their young age. In the guest editorial published by me "Chemistry: A European Journal" (vol. 16 (2010), pp. 13888–13893), I reported many details of the quality of the participants and of the whole Award Competition.

However, let me stress here some of points shown there. About 45% of the applicants have been chosen to give an oral presentation to the Nürnberg Congress. Among the participants one can find candidates with about 60 papers in peer-reviewed international journals and guiding a group of more than 20 PhDs and Post Docs, or candidates whose works got more than 1500 citations. The publication lists of most applicants proudly included the appearance of their work in the leading general science/chemistry journals such as *Science, Nature, Angewandte Chemie, Journal of the American Chemical Society*, and so on or the best niche journals in the fields of organic, inorganic, organometallic, physical, analytical, environmental, and medicinal chemistry. Several participants have been granted different prizes, have been invited to give different lectures, and achieved further recognitions such as front-end covers, hot articles, or highlights in top journals. Moreover, reading the application documents it comes out clearly that many of the competitors have different scientific interests and do have very exciting ideas for their future work. Further support for the applications, and a testament to the very high quality of the competitors, was apparent from the comments contained in the often very effusive recommendation letters coming from a number of eminent scientists. A flavor of these from the applications received can be found in the above cited guest editorial published by "Chemistry: A European Journal."

This is the pool from which I fished the contributors of the above two books, and of this book in particular.

In fact, this book gives an account of the most recent results of research in Organic, Inorganic, and Organometallic Synthesis as well as in Catalysis, based on a selection of work by leading young scientists. The authors provide the state of the art in their field of research and a perspective or preview of the future research directions.

The content covers some of the aspects of the international chemical research highlighted above. The book is divided into three parts dedicated to the Synthetic Methods, Catalysis, and Combinatorial and Chemical Biology, even if in some cases the content of one part may overlap that of another.

Part I begins with two new synthetic strategies.

The first deals with high-resolution time-of-flight mass spectrometry (TOF-MS), which, over the last decade, has been employed in an effort to throw light upon assembly–disassembly processes of polyoxometalate (POM) and coordination clusters and to identify novel reactive and intermediate species in reaction mixtures as well. The most recent developments of this type of application of mass spectrometry are discussed, showing that this strategy contributes to opening the door for further exploration, discoveries, and well-established designing methodologies toward materials with predefined functionalities.

The second synthetic strategy deals with automated synthesizers explaining how these may help synthetic chemists to prepare natural products. In particular, it is shown that the use of such systems eliminates wastage of time with the trivial, repetitive, and long steps of the traditional synthetic procedure and thereby gives the opportunity to expend more time on the advanced and challenging work of developing new synthetic routes for the total synthesis of natural products.

Ionic ozonides and amidoureas are the themes of two other chapters of Part I. In particular, Chapter 4 describes part of the investigations on ionic ozonides that have been performed over the last decades, focusing on the synthesis and structure–property relationships. In agreement with the authors, I hope that the chapter will motivate the reader to delve into this fascinating field of molecular inorganic chemistry.

Chapter 5, dealing with the very interesting field of the chemistry of amidinoureas, is entitled "Chemistry and biological properties of amidinoureas: strategies for the synthesis of original bioactive hit compounds" and concludes that "the race for amidinoureas has just started and their full potential has still to be discovered." There is again a clear invitation to the reader to shift some of his/her interest to this field.

Chapter 3 reports on the catalytic chemoselective reduction of amides and imides to the corresponding amines by using the hydrosilylation strategy and is therefore, in some respect, a bridge to the contributions reported in Part II.

Part II is specifically dedicated to catalysis. Again, the themes treated are very different and cover a wide area of the world of catalysis.

In particular, Chapter 8 details a contribution in the huge field of organocatalysis in general and chiral thioureas in particular. Different thiourea-catalyzed processes have been shown. Some of them are original and pioneering in this area and became a framework for further reactions to be explored.

Moving on to a different aspect of catalytic applications, Chapter 9 shows that electrosynthesis has several advantages for the coating of metallic supports. This study demonstrated the feasibility of the electrosynthesis of HT compounds (lamellar compounds with chemical formula $[M_2 + 1 - xM_3 + x(OH)_2](An - x/n)_n H_2O)$ to coat metallic foams leading to structured catalysts. The growth and composition of the coating are controlled by varying the deposition time, applied potential, and bath composition.

The overview of the state-of-the-art approach to the microkinetic analysis of complex chemical processes is provided in Chapter 10. At first, a general overview of the hierarchical multiscale approach for the microkinetic modeling and analysis of complex chemical processes at surfaces is given. Then, the results on the microkinetic modeling and analysis of the partial oxidation of methane on Rh catalysts are reviewed.

The focus of Chapter 11 is to review gold-catalyzed redox processes and their synthetic impact. Due to the fact that the field is rather wide, the chapter is just focused on homogeneous catalysis, skipping the heterogeneous methodologies. In agreement with the authors I hope that "even if key contributions have been left out, those that are included give the reader an overview of the targets already achieved as well as the new trends and challenges still ahead in the fascinating area of gold-catalyzed redox processes."

Transition Metal Complexes in Supported Liquid Phase (SLF) is the topic of Chapter 12. Dispersion of a concentrated catalyst solution on the surface of a porous substrates has the respective advantages of liquid- and solid-phase immobilization. In agreement with the authors it is possible to say that the examples summarized in this chapter shows that SLPs represent a versatile and successful approach to organometallic catalyst recovery and recycling. This is particularly interesting with the use of nanoscale or molecular catalysts under continuous-flow.

The last two chapters collected in this section are at least in some respect connected to the chemical biology and build therefore a bridge to the third part of this book.

The aim of Chapter 7 is to review the development of bioinspired iron catalysts for unactivated $Csp_3E{-}H$ oxidation over the last two decades, and to investigate the important points for the design of efficient and selective catalysts.

Chapter 6 deals with DNA catalysts for synthetic applications in biomolecular chemistry. The focus is on deoxyribozymes as only one partial feature of DNA's applications in current chemical research. In the chapter, a brief overview of the scope of reactions catalyzed by deoxyribozymes is followed by selected examples of DNA catalysts for the preparation of biomolecules that are challenging to be approached by other methods. The major emphasis is on DNA enzymes that enable the ligation of nucleic acid fragments in different topologies.

Part III is dedicated to Combinatorial and Chemical Biology. There are three chapters in this part.

Chapter 13 is just at the interface of chemistry and biology and deals with inhibiting pathogenic protein aggregation.

Chapter 14 entitled "Synthesis and Application of Macrocycles Using Dynamic Combinatorial Chemistry" reports on the ion transport across membranes mediated by a dynamic combinatorial library.

The book closes with a review illustrating representative examples of solid-phase-generated libraries including their evaluation against various biological targets in Chapter 15. As pointed out by the authors, the latter are mainly members of the G-protein coupled receptors family, to which about half of all new drugs belong. The selected syntheses are organized by their respective privileged substructures. An overview of different reaction conditions and linkers used in solid-phase chemistry is also given.

I feel that the book is of major interest to chemists in the organic, inorganic, and organometallic areas but also to material chemists.

It should be relevant for readers both from academia and industry since it deals with fundamental contributions and possible applications.

As I have done for the other books of this series, I cannot finish this preface without acknowledging all the authors and all the persons who helped and supported me in this project. In particular, I would like to thank Prof. Giovanni Natile, Prof. Francesco De Angelis, and Prof. Luigi Campanella, who, as Presidents of the Italian Chemical Society and/or EuCheMs representatives, strongly encouraged me during the years in this activity. And of course, I thank all those Societies (see the book cover) that motivated and supported the book.

Palermo
January 2012

Bruno Pignataro

List of Contributors

Francesco Basile
ALMA MATER
STUDIORUM-Università di
Bologna
Dip. Chimica Industriale e dei
Materiali
Viale Risorgimento 4
40136 Bologna
Italy

Patricia Benito
ALMA MATER
STUDIORUM-Università di
Bologna
Dip. Chimica Industriale e dei
Materiali
Viale Risorgimento 4
40136 Bologna
Italy

Stefan Bräse
Institute of Organic Chemistry
Karlsruhe Institute of Technology
(KIT)
Fritz-Haber-Weg 6
D-76131 Karlsruhe
Germany

Daniele Castagnolo
University of Siena
Dipartimento Farmaco Chimico
Tecnologico
Via Alcide de Gasperi 2
53100 Siena
Italy

Yi-Pin Chang
University of Oxford
Department of Chemistry
Mansfield Road
Oxford OX1 3TA
UK

Tamilselvi Chinnusamy
Institut für Technische und
Makromolekulare Chemie
RWTH Aachen University
Worringerweg 1
52074 Aachen
Germany

Leroy Cronin
The University of Glasgow
School of Chemistry
University Avenue
Glasgow G12 8QQ
UK

Shoubhik Das
Leibniz-Institut für Katalyse e.V.
an der Universität Rostock
Albert-Einstein-Str. 29a
18059 Rostock
Germany

Giuseppe Fornasari
ALMA MATER
STUDIORUM-Università di
Bologna
Dip. Chimica Industriale e dei
Materiali
Viale Risorgimento 4
40136 Bologna
Italy

Shinichiro Fuse
Tokyo Institute of Technology
Department of Applied Chemistry
2-12-1-H-101 Ookayama
Meguro-ku
Tokyo 152-8552
Japan

Laura Gómez
Universitat de Girona
Departament de Química
Campus de Montilivi
17071 Girona
Spain

Teresa de Haro
University of Zürich
Organic Chemistry Institute
Winterthurerstrasse 190
CH-8057 Zürich
Switzerland

Raquel P. Herrera
ARAID, Fundación Aragón I+D
E-50004 Zaragoza
Spain

and

CSIC-Universidad de Zaragoza
Departamento de Química
Orgánica
Instituto de Síntesis Química y
Catílisis Homogénea (ISQCH)
C/Pedro Cerbuna
12 E-50009 Zaragoza
Spain

Ulrich Hintermair
Institut für Technische und
Makromolekulare Chemie
RWTH Aachen University
Worringerweg 1
52074 Aachen
Germany

Claudia Höbartner
Research Group Nucleic Acid
Chemistry
Max Planck Institute for
Biophysical Chemistry
Am Fassberg 11
D-37077 Göttingen
Germany

Martin Jansen
Max Planck Institute for Solid
State Research
Inorganic Solid State Chemistry
Heisenbergstr. 1
70569 Stuttgart
Germany

Dagmar C. Kapeller
Institute of Organic Chemistry
Karlsruhe Institute of Technology (KIT)
Hermann-von-Helmholtz-Platz 1
D-76344
Eggenstein-Leopoldshafen
Germany

Walter Leitner
Institut für Technische und Makromolekulare Chemie
RWTH Aachen University
Worringerweg 1
52074 Aachen
Germany

Matteo Maestri
Politecnico di Milano
Dipartimento di Energia
Laboratory of Catalysis and Catalytic Processes
Piazza Leonardo da Vinci 32
20133 Milano
Italy

Eugenia Marqués-López
CSIC-Universidad de Zaragoza
Departamento de Química Orgánica
Instituto de Síntesis Química y Catálisis Homogénea (ISQCH)
C/Pedro Cerbuna
12 E-50009 Zaragoza
Spain

Haralampos N. Miras
The University of Glasgow
School of Chemistry
University Avenue
Glasgow G12 8QQ
UK

Kazuhiro Machida
ChemGenesis Incorporated
Development Department
4-10-2 Nihonbashi-honcho
Chuo-ku
Tokyo 103-0023
Japan

Marco Monti
ALMA MATER STUDIORUM-Università di Bologna
Dip. Chimica Fisica e Inorganica
Viale Risorgimento 4
40136 Bologna
Italy

Cristina Nevado
University of Zürich
Organic Chemistry Institute
Winterthurerstrasse 190
CH-8057 Zürich
Switzerland

Hanne Nuss
Max Planck Institute for Solid State Research
Inorganic Solid State Chemistry
Heisenbergstr. 1
70569 Stuttgart
Germany

P. I. Pradeepkumar
Indian Institute of Technology Bombay
Department of Chemistry
Powai
Mumbai 400076
India

Vittorio Saggiomo
Centre for Systems Chemistry
Stratingh Institute
Faculty of Mathematics and
Natural Sciences
University of Groningen
Nijenborgh 4
9747 AG Groningen
The Netherlands

Erika Scavetta
ALMA MATER
STUDIORUM-Università di
Bologna
Dip. Chimica Fisica e Inorganica
Viale Risorgimento 4
40136 Bologna
Italy

Domenica Tonelli
ALMA MATER
STUDIORUM-Università di
Bologna
Dip. Chimica Fisica e Inorganica
Viale Risorgimento 4
40136, Bologna
Italy

Takashi Takahashi
Tokyo Institute of Technology
Department of Applied Chemistry
2-12-1-H-101 Ookayama
Meguro-ku
Tokyo 152-8552
Japan

Angelo Vaccari
ALMA MATER
STUDIORUM-Università di
Bologna
Dip. Chimica Industriale e dei
Materiali
Viale Risorgimento 4
40136 Bologna
Italy

Part I
Synthetic Methods

1
Electrospray and Cryospray Mass Spectrometry: From Serendipity to Designed Synthesis of Supramolecular Coordination and Polyoxometalate Clusters

Haralampos N. Miras and Leroy Cronin

1.1
Introduction

Molecular self-assembly is an exciting occurrence which governs how simple building blocks [1] can be organized spontaneously into complex architectures [2]. Such self-assembly processes are highly dependent upon the experimental conditions [3] often to such a degree that total control is never easily achieved [4]. This can be frustrating since extremely small changes in reaction conditions can yield totally different results [5]. For instance, many research groups have reported the discovery of new building blocks, architectures, and materials that exhibit fundamentally new and interesting properties as a result of the self-assembly process that were simply not expected [6]. The serendipitous result in self-assembled chemical systems usually lacks the element of design. On the other hand, little progress would be made if serendipity (i.e., chance discoveries) was the only guide [7]. Future work usually uses valuable information that is extracted by observations made from earlier studies and used as a starting point for a more designed approach. Representative examples that fall into the category of self-assembled chemical systems are the polynuclear coordination compounds and the polyoxometalate (POM) clusters.

POM clusters represent an unparalleled range of architectures and chemical properties, acting as a set of transferable building blocks that can be reliably utilized in the formation of new materials. These key features are being exploited rapidly today after a rise in popularity of POMs over the last two decades [8, 9]. Today, POM chemistry is an important, emerging area that promises to allow the development of sophisticated molecule-based materials and devices with numerous applications ranging from electronics and catalysis to physics [10–13]. The reason for the explosion in the number of structurally characterized POM compounds is due to developments in instrumentation and novel synthetic approaches. In terms of technique development, fast and routine single-crystal data collection has allowed the area to accelerate to the point where the bottle neck has moved to structure refinement and crystallization of new compounds rather than the time taken for data collection and initial structure solution. However, despite all the promise, the

New Strategies in Chemical Synthesis and Catalysis, First Edition. Edited by Bruno Pignataro.

relentless increase in the number of structures and derivatives mean that it can be difficult to distinguish between the different cluster types and subtypes, whereas till now it seemed hard to be sure that the processes that govern the self-assembly of these complex architectural systems can be fully controlled and directed.

In a similar manner, supramolecular chemistry has been described by researchers as an information science in which libraries of building blocks that contain the necessary information self-assemble into larger architectures [14]. Complementary and precise interaction of individual species with the appropriate symmetry, geometry, and chemical information directs the product assembly. Also in this case, there is a long list of variables that triggers and directs the whole process by creating an experimental environment in which the "fittest" species prevail and evolve into the final product following the most energetically favorable route. Consequently, self-assembly has been recognized as a powerful tool for the construction of supramolecular scaffolds, as demonstrated by a plethora of exciting contributions [15].

The development of both classes of materials and exploitation of their potential require the ability to design novel synthetic procedures and the capacity to use results obtained by serendipity constructively in order to maximize the desirable outcome and the understanding of the selection rules that govern these chemical systems. In both the aforementioned cases, it is a great challenge and extremely important to understand and control the self-assembly. During this effort, the conventional spectroscopic techniques present significant drawbacks. For example, nuclear magnetic resonance (NMR) is of limited use when the symmetry of the assembled architecture is high [16], when the structures are labile or paramagnetic, and when the nuclei have poor receptivity. Additionally, the reaction mixtures are far too complicated to extract any useful information regarding the nature and availability of the building-block libraries, the selection rules that trigger their organization in a controlled fashion, as well as the preferable pathway that is selected for the formation of the polynuclear clusters.

Therefore, given the enormous challenge in understanding and controling the self-assembly process for a range of self-assembled, cluster-based architectures, high-resolution time-of-flight mass spectrometry (TOF-MS) has been employed over the last decade in an effort to shine light upon the assembly–disassembly process of high-nuclearity POMs and coordination clusters as well to identify novel reactive and intermediate species in reaction mixtures. Herein, we discuss the most recent developments and how electrospray ionization mass spectrometry (ESI-MS), cryospray ionization mass spectroscopy (CSI-MS), and variable temperature mass spectrometry (VT-MS) with a high-resolution TOF detector could offer a novel approach in overcoming the present difficulties, revealing important mechanistic information and allowing the identification of the building-block libraries present in solution under specific experimental conditions. Furthermore, this will allow us to take real control over the self-assembly processes and open the door for further exploration, discoveries, and, ultimately, well-established designing methodologies toward materials with predefined functionalities.

1.2 Background to ESI-MS

ESI-MS has been used extensively to investigate many types of POMs including vanadates [17], niobates [18], tantalates [19], chromates [20], molybdates [21], tungstates [22–24], and rhenates [25]. Howarth *et al.* investigated aqueous solutions of isopolytungstates, peroxotungstates, and heteropoly-molybdates, detecting the $[W_6O_{19}]^{2-}$ and $[W_2O_7]^{2-}$ species in aqueous solution for the first time [26]. Mixed-metal heteropolyanion clusters were also investigated, and the series of $[H_xPW_nMo_{12-n}O_{40}]^{(3-x)-}$ species reported. Further, it was shown that the ESI-MS-determined concentration of sensitive species (i.e., sensitive to changes in pH or the presence of other species) may differ from that determined in bulk measurements. Such studies have been extended to other clusters. For instance, it was shown that heteropolyoxomolybdate [27] compounds in acetonitrile could be successfully identified as intact anionic species, for example, $[S_2Mo_{18}O_{62}]^{2-}$. Also, mixed-metal polyoxomolybates and tungstates could be identified in solution [28], and studies on isopolyoxomolybdates [21], tungstates [22–24], and even isopolyoxochromates [20] have been reported. In each case, an aggregation process of additive polymerization involving $\{MoO_3\}$, $\{WO_3\}$, or $\{CrO_3\}$ moieties, respectively, was identified as giving rise to the larger POM aggregates observed. In addition, the wide applicability of the ESI-MS technique to complex systems and mixtures has been demonstrated in catalytic studies, where the real-time transformation of the substrate could be observed, helping the proposal of a mechanistic pathway [29a,b]. Studies of the potential-dependent formation of unknown multinuclear and mixed-valence polyoxomolybdate complexes when using online electrochemical flow cell electrospray mass spectrometry (EC-ESI-MS) have also been presented [29c].

1.2.1 Background to CSI-MS

CSI-MS was developed by Yamaguchi *et al.* [30] in order to investigate unstable organometallic complexes in which the presence of weak noncovalent interactions had precluded analysis by other ionization techniques such as fast-atom bombardment (FAB), matrix-assisted laser desorption ionization (MALDI), and ESI because of dissociation of the species. The technique is, therefore, of interest for investigations of labile POM systems because previous ESI-MS studies of such systems have been limited by the use of low-resolution detectors and the high temperatures utilized in the ESI process. The cryospray apparatus consists of, essentially, an electrospray source where the N_2 capillary and sprayer gases are maintained at very low temperatures ($\sim -100\,^{\circ}C$). The use of low-temperature gases promotes ionization of the target molecules, not by desolvation as in the conventional ESI process, but by increasing the polarizability of the target molecules at low temperature (i.e., the result of higher dielectric constant at low temperature). This allows the molecular ions of unstable species to be generated and transferred efficiently into the MS detector with minimal fragmentation effects [30]. An example of the potential of CSI-MS to investigate weakly hydrogen-bonded organic aggregates is the observation of large hydrogen-bonded chain structures of amino acids in solution, for example, L-serine, glycine, L-valine [31]. These observations are consistent with the single-crystal X-ray crystallographic data for some of these amino acids in the solid state. Furthermore, the use of CSI-MS studies in more complicated molecules of biological interest was crucial in an effort to investigate the conformational changes of human telomerase upon binding with naphthyridine dimmers [32]. The application of CSI-MS analysis to Pd and Pt cages and their host–guest chemistry has been successful as well as a result of the mild experimental conditions (low temperature). The use of NMR proved not to be very informative since it is difficult to determine precisely the number and structures of the guest molecules if plural guest molecules are encapsulated within the cage compound, whereas during the ESI-MS studies extensive fragmentation takes place because the majority of the Pt/Pd cages are unstable at the temperatures (100 °C) of the desolvation plate. Investigation of the constitution of Grignard reagents [33] in solution is extremely challenging. The use of CSI-MS studies, though, allowed the identification of the species in solution, which supports the idea that the major species of RMgCl Grignard-type reagents in tetrahydrofuran (THF) has the formula $RMg_2(\mu\text{-}Cl_3)$ [33]. CSI-MS is also applicable to the investigation of labile solution structures of various biomolecules. In the case of DNA, for example, various complexes of oligodeoxynucleotides were identified, and it was possible to observe very unstable species such as low-melting-temperature DNA duplexes which could not be detected by the conventional ESI-MS technique [34].

1.3
Application of High-Resolution ESI-MS and CSI-MS to Polyoxometalate Cluster Systems

In the context of this chapter, it is worth pointing out that POMs are ideal candidates for high-resolution studies since they have complex isotopic envelopes resulting from the high number of stable isotopes of tungsten (^{182}W, 26.5%; ^{183}W, 14.3%; ^{184}W, 30.6%; and ^{186}W, 28.4%) and molybdenum (^{92}Mo, 14.8%; ^{94}Mo, 9.3%; ^{95}Mo, 15.9%; ^{96}Mo, 16.7%; ^{97}Mo, 9.6%; ^{98}Mo, 24.1%; and ^{100}Mo, 9.6%), and are intrinsically charged. This allows complete determination of the cluster formula down to the last proton by matching the calculated versus experimentally obtained distribution envelopes. The difficulty associated with determining the protonation state of the cluster has thus far been a major drawback of standard crystallographic X-ray diffraction (XRD) studies, which often do not provide direct information on the protonation state of the cluster anions. Therefore, MS of POMs has the potential to become the standard analysis technique for complex cluster systems since it provides vital complementary information of the cluster composition in solution that cannot be deduced from crystallographic studies, as shown by the following examples.

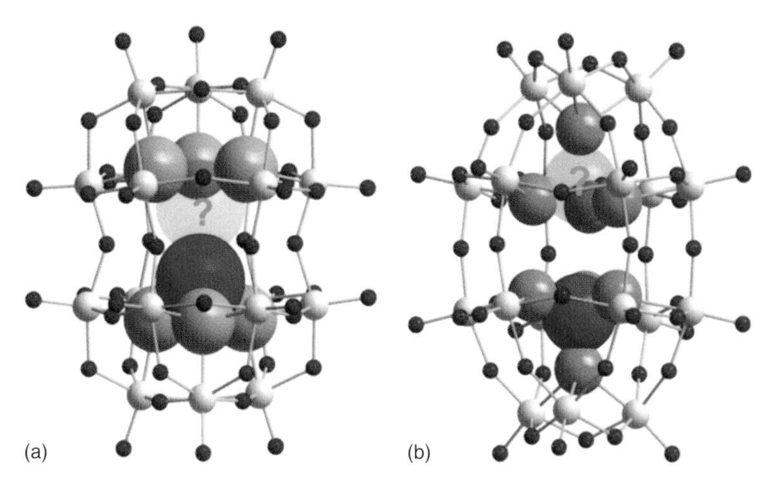

Figure 1.1 (a,b) Representation of the structures of $[H_mSb_nW_{18}O_{60}]^{y-}$ and $[H_mP_nW_{18}O_{62}]^{y-}$. Color scheme: W, gray; O, red; and Sb and P, purple.

1.3.1 Probing Protonation Versus Heteroatom Inclusion with ESI

The rugby-ball-shaped clusters [35] $[H_mW_{18}O_{60}X_n]^{y-}$ (where X = As, Sb, and Bi)[1] have been known for three decades, with an approximate formulation of $n = 1$, but their precise composition could not be confirmed unambiguously because of disorder of the heteroatoms over two positions in a single cluster. However, by using ESI-MS, we recently reported the direct probing of the D_{3d} symmetric isomer of the Sb-based heteropolyoxotungstate $[H_mSb_nW_{18}O_{60}]^{y-}$ [13c]. During the course of these studies, we discovered that the correct formulation is one in which the cluster contains one hetero-ion disordered over two positions (Figure 1.1a). This situation can be compared with the discovery of $[H_mP_nW_{18}O_{62}]^{y-}$, which was also reported with $n = 1$ (see Figure 1.1b). Both clusters appear to include only one heteroatom; in addition, the $[H_4SbW_{18}O_{60}]^{7-}$ is templated by a pyramidal SbO_3^{3-} anion, whereas $[H_4PW_{18}O_{62}]^{7-}$ contains one tetrahedral PO_4^{3-} anion. ESI-MS results were obtained on tetrabutylammonium (TBM) salts of the clusters in acetonitrile solution (see Figures 1.2 and 1.3).

The cation exchange process was used because the TBA^+ cations have a much higher mass than Na^+ or K^+ and give a larger separation between signals corresponding to differently charged or protonated cluster states while at the

1) It should be noted that, when using both the techniques of ESI-MS and CSI-MS, the determined concentration of sensitive species (i.e., sensitive to changes in pH or the presence of other species) can still differ to some extent from that determined in bulk measurements. This effect was investigated by Howarth *et al.* during ESI-MS studies [26], and is due to the interference in the equilibrium process by the drying agent (e.g., nitrogen gas) as the desolvation rapidly affects the pH and the concentrations of the solutes in the formation of the analytes.

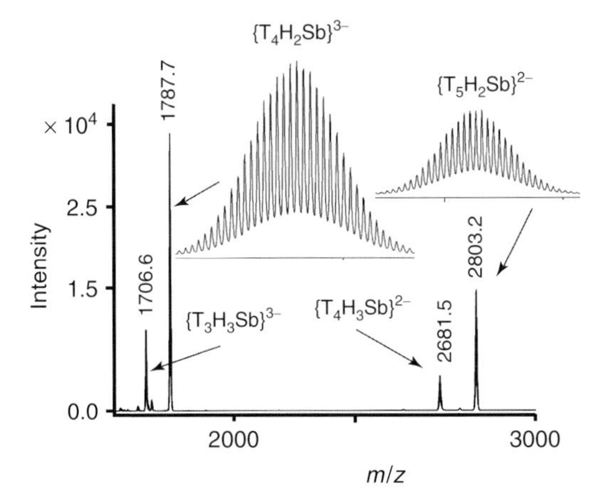

Figure 1.2 The negative ion mass spectrum of $[H_mSb_nW_{18}O_{60}]^{y-}$ showing the series of tri- and tetra-protonated forms of $(TBA)_y[H_xSb_1W_{18}O_{62}]^{(9-(x+y))-}$ in solution ($T = TBA^+$, $Sb = Sb_1W_{18}O_{60}$).

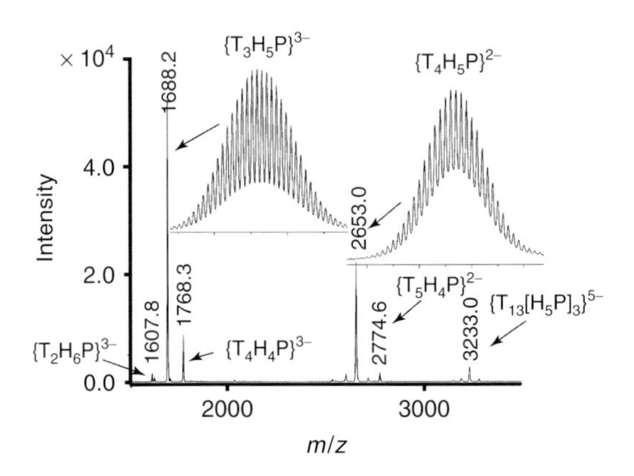

Figure 1.3 The negative ion mass spectrum of $[H_mP_nW_{18}O_{62}]^{y-}$ showing the series of tetra-, penta-, and hexa-protonated forms of $(TBA)_y[H_xP_1W_{18}O_{62}]^{(11-(x+y))-}$ in solution ($T = TBA^+, P = P_1W_{18}O_{62}$).

same time avoiding multiple overlapping distribution envelopes arising from different extents of protonation and/or hydration. Exhaustive analysis of the ESI-MS data shows that the compounds can be unambiguously identified in both positive and negative ion modes, and that the $[H_mSb_nW_{18}O_{60}]^{y-}$ species in solution to give the di- and tri-protonated forms (Figure 1.2) whereas $[H_mP_nW_{18}O_{62}]^{y-}$ can be observed as the hexa-, penta-, and tetra-protonated forms (Figure 1.3).

This study shows that heteropolyoxometalate clusters exist in solution in a range of accessible protonation states that can often not be identified by bulk analytical methods. For both compounds, the most intense single-ion signal or base peak

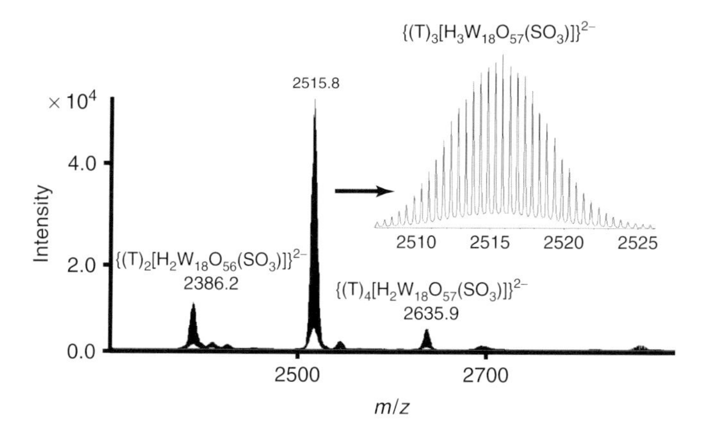

Figure 1.4 Negative ion mode mass spectra of $(TBA)_5[H_3W_{18}O_{57}(SO_3)]$ in acetonitrile ($T = TBA^+$).

was observed for the di- and penta-protonated species, respectively, and shows that it is possible to observe the protonation of a heteropoly acid as a function of the number of hetero-anions included within the cluster.

Also, using this data, it is possible to assign the protonation state of the cluster in the solid state. Furthermore, using a range of mixed-cluster ESI-MS experiments carried out as a function of concentration, we were able to quantify the transmitted ion intensity directly with the concentration of the cluster species in solution. This was done by comparison of the bis-phosphate Dawson cluster $(TBA)_6[P_2W_{18}O_{62}]$ with the monophosphate $(TBA)_6[H_5PW_{18}O_{62}]$. This gave a linear relationship between the relative ion intensity and the ratio of the concentration of the two compounds in solution. This means that in organic solvents it is possible to use ESI-MS studies to directly probe the solution equilibria.

In a similar manner, we recently reported the solution studies of the mono- and bi-sulfite tungsten-based capsules. The ESI-MS study of the monosulfite species in acetonitrile shows that the expected $\{W_{18}S\}$ cluster anion can be assigned unambiguously to different charge/cation states of this: *m/z* 2515.8 $\{(TBA)_3[H_3W_{18}SO_{60}]\}^{2-}$, 3434.8 $\{(TBA)_7[H_3W_{18}SO_{60}]_2\}^{3-}$, and 3001.4 $\{(TBA)_7[H_3W_{18}SO_{60}]\}^{2+}$ (see Figure 1.4 for the spectrum including an expansion of the main peak in the negative mode). The ESI-MS spectrum of the TBA salt of $\{W_{18}S\}$ clearly shows that the cluster is threefold protonated compared to the doubly protonated species found in its sodium analog; however, the composition of the main framework is otherwise identical as expected. It is interesting also to observe that the cluster is labile and that the parent threefold protonated cluster can decompose via water loss, as indicated by the peak at *m/z* 2386.2; this can be assigned to the $\{(TBA)_2[H_2W_{18}O_{56}(SO_3)]\}^{2-}$ species, which indicates that the decomposition process of the threefold protonated cluster may begin with the loss of one water molecule.

In contrast, the ESI-MS spectra of the bisulfite cluster anion in acetonitrile solvent are also recorded, and the expanded main peak is represented in

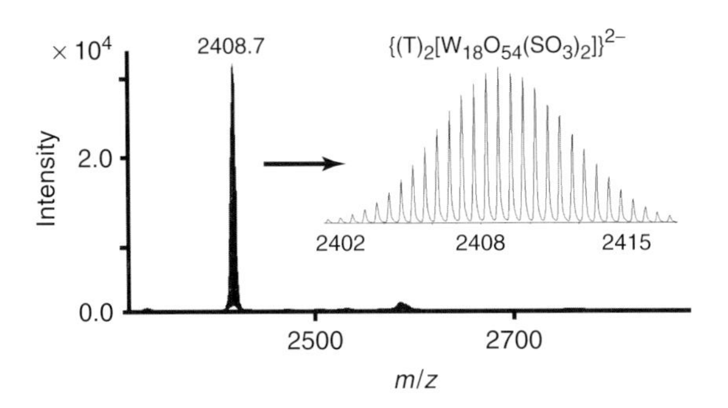

Figure 1.5 Negative ion mode mass spectra of $(TBA)_4[W_{18}O_{54}(SO_3)_2]$ in acetonitrile ($T = TBA^+$).

Figure 1.5, which gives totally different results. As reported previously [36], the $[W_{18}O_{56}(SO_3)_2(H_2O)_2]^{8-}$ cluster rearranges during the cation exchange process and forms the $[W_{18}O_{54}(SO_3)_2]^{4-}$ cluster. All the main peaks can be assigned, and related to the $\{W_{18}S_2\}$ species.

1.3.2
Solution Identification of Functionalized POMs

The functionalization of POM clusters via covalent grafting of organic functions allows the properties of the cluster to be tuned and novel functionalities to be introduced. In our recent work, an asymmetric, functionalized Mn-Anderson cluster was designed and synthesized, $(TBA)_3[MnMo_6O_{18}(C_4H_6O_3NO_2)(C_4H_6O_3NH_2)]$ [37], utilizing ESI-MS during the screening process of the reaction mixtures and fine adjustment of the experimental conditions of this labile system. Subsequently, the system was set up for crystallization, and the crystalline fractions were isolated from the mother liquor by filtration every 6 h. Each batch of the crystalline samples was then analyzed again using ESI-MS in order to determine the distribution and number of products present, allowing the correct batch to be "sorted" from the bulk, statistically defined mixture. The batch corresponding to the asymmetric compound was collected. Using this approach, ESI-MS studies not only helped the reaction mixtures to be "sorted" but also allowed confirmation of the intrinsic composition, as shown in Figure 1.6, and isolation of the pure crystalline phase of the desirable product. The peak at $m/z = 1669$ can be clearly assigned to the $[(TBA)_2\{MnMo_6O_{18}(C_4H_6O_3NO_2)(C_4H_6O_3NH_2)\}]^-$ anion. Additionally, the signal at $m/z = 1669$ is the only peak in the range of m/z 1300–2000, with no peaks being observed that would be expected from co-crystallization of the other possible species $[(TBA)_2\{MnMo_6O_{18}(C_4H_6O_3NH_2)_2\}]^-$ (m/z 1640) or $[(TBA)_2\{MnMo_6O_{18}(C_4H_6O_3NO_2)_2\}]^-$ (m/z 1700). The observation of such complex reaction mixtures under electrospray conditions is very important in the effort toward the design and synthesis of hybrid POM-based materials that possess the desirable functionality.

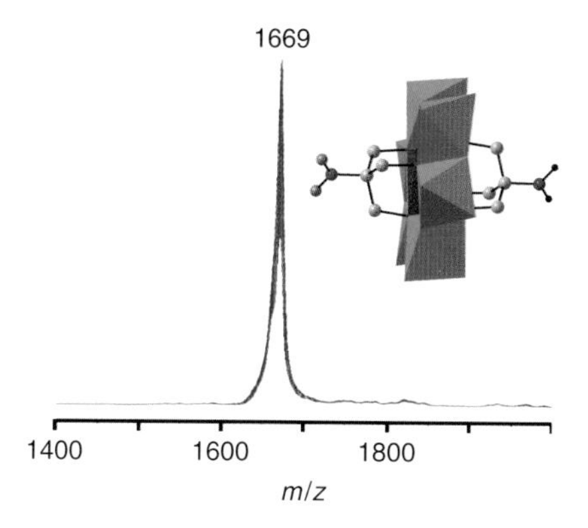

Figure 1.6 ESI-MS spectrum of the asymmetric compound in MeCN showing one signal at $m/z = 1669$ due to $[(TBA)_2\{MnMo_6O_{18}(C_4H_6O_3NO_2)(C_4H_6O_3NH_2)\}]^-$ species. Inset: representation of the X-ray crystal structure of $\{MnMo_6O_{18}(C_4H_6O_3NO_2)(C_4H_6O_3NH_2)\}^{2-}$. Color scheme: Mo, green polyhedra; Mn, pink polyhedron; C, gray; N, blue; O, orange; and H, black.

1.3.3
Solution Identification of New Isopolyoxotungstates and Isopolyoxoniobates

In our recent work, we targeted the exploration of aqueous solutions of isopolyoxotungstates using a combination of pH, ionic strength, and anion control to unravel the influence of these factors on the self-assembly of anionic isopolyoxotungstates. As such, we were able to demonstrate that it is possible to discover radically new cluster architectures in the isopolyoxotungstate family simply by acidifying solutions of sodium tungstate leading to the isolation and crystallization of an "S"-shaped $[H_4W_{22}O_{74}]^{12-}$ cluster (Figure 1.7) [12b].

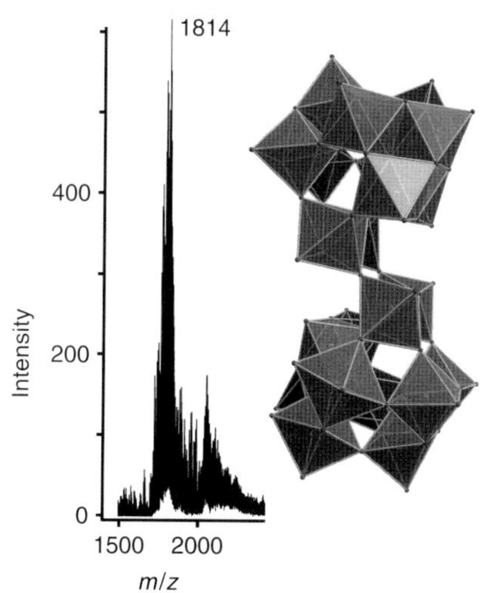

Figure 1.7 The crystal structure of the "S"-shaped cluster (b) and the negative ion mass spectrum for $\{(Na_9)[H_4W_{22}O_{74}]\}^{3-}$ in water (a). Simulation of the expected spectrum matches that of the observed peak at m/z 1814 [12b]. Color scheme: W, dark grey polyhedra and O, dark grey spheres.

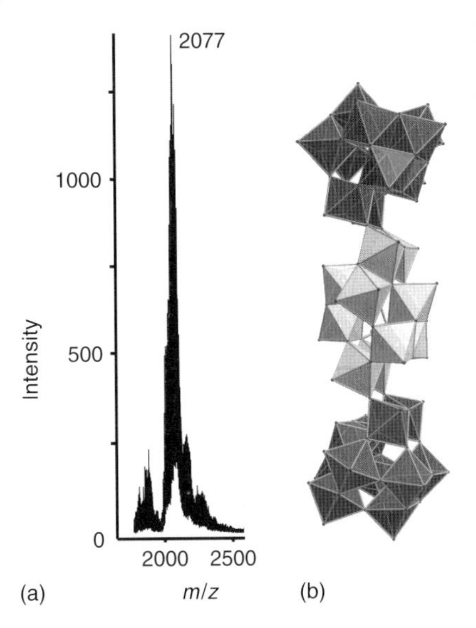

Figure 1.8 The crystal structure of the "ʃ"-shaped cluster (b) and the negative ion mass spectrum for $\{Na_8H_6[H_{10}W_{34}O_{116}]\}^{4-}$ in water (a). Simulation of the expected spectrum matches that of the observed peak at *m/z* 2077 [12b]. Color scheme: W_{22} unit, dark grey polyhedra; W_{34} unit, light grey polyhedra; and O, dark grey spheres.

The compound $Na_{12}[H_4W_{22}O_{74}] \cdot 31H_2O$ could not only be identified as containing the $[H_4W_{22}O_{74}]^{12-}$ anion in the solid state, but it was also identified in aqueous solution using ESI-MS analysis as the $\{(Na_9)[H_4W_{22}O_{74}]\}^{3-}$ anion at *m/z* 1814. The cluster framework $[H_4W_{22}O_{74}]$ can unambiguously be assigned as being present. The broad spectrum is simply due to the presence of several overlapping species arising from the many different envelopes. All these species are ions with a charge of 3− containing different amounts of sodium ions and protons. On lowering the pH to 2.4, a related "ʃ"-shaped $[H_{10}W_{34}O_{116}]^{18-}$ cluster is isolated as $Na_{18}[H_{10}W_{34}O_{116}] \cdot 47H_2O$ (Figure 1.8). The cluster framework $[H_{10}W_{34}O_{116}]$ can unambiguously be assigned as being present. Similarly, the broad spectrum is simply due to the presence of several overlapping species with different extents of protonation and hydration, which give many different distribution envelopes. The observation of such complex isopolytungstate-based clusters under electrospray conditions was not expected because of the number of possible species and the fragile nature of such clusters, but at the same time this offers many new possibilities for the systematic screening of reactions to discover new cluster architectures.

Another example recently reported by our group is the identification in solution of Nb_xO_y reaction mixtures. More specifically, we conducted the hydrothermal experiment over 20 h, using $\{Nb_6\}$ as the precursor, and then treated the mother liquor with hexadecyltrimethylammonium bromide. This procedure was intended to produce a material that could be dissolved in acetonitrile for analysis by high-resolution ESI-MS. These experiments resulted in the observation of a series of envelopes that could be assigned to $\{Nb_6\}$, $\{Nb_{10}\}$, $\{Nb_{20}\}$, and $\{Nb_{27}\}$. Although the $\{Nb_6\}$ was still present as the main component, it was possible to identify unambiguously a range of species that could be assigned as the $\{Nb_{27}\}$-based

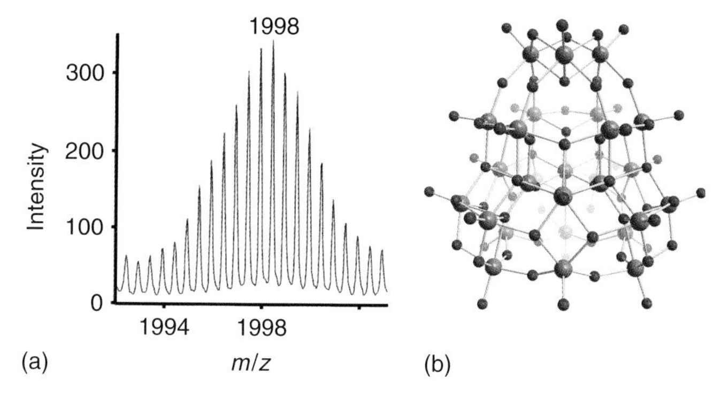

Figure 1.9 (a) Ball-and-stick representation of the structure of {Nb_{27}}. Color scheme: Nb, dark yellow and O, red spheres. (b) Distribution envelope of the {Nb_{27}} species identified unambiguously in the reaction mixture.

anions (Figure 1.9) [18a]. This finding clearly indicates that a gradual molecular growth takes place via the lower nuclearity species toward the final formation of the {Nb_{27}} cluster, which is the biggest isopolyoxoniobate species identified by ESI-MS in solution so far.

1.3.4
Solution Identification and Isolation of Mixed-Metal/Valence POMs with CSI-MS

The above examples show that ESI-MS studies have been extremely helpful in the identification of the composition, the extent of protonation, and the existence of other relatively stable species in the solution. However, these studies can be limited when we have to deal with labile clusters with complex compositions, or those that adopt large and unstable motifs. While a significant result was not obtained from the conventional ESI-MS of the subsequently discussed fragile species presumably due to the high desolvation plate temperature (180 °C), multiply charged high-nuclearity molecular ions were clearly observed without decomposition under CSI-MS conditions. Furthermore, given the extent of ionization and instability issues of such structures at high temperatures, it is often difficult to establish the presence of some cluster architectures using ESI-MS. In contrast, the low temperatures accessible (up to −100 °C) for use with a cryospray source minimize uncontrolled fragmentation and allow efficient transfer of very high nuclearity, yet labile, ionic species into the detector with minimal interference from the ionization and desolvation processes. Moreover, low temperatures allow the detection and entrapment of short-lived intermediate species that take part in a reaction mechanism, which is impossible to observe or to prove their existence using conventional methods. By employing this approach, it is then possible to transfer many of the labile and/or unstable species present in solution into the mass spectrometer and allow some correlation between the essentially gas-phase measurements and those

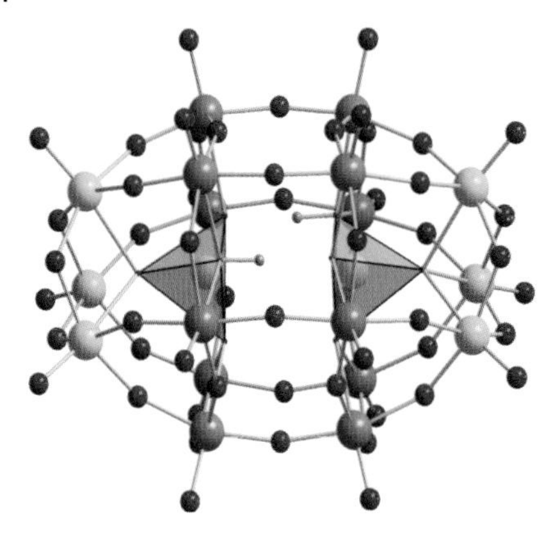

Figure 1.10 Representation of the isostructural anionic framework found in the structures of compounds with the general formula $[H_2V^{IV}M_{17}O_{54}(V^{V}O_4)_2]^{6-}$ (M = W or Mo). The $\{M_{17}V_1\}$ framework is shown in ball and stick and the V templates are shown by the polyhedra. Even though the "framework" V^{IV} ion cannot be formally located, theoretical calculations showed a preference for the two cap positions rather than the belt positions. Color scheme: Mo/W, dark green (belt), light green (cap); O, red; and V, dark green polyhedra [17a].

in the solution and solid state [26]. At this point, CSI-MS studies can offer the necessary environment for labile/unstable species to survive and collect important information which will help us understand the self-assembly mechanisms in supramolecular chemistry.

1.3.5
Mixed-Metal/Valence Hetero-POMs $V_2 \subset \{M_{17}V_1\}$

Over the last two years, we reported our first attempts to detect and study labile POM systems using CSI-MS. CSI-MS studies proved to be a tool of vital importance in first identifying these species using the correct combination of experimental parameters to minimize fragmentation allowing observation of new cluster architectures in solution, and then providing some indication of how the synthetic procedure can be optimized to yield the new architecture [17]. In this work, we aimed to replace the hetero-anion templates in the classical Dawson-like clusters, for example, $[W^{VI}{}_{18}O_{54}(PO_4)_2]^{6-}$, with transition metals; by scanning the reaction mixtures before crystallization, we were able to locate the reaction systems that produced the $V_2 \subset \{M_{17}V_1\}$ species, as shown in Figure 1.10.

The discovery studies were performed by precipitating solids from a range of candidate reaction systems under aqueous conditions. The precipitates were then transferred into the organic phase by ion exchange with TBA and "screened" using CSI-MS. These studies showed that this family of clusters was present in solution prior to their structural analysis. The studies also showed that the TBA salts of the $\{M_{17}V_3\}$(M = W, Mo) clusters were stable in solution (Figure 1.11). A range of charge (−1 to−3) and protonation (0–2) states were observed. Also, the direct

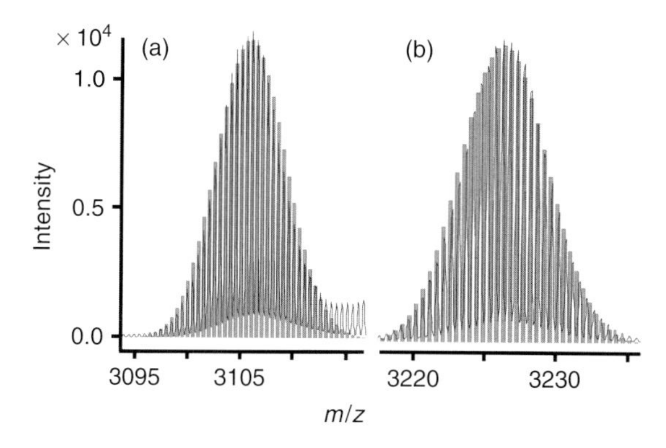

Figure 1.11 Positive ion mass spectrum showing the $\{(TBA)_{10-n}[H_nV_3W_{17}O_{62}]\}^{2+}$ in acetonitrile solution. (a) where $n = 2$ at m/z about 3106. (b) where $n = 1$ m/z about 3227. The black line shows the actual spectrum and the grey bar graph is the predicted envelop.

observation of $\{(TBA)_8[H_2V^V{}_2V^{IV}W_{17}O_{62}]\}^{2+}$ allowed us to confirm that the clusters observed in the solid state all have six cations associated and are di-protonated. This observation is useful since this gives unambiguous proof that the Dawson capsules are present in solution, establishes the existence of exactly one vanadium metal center on the shell of the Dawson framework (which is almost impossible to distinguish crystallographically), and also confirms the extent of protonation of the cluster, which is extremely difficult to determine directly.

Another challenging case is the mixed-metal/valence sulfite POM systems. The discovery of such species in solution, characterization, and final optimization of the synthetic conditions proved to be an intriguing task. In this investigation, CSI-MS proved to be an important technique in our efforts to discover new Dawson-like capsules in solution, allowing the compound to be identified prior to structural analysis [17b]. CSI-MS studies of the tetrapropylammonium (TPA) salts of the cluster (Figure 1.13) dissolved in acetonitrile confirmed that the sulfite capsule retained its integrity in solution, and the peaks seen were assigned to $\{(TPA)_4[H_{1-n}V^V{}_{5+n}V^{IV}{}_{2-n}Mo_{11}O_{52}(SO_3)]\}^{2-}$ where $n = 1$ (with only one vanadium ion in oxidation state IV), giving an envelope centered at m/z about 1534.5, and where $n = 0$ (with two vanadium ions in oxidation state IV, requiring one proton), giving an envelope centered at m/z about 1535.0 (Figure 1.12).

In the majority of the cases, effort to increase the cluster's solubility in organic solvents via cation exchange process is desirable, since the use of aqueous solutions can cause ion-transfer problems resulting in very low-intensity signals, which can be difficult to analyze in detail. Furthermore, aqueous reaction mixtures often results in the observation of a plethora of species resulting from the multitude of possibilities arising from the clusters transferring into the MS detector with sodium and potassium cations with multiple water ligands.

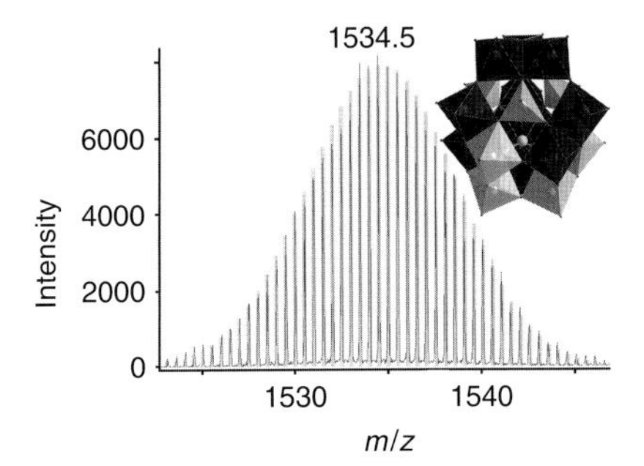

Figure 1.12 Negative ion mass spectrum in acetonitrile solution of $\{(Pr_4N)_4[H_{1-n}V^V{}_{5+n}V^{IV}{}_{2-n}Mo_{11}O_{52}(SO_3)]\}^{2-}$. Two envelopes can be seen where $n = 1$ (with only one vanadium ion in oxidation state IV) giving an envelope centered at m/z about 1534.5, and where $n = 0$ (with two vanadium ions in oxidation state IV, requiring one proton) giving an envelope centered at m/z about 1535.0. Black line: experimental data, light grey bars: simulation of isotope pattern. Inset: polyhedral representation of the anion. Color scheme: Mo, dark grey polyhedra; V, light grey polyhedra; and S, light grey sphere.

In an effort to extend the work on mixed-metal POM-based sulfite capsules, we managed to adjust appropriately the experimental conditions, which led to a molecular evolution of the $\{Mo_{11}V_7\}$ Dawson-like capsule to a "crowned" version of it with the formula $[Mo^{VI}{}_{11}V^V{}_5V^{IV}{}_2O_{52}(\mu_9 - SO_3)(Mo^{VI}{}_6V^VO_{22})]^{10-}$ [17d]. Following a similar approach and using as starting material the $\{Mo_{11}V_7\}$ capsule, it was possible to identify the new species in solution by CSI-MS. By fine adjustment of the solution's concentration, we managed to optimize the synthetic procedure and characterize the material in solid state by XRD analysis (Figure 1.13).

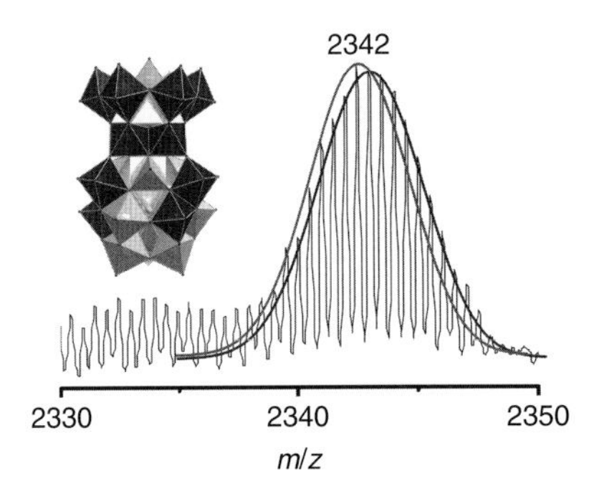

Figure 1.13 Negative ion mass spectrum in acetonitrile solution of $\{(Pr_4N)_7H_n[Mo^{VI}{}_{11}V^V{}_{5-n}V^{IV}{}_{2+n}O_{52}(\mu_9 - SO_3)(Mo_6^{VI}V^VO_{22})](NH_4)_2(CH_3CN)\}^{2-}$. Two envelopes can be seen where $n = 0$ (with only two vanadium ions in oxidation state IV) giving an envelope centered at m/z about 2342.1, and where $n = 1$ (with three vanadium ions in oxidation state IV, requiring three protons) giving an envelope centered at m/z about 2342.6. Black line: experimental data, Black/grey lines: profile lines of the simulated isotope patterns. Inset: polyhedral presentation of $\{Mo_{17}V_8\}$ capsule. Color scheme: Mo, dark grey polyhedra; V, light grey polyhedra; and S, light grey sphere.

The above examples demonstrate that the application of CSI-MS to investigate and discover labile or unstable cluster species in solution was used to help discover and then optimize the synthetic approach to isolate the clusters, and this proved to be of major significance. That became obvious when we carried out the same studies in the ESI mode on the same series of solutions, where in most of the cases the results were inconclusive owing to excessive fragmentation and led to a mixture of many different kinds of species revealing different compositions, oxidation states, and possible structures, making the assigning process impossible. The fine adjustment of the experimental conditions (concentrations, pH, and temperature) allows the direct observation of the species present in the reaction system using the CSI-MS technique: an approach that we have found useful to observe reactive building blocks and high-nuclearity clusters. In this context, we are utilizing CSI-MS to identify new POM clusters with novel templates and architectures that would not be easily isolated without prior detailed knowledge of the clusters present in the reaction solutions.

1.3.6 Periodate-Containing POMs

In an extension of this technique to look at reactive clusters, we utilized CSI-MS studies to identify and unambiguously reveal the composition of a new cluster species in solution and investigate the relationship between the solution and solid states. In this work, we explored the insertion of new high-oxidation-state guests into the "Dawson capsule" to allow modulation of the cluster's physical properties, for example, redox and catalytic properties, and acidity. The encapsulation of such templates should significantly affect the acidic, catalytic, and redox properties of the resulting cluster systems. In this work, we examined the tungsten Dawson capsule, $K_6[H_3W_{18}O_{56}(IO_6)] \cdot 9H_2O$, which we have both synthesized and characterized in the solid state [38]. To rule out the possibility that the aforementioned compound could be assigned as $[H_4W_{19}O_{62}]^{6-}$ (i.e., I substituted for W or other hetero-ions of comparable ionic radius), we utilized high-resolution CSI and ESI-MS to identify the exact elemental composition of the cluster anion. To simplify the MS experiments, the potassium salt of the aforementioned compound was ion-exchanged with TPA^+ cations, yielding $(TPA)_6[H_3W_{18}O_{56}(IO_6)]$. Figure 1.14 shows the mass spectrum of $(TPA)_6[H_3W_{18}O_{56}(IO_6)]$ in acetonitrile in which all major peaks are related to the $\{W_{18}I\}$ capsule. In contrast, the TPA^+ salt of $[H_4W_{19}O_{62}]^{6-}$ [39] capsule in acetonitrile gives a different mass spectrum in the same m/z region. Therefore, CSI mass spectral studies, in combination with elemental analysis and crystallography, have given us unambiguous proof of the existence and integrity of $\{W_{18}I\}$ species.

1.3.7 Probing the Formation of POM-Based Nano-Structures

The extension of MS studies to supramolecular architectures poses several important questions regarding the nature of the assemblies present in solution

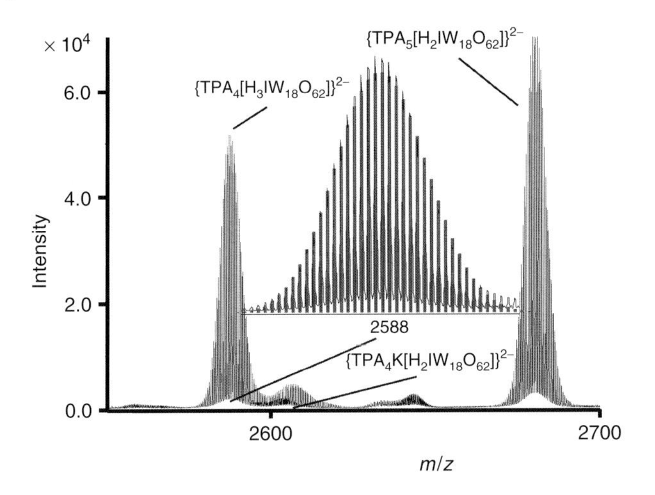

Figure 1.14 Negative-mode mass spectra of $(TPA)_6[H_3IW_{18}O_{62}]$ in acetonitrile. An expansion of the peak at $m/z = 2588$ is shown along with the calculated isotopic pattern (light grey bars).

versus those transmitted and observed in ESI-MS experiments. However, we have also made progress here as shown by the synthesis and characterization of a nano-structured POM-based assembly in the solid state [40]. By grafting of a H-bonding donor "cap" onto a POM cluster, we could control the supramolecular self-assembly of cluster species in solution, as well as in the solid state, leading to the formation of macromolecular H-bonded nano-assemblies of hybrid POM clusters. Most importantly, it was observed that the formation of these supramolecular architectures could be externally controlled by simply changing the grafted H-bonding organic cap (Figure 1.15).

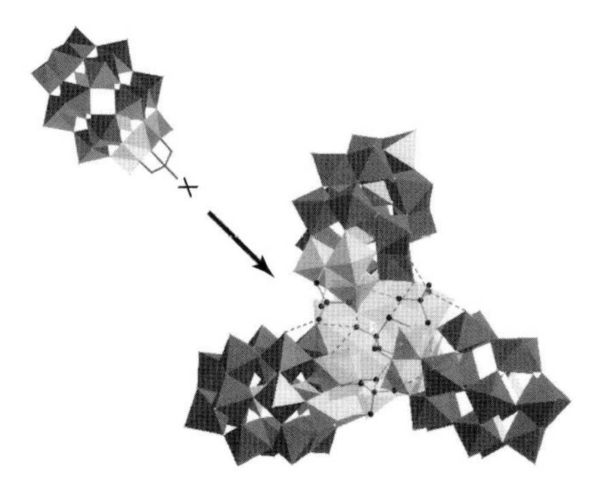

Figure 1.15 Hydrogen-bonded distorted tetrahedral nano-assembly of $[H_2NC(CH_2O)_3P_2V_3W_{15}O_{59}]^{6-}$ cluster. One lobe of the tetrahedron – pointing toward the observer – is made transparent for easy visualization of the H-bonding interactions. Color scheme: O, dark grey spheres; W, dark grey polyhedra; V, light grey polyhedra; and PO_4, grey polyhedral within the capsules, where X: $-NH_2$, $-NO_2$, and $-CH_3$.

In this case, CSI-MS studies did not only help us to identify and establish the integrity of the cluster but also to observe the formation of hydrogen-bonded nanostructures in the solution as well as to reveal information about the mechanism of this self-assembly process. By using cryospray experiments at $-40\,^{\circ}C$, we were able to examine the self-assembly processes in solution in great detail. For instance, CSI-MS studies of $(TBA)_4H_2[H_2NC(CH_2O)_3P_2V_3W_{15}O_{59}]$ in dilute acetonitrile solutions revealed that the tris(hydroxymethyl)-aminomethane-grafted POM cluster exists in solution phase as monomers, dimers, and trimers, and also the tetramer could be clearly observed. The observed mass and charge corresponding to each of these peaks clearly matched with the assigned formulae as well as the simulated spectra. The role of the H-bonding ability of the organic cap of the cluster anions in forming the multiple aggregation of cluster species in solution and gas phase was examined by comparing the CSI-MS of cluster species $[H_2NC(CH_2O)_3P_2V_3W_{15}O_{59}]^{6-}$ (**1**), $[O_2NC(CH_2O)_3P_2V_3W_{15}O_{59}]^{6-}$ (**2**), and $[H_3CC(CH_2O)_3P_2V_3W_{15}O_{59}]^{6-}$ (**3**) under identical experimental conditions. It was observed that the $-NO_2$-tipped cluster (**2**) also forms multiple aggregates in solution and gas phases, quite similar to cluster (**1**). Interestingly, it was found that the $-NO_2$ capped cluster is more efficient in forming such multiple aggregates in solution since assemblies of five (pentamer) and even six (hexamer) alkoxy POM clusters are observed. The weakly hydrogen-bonding $-CH_3$-tipped cluster forms mainly monomeric species in solution, and no higher aggregates are observed (Figure 1.16).

The aforementioned CSI-MS studies presented an approach to observe the formation of nano-assemblies of POM clusters in solution and in the gas phase and have given us important missing information on the "mechanism" that governs this assembly process. The outcome of this study allowed us to modify accordingly the extent of the H-bond network and consequently to control the interaction between nano-sized molecules. Furthermore, the observation of the gigantic hydrogen-bonded tetrahedral nano-assembly of the alkoxy cluster anions of $[H_2NC(CH_2O)_3P_2V_3W_{15}O_{59}]^{6-}$ along with 19 TBA counterions in solution phase by CSI-MS analysis underlines the potential of this technique to investigate the role of weak interactions such as hydrogen bonding in understanding the self-organization processes involved in the synthesis of novel POM-based functional nanomaterials.

1.3.8
Mechanistic Insights into POM Self-Assembly Using ESI- and CSI-MS

Given the power of CSI-MS to look in detail at labile POM solutions, we envisaged utilizing this technique to shed light upon the mechanism of self-organization of individual POM molecules from its reagents in solution. The study of the self-assembly processes of POMs has previously been pursued by the use of ligands with well-defined binding sites. This allows easier elucidation of the assembly processes, as the POM units effectively behave like multidentate organic ligands and bind readily to secondary transition metals [41]. It was for these reasons that the use of silver(I) cations as linking groups between POM units has been investigated by our group. It was found that, in the reaction of $[TBA_2(Mo_6O_{19})]$ with silver(I)

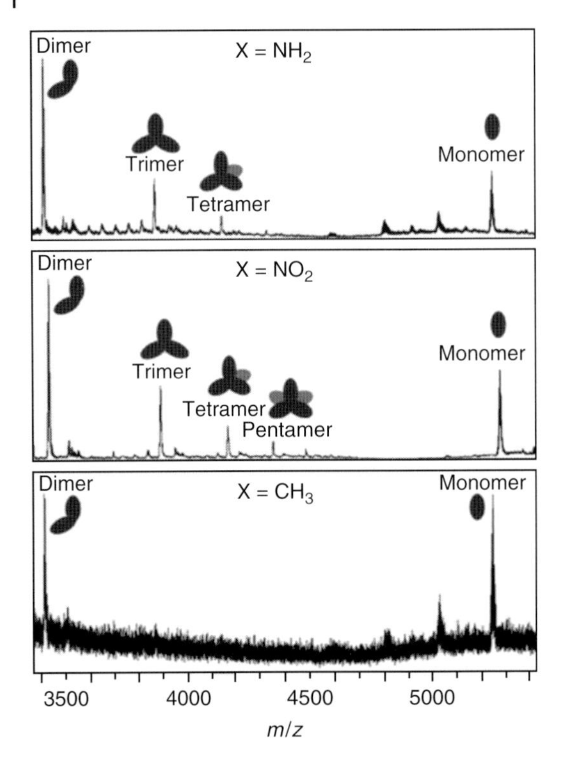

Figure 1.16 Comparison of the CSI-MS spectra and supramolecular assemblies of clusters **1**($-NH_2$), **2**($-NO_2$), and **3**($-CH_3$) as a function of different substituent (X) on the organic cap. Spectra are on the same m/z scale.

fluoride in a mixed acetonitrile–methanol solution, a unique one-dimensional chain structure of the composition ($TBA_{2n}[Ag_2Mo_8O_{26}]_n$) is produced. Furthermore, many other architectures, involving specifically the aggregation of {Ag(Mo_8)Ag} synthons, have been produced [42]. In this work, the in-solution interconversion of Lindqvist into β-octamolybdate anions and subsequent self-assembly into the silver-linked POM structure $[(n-(C_4H_9)_4N)_{2n}(Ag_2Mo_8O_{26})_n]$ was investigated using CSI-MS and electronic absorbance spectroscopy [43], allowing the direct monitoring of real-time rearrangements in the reaction mixture. Only mono-anionic and di-anionic series were observed in these results, from approximately m/z 285 to as high as approximately m/z 3800, indicating the efficient transfer of very high nuclearity, yet labile, ionic species into the detector with minimal interference from the ionization and desolvation processes. Of these identified anion series (see Figure 1.17), only $[HMo_mO_{3m+1}]^-$ series had been observed in previous ESI-MS studies on polyoxomolybdate systems [21], which underpins the advance in understanding that can be made with CSI-MS studies for detecting molecular building blocks. The $[H_7Mo_mO_{3m+3}]^-$ series is of special interest with regard to this POM reaction system. This is because the role of the AgI moiety in the assembly of the

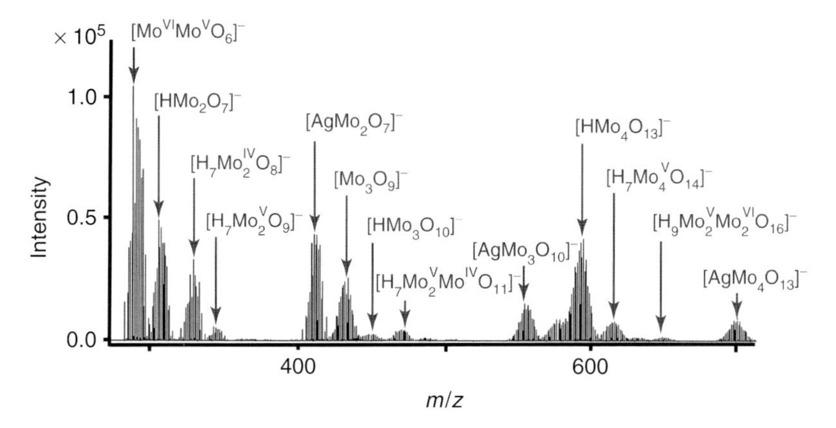

Figure 1.17 CSI-MS data collected for the reaction solution. The six mono-anionic series identified within these results are highlighted. The steps toward the assembly of the {Ag(Mo$_8$)Ag} synthon units can be observed by examination of anion series (vi), which highlights the role of the Ag^+ in the rearrangement process of these clusters. Of particular note from this series are the peaks at *m/z* 410.7 and 700.5, which are attributed to the species $[AgMo_2O_7]^-$ and $[AgMo_4O_{13}]^-$, respectively.

stable silver-linked octamolybdate species has been established by mass spectral methods for the first time and is shown to be crucial for the formation of the larger cluster fragments.

Detection of the $[AgMo_2O_7]^-$ fragment (peak at *m/z* 410) of the (Ag{Mo$_8$}Ag) synthon from the reaction solution supports the theory of rearrangement of the Lindqvist anion into $[AgMo_2O_7]^{-n}$, which is the smallest stable unit of the silver-linked POM chain. Indeed, the stable nature of this fragment of the (Ag{Mo$_8$}Ag) synthon unit allowed the isolation in the solid state of $Ag_2Mo_2O_7$ clusters linked into 1D chains, which has been reported by Gatehouse and Leverett [44]. Detection of the $[AgMo_4O_{13}]^-$ species (peak at *m/z* 700.5), being half of the (Ag{Mo$_8$}Ag) synthon unit, represents the next stepping stone in the final rearrangement to the stable silver-linked octamolybdate species.

In the higher mass range of the CSI-MS analyses carried out, the structure-directing effect of the organic cations has been illustrated for the first time. Detection of the species $[(AgMo_8O_{26})TBA_2]^-$ (peak at *m/z* 1776.6), $[(Ag_2Mo_8O_{26})(Mo_4O_{13})TBA_3]^-$ (peak at *m/z* 2718.3), and $[(Ag_2Mo_8O_{26})(Mo_8O_{26})TBA_5]^-$ (peak at *m/z* 3796.5), each with an increasing organic cation contribution, shows the increasing metal nuclearity of the chain of polymer compound concomitant with the associated increase in organic cations present (Figure 1.18). This observation can be interpreted as the start of the self-assembly aggregation process where "monomeric" units assemble into larger fragments, which eventually leads to the formation of crystals of the compound $(TBA_{2n}[Ag_2Mo_8O_{26}]_n)$. Kinetics of the rearrangement process of Lindqvist anions into (Ag{Mo$_8$}Ag) synthons in the reaction solution was studied using electronic absorbance measurements. The relationship between decreasing Lindqvist anion concentration and concomitant

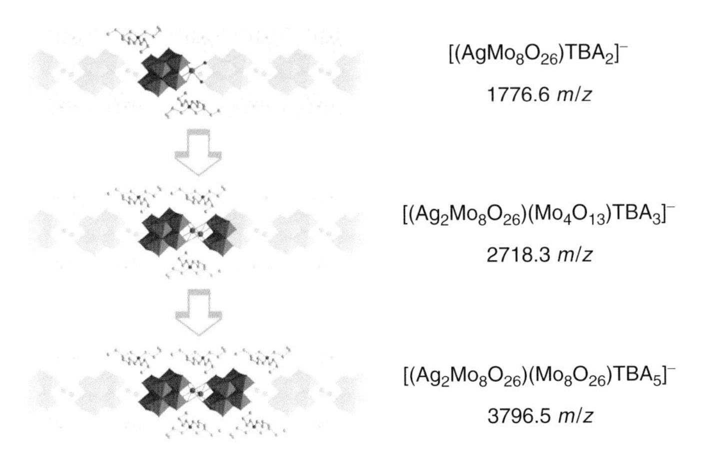

Figure 1.18 Structural representation of the higher mass fragments (highlighted) identified within the CSI-MS analyses of a reaction solution. This diagram illustrates the increasing metal nuclearity of the chain concomitant with the associated increase in organic cations present. Color scheme: Mo, green polyhedra; Ag, pink; O, red; C, gray; and N, blue.

increase in $\{Mo_8\}$ anion concentration was further supported by monitoring the reaction solution over time using CSI-MS experiments.

In summary, the use of CSI-MS in this way to monitor in real-time, in-solution rearrangements in a POM reactant solution is, to our knowledge, unprecedented. This approach can be extended to investigate the bottom-up, in-solution processes governing the formation of other POM systems, enhancing our understanding and giving us the potential to control the building-block principles involved.

In an effort to monitor the reaction mixture and investigate the formation mechanism of a more complicated POM-based chemical system, we utilized the ESI-MS technique to investigate, in real time, the formation of a complex and organic–inorganic POM hybrid system. The reaction system chosen for investigation in this case was found by Hasenknopf *et al.* [45] and involves the rearrangement of $[\alpha\text{-}Mo_8O_{26}]^{4-}$, coordination of Mn^{III}, and coordination of two tris(hydroxylmethyl)aminomethane (TRIS) molecules, to form the symmetrical Mn-Anderson cluster $TBA_3[MnMo_6O_{18}((OCH_2)_3CNH_2)_2]$. This piece of work [46] is very important since the organic–inorganic POM hybrids can be used as synthons in the design of nanoscale hybrid POM architectures, thereby increasing our knowledge of the aggregation processes that form such building blocks and making the field of nanoscale functional materials more accessible for further exploration.

The speciation and fragment rearrangements were investigated, and the first spectrum was dominated by peaks that could be assigned to isopolyoxomolybdate fragments of the rearranging $[\alpha\text{-}Mo_8O_{26}]^{4-}$ anion (Figure 1.19). The dominance of these isopolyoxomolybdate fragments indicates that the $[\alpha\text{-}Mo_8O_{26}]^{4-}$ anion rearranges into these smaller fragment ions prior to further coordination with

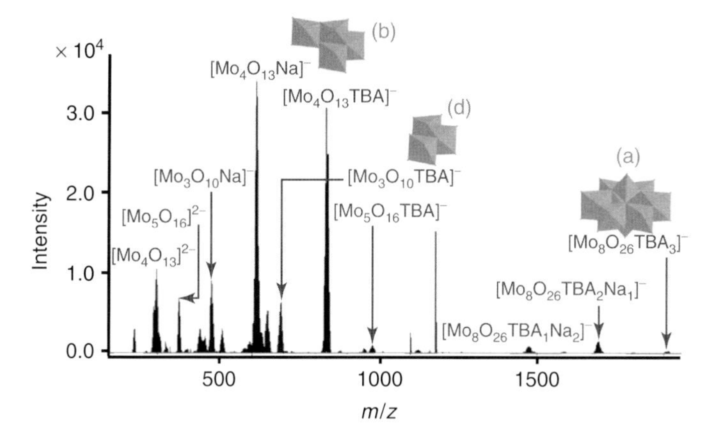

Figure 1.19 ESI-MS spectra collected of the reaction solution of $[\alpha\text{-}Mo_8O_{26}]^{4-}$, recorded after stirring at room temperature for 13 min. The spectrum is dominated by isopolyoxomolybdate fragment peaks. Of particular note are the two major peaks in this spectrum, at *m/z* 614.6 (the base peak) and 833.8, which are attributed to the species $[Mo_4O_{13}Na]^-$ and $[Mo_4O_{13}TBA]^-$ respectively, that is, half of the parent $(Mo_8O_{26})^{4-}$ cluster anion. Some of the intermediate prominent fragment ions are also shown here and are labeled similarly. Color scheme: Mo, dark grey polyhedra; and O, dark grey spheres. Structural representations: (a) $[\alpha\text{-}Mo_8O_{26}]^{4-}$, (b) $[Mo_4O_{13}TBA]^-$, and (d) $[Mo_3O_{10}TBA]^-$.

the Mn cations and TRIS groups. Indeed, the first indications of this further coordination are illustrated by the presence of very low-intensity peaks containing TRIS groups and manganese cations, for example, $[Mo_2O_5((OCH_2)_3CNH_2)]^-$ (*m/z* 387.8) and $[Mn^{III}Mo_3O_8((OCH_2)_3CNH_2)_2]^-$ (*m/z* 706.7).

The complexity and ion series observed in this spectrum remain observable through to the final spectrum recorded after refluxing for approximately 30 h (see Figure 1.20). It is interesting to note at this point the presence of Mn^{II} ions, particularly in the smaller *m/z* fragment ions, and mixed oxidation state species where molybdenum is found to exist in oxidation states IV–VI. Observation of molybdenum and manganese centers in reduced oxidation states is not entirely unexpected because of the high voltages utilized in the mass spectrometer during the ion-transfer process [47]. Also, a singly reduced molybdate species $[Mo^VO_3]^-$, and the corresponding singly reduced tungstate species $[W^VO_6]^-$, has been observed in previous studies, along with mixed oxidation state fragments of polyoxomolybdate ions [47].

In this work, it is rather important to note that further information about the rearrangement processes taking place in solution from $[\alpha\text{-}Mo_8O_{26}]^{4-}$ through to the formation of the product cluster anion $[MnMo_6O_{18}((OCH_2)_3CNH_2)_2]^{3-}$ can be reliably extracted from the ESI-MS by monitoring the reaction as a function of time. These observations, when considered as a whole, are crucial in the effort to increase our understanding of the rearrangement processes taking place in the reaction solution. First, the exponential decrease in peak intensity of the $[Mo_8O_{26}TBA_3]^-$ and the $[Mo_4O_{13}TBA]^-$ ions could

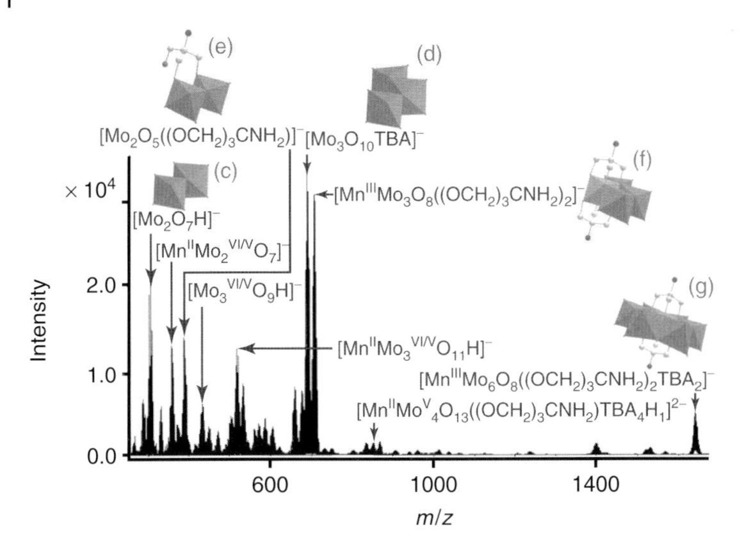

Figure 1.20 ESI-MS spectra collected of the reaction solution of $[\alpha\text{-}Mo_8O_{26}]^{4-}$, recorded after refluxing at 80 °C for approximately 30 h. Some of the intermediate prominent fragment ions are also shown here and are labeled similarly. Color scheme: Mo, dark grey polyhedra; Mn, grey polyhedra; O, grey spheres; N, dark grey spheres; and C, light grey. H atoms are omitted for clarity. Structural representation: (c) $[Mo_2O_7H]^-$, (d) $[Mo_3O_{10}TBA]^-$, (e) $[Mo_2O_5((OCH_2)_3CNH_2)]^-$, (f) $[Mn^{III}Mo_3O_8((OCH_2)_3CNH_2)_2]^-$, and (g) $[MnMo_6O_{18}((OCH_2)_3CNH_2)_2]^{3-}$.

suggest an initial rapid decomposition and rearrangement of the $[\alpha\text{-}Mo_8O_{26}]^{4-}$ via the formation of the $[Mo_4O_{13}]^{2-}$ cluster species, that is, $[Mo_4O_{13}Na_1]^-$ (m/z 614.6, base peak) and $[Mo_4O_{13}TBA]^-$ (m/z 833.8), which are half fragments of the $\{\alpha\text{-}Mo_8\}$ clusters and the most prominent peaks in the first spectrum recorded (see Figure 1.20), into the smaller, stable dinuclear, that is, $[Mo_2O_7H]^-$, and trinuclear, that is, $[Mo_3O_{10}TBA]^-$, isopolyoxomolybdate fragments. The concomitant increase in peak intensity of these small isopolyoxomolybdate ions over the time of reaction supports this proposal. Then the identification of similar dinuclear and trinuclear molybdate fragments possessing further coordination with TRIS and manganese ions, that is, $[Mo_2O_5((OCH_2)_3CNH_2)]^-$ and $[Mn^{III}Mo_3O_8((OCH_2)_3CNH_2)_2]^-$, and whose concentrations also increase over the time of reaction, suggests subsequent coordination of these building blocks with TRIS, manganese ions, and further molybdate anionic units, thereby building up the final Mn-Anderson-TRIS product ion.

The aforementioned cases demonstrate that the utilization of real-time mass spectrometry opens the door to investigations of even more complex chemical systems. Generally, the application of this approach to a wide range of nanomolecular systems will improve our understanding of the self-assembly processes and allow us to design efficient synthetic procedures which would lead to the formation of complex functional architectures.

1.4 Species Identification and Probing Structural Transformations in Multi-Metallic Systems

The diversity and effectiveness of CSI-MS studies has also been proven in the case of multi-metallic coordination systems. Careful scanning of $\{Co_{(12-x)}Ni_x\}$–tach (where tach is *cis,trans*-1,3,5-triaminocyclohexane) reaction mixtures of different concentration ratios allowed us to isolate the correct reaction conditions, which led to the identification, characterization, and finally the formation of pure desirable products [48]. The existence of a combinatorial library of Co–Ni species makes the system extremely difficult to characterize, yet allows us to design multinuclear bimetallic magnetic materials. Close examination of the ESI-MS data of the identified compounds clearly shows envelopes corresponding to the $\{Ni_{12}\}$ and $\{Co_{12}\}$ intact cluster species, which can each be assigned as $[Ni_{12}(CH_3O)_{12}(CH_3CO_2)_9(CO_3)(tach)_6]^+$ and $[Co_{12}(OH)_{12}(CH_3CO_2)_{10}(CO_3)(H_2O)_6(tach)_6(H)]^+$ (Figure 1.21). Furthermore, the $\{Co_6Ni_6\}$ species gives a CSI-MS envelope which matches that expected for a discrete mixed species (containing $(OH)_6$ and $(CH_3O)_6$ groups), rather than the supposition of many possible outcomes. Moreover, we found that isostructural mixed Ni–Co clusters could be accessed by controlling the pH–metal ion ratio of the reaction solution after studying the system by CSI-MS, and these mixed systems can be observed using MS and isolated in the pH range of 3.5–10.5 (Figure 1.22) as a function of pH and concentration ratios.

Using the same conditions and solution identification technique, we managed to isolate the $\{Co_{13}\}$ species (Figure 1.23, left), which represents the same structural motif with the aforementioned $\{M_{12}\}$ compounds and consists of three Co-cubanes that are bridged together through a $[CoO_3(OAc)]^{5-}$ unit. Utilizing CSI-MS studies,

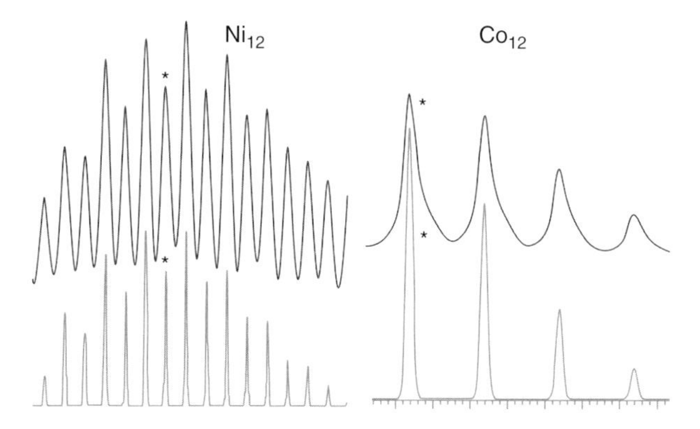

Figure 1.21 Cryospray mass spectrum for the $\{M_{12}\}$ clusters at −40 °C. Experimental (top) and simulated data (below). (b) $[Co_{12}(OH)_{12}(CH_3CO_2)_{10}(CO_3)(H_2O)_6(tach)_6H]^+$;* = m/z 2445. (a) $Ni_{12}(CH_3O)_{12}(CH_3CO_2)_9(CO_3)(tach)_6]^+$; * = m/z 2442 (simulated and experimental within 1 Da).

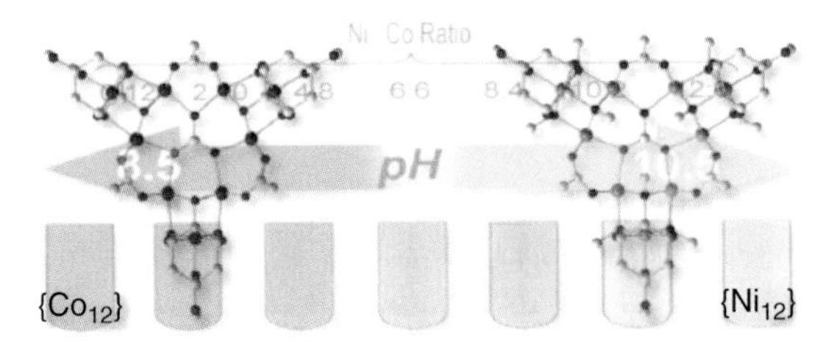

Figure 1.22 Structures of the $\{Co_{12}\}$ and $\{Ni_{12}\}$ complexes and a schematic showing the colors of mixed $\{Ni_{12-n}Co_n\}$ ($n = 1, 2, \ldots, 11$) intermediates. Color scheme: Ni, green; Co, purple; C, gray; N, blue; and O, red.

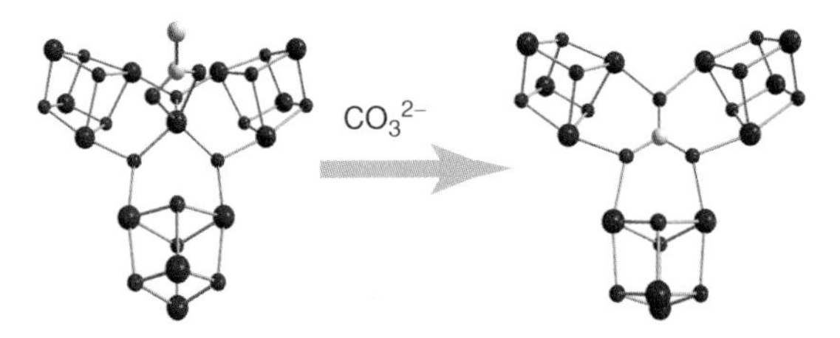

Figure 1.23 Comparison of cores of $\{Co_{13}\}$ (left) and $\{Co_{12}\}$ (right), capping *trans*-tach ligands omitted for clarity.

we observed the real-time structural transformation in solution of $\{Co_{13}\}$ to $\{Co_{12}\}$ daughter products by exchanging the central templating core $[CoO_3(OAc)]^{5-}$ for the carbonate anion (Figure 1.23) [48b]. The aforementioned observation also demonstrates that the $\{Co_{12}\}$ cluster core in solution originates from the $\{Co_{13}\}$ cluster core that forms in the absence of "external" ligand templates, giving us important hidden information about what takes place during the self-assembly process. The ability to induce systematic transformations to a given molecular framework or supramolecular architecture, that is, to fine-tune electronic or magnetic properties, is crucial in the design of functional molecule-based materials.

In another piece of work [49], the use of CSI-MS studies has allowed the identification of a multi-metallic Cr-based species in solution formed by reactive $\{Cr_6\}$ species, where the $\{Cr_6\}$ forms a larger complex via sodium fluoride interactions to give $[(NH_2Et_2)\{CrNa_{14}F_6(H_2O)_{10}\}\{Cr_6F_{11}(O_2CBu)_{10}\}_4]$. Although ESI-MS measurements did not allow the observation of the intact cluster, CSI-MS experiments on a $\{Cr_{25}\}$ (Figure 1.24) compound in THF:CH_3CN (70 : 30) at $-40\,^{\circ}C$ showed that the supramolecular assembly is present as a di-cation. The parent cluster ion can be found in two overlapping envelopes centered at an m/z of about 3410.2. Two species can be identified that are related to the parent cluster. Both species are doubly charged and the observed pattern corresponds to (i) $C_{204}H_{390}NCr_{25}F_{48}Na_{14}O_{89}$ and (ii) $C_{204}H_{388}NCr_{25}F_{50}Na_{14}O_{87}$, respectively. The most unusual aspect of the aforementioned material is that the $\{Cr_6\}$ "horseshoe" acts as a poly-nucleating fluoride donor and that the resulting supramolecular assembly has significant

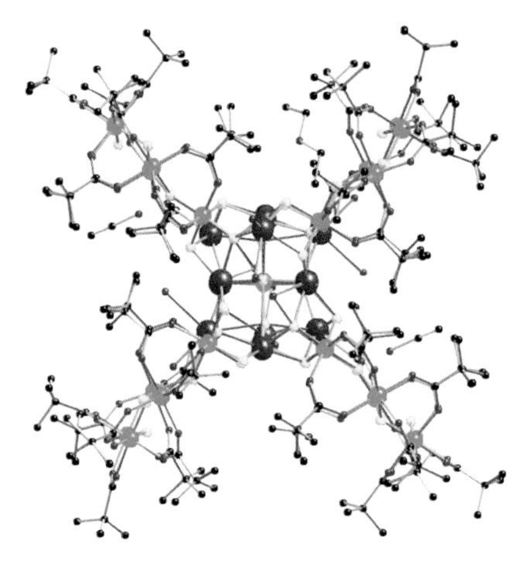

Figure 1.24 Crystal structure of $[(NH_2Et_2)\{CrNa_{14}F_6(H_2O)_{10}\}\{Cr_6F_{11}(O_2CBu)_{10}\}_4]$. Color scheme: Cr, green; O, red; F, yellow; N, blue; C, black; Na, purple; and H, not shown.

solution stability. CSI-MS is possibly the only tool currently available that can unambiguously demonstrate the structural integrity of this complex in solution.

CSI-MS can also be used to observe intact multi-metallic coordination complexes in solution, and this has been illustrated through the work investigating a novel pentanuclear palladium(II) complex [50]. For example, we have designed ligand L1 (see Figure 1.25) to interact with anionic species and to coordinate to metal ions. Interestingly, ligand L1 is transformed to L2 upon complexation. On reaction of compound L1 with palladium(II) acetate in dichloromethane (DCM), a pentanuclear Pd(II) complex is formed in which the Pd(II) mediates a ligand transformation of the aforementioned compound into a phenanthridone-based ligand (L2). The Pd(II) coordination compound has been characterized in the solid state via single-crystal X-ray crystallography and has been formulated as $[Pd_5(L_2)_2(OAc)_8](Br)_2$ (see Figure 1.26). CSI-MS investigations have also shown that the Pd_5L_2 core can be observed in the solution state by the identification of the $[Pd_5(C_{40}H_{34}N_{12}O)_2(CH_3CO_2)_5Cl_6]^-$ anion at -40 °C.

1.5
Future Challenges and Conclusions

In recent years, there has been an unprecedented rise in the number and size of interesting supramolecular clusters that have been structurally characterized by single-crystal X-ray crystallography. We showed that POMs and multi-metallic coordination compounds are ideal candidates for the development of a new type of supramolecular chemistry based upon the building-block ideas already established; using these ideas it should be possible to work toward designing nanomolecules

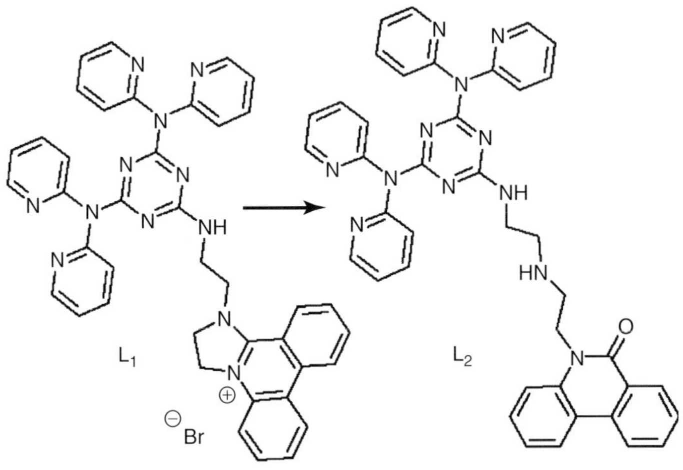

Figure 1.25 The transformation of compound L1 into L2 upon complexation with Pd(OAc)2 in DCM via a ring-opening process; this occurs in a 33% yield.

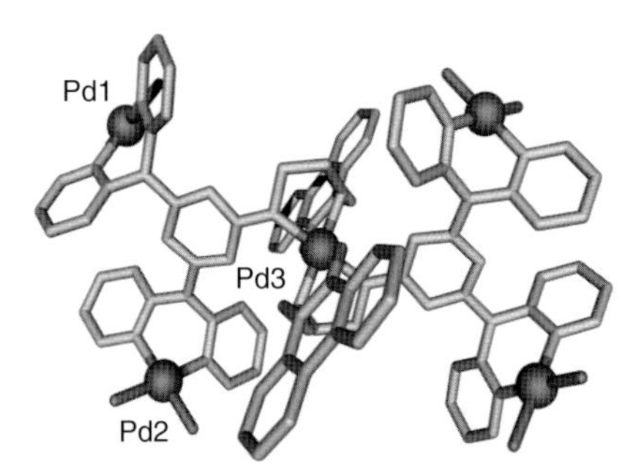

Figure 1.26 Ball-and-stick representation of $[Pd_5(L_2)_2(OAc)_8]^{2+}$. The solvent, anions of crystallization, and terminal acetate groups as well as the H atoms, are omitted for clarity. Color scheme: C, light grey sticks; N, grey sticks; O, dark grey sticks; and Pd, dark grey spheres. Pd_3 rests on an inversion center.

of increasing size and complexity. On the other hand, serendipity prevents us from designing the desirable functionality and manipulating the correct set of fundamental synthons under specific experimental conditions. In this chapter, we showed that, by utilizing the power of high-resolution ESI and CSI-MS to investigate very large labile frameworks, it is possible to overcome such drawbacks and promote further our understanding of the building-block principles and selection rules that govern the bottom-up self-assembly processes of inorganic complexes, supramolecular architectures, and cluster formation processes in solution. Although it is true that environmental conditions found in MS experiments differ from ordinary synthetic procedures, we have found them to be increasingly able to correlate the data from the MS studies with solution data. This is especially relevant to intrinsically charged clusters/supramolecular architectures, but it is important to note that labile systems may be easily disturbed by MS experiment and only experiments with good controls can be used with confidence to directly probe the solution phase. Indeed, the validation of MS studies for direct probing of the species in solution is one important part of our ongoing research.

Furthermore, the extension of the use of ESI-MS and CSI-MS for identification of intermediate reactive species in many different reaction mechanisms could reveal an unprecedented level of information, allowing the monitoring and classification of the species that prevail under specific conditions and directing the bottom-up self-assembly of fundamental building blocks toward higher nuclearity species through a molecular evolution process. Moreover, we will soon introduce the new field of VT-MS by presenting the first results obtained from complex chemical systems, which will open the door for additional insight to the assembly processes by "trapping" the key reactive intermediates that are responsible for the final structural features and properties of the materials. The above extracted information will allow initially a correlation to be made between the isolated architecture and potentially the common intermediate species that exist in solution, of conceptually related chemical systems, which will give us the opportunity to create a library of these key intermediates for future reference and design. This is an important area for research, as our increasing knowledge of these formation mechanisms will allow us to design procedures and direct the synthesis of new POM-based and multinuclear coordination clusters that adopt specific architectures and exhibit desirable functionality.

Finally, the discoveries that have been discussed in this chapter clearly show that the first significant steps toward this goal have been accomplished, while the ultimate achievement for controlled and directed self-assembly of complex functional nanoscale chemical systems is brought one step closer by employing these techniques. We may even be able to directly use mass spectrometry to guide synthetic routes to isolate desired products, plan more effective synthetic procedures based on the extracted information by ESI and CSI-MS studies, and design novel procedures for synthesizing functional materials, thereby bridging the gap between self-assembly, serendipity, and designed synthesis.

References

1. Steed, J.W. and Atwood, J.L. (2000) *Supramolecular Chemistry*, John Wiley & Sons, Ltd, Chichester.
2. Sanchez, G., de Soler-Illia, G.J., Ribot, F., Lalot, T., Mayer, C.R., and Cabuil, V. (2001) *Chem. Mater.*, **13**, 3061.
3. Long, D.-L., Kögerler, P., Farrugia, L.J., and Cronin, L. (2003) *Angew. Chem. Int. Ed.*, **42**, 4180.
4. Long, D.-L., Tsunashima, R., and Cronin, L. (2010) *Angew. Chem. Int. Ed.*, **49**, 1736.
5. (a) Long, D.-L., Abbas, H., Kögerler, P., and Cronin, L. (2004) *J. Am. Chem. Soc.*, **126**, 13880; (b) Long, D.-L., Kögerler, P., and Cronin, L. (2004) *Angew. Chem. Int. Ed.*, **43**, 1817.

6. (a) Cronin, L., Kögerler, P., and Müller, A. (2000) *J. Solid State Chem.*, **152**, 57; (b) Hussain, F., Gable, R.W., Speldrich, M., Kögerler, P., and Boskovic, C. (2009) *Chem. Commun.*, 328; (c) Kögerler, P., Tsukerblat, B., and Müller, A. (2010) *Dalton Trans.*, **39**, 21; (d) Mal, S.S. and Kortz, U. (2005) *Angew. Chem. Int. Ed.*, **44**, 3777; (e) Bassil, B.S., Ibrahim, M., Sankar Mal, S., Suchopar, A., Ngo Biboum, R., Keita, B., Nadjo, L., Nellutla, S., van Tol, J., Dalal, N.S., and Kortz, U. (2010) *Inorg. Chem.*, **49**, 4949.
7. Winpenny, R.E.P. (2002) *J. Chem. Soc. Dalton Trans.*, 1.
8. (a) Pope, M.T. and Müller, A. (1991) *Angew. Chem. Int. Ed. Engl.*, **30**, 34; (b) Müller, A. and Roy S. (2004) in *The Chemistry of Nanomaterials: Synthesis, Properties and Applications* (eds C.N.R. Rao, A. Müller, and A.K. Cheetham), Wiley-VCH Verlag GmbH, Weinheim, pp. 452–475.
9. Hill, C.L. (1998) *Chem. Rev.*, **98**, 1.
10. Long, D.-L., Burkholder, E., and Cronin, L. (2007) *Chem. Soc. Rev.*, **36**, 105.
11. Hasenknopf, B. (2005) *Front. Biosci.*, **10**, 275.
12. (a) Long, D.-L., Abbas, H., Kögerler, P., and Cronin, L. (2004) *J. Am. Chem. Soc.*, **126**, 13880; (b) Miras, H.N., Yan, J., Long, D.-L., and Cronin, L. (2008) *Angew. Chem. Int. Ed.*, **47**, 8420; (c) Ritchie, C., Ferguson, A., Nojiri, H., Miras, H.N., Song, Y.-F., Long, D.-L., Burkholder, E., Murrie, M., Kögerler, P., Brechin, E.K., and Cronin, L. (2008) *Angew. Chem. Int. Ed.*, **47**, 5609.
13. (a) Miras, H.N., Cooper, G.J.T., Long, D.-L., Bögge, H., Müller, A., Streb, C., and Cronin, L. (2010) *Science*, **327**, 72; (b) Ritchie, C., Streb, C., Thiel, J., Mitchell, S.G., Miras, H.N., Long, D.-L., Boyd, T., Peacock, R.D., McGlone, T., and Cronin, L. (2008) *Angew. Chem. Int. Ed.*, **47**, 6881; (c) Long, D.-L., Streb, C., Song, Y.-F., Mitchell, S.G., and Cronin, L. (2008) *J. Am. Chem. Soc.*, **130**, 1830.
14. (a) Lehn, J.-M. (2007) *Chem. Soc. Rev.*, **36**, 151; (b) Lehn, J.-M. (2002) *Science*, **295**, 2400; (c) Lehn, J.-M. (1999) *Chem. Eur. J.*, **5**, 2455; (d) Lehn, J.-M. (1995) *Supramolecular Chemistry: Concepts and Perspectives*, Wiley-VCH Verlag GmbH, Weinheim; (e) Lehn, J.-M. (1978) *Acc. Chem. Res.*, **11**, 49.
15. (a) Pluth, M.D. and Raymond, K.N. (2007) *Chem. Soc. Rev.*, **36**, 161; (b) Saalfrank, M.R.W. and Demleitner, B. (1999) in *Perspectives in Supramolecular Chemistry*, vol. **5** (ed. J.-P. Sauvage), John Wiley & Sons, Ltd, Chichester, p. 1; (c) Holliday, B.J. and Mirkin, C.A. (2001) *Angew. Chem. Int. Ed.*, **40**, 2022; (d) Winpenny, R.E.P. (1999) in *Perspectives in Supramolecular Chemistry*, vol. **5** (ed. Sauvage, J.-P.), John Wiley & Sons, Ltd, Chichester, p. 193; (e) Cutland, A.D., Malkani, R.G., Kampf, J.W., and Pecoraro, V.L. (2000) *Angew. Chem. Int. Ed.*, **39**, 2689; (f) Miras, H.N., Chakraborty, I., and Raptis, R.G. (2010) *Chem. Commun.*, 2569; (g) Oshio, H., Hoshino, N., Ito, T., Nakano, M., Renz, F., and Gütlich, P. (2003) *Angew. Chem. Int. Ed.*, **42**, 223; (h) Watton, S.P., Fuhrmann, P., Pence, L.E., Caneschi, A., Cornia, A., Abbati, G.L., and Lippard, S.J. (1997) *Angew. Chem. Int. Ed. Engl.*, **36**, 2774.
16. (a) Seeber, G., Kögerler, P., Kariuki, B.M., and Cronin, L. (2004) *Chem. Commun.*, 1580; (b) Seeber, G., Kögerler, P., Kariuki, B.M., and Cronin, L. (2002) *Chem. Commun.*, 2912.
17. (a) Miras, H.N., Long, D.-L., Kögerler, P., and Cronin, L. (2008) *Dalton Trans.*, 214; (b) Miras, H.N., Nieves Corella Ochoa, M., Long, D.-L., and Cronin, L. (2010) *Chem. Commun.*, 8148; (c) Walanda, D.K., Burns, R.C., Lawrance, G.A., and Von Nagy-Felsobuki, E.I. (1999) *Inorg. Chem. Commun.*, **10**, 487; (d) Miras, H.N., Stone, D.J., McInnes, E.J.L., Raptis, R.G., Baran, P., Chilas, G.I., Sigalas, M.P., Kabanos, T.A., and Cronin, L. (2008) *Chem. Commun.*, 4703.
18. (a) Tsunashima, R., Long, D.-L., Miras, H.N., Gabb, D., Pradeep, C.P., and Cronin, L. (2010) *Angew. Chem. Int. Ed.*, **49**, 113; (b) Ohlin, C.A., Villa, E.M., Fettinger, J.C., and Casey, W.H. (2008) *Angew. Chem. Int. Ed.*, **47**, 5634.
19. Sahureka, F., Burns, R.C., and von Nagy-Felsobuki, E.I. (2003) *Inorg. Chim. Acta*, **351**, 69.

20. (a) Aubriet, F., Maunit, B., Courrier, B., and Muller, J.F. (1997) *Rapid Commun. Mass Spectrom.*, **11**, 1596; (b) Molek, K.S., Reed, Z.D., Ricks, A.M., and Duncan, M.A. (2007) *J. Phys. Chem. A*, **111**, 8080.
21. (a) Walanda, D.K., Burns, R.C., Lawrance, G.A., and Von Nagy-Felsobuki, E.I. (1999) *J. Chem. Soc. Dalton Trans.*, 311; (b) Alyea, E.C., Craig, D., Dance, I., Fisher, K., Willett, G., and Scudder, M. (2005) *CrystEngComm*, **7**, 491.
22. (a) Bonchio, M., Bortolini, O., Conte, V., and Sartorel, A. (2003) *Eur. J. Inorg. Chem.*, **4**, 699; (b) Walanda, D.K., Burns, R.C., Lawrance, G.A., and Von Nagy-Felsobuki, E.I. (2000) *J. Cluster Sci.*, **1**, 5.
23. Boglio, C., Lenoble, G., Duhayon, C., Hasenknopf, B., Thouvenot, R., Zhang, C., Howell, R.C., Burton-Pye, B.P., Francesconi, L.C., Lacote, E., Thorimbert, S., Malacria, M., Afonso, C., and Tabet, J.C. (2006) *Inorg. Chem.*, **3**, 1389.
24. (a) Mayer, C.R., Roch-Marchal, C., Lavanant, H., Thouvenot, R., Sellier, N., Blais, J.C., and Sécheresse, F. (2004) *Chem. Eur. J.*, **21**, 5517; (b) Dablemont, C., Proust, A., Thouvenot, R., Afonso, C., Fournier, F., and Tabet, J.C. (2004) *Inorg. Chem.*, **43**, 3514.
25. (a) Sahureka, F., Burns, R.C., and von Nagy-Felsobuki, E.I. (2002) *Inorg. Chem. Commun.*, **5**, 23; (b) Sahureka, F., Burns, R.C., and von Nagy-Felsobuki, E.I. (2001) *J. Am. Soc. Mass Spectrom.*, **10**, 1136.
26. Deery, M.J., Howarth, O.W., and Jennings, K.R.J. (1997) *Chem. Soc. Dalton Trans.*, 4783.
27. Colton, R. and Traeger, J.C. (1992) *Inorg. Chim. Acta*, **201**, 153.
28. Tuoi, J.L. and Muller, E. (1994) *Rapid Commun. Mass Spectrom.*, **9**, 692.
29. (a) Waters, T., O'Hair, R.A.J., and Wedd, A.G. (2003) *J. Am. Chem. Soc.*, **125**, 3384; (b) Waters, T., O'Hair, R.A.J., and Wedd, A.G. (2003) *Int. J. Mass Spectrom.*, **228**, 599; (c) Gun, J., Modestov, A., Lev, O., Saurenx, D., Vorotyntsev, M.A., and Poli, R. (2003) *Eur. J. Inorg. Chem.*, **3**, 482.
30. (a) Sakamoto, S., Fujita, M., Kim, K., and Yamaguchi, K. (2000) *Tetrahedron*, **56**, 955; (b) Saito, K., Sei, Y., Miki, S., and Yamaguchi, K. (2008) *Toxicon*, **51**, 1496.
31. Kunimura, M., Sakamoto, S., and Yamaguchi, K. (2002) *Org. Lett.*, **4**, 347.
32. Nakatani, K., Hagihara, S., Sando, S., Sakamoto, S., Yamaguchi, K., Maesawa, C., and Saito, I. (2003) *J. Am. Chem. Soc.*, **125**, 662.
33. Yamaguchi, K. (2003) *J. Mass Spectrom.*, **38**, 473.
34. Sakamoto, S. and Yamaguchi, K. (2003) *Angew. Chem. Int. Ed. Engl.*, **42**, 905.
35. (a) Jeannin, Y., and Martin-Frére, J. (1979) *Inorg. Chem.*, **18**, 3010; (b) Ozawa, Y., and Sasaki, Y. (1987) *Chem. Lett.*, **185**, 923. (c) Rodewald, D., and Jeannin, Y. (1999) *C. R. Acad. Sci. Ser. IIc: Chim.*, **2**, 161.
36. Song, Y.-F., Long, D. -L., Kelly, S.E., and Cronin, L. (2008) *Inorg. Chem.*, **47**, 9137.
37. (a) Long, D.-L., Abbas, H., Kögerler, P., and Cronin, L. (2005) *Angew. Chem. Int. Ed.*, **44**, 3415; (b) Yan, J., Long, D.-L., Miras, H.N., and Cronin, L. (2010) *Inorg. Chem.*, **49**, 1819.
38. Long, D.-L., Song, Y.-F., Wilson, E.F., Kögerler, P., Guo, S.-X., Bond, A.M., Hargreaves, J.S.J., and Cronin, L. (2008) *Angew. Chem. Int. Ed.*, **47**, 4384.
39. Long, D.-L., Kögerler, P., Parenty, A.D.C., Fielden, J., and Cronin, L. (2006) *Angew. Chem. Int. Ed.*, **45**, 4798.
40. Pradeep, C.P., Long, D.-L., Newton, G.N., Song, Y.-F., and Cronin, L. (2008) *Angew. Chem. Int. Ed.*, **23**, 4388.
41. Zheng, S.T., Chen, Y.M., Zhang, J., Xu, J.Q., and Yang, G.Y. (2006) *Eur. J. Inorg. Chem.*, **2**, 397.
42. Abbas, H., Pickering, A.L., Long, D.-L., Kögerler, P., and Cronin, L. (2005) *Chem. Eur. J.*, **4**, 1071.
43. Wilson, E.F., Abbas, H., Duncombe, B.J., Streb, C., Long, D.-L., and Cronin, L. (2008) *J. Am. Chem. Soc.*, **130**, 13876.
44. Gatehouse, B.M. and Leverett, P. (1976) *J. Chem. Soc. Dalton Trans.*, 1316.
45. Marcoux, P.R., Hasenknopf, B., Vaissermann, J., and Gouzerh, P. (2003) *Eur. J. Inorg. Chem.*, **13**, 2406.

46. Wilson, E.F., Miras, H.N., Rosnes, M.H., and Cronin, L. (2011) *Angew. Chem. Int. Ed.*, **50**, 3720.
47. (a) Gun, J., Modestov, A., Lev, O., Saurenx, D., Vorotyntsev, M.A., and Poli, R. (2003) *Eur. J. Inorg. Chem.*, **3**, 482; (b) Sahureka, F., Burns, R.C., and von Nagy-Felsobuki, E.I. (2002) *Inorg. Chim. Acta*, **332**, 7.
48. (a) Cooper, G.J.T., Newton, G.N., Kögerler, P., Long, D.-L., Engelhardt, L., Luban, M., and Cronin, L. (2007) *Angew. Chem. Int. Ed*, **46**, 1340; (b) Newton, G.N., Cooper, G.J.T., Kögerler, P., Long, D.-L., and Cronin, L. (2008) *J. Am. Chem. Soc*, **130**, 790.
49. Rancan, M., Newton, G.N., Muryn, C.A., Pritchard, R.G., Timco, G.A., Cronin, L., and Winpenny, R.E.P. (2008) *Chem. Commun.*, 1560.
50. Kitson, P.J., Song, Y.-F., Gamez, P., de Hoog, P., Long, D.-L., Parenty, A.D.C., Reedijk, J., and Cronin, L. (2008) *Inorg. Chem.*, **47**, 1883.

2
Efficient Synthesis of Natural Products Aided by Automated Synthesizers and Microreactors

Shinichiro Fuse, Kazuhiro Machida, and Takashi Takahashi

2.1
Efficient Synthesis of Natural Products Aided by Automated Synthesizers

The most exciting aspect of the total synthesis of natural products is the construction of new synthetic routes. Synthetic chemists try to find the precise reaction conditions that activate only the desired functional groups (Figure 2.1). Expert thought, knowledge, and skills are required for this task. It is preferable that synthetic chemists spend most of their time and effort in constructing new synthetic routes for efficient total synthesis of natural products. However, a great deal of effort is necessary to perform simple, repetitive processes, including the supply of synthetic intermediates using known synthetic routes, especially in the total synthesis of complicated natural products, because many synthetic steps are usually necessary for the preparation of synthetic intermediates (Figure 2.1). Synthetic chemists sometimes encounter difficulties in reproducing results. The lack of reproducibility usually stems from the variability of synthetic manipulation. For example, in the case of the dropwise addition of a reagent into a solution of a substrate via syringe over 5 min, for the first attempt, a synthetic chemist might add the reagent very slowly in the early stage and accelerate the rate of addition in the later stages. However, during a second attempt, the same chemist might add the reagent at the same rate over 5 min. The subtle difference in the experimental manipulation may affect the temperature and concentration of the reaction mixture, which might change the results of the reaction, especially during the syntheses of complicated natural products that have multiple reaction points. It is sometimes difficult to detect these subtle differences in experimental manipulation from published papers and experimental notebooks.

The deployment of automated synthesizers might be a solution to these problems [1]. If synthetic chemists run the reactions in automated synthesizers and digitally store the experimental procedures, anyone could reproduce the results by using the same apparatus and compounds, regardless of the time and place. As a result, synthetic chemists could expend more effort on the advanced and challenging work of construction of new synthetic routes. However, the automation of general solution-phase syntheses is limited by the wide variety of experimental operations

New Strategies in Chemical Synthesis and Catalysis, First Edition. Edited by Bruno Pignataro.

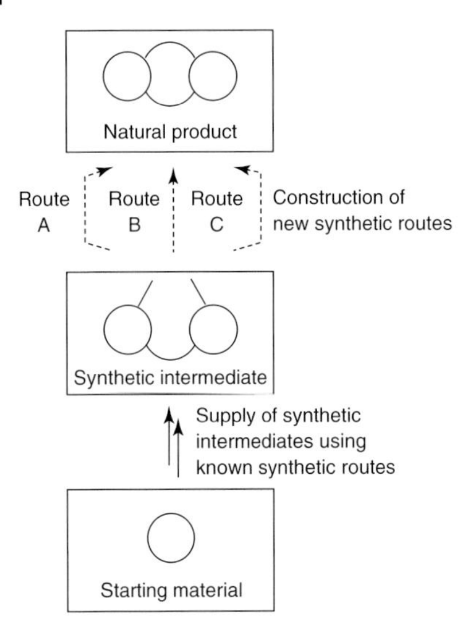

Figure 2.1 Process of total synthesis of natural products.

required, compared to those used for solid-phase synthesis. The automation of liquid–liquid extractions in particular is not easy [2]. Moreover, it is sometimes necessary to modify the reaction conditions and/or experimental operations to utilize an automated synthesizer [3, 4]. Therefore, the application of an automated process to solution-phase synthesis is generally challenging. It is nevertheless important to apply various solution-phase reactions to an automated synthesizer. This chapter describes the efficient synthesis of taxol [5] and nine-membered masked enediyne [6], aided by automated synthesizers.

2.1.1 The Process of Automating the Supply of Synthetic Intermediates

Automated syntheses of synthetic intermediates were performed via the four stages as follows:

1) Understand the properties of automated synthesizers.
2) Perform reactions manually and classify them into two groups. Reactions in the first group can be performed by automated synthesizers without altering the reaction conditions and/or synthetic operations, while reactions in the second group require alteration of the conditions and/or the operations before utilizing the apparatus. Decide which reaction conditions and/or operations must be changed in the second group.
3) Examine the influence of the changes in conditions and/or operations for the reaction results. Usually, these examinations are performed manually. If a decrease in conversion and/or purity is observed, the conditions and/or operations should be changed again.
4) Attempt to perform reactions in automated synthesizers.

In this chapter, we explain the detailed process for carrying out a "general reaction" in Figure 2.3, utilizing our original automated synthesizer, ChemKonzert® (Figure 2.2) [7]. ChemKonzert shares many common features with other solution-phase automated synthesizers. Therefore, the process of automating the supply of synthetic intermediates should also be useful for chemists attempting to automate synthetic operations when using other automated synthesizers.

1) ChemKonzert [7] consists of 2 reaction vessels (RF1, RF2, 100–1000 ml), a centrifugal separator (SF, 700 ml), 2 receivers (SF1, SF2) (500 ml), 2 glass filters (FF1, FF2) (500 and 100 ml), 12 substrate and reagent reservoirs (RR1–RR12) (100–200 ml), 6 solvent and wash solution bottles (RS1–RS6) (500 ml), 3 drying pads (DT1–DT3), a round flask (CF), 2 solvent tanks (WT1, WT2), and a computer controller. The glassware is interconnected with Teflon® tubes, and solutions are transferred under reduced pressure using a diaphragm pump. Separation of the organic and aqueous layers is performed by measuring the electroconductivity of the two phases with a sensor, and the liquid flow is regulated by solenoid valves controlled with our original Windows software (KonzertMeister). The users upload the procedures into a computer, add substrates and reagents to the reservoirs, and fill the solvent bottles. The synthesizer carried out the reaction procedures as follows. The substrate and reagents in the RR were added to the RF at a controlled reaction temperature (20–130 °C) under a nitrogen atmosphere. After the reaction was complete, a quenching reagent in the RR or the RS was added to the RF and the mixture was transferred to the SF, with removal of the precipitate through the FF. After centrifugal separation, the two resulting phases were separated, their electroconductivity was measured with a sensor, and they were transferred to the SF1 and SF2. The aqueous solution in SF1 was returned to the RF. After addition of the extraction solvent from the RS, the mixture was stirred and then transferred to the SF. After two or three extractions, the combined organic solution in the SF2 receiver was washed with aqueous solutions of sodium bicarbonate and sodium chloride from the RS in the RF. The organic layer was separated in the SF and transferred to SF2. The organic layer in SF2 was then passed through a $MgSO_4$ or a Na_2SO_4 plug, DT, for drying. The filtrate was stored in the CF for purification after evaporation of the solvent (manual). Unless purification was necessary, the filtrate was directly transferred to another RF, concentrated under reduced pressure, and the next reaction was carried out sequentially. Finally, the whole apparatus was washed with water and acetone from the solvent tanks and dried under reduced pressure.
   Certain properties of the automated synthesizer, shown below, must be considered before performing a reaction.
   a. Solutions are transferred through the Teflon tube; therefore, the transfer of a large amount of viscous insoluble solids is difficult. Many solution-phase automated synthesizers other than ChemKonzert also transfer solutions by tubes, thus the transfer of solids is not easy.
   b. Separation of two layers is performed by measuring their electroconductivity with a sensor.

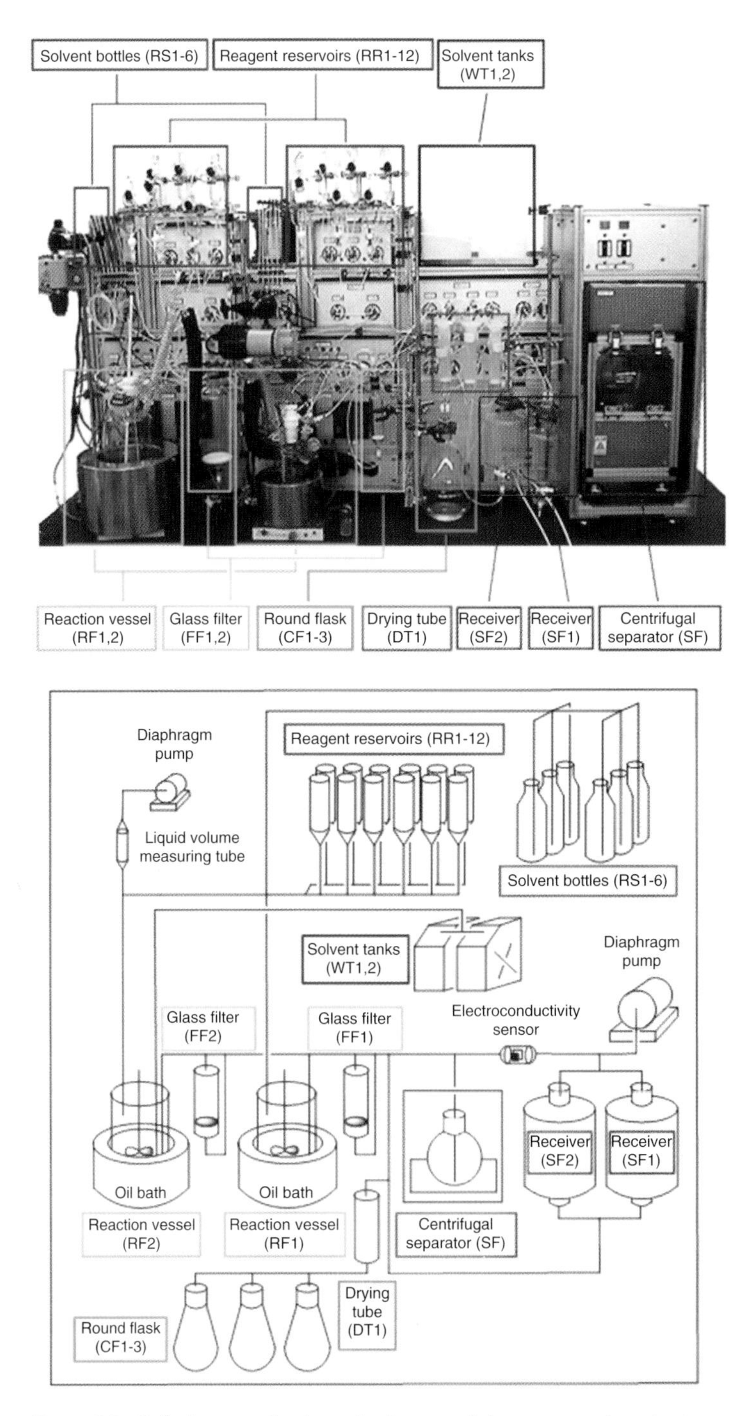

Figure 2.2 Full picture and schematic diagram of the automated synthesizer, ChemKonzert.

If a large amount of organic solvent capable of dissolving salts, such as MeOH, is used, the difference in the electroconductivity of the two layers is reduced, so the automated synthesizer cannot detect the interface of the two layers. This is also a common feature in many solution-phase automated synthesizers that use electroconductivity sensors for phase separation.

2) The "general reaction" in Figure 2.3 must be performed manually, with the previously described properties of the automated synthesizer in mind. The solubility of reagents and substrates, as well as the precipitation of solids, should be checked at this point. If the experimental operations that were carried out manually cannot be performed by the automated synthesizer, the experimental operations and/or reaction conditions must be changed (Figure 2.3).
   a. **Operation A**: MeOH is used as a solvent in the "general reaction," which will cause a phase separation error (Figure 2.3). MeOH should be changed to mixed solvents, such as MeOH/THF or MeOH/THF/H_2O. Researchers must select a suitable solvent that will not decrease the yield of the product sought.
   b. **Operation B**: In the manual operation of the "general reaction," the liquid reagent b was added via syringe into the reaction mixture, which was cooled via an ice bath (Figure 2.3). The temperature of the reaction mixture during the addition of reagent b can be precisely controlled using a sensor and a drop-rate controller.
   c. **Operation C**: In the manual operation of the "general reaction," a solid reagent c was added directly into the reaction mixture (Figure 2.3). Many automated synthesizers, including ChemKonzert, cannot transfer solids. Therefore, the solid reagent c must be dissolved in a solvent and the resulting solution of reagent c added to the reaction mixture.
   d. **Operation D**: In the manual operation of the "general reaction," the reaction mixture was directly added to an Erlenmeyer flask using a funnel; thus, the precipitation of solids was not a big problem (Figure 2.3). However, automated synthesizers, including ChemKonzert, cannot transfer solids; therefore, a suitable solvent that can dissolve the solids must be added before the transfer of the reaction mixture.
   e. **Operation E**: In the manual operation of the "general reaction," extractions were performed by shaking the mixture of two layers using a separation funnel, and they were separated by detecting the interface of the two layers visually (Figure 2.3). Extractions in the automated synthesizer were performed by vigorous stirring of the two layers in the reaction flask, and then separating them by measurement of their electroconductivity with a sensor.
   f. **Operation F**: In general, the organic layer is dried by adding drying agents such as $MgSO_4$ or Na_2SO_4 to the organic layer (Figure 2.3). Automated synthesizers cannot transfer solids, so the organic layer is dried by passing it through a dry pad column.

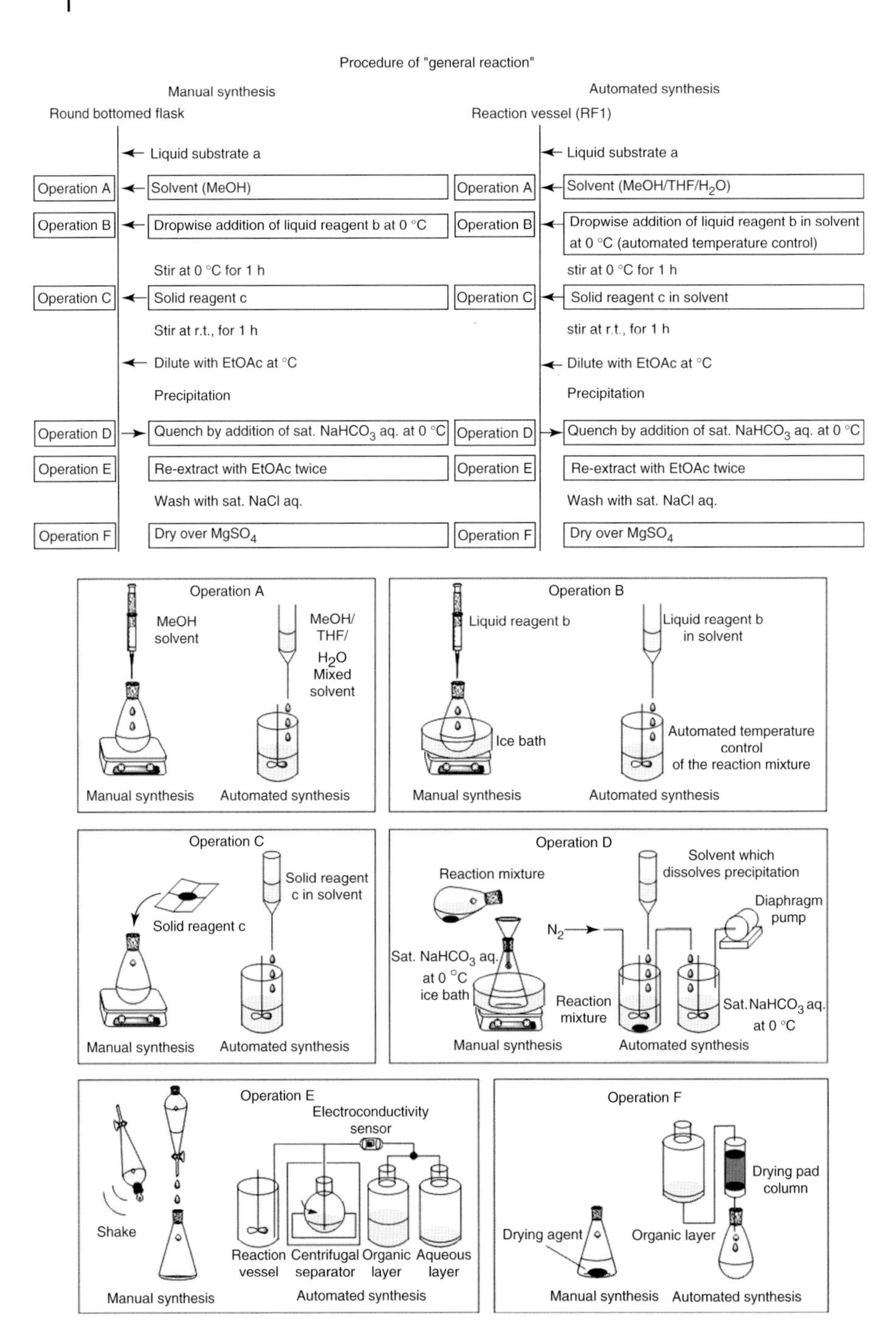

Figure 2.3 Manual and automated procedures for the "general reaction".

3) The influence of the changes in reaction conditions and/or synthetic operations (described in point 2) on the results of the reactions must be checked.
   a. **Operation A**: It is necessary to check the influence of the alteration of the solvent, from MeOH to the mixed solvent, on the results of the reaction.
   b. **Operation B**: Selectivity in the reaction might be improved by altering the temperature automatically. However, this alteration might change the reaction time. The reaction should be followed by carefully monitoring the temperature of the reaction mixture to determine the suitable reaction time.
   c. **Operation C**: The solubility of solid reagent c should be examined first. The need for additional examinations is dependent on the solubility of reagent c, as shown in Figure 2.4.
   d. **Operation D**: As with operation A, it is necessary to add the proper amount of a suitable solvent to dissolve the precipitate. It is sometimes possible for ChemKonzert to transfer solids, if they can be finely milled by vigorous stirring.
   e. **Operation E**: The extraction operation of the automated synthesizer is different from that of the manual operation, and this might change the extraction efficiency. The extraction operation of the automated synthesizer should be reproduced manually, as precisely as possible, to determine the proper stirring speed and time to extract the desired compound fully.
   f. **Operation F**: It is necessary to determine the proper amount of a suitable drying agent, depending on the amount of the organic layer.
4) The automated synthesizer should be tested using water and an organic solvent, without using reagents and substrates. Clogs in tubes or valves, leaks from joints, and movement of valves, pumps, and sensors should be checked. If no problem is observed, the "general reaction" in Figure 2.3 should be attempted in the automated synthesizer, using reagents and substrates. Generally, it is best for the operator to check all automated

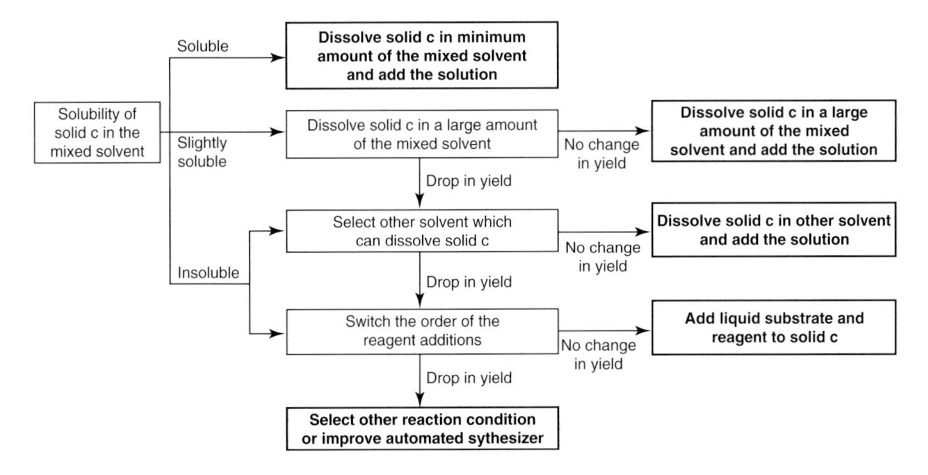

Figure 2.4 Examinations for the addition of solid reagent c, utlizing ChemKonzert.

operations visually during the first trial. In particular, the separation of two layers by the electroconductivity sensor should be carefully checked, because the electroconductivity of the two layers would differ from the test operation because of the use of reagents and substrates.

The process described above is for automated synthesis using ChemKonzert. The detailed issues that need to be tested before automated synthesis differ with the synthesizer used. It is very important to understand the properties of the automated synthesizer and decide which reaction conditions and/or synthetic operations must be changed.

2.1.2
Efficient Synthesis of a Cyanohydrin Key Intermediate for Taxol Using Automated Synthesizers

Taxol possesses very potent antitumor activity, and it is used clinically for various cancers [8]. Taxol is notable for its difficult synthesis [9]. In fact, more than 200 research groups have tried to synthesize the compound since 1971, when its structure was determined by Wall *et al.* [10]. It took more than 20 years for the first total synthesis, accomplished by Holton *et al.* [11]. Only seven synthesis routes, including that by our group, have been reported [11–16]. Generally, teams consisting of many expert synthetic chemists are formed for the total synthesis of complicated natural products such as taxol. Generally, chemists focus on constructing new routes, and must depend on others to provide a sufficient amount of synthetic intermediates using an established synthetic route. Perhaps even one student could achieve the total synthesis of a complex natural product if the supply of synthetic intermediates could be provided through automated synthesizers. The description of a successful racemic formal total synthesis of taxol follows. The 36 steps in the synthesis of the key intermediate were performed by automated synthesizers [5].

The strategy for the synthesis of taxol (**1**) is shown in Scheme 2.1. The 13-OH group is located near the C ring, owing to the endo structure of the taxane skeleton (see the 3D structure of the taxane skeleton in Scheme 2.1), and the 13-OH group might decrease the reactivity of the functional groups on the C ring. The 7-OH group is easily epimerized under the influence of the 9-keto group through the retro-aldol and aldol sequences [17]. In consideration of these issues, 9,13-dideoxy enone **3** was designed as a precursor of taxol. The intention was to introduce the 9- and 13-OH groups by using the 10-keto group and the Δ^{11} alkene of **3** at the final stage of the total synthesis. The oxetane ring will be attacked by an adjacent hydroxy group under both acidic and basic conditions [18]. Therefore, the plan was to construct the oxetane ring at a later stage of the total synthesis using $\Delta^{4,(20)}$-exo methylene of **4**. The crucial issue in the total synthesis of taxol is the construction of the B ring. This eight-membered ring is highly strained by steric repulsions between the 16-, 17-, and 19-methyl groups, and also by the existence of the bridge head Δ^{11}-alkene. We planned to perform the intramolecular alkylation of protected cyanohydrin **5** because this alkylation reaction is not reversible [19, 20]. In addition,

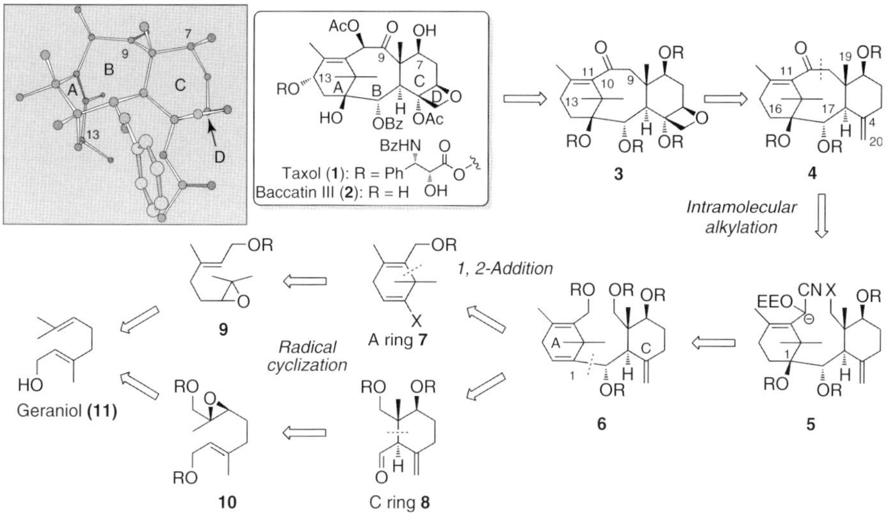

Scheme 2.1 Synthetic strategy for taxol.

the anion generated from the protected cyanohydrin is thermally stable. We can, therefore, perform the cyclization at a high temperature to make up for the entropic and enthalpic disadvantages in cyclization. The protected cyanohydrin **5** can be obtained by the stereo- and regio-controlled introduction of the 1-OH group into allylic alcohol **6**, followed by the protecting group changes. The allylic alcohol **6** can be obtained by the stereoselective 1,2-addition of the alkenyl anion, generated from the A ring **7**, to the C ring **8**. The plan was to prepare both A and C rings from the epoxy alkenes **9** and **10** by way of Ti(III)-mediated radical cyclization [21, 22]. These epoxy alkenes can be prepared from the common starting material geraniol (**11**). This synthetic strategy offers an enantioselective approach to taxol (**1**), starting from the optically active epoxide **10**, which is readily prepared by the Sharpless asymmetric epoxidation of the allylic alcohol.

A ring **7** and C ring **8** were prepared utilizing the commercially available automated synthesizer, Sol-capa, after some modifications (Figure 2.5). Sol-capa consists of one reaction vessel (1000 ml, −5 to 150 °C), one extraction vessel (1000 ml), two receivers (1000 ml), seven solvent and wash-solution bottles (200 ml), one solvent tank, one drying pad, and a computerized controller. The glassware is interconnected with Teflon tubes, and the solutions are transferred using a pump. Separation of the organic and aqueous layers is performed by measuring the electroconductivity of the two phases with a sensor, and the liquid flow is regulated by solenoid valves controlled with Windows software. The users upload the procedures into a computer, fill the solvent and wash solutions, and manually add substrates and reagents to the reaction vessel. The synthesizer carried out the automated workup as follows. After the reaction was complete, a quenching reagent in the wash solution bottle was added to the reaction vessel and the mixture was transferred to the extraction vessel, with removal of the precipitate through a small

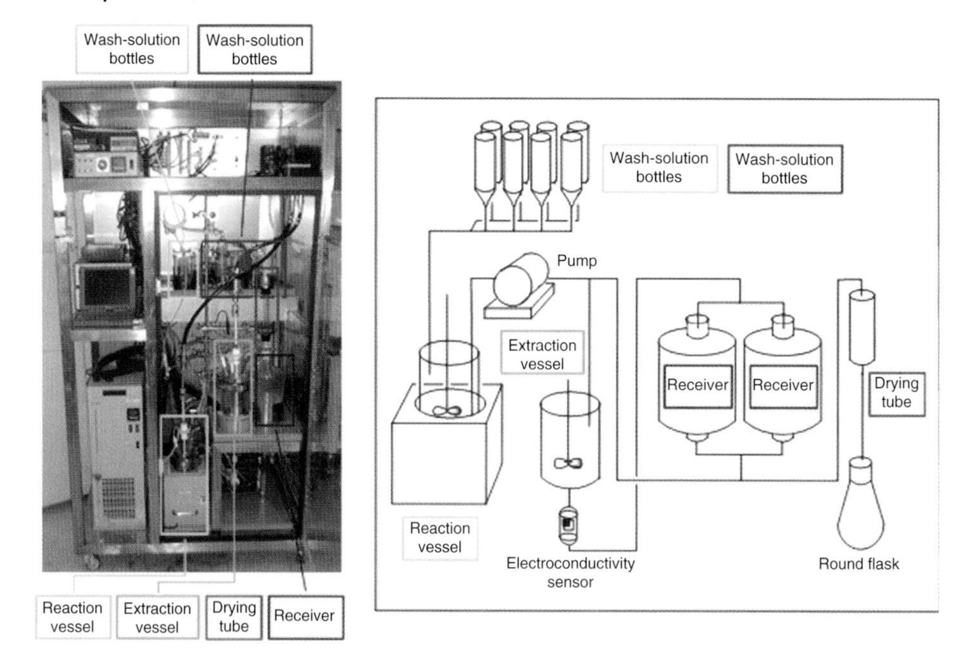

Figure 2.5 Full picture and schematic diagram of the automated synthesizer Sol-capa.

glass filter. The two phases were separated, their electroconductivity was measured with a sensor, and they were transferred to the two receivers. The aqueous solution in the receiver was returned to the extraction vessel. After addition of the extraction solvent from the solvent bottle, the mixture was stirred and separated. After two or three extractions, the combined organic solution in another receiver was washed with aqueous solutions of sodium bicarbonate and sodium chloride from the wash solution bottles. The organic layer was separated in the extraction vessel and transferred to the receiver. The organic layer in the receiver was then passed through a $MgSO_4$ or a Na_2SO_4 plug for drying. The filtrate was transferred to a flask for purification after evaporation of the solvent (manual). Finally, the entire apparatus was washed with organic solvent and/or water from the solvent tank.

This apparatus could not perform three types of reactions:

- reactions that accompany a large amount of precipitation;
- reactions that accompany the generation of an emulsion during the phase-separation process; and
- reactions in which a large amount of an organic solvent capable of dissolving salts, such as MeOH, is used.

Considering these properties of Sol-capa, reactions a–l (Scheme 2.2) were examined. As a result, the reactions indicated in italics required alteration of the reaction conditions. In reactions d and q, at least 3 equiv. of Cp_2TiCl_2 was required to complete the reaction; thus, a large amount of titanium salts was precipitated after the reaction was quenched. The Cp_2TiCl_2-catalyzed radical cyclization conditions

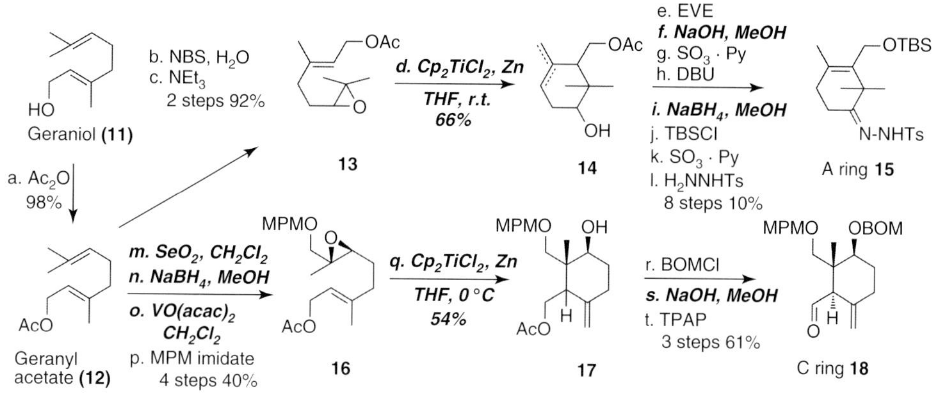

Scheme 2.2 Synthesis of A ring **15** and C ring **18**. NBS: *N*-bromosuccinimide; Cp: cyclopentadienyl; DBU: 1,8-diazabicyclo[5.4.0]undecene; TBS: *tert*-butyldimethylsilyl; EE: ethoxy ethyl; Ts: *p*-toluenesulfonyl; MPM: 4-methoxyphenylmethyl; BOM: benzyloxymethyl; TPAP: tetrapropylammonium perruthenate.

were examined. As a result, reactions d and q were successfully performed in automated synthesizers, utilizing novel catalytic radical cyclization conditions (d: 0.1 equiv. Cp_2TiCl_2, Mn, Et_3B, lutidine·HCl; q: 0.1 equiv. Cp_2TiCl_2, Mn, Et_3B, TMSCl) [21]. In reactions f, i, n, and s, a large amount of MeOH was used as solvent. A decision was made to use the mixed solvent $MeOH/THF/H_2O$ instead of MeOH. In reactions m and o, a large amount of a heavy organic solvent, CH_2Cl_2, was used. This caused the generation of an emulsion during the phase separation. Hexane in reaction m and toluene in reaction o were used instead of CH_2Cl_2. As a result of these alterations, a total of 18 steps were successfully performed in the Sol-capa in the 20–300 g scale. Purities and yields were almost the same as those obtained by manual experiments.

The preparation of the key synthetic intermediate **24** was performed in ChemKonzert (Scheme 2.3). The synthetic scheme included reactions that employed anhydrous *t*-BuLi, alkenyl lithium, and $LiAlH_4$. In addition, reactions d and n were accompanied by the evolution of a large amount of gas during the work-up process. These reactions could not be performed using Sol-capa. Thus, we utilized our original automated synthesizer, ChemKonzert (Figure 2.2). A total of 15 steps were successfully performed utilizing ChemKonzert on a scale that ranged from 100 mg to 1 g. Purities and yields were almost the same as those obtained by manual experiments. ChemKonzert removed the solvent by heating the solutions under reduced pressure. This function enabled the sequential performance of multiple reactions in ChemKonzert. In fact, several two-step reactions, namely, g and h, i and j, and k and l, were sequentially performed without taking compounds out of ChemKonzert after the first reaction. This was very valuable, because it allowed setting up of the substrate, solvents, and reagents and overnight operation of the programmed procedure to obtain the product of the two-step reactions.

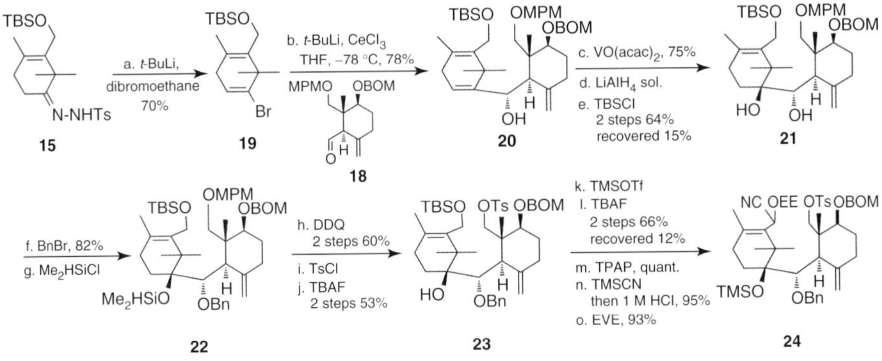

Scheme 2.3 Synthesis of protected cyanohydrin **24**. DDQ: 2,3-dichloro-5,6-dicyano-1,4-benzoquinone; TBAF: tetrabutylammonium fluoride; TMS: trimethylsilyl; Tf: trifluoromethanesulfonyl.

The crucial eight-membered ring formation from **24** to **25** was effectively assisted by microwave irradiation in the presence of excess $LiN(TMS)_2$ (Scheme 2.4). The reaction period was dramatically decreased from 10 h (conventional heating condition) to 15 min (microwave irradiation condition) [5].

Further transformations were carried out manually (Scheme 2.4). Hydrolysis of cyanohydrin ether **25** afforded the corresponding ketone **26**. Regio- and stereoselective allylic oxidation, dihydroxylation of $\Delta^{4,(20)}$-exo methylene with OsO_4–quinuclidine complex, and selective acetylation of the primary alcohol, followed by mesylation of the secondary alcohol in **26**, afforded **27**. Prior to the formation of the oxetane ring, **27** was converted into 1,2-carbonate **28** (Scheme 2.4). Hydrogenolysis of the benzyl and benzyloxymethyl ethers in **27**, followed by treatment with triphosgene, provided the carbonate. Triethylsilyl (TES) protection of 7-OH and removal of the acetyl group afforded **28**. Oxetane formation without deconjugation of Δ^{11}-alkene to $\Delta^{12(18)}$-exo methylene was crucial. Treatment of **28** with *N,N*-diisopropylethylamine (DIEA) in hexamethylphosphoric triamide (HMPA) at 100 °C provided the desired **29** in 77% yield with recovery of the starting material **28** (20%). Acetylation and addition of phenyl lithium to the 1,2-carbonate group of **29** provided the benzoate, which is a racemic form of the Danishefsky intermediate (m.p. 214–216 °C) for the synthesis of taxol (**1**). Some modification of the reported six-step procedure led to the total synthesis of (±)-baccatin III (**2**) (m.p. 221–223 °C), with spectral data identical to that previously reported. Thus, the racemic formal total synthesis of taxol was accomplished aided by automated synthesizers [5].

2.1.3
Efficient Synthesis of a Cyclic Ether Key Intermediate for Nine-Membered Masked Enediyne, Using an Automated Synthesizer

Naturally occurring antibiotics, such as kedarcidin chromophore (**30**) [23, 24] and C-1027 chromophore (**31**) [25, 26], that contain a nine-membered enediyne skeleton

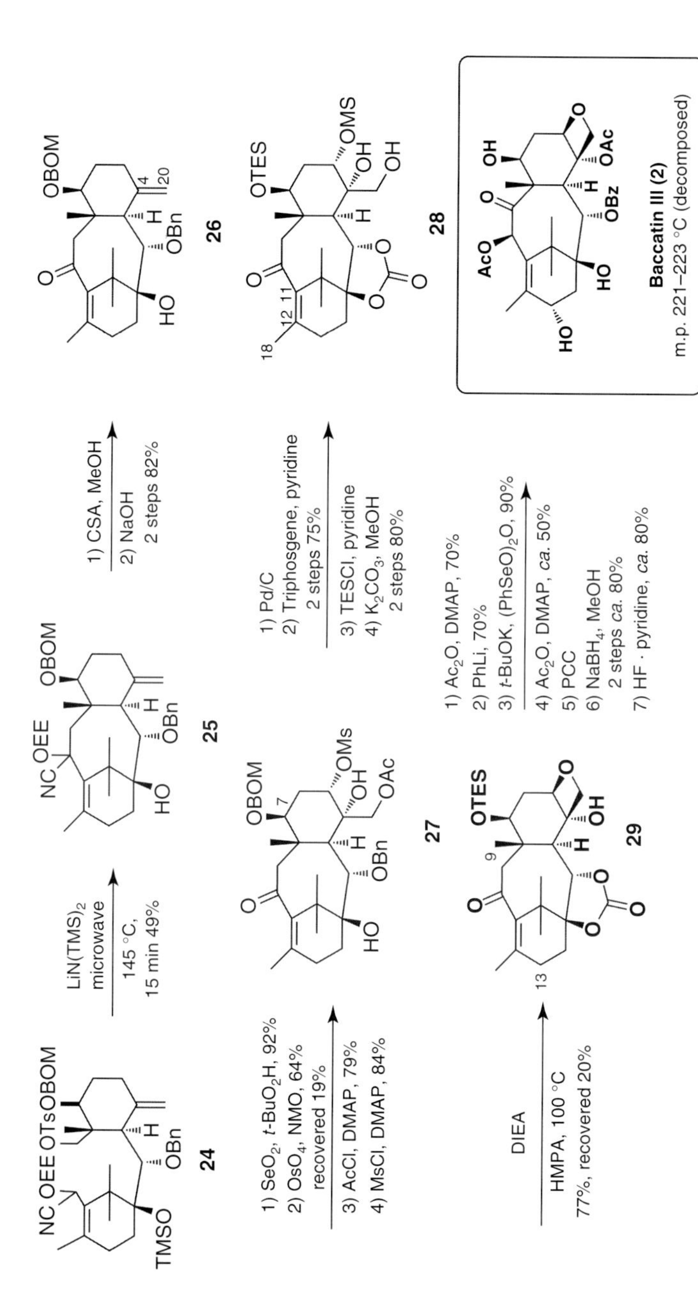

Scheme 2.4 Racemic formal total synthesis of taxol. CSA: camphorsulfonic acid; NMO: 4-methylmorpholine *N*-oxide; Ms: methanesulfonyl; DMAP: 4-(dimethylamino)pyridine; TES: triethylsilyl; DIEA: *N*,*N*-diisopropylethylamine; HMPA: hexamethylphosphoric triamide; PCC: pyridinium chlorochromate.

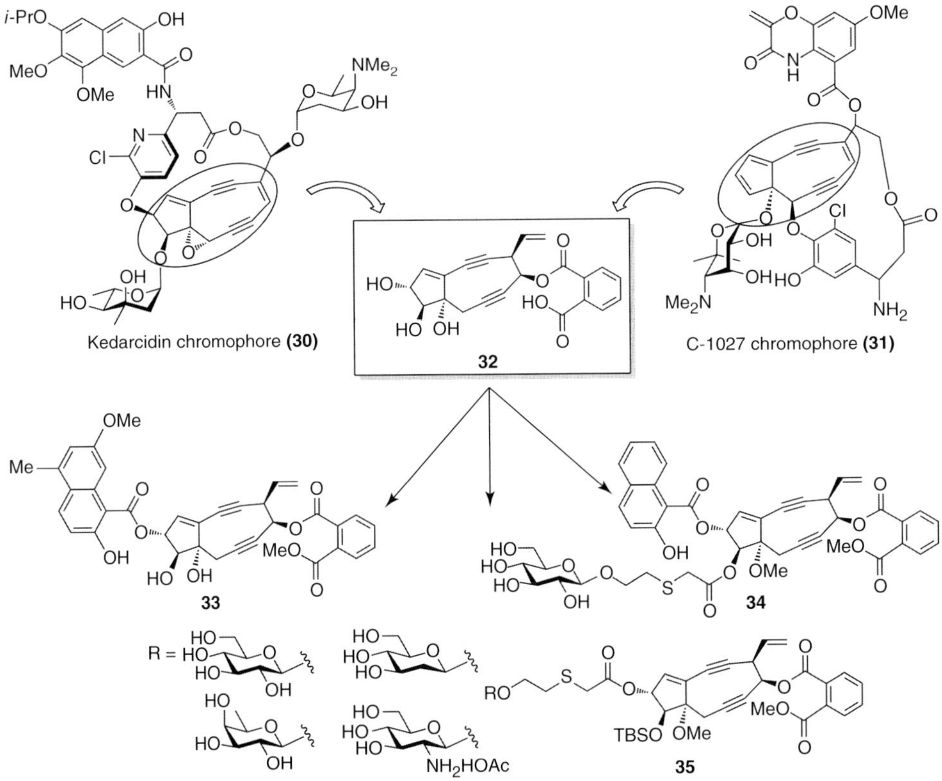

Figure 2.6 Chemical structures of kedarcidin chromophore (**30**), C-1027 chromophore (**31**), the masked nine-membered enediyne **32**, and their analogs **33–35**.

have displayed potent antitumor activity (Figure 2.6). In addition to their uses in pharmaceuticals, the nine-membered enediynes are powerful tools for chemical biology because they cleave DNA strands via the formation of a highly active biradical. We previously demonstrated that the masked nine-membered enediyne **32** exhibits DNA-cleaving activity. In addition, its derivatives **33** [27], **34**, and **35** [28, 29], which contain sugars and/or a naphthoate moiety, possess base-selective DNA-cleaving activity (Figure 2.6). Despite the importance of the nine-membered enediynes, the applications of these molecules are limited by the difficulty of their syntheses [23, 24, 26, 30–34]. Synthesis of the nine-membered enediynes, which are structurally complex and unstable, typically requires additional synthetic steps and careful manipulation. Thus, only a few laboratories can synthesize, modify, and utilize these "molecular scissors" to cleave DNA. Here, we describe the synthesis of the 12-membered cyclic ether **36** (Scheme 2.5), which is a chemically stable key synthetic intermediate of **32–35**. To improve the reproducibility and efficiency of the synthesis, an automated synthesizer originally developed by our group, the

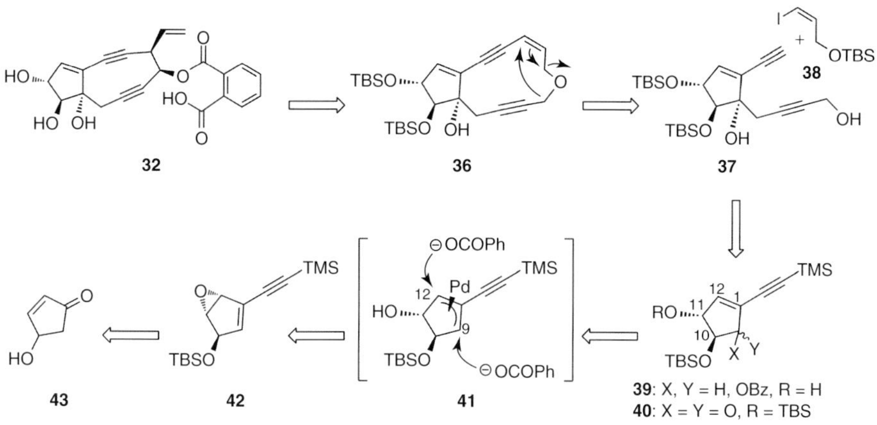

Scheme 2.5 Synthetic strategy for the masked nine-membered enediyne **32**.

ChemKonzert (Figure 2.2), was used in a 16-step process with some modifications of the reaction conditions [6].

The synthetic route for the masked nine-membered enediyne **32** is illustrated in Scheme 2.5. The transannular [2,3]-Wittig rearrangement [35] of the 12-membered cyclic ether **36** is the most crucial step in the construction of the highly strained nine-membered diyne, in which the newly formed secondary alcohol can be used to introduce a phthalic acid. β-Elimination of the phthalate "triggers" biradical formation. Cyclic ether **36** can be prepared via the Sonogashira coupling reaction between **37** and **38**, followed by intramolecular etherification. Selection of an appropriate protecting group for the allylic alcohol **38** is important. The protecting group must be stable under palladium-catalyzed reaction conditions. In addition, after the coupling reaction, it must be removed without affecting either the siloxy groups at the 10 and 11 positions or the unstable conjugated dienyne moiety. Enediyne **37** was constructed by stereoselective 1,2-addition to ketone **40**. The second key reaction is the palladium-catalyzed addition of BzOH to the diene monoepoxide **42** [36, 37]. This reaction induces trans stereochemistry between the hydroxy groups at the 10 and 11 positions and the $\Delta^{1,(12)}$ double bond in **39**. Although the palladium-catalyzed reaction produces four possible regio- and stereoisomers by BzOH displacement from the α and β faces at the 9 and 12 positions via the π-allylpalladium complex **41**, all products could be converted into the enone **40** in three steps. This methodology offers an enantioselective approach to **32**, starting from the optically active epoxide **42**, which is readily prepared from (*S*)-4-hydroxy-2-cyclopentenone [38]. The efficient, automated synthesis of racemic **36** is described below.

In consideration of the properties of ChemKonzert, reactions a–p were manually (Scheme 2.6) examined. The reactions indicated in italics required alteration of their reaction conditions. In the key palladium-catalyzed addition reaction e, the solid reagent benzoic acid was added directly to the reaction mixture in the manual experiment. The transfer of solids cannot be performed using the ChemKonzert,

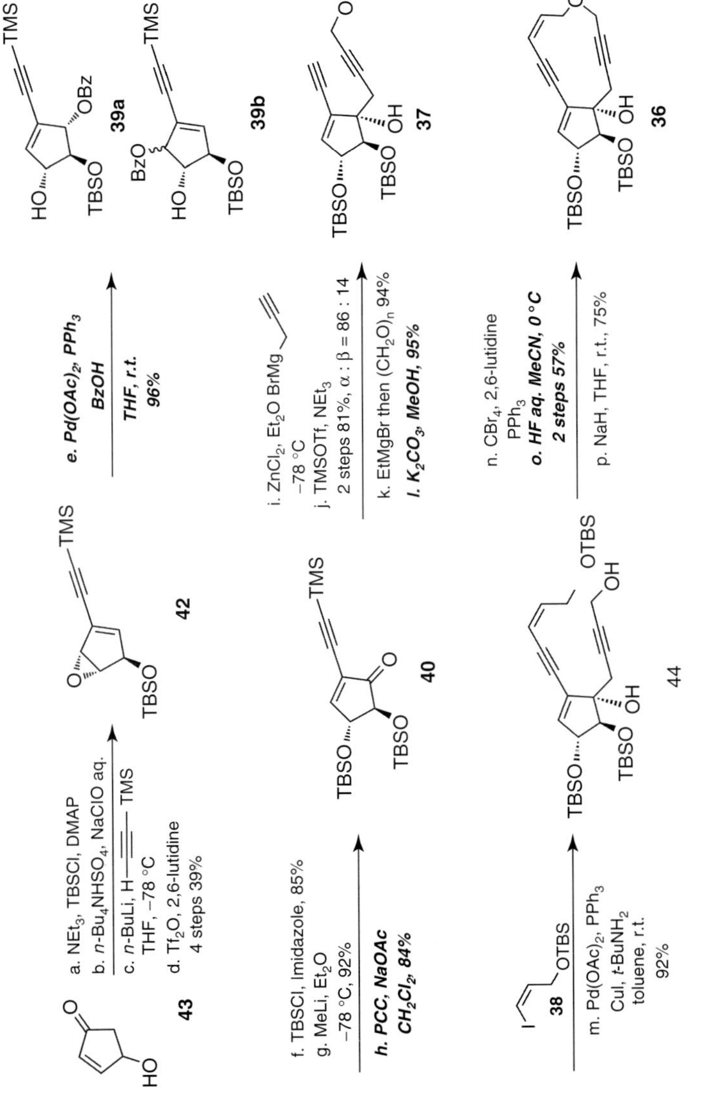

Scheme 2.6 Synthesis of the cyclic ether key intermediate **36**.

so the THF solution of benzoic acid was used in the automated synthesis. During oxidation h, a large amount of highly viscous chromium salts was precipitated. Therefore, various kinds of oxidation conditions were examined, including Swern oxidation, Parikh–Doering oxidation, Dess–Martin oxidation, and TEMPO–NaClO oxidation. All these oxidation conditions gave unsatisfactory results, while the IBX (*o*-iodoxybenzoic acid) oxidation in dimethylsulfoxide (DMSO) afforded the desired product without generating precipitation. This oxidation reaction could be performed using the ChemKonzert. In reaction l, namely, removal of the trimethylsilyl (TMS) group, a large amount of MeOH was used as a solvent. Therefore, the decision was made to use the mixed solvent MeOH/THF instead of MeOH.

In reaction o, the selective removal of the *tert*-butyldimethylsilyl (TBS) group using HF, it is necessary to monitor the reaction continuously during the selective deprotection and to adjust the reaction time to prevent over-desilylation. However, ChemKonzert begins work-up operations after the programmed reaction time has elapsed. This reaction cannot, therefore, be performed using the automated synthesizer. The palladium-catalyzed coupling reaction of the vinyl iodides was examined, which showed various protecting groups at the allylic hydroxyl group. As a result, vinyl iodide **45**, which has a 4-methoxyphenylmethyl (MPM) group, afforded the desired product in 65% yield (Scheme 2.7). The MPM group was chemoselectively deprotected by 2,3-dichloro-5,6-dicyano-1,4-benzoquinone (DDQ) without affecting the other functional groups, to give the cyclization precursor **47** in 70% yield from **46**. This deprotection condition is more suitable for use with the automated synthesizer than the previously described desilylation conditions, because the risk of over-desilylation during the deprotection step is much lower. Macro-etherification was performed successfully utilizing the automated synthesizer (Scheme 2.7). When workup of this cyclization reaction was performed manually, the addition of a large amount of EtOAc enabled liquid–liquid extraction. However, the addition of

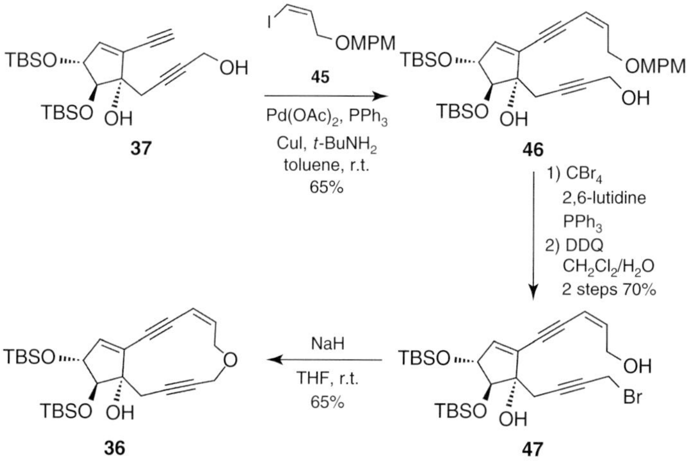

Scheme 2.7 Improved synthesis of the cyclic ether key intermediate **36** from **37** aided by ChemKonzert.

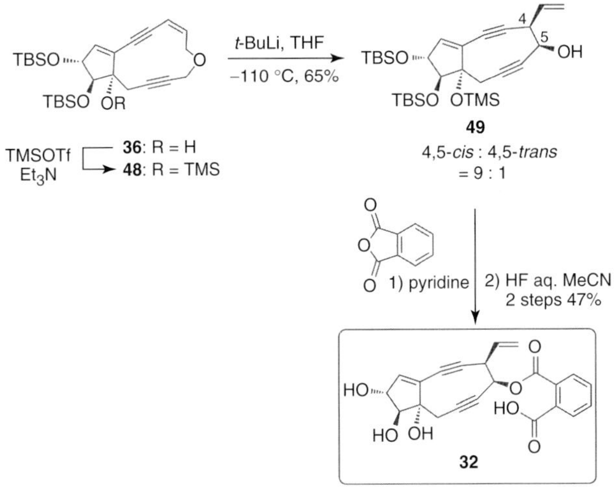

Scheme 2.8 Synthesis of the masked nine-membered enediyne **32**.

EtOAc was not suitable for the synthesizer because the maximum solution volume of the centrifugal separation was 700 ml. Therefore, the reaction mixture was heated to 50 °C under reduced pressure to remove THF. Automated workup and manual silica gel column purification furnished the desired 12-membered cyclic ether **36** as a chemically stable solid, which could be stored at room temperature for long periods. For this cyclization, the addition of a catalytic amount of water was crucial to obtain reproducible results. A catalytic amount of EtOH was also effective.

Further transformations were carried out manually (Scheme 2.8). After the protection of the alcohol **36** with a TMS group, the key [2,3]-Wittig rearrangement was carried out to afford the desired product **49**. Treatment of **49** with phthalic anhydride, followed by removal of the TBS groups, afforded the desired masked nine-membered enediyne **32**.

2.1.4
List of Reactions Successfully Performed in Automated Synthesizers

The reactions that were successfully performed in automated synthesizers are shown in Table 2.1. The list includes various kinds of reactions, including C–C bond formations, oxidation and reduction reactions, protection and deprotection sequences, and other transformation reactions. Toxic, anhydrous, foul-smelling, and lachrymatory compounds are used in many of these reactions.

Recently, we accomplished the total synthesis of the histone deacetylase (HDAC) inhibitor, spiruchostatin B [39]. A five-step synthesis that included an asymmetric aldol reaction was carried out in an automated synthesizer to provide an (*E*)-(*S*)-3-hydroxy-7-thio-4-heptenoic acid segment which is the crucial structure of cysteine-containing depsipeptidic natural products such as spiruchostatins, FK228, FR901375, and largazole for their inhibitory activity against HDACs.

Table 2.1 Reactions performed by automated synthesizer.

C–C bond formation	
Nucleophilic addition of alkenyl lithium	*t*-BuLi, $CeCl_3$, alkenyl bromide, THF, −78 °C
Radical cyclization	Cp_2TiCl_2, Mn, TMSCl, K_2CO_3, 0 °C
Radical cyclization	Cp_2TiCl_2, Mn, BEt_3, lutidine·HCl, THF, 25 °C
Cyanohydrin formation	TMSCN, KCN, 18-crown-6, 25 °C, then 1 M HCl
Nucleophilic addition of alkenyl lithium	Trimethylsilylacetylene, *n*-BuLi, THF, −78 to − 40 °C
Nucleophilic addition of Grignard reagent	$ZnCl_2$, propargyl magnesium bromide, Et_2O, −78 °C
Nucleophilic addition of Grignard reagent	EtMgBr, THF, $(CH_2O)n$, 50–25 °C
Sonogashira coupling	$Pd(OAc)_2$, PPh_3, CuI, *t*-$BuNH_2$, toluene, 25 °C
Oxidation	
Bromohydrin formation	NBS, H_2O, *t*-BuOH, 25 °C
Parikh–Doering oxidation	SO_3· Py, DMSO, Et_3N, CH_2Cl_2, 0 °C
Ley oxidation	TPAP, NMO, CH_2Cl_2, 25 °C
Epoxidation	$VO(acac)_2$, TBHP, benzene, 6 °C
Epoxidation	*n*-Bu_4NHSO_4, NaClO aq., CH_2Cl_2, 0 °C
SeO_2 oxidation	SeO_2, TBHP, salicylic acid, hexane, 60 °C
IBX oxidation	IBX, DMSO, 50 °C
Reduction	
$LiAlH_4$ reduction	$LiAlH_4$, Et_2O sol., 40 °C
$NaBH_4$ reduction	$NaBH_4$, $MeOH/H_2O/THF$, 0 °C
Protection of alcohol	
Ac protection	Ac_2O, Et_3N, DMAP, 25 °C
EE protection	EVE, CSA, CH_2Cl_2, 0 °C
MPM protection	MPM imidate, TfOH, Et_2O, 25 °C
BOM protection	BOMCl, *i*-Pr_2NEt, CH_2Cl_2, 25 °C
Bn protection	BnBr, *n*-Bu_4NHSO_4, 50% aq. KOH, 25 °C
TBS protection	TBSCl, Et_3N, CH_2Cl_2, 25 °C
TBS protection	TBSCl, Et_3N, DMAP, CH_2Cl_2, 25 °C
TBS protection	TBSCl, imidazole, CH_2Cl_2, 25 °C
TBS protection	TBSCl, imidazole, DMF, 25 °C
TMS protection	TMSOTf, 2,6-lutidine, *i*-Pr_2NEt, 0 °C
TMS protection	TMSOTf, NEt_3, CH_2Cl_2, 0 °C
Ts protection	TsCl, DMAP, $CHCl_3$, 50 °C
Deprotection	
Ester solvolysis	NaOH, $MeOH/H_2O/THF$, 25 °C
Removal of silyl group	TBAF, THF, 25 °C
Removal of Bz group	MeLi, Et_2O, −78 to 0 °C
Removal of MPM group	DDQ, CH_2Cl_2/buffer, 25 °C
Miscellaneous	
Shapiro reaction – bromination	*t*-BuLi, 1,2-dibromoethane, THF, −78 °C

(*continued overleaf*)

Table 2.1 (continued)

Epoxide formation	Et_3N, toluene, reflux
Alkene isomerization	DBU, CH_2Cl_2, 25 °C
β-Elimination of hydroxy group	Tf_2O, 2,6-lutidine, CH_2Cl_2, −78 to 0 °C
Pd-catalyzed addition of BzOH	$Pd(OAc)_2$, PPh_3, BzOH, THF, 25 °C
Removal of TMS group	K_2CO_3, MeOH/THF, 25 °C
Bromination	CBr_3, PPh_3, MeCN, 0–25 °C
Etherification	NaH, THF, 25 °C

2.2 Continuous-Flow Synthesis of Vitamin D_3

Another problem in the preparation of the synthetic intermediates is the scaling up of the reactions. Strict temperature control is sometimes difficult in medium- to large-scale reactions, and hence undesired products may be generated. Microflow synthesis has attracted much attention in the past decade. Microreactors have two advantages over conventional batch reactors: (i) reaction conditions can be precisely controlled through rapid heat transfer and mixing; and (ii) the syntheses can be scaled up by running a single reactor for extended periods of time, or by the addition of more identical flow reactors in parallel. Recently, there has been an increase in the utilization of integrated flow microreactor systems [40–42], such as the continuous-flow, multistep synthesis of organic compounds that do not require the purification of unstable intermediates.

The continuous-flow synthesis of vitamin D_3 (**50**) is described below [43]. Photo and thermal isomerization of provitamin D_3 (**51**) are conventional methods for the production of both vitamin D_3 and its analogs, because of the ready availability of the starting compounds (Scheme 2.9). In fact, vitamin D_3 is commercially produced by this sequence. However, the overall yields of vitamin D_3 and its analogs obtained using photo and thermal isomerization are generally low (<20%) [44–46] as a result of the poor selectivity of the photoisomerization step. The photoisomerization of provitamin D_3 (**51**) into previtamin D_3 (**52**) is not selective (Scheme 2.9, reaction I). Because previtamin D_3 (**52**) has an absorption wavelength similar to that of provitamin D_3 (**51**), undesired products, such as lumisterol (**53**) and tachysterol (**54**), result from the equilibrium between the products.

Therefore, with the present industrial method, it is necessary to interrupt the irradiation after relatively low conversion (10–20%) of provitamin D_3 (**51**) to previtamin D_3 (**52**). The unconverted provitamin D_3 (**51**) is recycled, while the previtamin D_3 (**52**) must be purified in an expensive work-up procedure. Various other sources of UV irradiation have been considered to improve the yield of previtamin D_3 (**52**). In fact, the use of excimer or exciplex lasers has reportedly been effective in the photoisomerization of provitamin D_3 (**51**) into previtamin D_3 (**52**) [47–49]. However, the use of narrow-band, high-intensity light sources, which

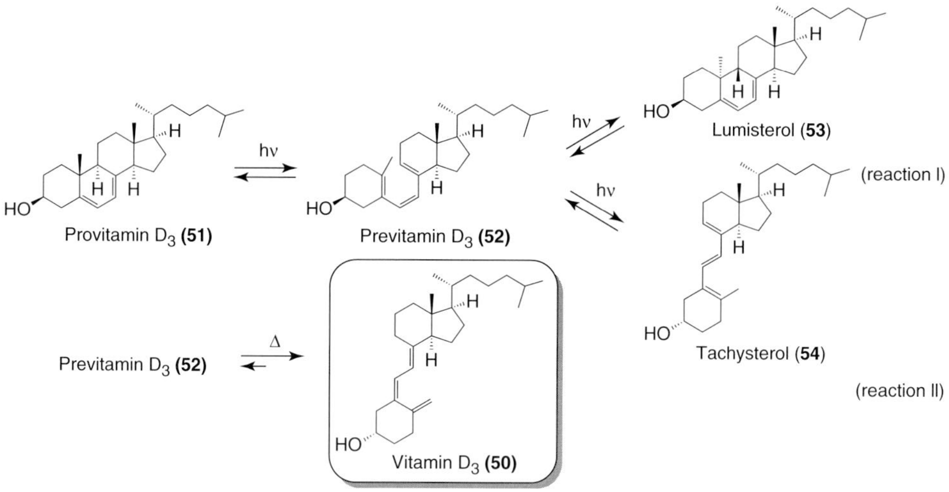

Scheme 2.9 Two-step conversion of provitamin D_3 (**51**) to vitamin D_3 (**50**).

is essential for larger scale application, is prohibitively expensive and requires specialized equipment setup. The use of a solution filter with an economical light source to generate narrow-band spectra has been reported [50]. However, the need to dispose of a large amount of waste is problematic in this case. The use of a sensitizer [51] or a filter compound [52] has been reported. However, an expensive work-up procedure to remove these compounds from the reaction mixture is necessary in these cases.

Vitamin D_3 and its analogs have a broad spectrum of biological activity, such as cell differentiation, regulation of calcium metabolism, and immune function [53, 54]. Moreover, antitumor activity of vitamin D_3 has recently been reported. Therefore, the development of a facile method for the preparation of vitamin D_3 and its analogs is highly important [55–57].

Our group focused on the development of an efficient synthetic methodology for vitamin D_3 utilizing microreactors and a high-intensity, economical light source, that is, a high-pressure mercury lamp [43]. Photomicroreactors have several advantages over conventional batch reactors. Specifically, photomicroreactors exhibit improved light-penetration efficiency owing to the thinness of the reaction mixture in the microspace. As previously described, microreactors also have advantages in thermal reactions due to rapid heat transfer.

After the examination of many reaction conditions and various combinations of microreactors, the two-stage irradiation method shown in Figure 2.7 afforded the best results [43]. The details of microflow synthesis follow. We anticipated that the mixture of previtamin D_3 (**52**) and tachysterol (**54**) prepared from provitamin D_3 (**51**) using the photomicroreactor (313–578 nm) would be converted into the desired vitamin D_3 (**50**) using the photo and thermal microreactor (360 nm, 100 °C). Consequently, the equilibrium for the photoisomerization of tachysterol (**54**) to previtamin D_3 (**52**) shifted to produce more previtamin D_3 (**52**).

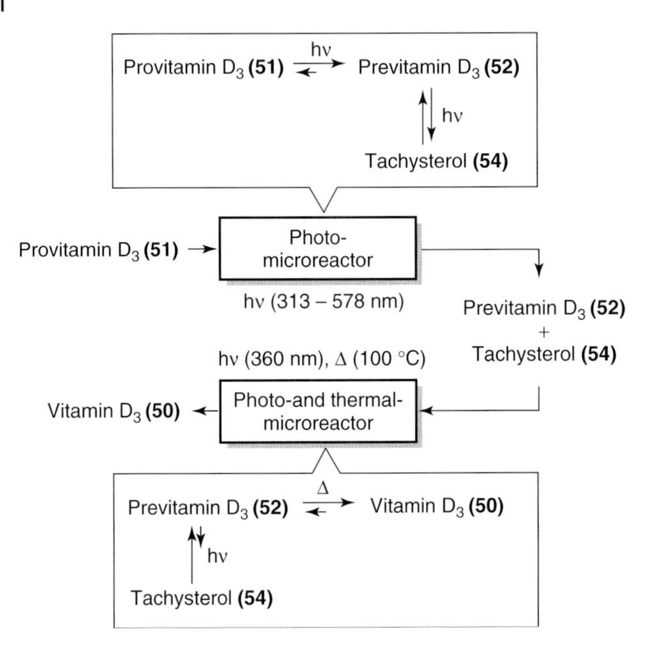

Figure 2.7 Microflow synthesis of vitamin D_3 (**50**) using a two-stage method.

A continuous-flow system was prepared (Figure 2.8) [43]. Two microreactors were connected with poly(ethylene ether ketone) (PEEK) tubing. The second microreactor was put on hot oil (100 °C). Then, 30 mM solutions of provitamin D_3 (**51**) in 1,4-dioxane were introduced with a syringe pump at flow rates of 5 μl/min. The desired vitamin D_3 (**50**) was obtained in excellent yield (HPLC-UV: 60%). This was the first microflow system to be applied to the synthesis of vitamin D_3 from

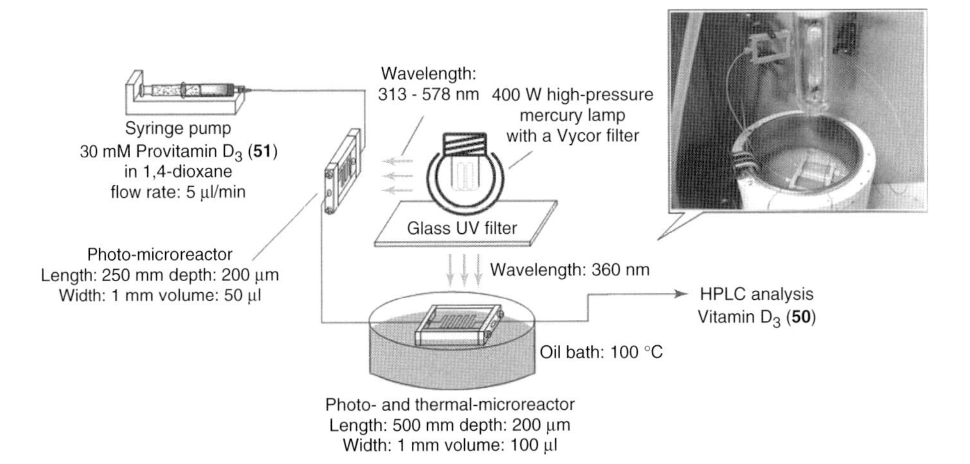

Figure 2.8 Two-stage, continuous-flow synthesis of vitamin D_3 (**50**) using two microreactors.

provitamin D_3. Our report is also the first demonstration of photo and thermal reactions in a single microreactor. Using the continuous-flow system, the desired vitamin D_3 was obtained in excellent yield (HPLC-UV: 60%, isolated: 32%). To the best of our knowledge, this yield is the highest ever achieved without the use of a laser, a sensitizer, or a filter compound. It should be noted that the continuous microflow synthesis of vitamin D_3 does not require the purification of intermediates or high dilution conditions, so waste is reduced. The protocol described here should be applicable to the synthesis of a variety of vitamin D_3 analogs.

2.3 Conclusions

There is concern that increased automation in synthetic organic chemistry might result in loss of jobs. In our opinion, many talented synthetic chemists would prefer to try advanced, challenging tasks such as the development of new synthetic routes and the discovery of new reaction conditions. However, they likely would prefer to avoid simple, routine work such as the repetitive preparation of synthetic intermediates using known synthetic schemes or the optimization of reaction conditions by changing the types of solvents, concentrations, temperatures, and so on. It is usually not profitable to perform the former types of tasks utilizing automated synthesizers. In such advanced work, synthetic operations vary on a daily basis because researchers attempt to examine various types of reactions. As described in Section 2.1, time and effort is necessary for the automation of synthetic procedures. Therefore researchers are forced to spend their time and effort on the automation of synthetic procedures every time they change reactions. However, the utilization of automated synthesizers is suitable for simple and repetitive work such as the preparation of synthetic intermediates or the optimization of reaction conditions [58]. Because little time and effort is needed for automation, researchers can benefit from it. Researchers should perform a labor–benefit analysis when they are considering whether a synthetic operation should be automated. The proper use of automated synthesizers can relieve synthetic chemists from tedious, simple, and repetitive work. They can, therefore, spend more time and effort on advanced and challenging tasks. The automated synthesizer is a strong partner rather than a rival. In recent years, risk management and hazard avoidance are strongly desired, not only in companies but also in educational institutions. The use of automated synthesizers contributes not only to the improvement of efficiency and reproducibility but also to the protection of researchers from potentially harmful compounds.

Acknowledgments

The authors thank Prof. Takayuki Doi (Tohoku University) for his support in completing the research described in this chapter. They also thank Dr. Tohru Sugawara, Mr. Yoichiro Hirose (ChemGenesis Inc.), and Dr. Shigetoshi Sekiyama

(SIC Co.) for the development of ChemKonzert. Thanks are also due to Sumitomo Chemical Co., Ltd., Kuraray Co., Ltd., EIWEISS Chemical Corporation, and Central Grass Co., Ltd., for a generous supply of 4-hydroxy-2-cyclopentenone, geraniol, EDCI, and Tf_2O. The continuous-flow synthesis of vitamin D_3 was supported by the Project "Micro-Chemical Technology for Production, Analysis, and Measurement Systems" from NEDO, Japan.

References

1. Cork, D.G. and Sugawara, T. (2002) *Laboratory Automation in the Chemical Industries*, Marcel Dekker Inc., New York.
2. Jordan, S., Moshiri, B., and Durand, R. (2002) *J. Assoc. Lab. Autom.*, **7**, 74–77.
3. Kuroda, N., Hattori, T., Kitada, C., and Sugawara, T. (2001) *Chem. Pharm. Bull.*, **49**, 1138–1146.
4. Shih, H.W., Guo, C.W., Lo, K.H., Huang, M.Y., and Cheng, W.C. (2009) *J. Comb. Chem.*, **11**, 281–287.
5. Doi, T., Fuse, S., Miyamoto, S., Nakai, K., Sasuga, D., and Takahashi, T. (2006) *Chem. Asian J.*, **1**, 370–383.
6. Tanaka, Y., Fuse, S., Tanaka, H., Doi, T., and Takahashi, T. (2009) *Org. Process Res. Dev.*, **13**, 1111–1121.
7. Machida, K., Hirose, Y., Fuse, S., Sugawara, T., and Takahashi, T. (2010) *Chem. Pharm. Bull.*, **58**, 87–93.
8. Chabner, B. and Longo, L.D. (2006) *Cancer Chemotherapy and Biotherapy: Principles and Practice*, Lippincott Williams & Wilkins, Philadelphia.
9. Nicolaou, K.C., Dai, W.M., and Guy, R.K. (1994) *Angew. Chem. Int. Ed.*, **33**, 15–44.
10. Wani, M.C., Taylor, H.L., Wall, M.E., Coggon, P., and McPhail, A.T. (1971) *J. Am. Chem. Soc.*, **93**, 2325–2327.
11. Holton, R.A., Somoza, C., Kim, H.B., Liang, F., Biediger, R.J., Boatman, P.D., Shindo, M., Smith, C.C., Kim, S.C., Nadizadeh, H., Suzuki, Y., Tao, C.L., Vu, P., Tang, S.H., Zhang, P.S., Murthi, K.K., Gentile, L.N., and Liu, J.H. (1994) *J. Am. Chem. Soc.*, **116**, 1597–1598.
12. Morihira, K., Hara, R., Kawahara, S., Nishimori, T., Nakamura, N., Kusama, H., and Kuwajima, I. (1998) *J. Am. Chem. Soc.*, **120**, 12980–12981.
13. Wender, P.A., Badham, N.F., Conway, S.P., Floreancig, P.E., Glass, T.E., Houze, J.B., Krauss, N.E., Lee, D.S., Marquess, D.G., McGrane, P.L., Meng, W., Natchus, M.G., Shuker, A.J., Sutton, J.C., and Taylor, R.E. (1997) *J. Am. Chem. Soc.*, **119**, 2757–2758.
14. Mukaiyama, T., Shiina, I., Iwadare, H., Sakoh, H., Tani, Y., Hasegawa, M., and Saitoh, K. (1997) *Proc. Jpn. Acad. Ser. B*, **73**, 95–100.
15. Masters, J.J., Link, J.T., Snyder, L.B., Young, W.B., and Danishefsky, S.J. (1995) *Angew. Chem. Int. Ed.*, **34**, 1723–1726.
16. Nicolaou, K.C., Yang, Z., Liu, J.J., Ueno, H., Nantermet, P.G., Guy, R.K., Claiborne, C.F., Renaud, J., Couladouros, E.A., Paulvannan, K., and Sorensen, E.J. (1994) *Nature*, **367**, 630–634.
17. Kingston, D.G.I., Molinero, A.A., Rimoldi, J.M., Kirby, R.E., Moore, R.E., Steglich, W., and Tamm, C. (eds) (1993) *The Taxane Diterpenoids*, Springer-Verlag, New York.
18. Nicolaou, K.C., Nantermet, P.G., Ueno, H., Guy, R.K., Couladouros, E.A., and Sorensen, E.J. (1995) *J. Am. Chem. Soc.*, **117**, 624–633.
19. Miyamoto, S., Doi, T., and Takahashi, T. (2002) *Synlett*, 97–99.
20. Takahashi, T., Iwamoto, H., Nagashima, K., Okabe, T., and Doi, T. (1997) *Angew. Chem. Int. Ed. Engl.*, **36**, 1319–1321.
21. Fuse, S., Hanochi, M., Doi, T., and Takahashi, T. (2004) *Tetrahedron Lett.*, **45**, 1961–1963.
22. Nakai, K., Kamoshita, M., Doi, T., Yamada, H., and Takahashi, T. (2001) *Tetrahedron Lett.*, **42**, 7855–7857.

23. Ren, F., Hogan, P.C., Anderson, A.J., and Myers, A.G. (2007) *J. Am. Chem. Soc.*, **129**, 5381–5383.
24. Ogawa, K., Koyama, Y., Ohashi, I., Sato, I., and Hirama, M. (2009) *Angew. Chem. Int. Ed.*, **48**, 1110–1113.
25. Otani, T., Minami, Y., Marunaka, T., Zhang, R., and Xie, M.Y. (1988) *J. Antibiot.*, **41**, 1580–1585.
26. Inoue, M., Ohashi, I., Kawaguchi, T., and Hirama, M. (2008) *Angew. Chem. Int. Ed.*, **47**, 1777–1779.
27. Takahashi, T., Tanaka, H., Yamada, H., Matsumoto, T., and Sugiura, Y. (1997) *Angew. Chem. Int. Ed. Engl.*, **36**, 1524–1526.
28. Takahashi, T., Tanaka, H., Matsuda, A., Doi, T., and Yamada, H. (1998) *Bioorg. Med. Chem. Lett.*, **8**, 3299–3302.
29. Takahashi, T., Tanaka, H., Matsuda, A., Doi, T., Yamada, H., Matsumoto, T., Sasaki, D., and Sugiura, Y. (1998) *Bioorg. Med. Chem. Lett.*, **8**, 3303–3306.
30. Wender, P.A., McKinney, J.A., and Mukai, C. (1990) *J. Am. Chem. Soc.*, **112**, 5369–5370.
31. Doi, T. and Takahashi, T. (1991) *J. Org. Chem.*, **56**, 3465–3467.
32. Kobayashi, S., Reddy, R.S., Sugiura, Y., Sasaki, D., Miyagawa, N., and Hirama, M. (2001) *J. Am. Chem. Soc.*, **123**, 2887–2888.
33. Myers, A.G., Glatthar, R., Hammond, M., Harrington, P.M., Kuo, E.Y., Liang, J., Schaus, S.E., Wu, Y.S., and Xiang, J.N. (2002) *J. Am. Chem. Soc.*, **124**, 5380–5401.
34. Komano, K., Shimamura, S., Inoue, M., and Hirama, M. (2007) *J. Am. Chem. Soc.*, **129**, 14184–14186.
35. Nakai, T. and Mikami, K. (1986) *Chem. Rev.*, **86**, 885–902.
36. Takahashi, T., Kataoka, H., and Tsuji, J. (1983) *J. Am. Chem. Soc.*, **105**, 147–149.
37. Deardorff, D.R., Myles, D.C., and Macferrin, K.D. (1985) *Tetrahedron Lett.*, **26**, 5615–5618.
38. Suzuki, M., Kawagishi, T., Suzuki, T., and Noyori, R. (1982) *Tetrahedron Lett.*, **23**, 4057–4060.
39. Fuse, S., Okada, K., Iijima, Y., Munakata, A., Machida, K., Takahashi, T., Takagi, M., Shin-ya, K., and Doi, T. (2011) *Org. Biomol. Chem.*, **9**, 3825–3833.
40. Yoshida, J. (2010) *Chem. Rec.*, **10**, 332–341.
41. Hartman, R.L., Naber, J.R., Buchwald, S.L., and Jensen, K.E. (2010) *Angew. Chem. Int. Ed.*, **49**, 899–903.
42. Brasholz, M., Macdonald, J.M., Saubern, S., Ryan, J.H., and Holmes, A.B. (2010) *Chem. Eur. J.*, **16**, 11471–11480.
43. Fuse, S., Tanabe, N., Yoshida, M., Yoshida, H., Doi, T., and Takahashi, T. (2010) *Chem. Commun.*, **46**, 8722–8724.
44. Kubodera, N., Watanabe, H., and Miyamoto, K. (1994) JP80626.
45. Katoh, M., Mikami, T., and Watanabe, H. (1994) JP72994.
46. Kubodera N. and Watanabe, H. (1991) JP188061.
47. Dauben, W.G. and Phillips, R.B. (1982) *J. Am. Chem. Soc.*, **104**, 5780–5781.
48. Dauben, W.G. and Phillips, R.B. (1982) *J. Am. Chem. Soc.*, **104**, 355–356.
49. Malatesta, V., Willis, C., and Hackett, P.A. (1981) *J. Am. Chem. Soc.*, **103**, 6781–6783.
50. Sato, T., Yamauchi, H., Ogata, Y., Kunii, T., Kagei, K., Katsui, G., Toyoshima, S., Yasumura, M., and Kobayashi, T. (1980) *J. Nutr. Sci. Vitaminol.*, **26**, 545–556.
51. Eyley, S.C. and Williams, D.H. (1975) *J. Chem. Soc., Chem. Commun.*, 858–858.
52. Okabe, M., Sun, R.C., Scalone, M., Jibilian, C.H., and Hutchings, S.D. (1995) *J. Org. Chem.*, **60**, 767–771.
53. Ettinger, R.H. and DeLuca, H.F. (1996) *Adv. Drug Res.*, **28**, 269–312.
54. Bouillon, R., Okamura, W.H., and Norman, A.W. (1995) *Endocr. Rev.*, **16**, 200–257.
55. Doi, T., Hijikuro, I., and Takahashi, T. (1999) *J. Am. Chem. Soc.*, **121**, 6749–6750.
56. Hijikuro, I., Doi, T., and Takahashi, T. (2001) *J. Am. Chem. Soc.*, **123**, 3716–3722.
57. Doi, T., Yoshida, M., Hijikuro, I., and Takahashi, T. (2004) *Tetrahedron Lett.*, **45**, 5727–5729.
58. McMullen, J.P., Stone, M.T., Buchwald, S.L., and Jensen, K.E. (2010) *Angew. Chem. Int. Ed.*, **49**, 7076–7080.

3
Chemoselective Reduction of Amides and Imides

Shoubhik Das

3.1
Introduction

Catalytic methods offer efficient and versatile strategies for the synthesis of important classes of organic compounds and represent in general a key technology for the advancement of "green chemistry," specifically for preventing wastage, reducing energy consumption, achieving high atom efficiency, and generating advantageous economics. The reduction of ketones, aldehydes, imines, nitriles, and carboxylic acid derivatives to alcohols and amines is one of the most fundamental and widely employed reactions in synthetic organic chemistry [1]. Sodium borohydride, lithium aluminum hydride, and other stoichiometric reducing agents are generally used for laboratory-scale syntheses [2]. However, the increased demands for atom economy, cleaner synthesis, and straightforward work-up procedures make the use of these reagents disadvantageous. Reduction procedures that make use of catalytic amounts of metals show better ecology, and are more cost effective and potentially easier to operate than those that require the cleanup of boron or aluminum waste at the end of the reaction. For this purpose, heterogeneous catalysts such as Pd/C and Pt/C might be used because they represent the most economical means to carry out these reductions [1a]. However, especially in cases where chemoselectivity, milder conditions, and functional group tolerance are required, homogeneous catalysts can be better suited for the task. For this reason, there have been consistent research efforts toward the development of new and improved homogeneous catalytic systems.

A number of amine compounds are well known in chemistry and biology. They are widely used from the agrochemical to the pharmaceutical industry. Notable examples (Scheme 3.1) of established drugs and agrochemicals and their areas of application include abacavir (HIV), albuterol (asthma), atrazine (pesticide), carfentanil (tranquilizer), bupropion (depression), dicycloverine (irritable bowel syndrome), serotonin (central nervous system), imipramine (depression), Flomax (prostate), indoprofen (inflammation), and donepezil (Alzheimer). Direct reduction of amides, nitriles, nitro compounds, and imines (Scheme 3.2) is the most viable route for the synthesis of chiral and nonchiral amines. Owing to its simplicity

New Strategies in Chemical Synthesis and Catalysis, First Edition. Edited by Bruno Pignataro.

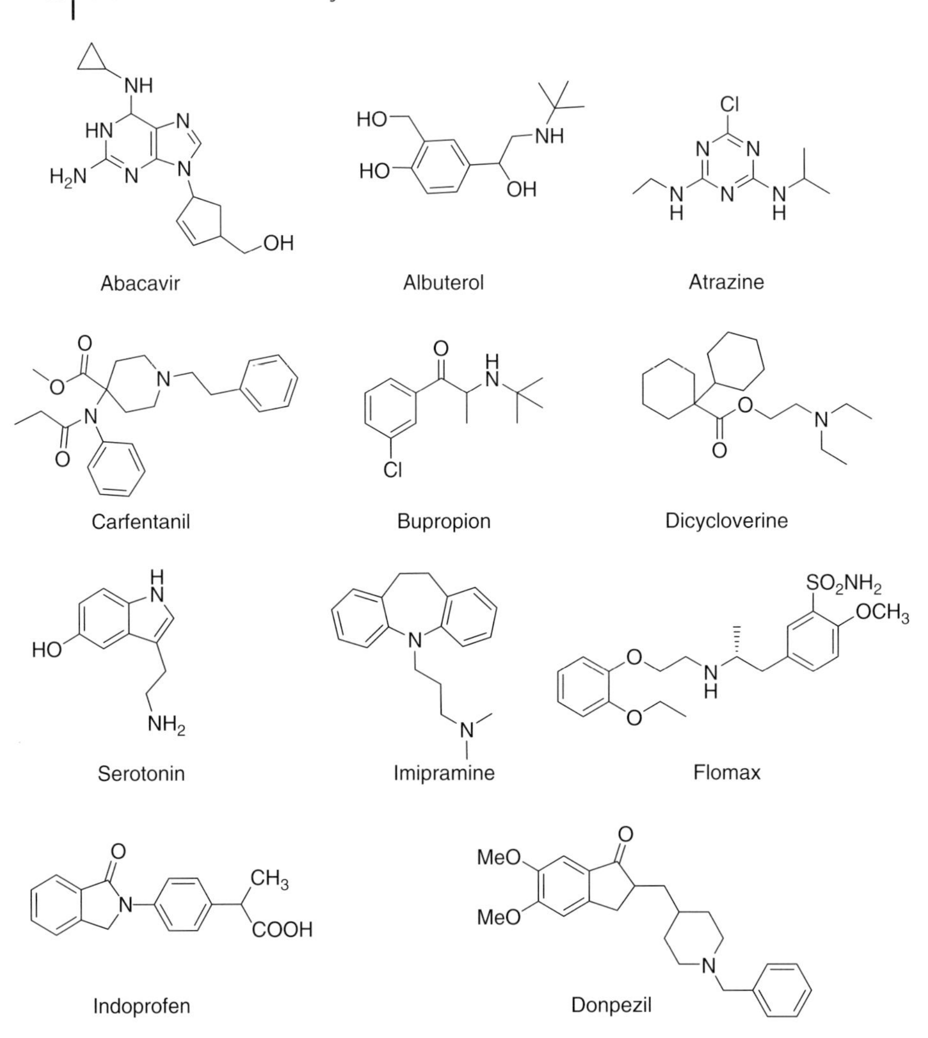

Scheme 3.1 Different amine structures used in pharmaceuticals and agrochemicals.

and apparent environment friendliness, hydrogenation is often a method of choice especially in view of the fact that H_2 is the only stoichiometric reagent used here [3]. However, the absence of suitable hydrogenation methods, especially for amides, as well as the necessity for high-pressure hydrogenation and metal leaching to the products is the major drawback to date.

In this respect, hydrosilylation is a well-accepted chemical tool which can bring chemoselectivity and regioselectivity together with different metal complexes as well as organocatalysts under mild conditions [4]. In addition, operational simplicity, safety, use of benign and low-cost reducing agents, and reduction of safety constraints associated with the use of gaseous and hazardous hydrogen or of metal hydrides of stoichiometric quantity make hydrosilane an alternative route. In addition, reactivity of silane can be tuned by different substituents on the silicon

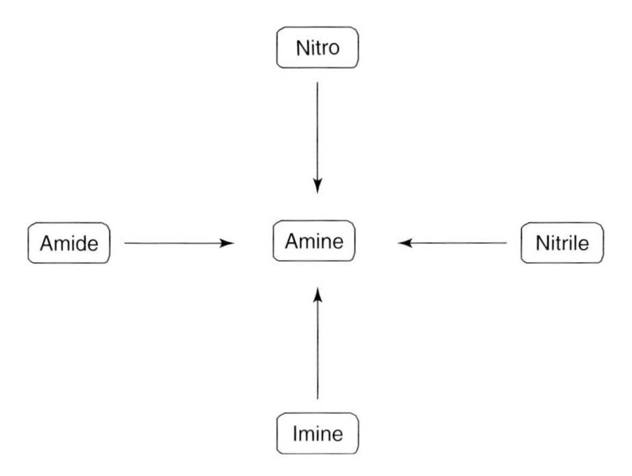

Scheme 3.2 Direct route for the synthesis of amines.

Table 3.1 Cost of different silanes.

Entry	Hydrosilane	Price (€/mmol)
1	$Me(OEt)_2SiH$	0.22
2	$TMSOSiMe_2H$	0.16
3	$PhSiH_3$	0.72
4	Et_2MeSiH	0.10
5	Et_3SiH	0.15
6	$(EtO)_3SiH$	0.26
7	PMHS	0.01
8	$Me_2SiHSiHMe_2$	16.9
9	Ph_2SiH_2	0.38

atom. The general order of reactivity of different silanes is as follows: PMHS (polymeric hydrosiloxanes) $<$ $Et_2OMeSiH$ $<$ $(OEt)_3SiH$ $<$ Ph_3SiH $<$ Ph_2SiH_2 $<$ $PhSiH_3$. Moreover, different silanes show different reactivities toward a particular reaction and also differ in prices (Table 3.1). Cheaper silanes such as PMHS and 1,1,3,3-tetramethyldisiloxane (TMDS) permit their application in industry, whereas expensive silanes are used only for the laboratory-scale syntheses. So, herein we will focus on our recent methods for reduction of amides and imides to the corresponding amines.

3.2 Reduction of Tertiary Amides

The synthesis of amines continues to be an important and current topic in organic synthesis. With respect to their preparation, the reduction of carboxamides is one of the classical and frequently applied procedures. The ease of the amide-bond

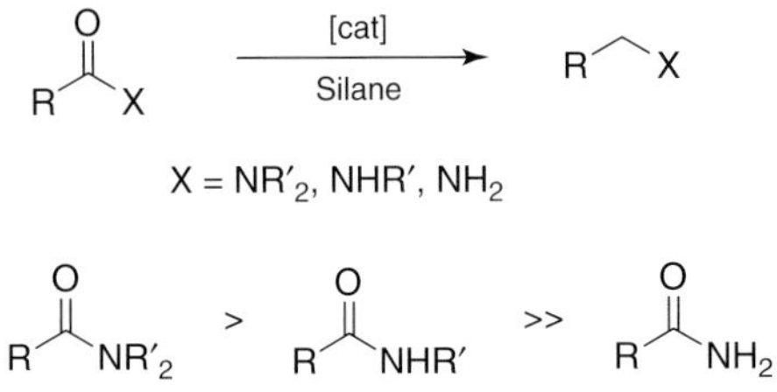

Scheme 3.3 Reactivity order of different amides.

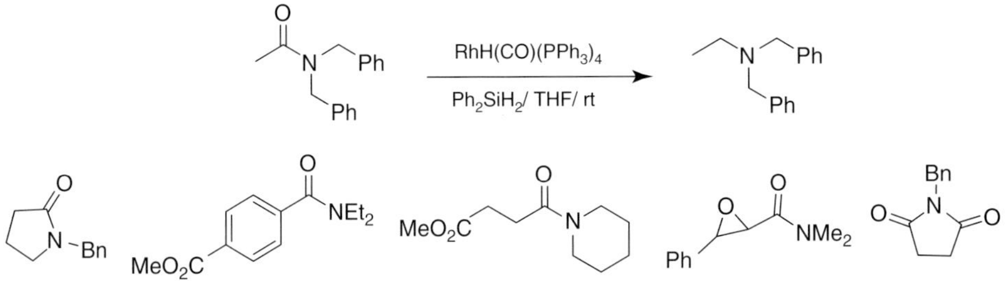

Scheme 3.4 Rhodium-catalyzed reduction of tertiary amides.

formation makes amides highly attractive starting materials for the synthesis of secondary and tertiary amines, although, reactivity-wise, tertiary amides are reduced more easily compared to secondary amides while primary amides tend to dehydrate to nitriles (Scheme 3.3) [5]. At the beginning of this decade, many reports were published regarding the catalytic hydrosilylation of secondary and tertiary amides. By late 1990s, Ito and coworkers, for the first time, developed an efficient method for the reduction of various tertiary amides in presence of 10 mol% $RhH(CO)(PPh_3)_3$ and diphenyl silane (Scheme 3.4) [6]. Chemoselectivity under mild condition was the main attraction of this protocol: mainly reduction of tertiary amides in presence of ester and epoxy functional groups. Later, Fuchikami and Igarashi showed that the reaction of amides with hydrosilanes can also be catalyzed by a variety of transition-metal complexes in the presence or absence of halides and amines as co-catalysts to afford the corresponding amines in good yields [7]. Notably, they were also able to reduce secondary and primary amides using their catalytic protocol (Scheme 3.5).

A substantial and detailed contribution on the catalytic hydrosilylation of amides came from the group of Nagashima (Scheme 3.6) [8]. A notable improvement of their first protocol on the hydrosilylation of tertiary amides has been the use of inexpensive PMHS as the hydrogen source. At the end of the reaction, amines were easily separated by washing with ether, while the ruthenium remained trapped in the resin. In the course of their studies, they looked for similar reduction of amides with PMHS involving self-encapsulation of the metal species by using other transition-metal catalysts. Commercially available H_2PtCl_6 and

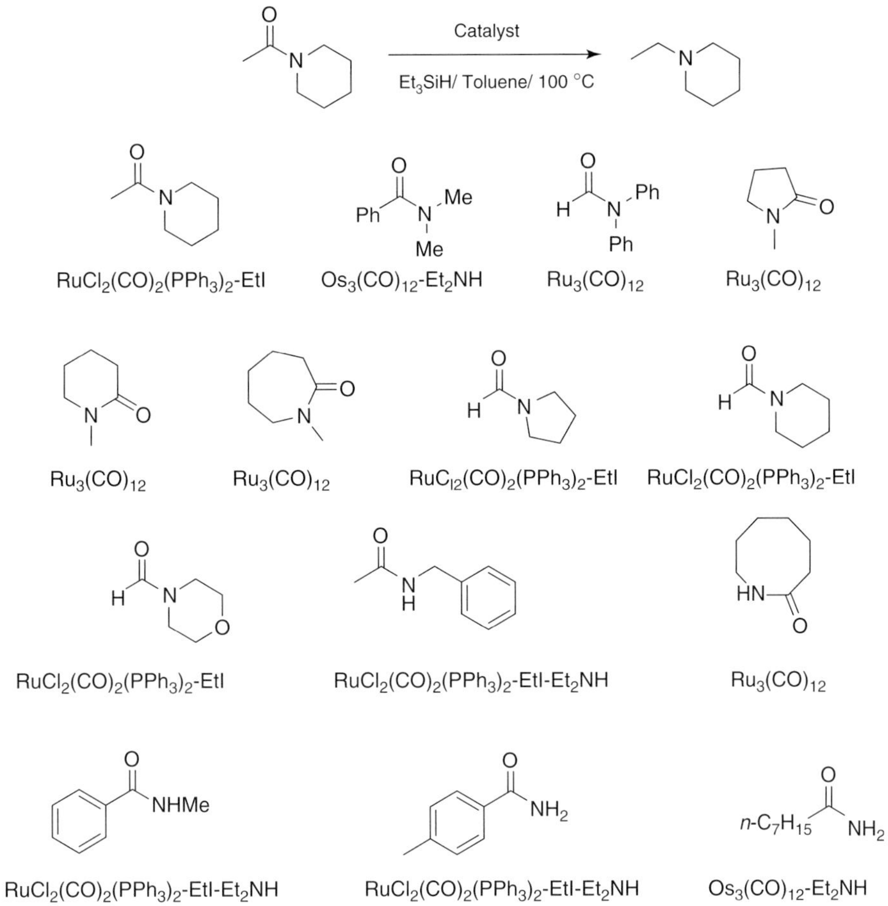

Scheme 3.5 Ruthenium-catalyzed reduction of amides.

other platinum compounds are widely used for catalytic hydrosilylation of alkenes on both the laboratory and industrial scale. Remarkably, this catalyst reduced the amide group chemoselectively even in the presence of double bonds at relatively low temperatures (generally 50–60 °C). Recently, Nagashima and coworkers used again platinum catalysts to show the synergistic effect of two Si–H groups in the reduction of carboxamides to amines under mild conditions. The rate of the reaction was dependent on the distance between two Si–H groups; TMDS and 1,2-bis(dimethylsilyl)benzene were found to be effective reducing reagents. It should be noted that the reduction of amides having other reducible functional groups such as NO_2, CO_2R, CN, C=C, Cl, and Br moieties proceeded selectively, providing a reliable method for the access to functionalized amine derivatives.

In addition to their platinum catalyst, Nagashima and coworkers showed that tertiary amides can be even reduced in presence of keto or ester groups by using ruthenium catalysts in presence of an equimolar amount of triethylamine and

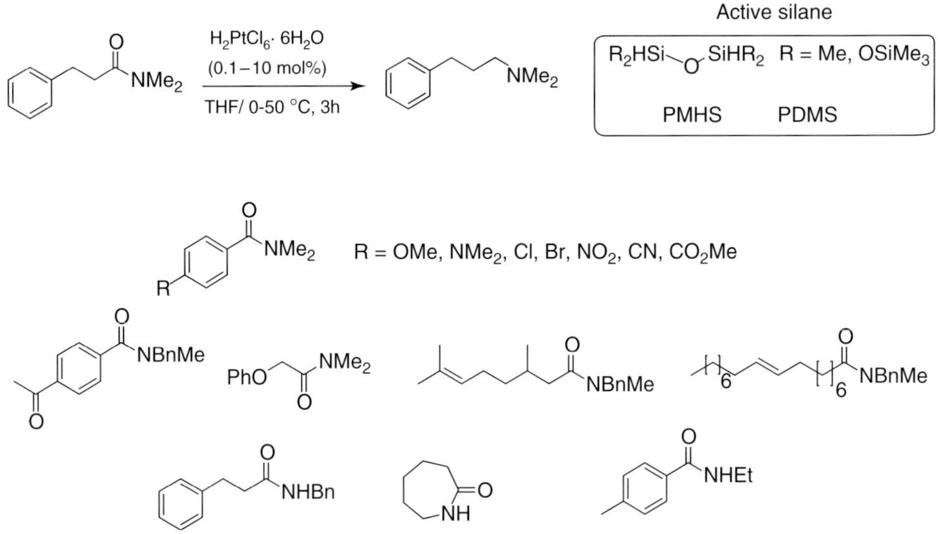

Scheme 3.6 Hydrosilylation of tertiary amides using the platinum complex.

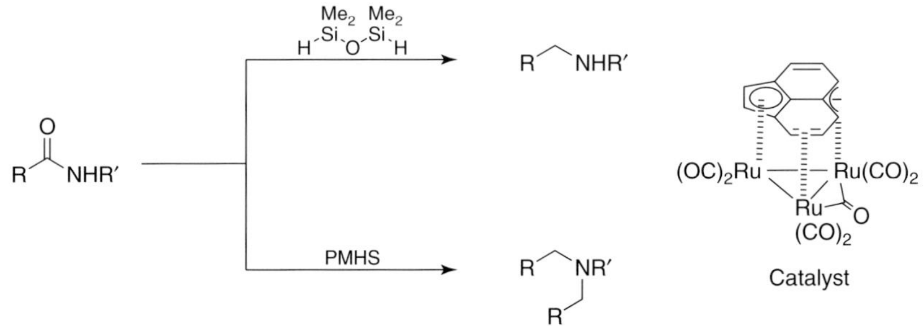

Scheme 3.7 Hydrosilylation of secondary amides using the ruthenium cluster.

$PhMe_2SiH$ (Scheme 3.7). Triethylamine inhibited the reduction of keto and ester groups, and also the competitive experiment where *N,N*-dimethylhexanamide and 2-heptanone or methyl hexanoate were present in a ratio of 1 : 5 showed selective hydrosilylation of the amide without reducing the ketone or the ester. In the same year, they accomplished also the more difficult reduction of secondary amides by using two different protocols (Scheme 3.7). In contrast, tertiary amines were formed with high selectivity by using a lower concentration of the catalyst **1** (1 mol%) and PMHS as reducing agent.

Later, Fernandes *et al.* reported the use of a simple molybdenum catalyst for the hydrosilylation of various amides at reflux temperature in toluene, with phenylsilane as hydrosilylation agent [9]. The yields were in general good to excellent in the case of tertiary amides and moderate in the case of secondary amides, following the general reactivity order. A simple and practical procedure for a direct reductive conversion from a variety of tertiary amides to the corresponding

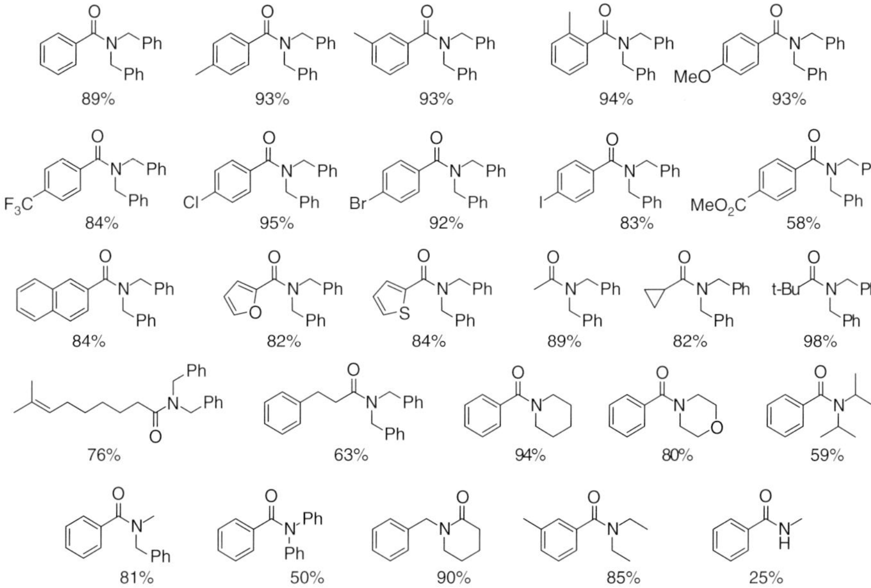

Scheme 3.8 Substrates scope for iron-catalyzed reduction of tertiary amides.

tertiary amines has been developed by Sakai and coworkers, which showed the potential of the simple $InBr_3/Et_3SiH$ reducing system [10].

Despite of the presence of numerable metal catalysts for the reduction of tertiary amides, chemoselective reduction of amides with biomimetic, cheap metals was still highly demanding [11]. In this respect, we thought iron could be a suitable alternative because of its abundant availability, lower price, nontoxicity, and lower environmental hazards. Initially, we took *N,N*-dimethyl benzamide as a model substrate. In the presence of $Fe_3(CO)_{12}$ and $PhSiH_3$, reduction took place at 100 °C and the yield reached to 97% within 24 h [12]. Other iron sources such as $Fe_2(CO)_9$, $Fe(OAc)_2$, $Fe(acac)_3$, and Fe_3O_4 were also applied under the same reaction condition but gave only poor yields. Next, our attention was focused on different substitutional silanes, which resulted in the observation that the inexpensive, nontoxic PMHS was the most effective out of other reactive silanes such as Et_3SiH, $(OEt)_3SiH$, TMDS, and so on. The general applicability of the method and functional group tolerance of the iron catalyst system were demonstrated in the reduction of various amides (Scheme 3.8). In general, aromatic, aliphatic, heteroaromatic, and heterocyclic amides were reduced smoothly under our reaction conditions. To our delight, the electron-withdrawing and electron-donating group in the aromatic ring could not hamper the reduction procedure, but steric crowding on the carbonyl carbon as well as on the nitrogen part imparted slower reduction as well as lower yield. Notably, this catalytic system also tolerated different functional groups such as halides, ethers, esters, and alkenes. Moreover, the reaction was easily scaled up to the 100 mmol scale without any change in the yield of the reaction.

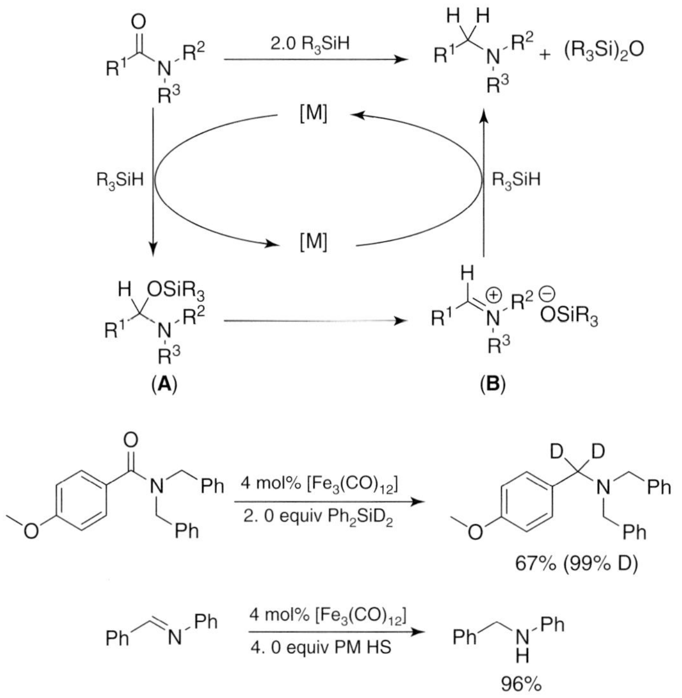

Scheme 3.9 Mechanism for iron-catalyzed reduction of tertiary amides.

With respect to the mechanism, we believe, initially an activation of the iron precursor with silane took place to give an activated anionic silyl hydride. Similar activations are known in the presence of Lewis base as well as transition-metal complexes. As shown in Scheme 3.9, this active species hydrosilylates one of the carbonyl group of the amide to produce the corresponding *O*-silylated *N,O*-acetal **A**. Further reduction of **A** with another equivalent of silane, probably via formation of the iminium species **B**, leads to the final product. This mechanistic proposal was well supported by the reaction of amide with Ph_2SiD_2 where both deuterium atoms were incorporated on the former carbonyl carbon. In addition, the iron catalyst was proved to be successful for the reduction of aldimine in the presence of 4 mol% catalyst and PMHS, which envisages the possibility for the iron-catalyzed stereoselective reduction of imines in the near future (Scheme 3.9).

In our further investigations to find milder conditions as well as to improve selectivity for the hydrosilylation of amides, a system combining zinc acetate and $(EtO)_3SiH$ was brought into the focus [13]. Other zinc salts such as zinc triflate, zinc fluoride, zinc bromide, and zinc iodide were less active, and also, among other silanes, phenyl silane, diphenyl silane, and methyldiethoxysilane were active only at a higher temperature (65 °C). As shown in Scheme 3.10, a variety of amides (aromatic, aliphatic, heterocyclic, and alicyclic) were reduced completely to the desired amines. After showing the general applicability of our zinc catalyst for

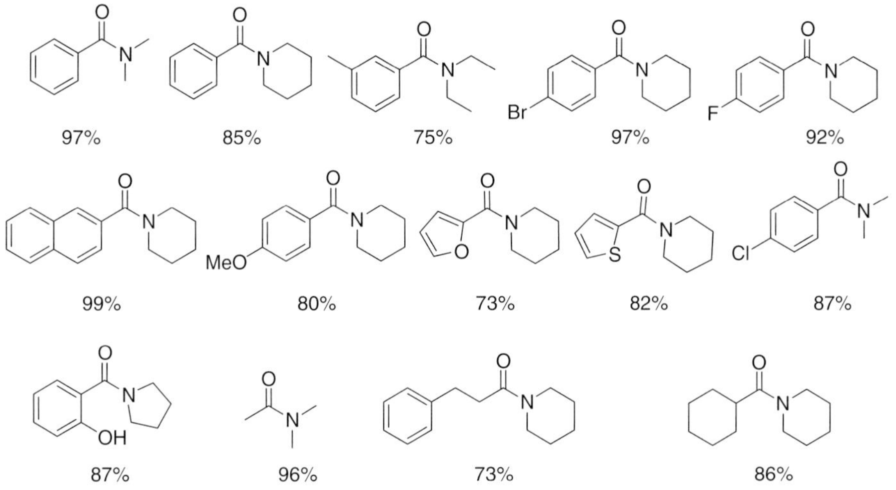

Scheme 3.10 Substrates scope for zinc acetate-catalyzed reduction of amide.

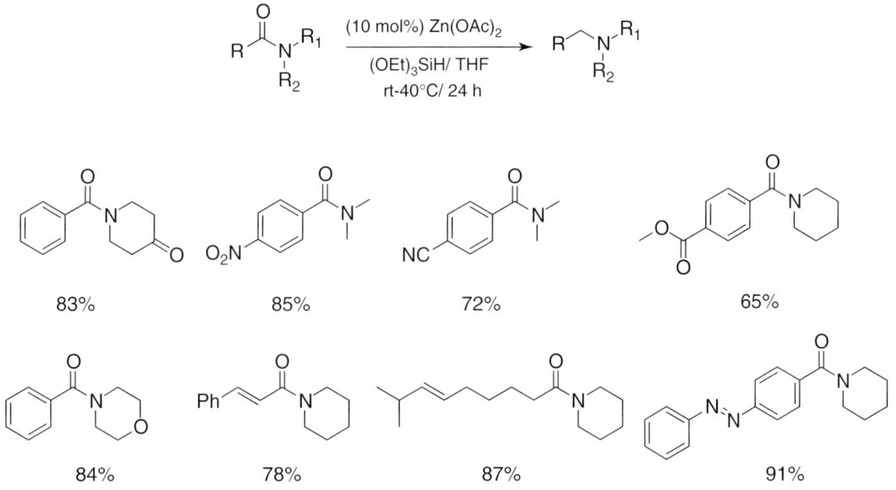

Scheme 3.11 Functional group tolerance.

different substrates, we concentrated on studying the functional group tolerance. As summarized in Scheme 3.11, our system worked excellently for amides including ester, ether, keto, nitro, cyano, and azo groups, as well as for isolated and conjugated double bonds. To our delight, in none of the cases any functionalities were reduced, which rendered novelty to this method. Chemoselectivity was also verified by exploring 1 : 1 mixtures of acetophenone, benzonitrile, *N,N*-dimethylbenzamide, methylbenzoate, and nitrobenzene together under this reaction condition, but only benzyl amine was obtained and other functional groups remained unreacted (Scheme 3.12).

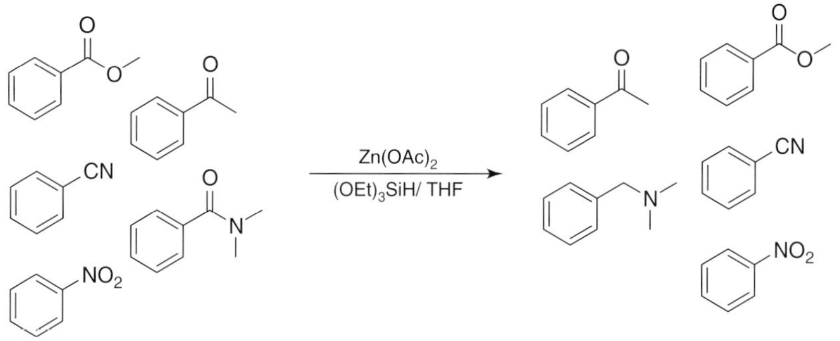

Zn(OAc)$_2$

(OEt)$_3$SiH/ THF

Scheme 3.12 Comparative reduction of tertiary amides in presence of different functional groups.

For mechanistic investigation, we stirred a 1 : 1 mixture of zinc acetate and amide (for 2 h) in tetrahydrofuran (THF), but no change of C=O stretching frequency was observed in the IR spectra, leading to the conclusion that the role of zinc acetate in this reaction was not to activate the carbonyl substrate but rather to activate the silane. Later, a 1 : 1 mixture of zinc acetate and triethoxysilane in THF-d_8 was monitored by ^{1}H NMR spectroscopy, and a small perturbation in the integration of resonances for silane hydrogen was observed, which was strong indication that, in the absence of a substrate, zinc acetate/silane adduct formation is partial and its concentration is low for ^{1}H NMR spectroscopy. Later, the chemical shift of ^{29}Si NMR spectroscopy for this mixture gave a single peak at 82.2 ppm (pure HSi(OEt)$_3$ value is 59.24), which is in the range for –Si(OEt)$_3$, affording strong proof for the activation of silane hydrogen by zinc acetate.

At the same time, the absence of any signal neither at low temperature nor between +30 and −30 ppm in ^{1}H NMR spectroscopy strengthened our proposal on the activation of Si–H bond by zinc acetate. However, the basicity of substrate was relevant in competitive reactions. When a 1 : 1 mixture of *N*-(4-methoxy-benzoyl)piperidine and *N*-(4-fluorobenzoyl)piperidine was subjected to standard hydrosilylation condition for 10 h, the mixture contained 23% of the corresponding fluorinated amine compared to 8% methoxy amine (gas chromatography (GC) analysis of the mixture). This discriminative behavior of the more basic substrates favored coordination of the carbonyl oxygen to silicon as well as nucleophilic attack at the carbonyl carbon atom. After getting this information, we proposed the reaction mechanism shown in Scheme 3.13. Zinc acetate activates the silane and forms an active species **A**.

Basicity of the amide also helps in the activation of the silane toward zinc acetate. Then the active species **A** attacks the carbonyl group of the amide and generates an *N,O*-acetal intermediate **C** via **B**, which, after releasing anionic zinc ether (**D**), converts itself to the iminium species (**E**). Then, the anionic zinc ether (**D**) incorporated with another equivalent of silane converts the iminium ion to the corresponding amine and siloxane. This mechanism is well established with the deuterated reaction with Ph_2SiD_2.

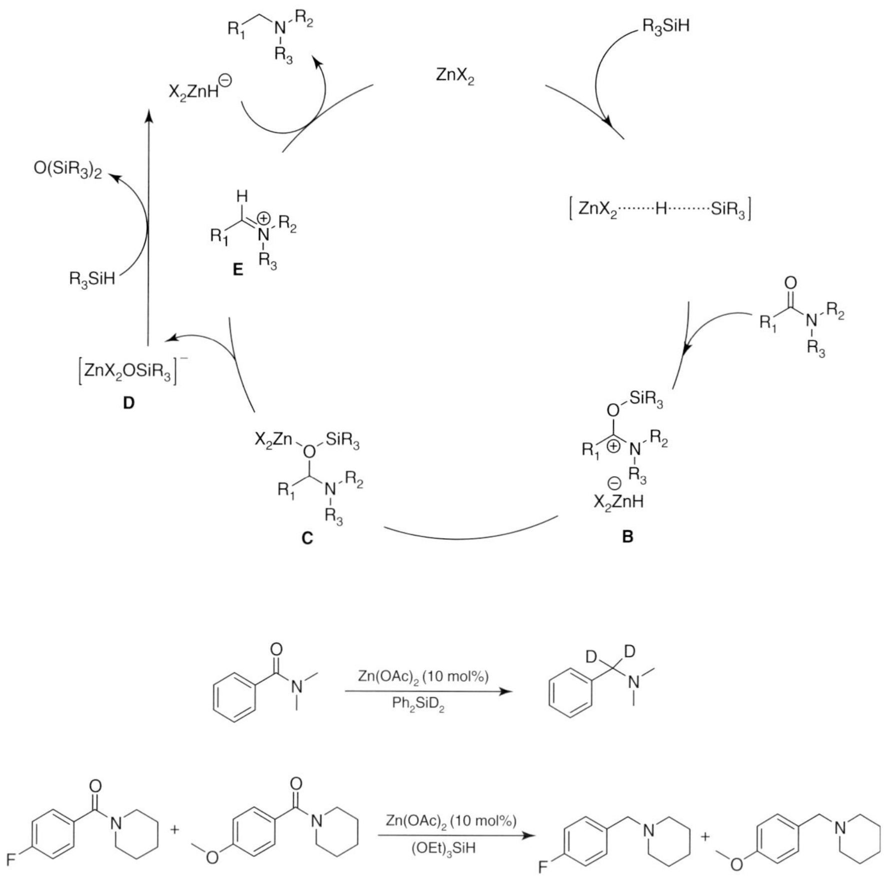

Scheme 3.13 Proposed mechanism of Zn-catalyzed hydrosilylation.

Regarding the triethoxy silane, it should be noted that this silane can easily generate a pyrophoric gas (probably SiH_4), which can result in fire and explosion [14]. However, we never faced any kind of fire or explosion even in the case of multigram-scale reaction. Nevertheless, safety issue regarding triethoxysilane prompted us to explore an alternative silane that would also work under mild condition and could tolerate different functional groups. In this respect, methyldiethoxysilane was a suitable choice. Although the reaction proceeded at a higher temperature (65 °C), it tolerated different functional groups. Once the optimized conditions were achieved, the scope and limitations of tertiary amides were explored. A variety of amides, including aromatic and heteroaromatic ones, were hydrosilylated smoothly with high yields up to 93% (Scheme 3.14). On the other hand, steric hindrance on the amine part of the amide bond had a significant effect. Surprisingly, the reaction proceeded well when there was an aliphatic *N*-piperidine or *N*-cyclopentyl moiety, but no reaction was observed when the *N*-piperidine moiety was replaced by a benzyl or phenyl group. A number of functional groups such as double bond, nitro,

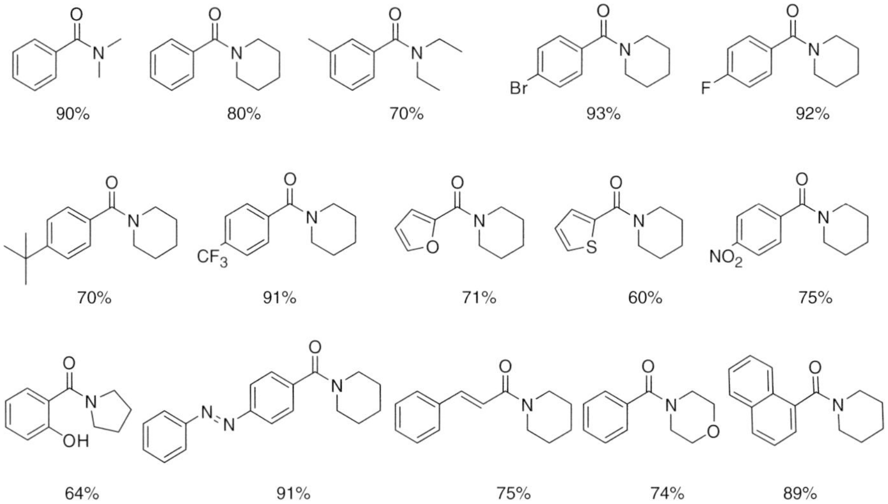

90% 80% 70% 93% 92%

70% 91% 71% 60% 75%

64% 91% 75% 74% 89%

Scheme 3.14 $Zn(OAc)_2$- and $(OEt)_2MeSiH$-catalyzed hydrosilylation of tertiary amides [15].

azo, and ether were also well tolerated under our reaction conditions. The utility of this reaction is unequivocally supported by scaleup of the experiments [15].

3.3 Reduction of Secondary Amides

After the successful chemoselective reduction of tertiary amides, our interest grew on the reduction of secondary amides. It is well known that reductions of secondary amides are very difficult compared to tertiary amides and catalytic hydrosilylation behaves in a similar manner. Earlier success from the group of Nagashima on the catalytic reduction of amides by combination of the ruthenium [8e] and platinum catalysts [8g] with dual Si–H moieties triggered us to reconsider silanes having dual proximate Si–H groups. Initially, we took *N*-benylbenzamide as a model substrate for the reduction of secondary amides. Next, we started to look for better catalysts and, to our surprise, found that in presence of 20 mol% zinc trilfate and 3 mmol of TMDS the reaction got full conversion with excellent yield [15]. Notably, monosilanes such as phenyl silane, diphenyl silane, methyldiethoxy silane, and triethoxy silane were very less active. Contrary to the finding of Nagashima *et al.*, 1,1,1,3,3-pentamethyldisiloxane, which also has two silicon centers but only one Si–H group, was much less active together with 1,2-bis(dimethylsilyl)benzene to afford *N*-benzyl benylamine. It implied that the presence of oxygen atom in between dual the Si–H moiety forced the reaction in case of the secondary amides.

Later, we started to investigate the effect of various metal triflates and the zinc precursor. It was quite obvious that the reaction did not move without the presence of any catalyst and, to our surprise, no other metal triflate such as scandenium triflate, iron (II) triflate, iron (III) triflate, ytterbium triflate, and zirconium triflate showed any activity at higher temperatures. Other zinc sources such as zinc chloride, zinc bromide, and zinc iodide showed much less reactivity. A lower catalyst loading led to a lower yield of the corresponding amine. Appropriate selection of the solvent was also necessary in this reductive hydrosilylation.

Expediently, our reduction protocol proceeded smoothly in case of multi-gram-scale reaction. The removal of siloxane that was concomitantly produced was easily achieved by treating the reaction mixture with a sodium hydroxide solution overnight followed by easy aqueous workup and purification by column chromatography. Once the optimized conditions were achieved, the scope and limitations of zinc triflate-catalyzed reductive hydrosilylation of secondary amides were explored. A variety of amides including aromatic, heteroaromatic, aliphatic, and alicyclic were smoothly reduced with high yields of up to 90% under our reaction condition (Scheme 3.15).

Both electron-donating and electron-withdrawing substituents on the aromatic ring at the para or meta position had little effect on the yield of the reaction, whereas steric crowding on both sides of the amide bond had a significant effect on the activities. The reactivity decreased when the benzyl group in the amine part of the amide group was replaced by a phenyl ring and a tertiary butyl group. In case of *para* and *meta* methyl in the benzene ring, yields

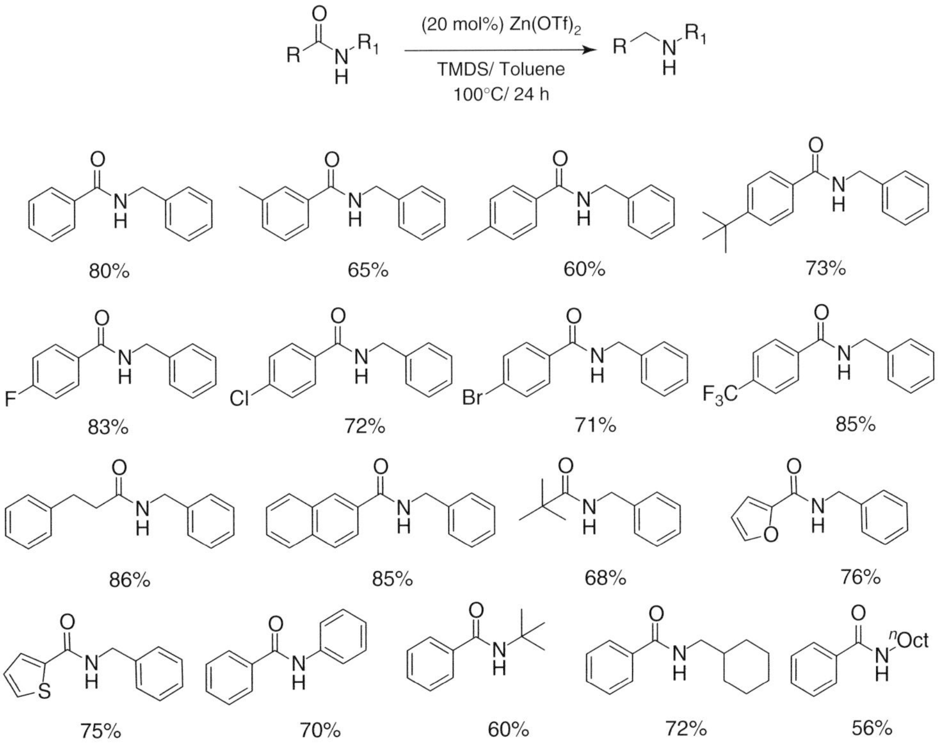

(20 mol%) $Zn(OTf)_2$

TMDS/ Toluene

100°C/ 24 h

80% 65% 60% 73%

83% 72% 71% 85%

86% 85% 68% 76%

75% 70% 60% 72% 56%

Scheme 3.15 Zn-catalyzed hydrosilylation of secondary amides.

were lower because of the distillation method for purification. Furthermore, amides containing para-substituted halogens (fluoro, chloro, and bromo) took a longer time for completion, but in none of the cases we observed any reductive dehalogenated product. In general, our catalyst showed a wide substrate scope for the reduction of secondary amides, and these results clearly demonstrated that the procedure offers a practical method for the synthesis of various secondary amines with high purity in moderate to excellent yield.

Chemoselective reduction of secondary amides in presence of other functional groups is a challenge to be solved in natural-products synthesis or multifunctional molecules. Our group, along with the groups of Barbe and Charette [16] and Nagashima [8g], has reported various reducible functional group tolerances in the case of tertiary amides but catalytic chemoselective reduction of secondary amides was still a challenge to be overcome. Next, we evaluated the chemoselectivity of the reduction process by grafting different functionalities on the *N*-benzyl benzamide or amine part of the amide group. By positioning these functionalities either at the carbonyl part or at the amine part of amide moiety, we minimized both the steric and electronic influence of the functional group on the reaction. To our delight, under our reaction conditions nitrile, azo, double-bond, triple-bond, nitro, ether, and ester groups were well tolerated, affording the corresponding amines in good to excellent yield (Scheme 3.16). It should be noted that, in all of the cases, none of the reduced functionalities was observed, thereby demonstrating the excellent chemoselectivity of the reduction process.

In summary, we established for the first time a highly chemoselective reduction of amide to amine with cheap zinc catalysts and silanes under very mild conditions. The features of cheap commercially available catalysts and silanes, operational safety, simple and mild conditions, and easier purification methods make these new protocols attractive. Our catalytic systems tolerated a variety of functional groups including ester, nitrile, nitro, ether, double-bond, and azo groups, which is completely unique compared to the conventional hydride reduction by aluminum and boron hydride, which involve not only poor functional group tolerance but also costlier purification problems.

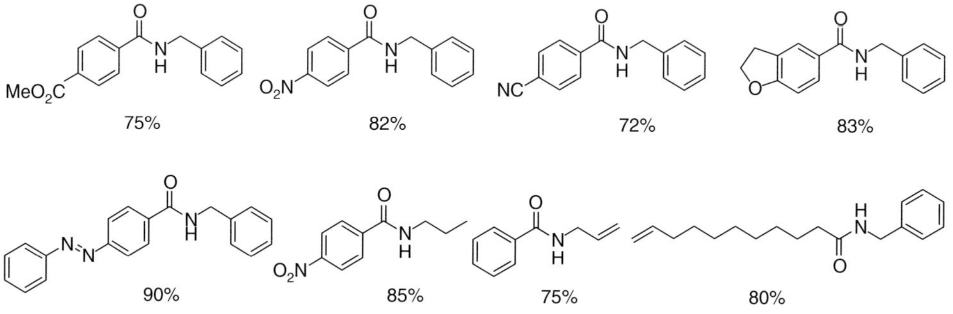

Scheme 3.16 Functional group tolerances for secondary amides.

3.4 Dehydration of Primary Amides

The long demand for primary amine in organic synthesis prompted us to look for the chemoselective reduction of primary amides. But successful reduction of secondary and tertiary amides did not continue there. Initially, we took benzamide as a model substrate for the hydrosilylation with different iron complexes as well as fluoride catalysts. To our surprise, we always obtained benzonitrile instead of benzyl amine (Scheme 3.17). Although traditionally this transformation can also be performed either in the presence of strongly acidic dehydrating agents [17], such as P_2O_5, $POCl_3$, $SOCl_2$, or $TiCl_4$, or in basic reagents sodium borohydride [18e] or more sophisticated reagents such as a combination of ethyl dichlorphosphate and DBU [18f], pivaloyl chloride–pyridine [18g], or trichloro acetyl chloride [18h], our synthetic protocol meets the criteria for today's sustainable synthesis.

Different iron precursors such as $Fe(OAc)_2$, $Fe(CO)_5$, $Fe_2(CO)_9$, $[CpFe(CO)_2]_2$, $Fe_3(CO)_{12}$, $[Et_3NH][HFe_3(CO)_{11}]$, $FeCl_2$, and $Fe(acac)_3$ were applied in the presence of methyldiethoxy silane, and the best activity was found using 2 mol% of $[Et_3NH][HFe_3(CO)_{11}]$ [18a]. Parallel to this iron complex, different other bases and copper sources were also applied, but surprisingly tetra-*n*-butylammonium fluoride (TBAF) gave the best yield within 30 min [18b]. Different other silanes were also screened to optimize the reaction condition, and surprisingly diphenyl silane also showed similar activity as methyldiethoxy silane, but the cheaper and lower reactivity methyldiethoxy silane was taken as a dehydrating agent. Notably, a number of amides were dehydrated to nitrile during the reaction conditions at 100 °C; especially, aromatic, aliphatic, and heterocyclic amides showed high activity (Scheme 3.18). Different functional groups such as the double bond, ester, and halogens were well tolerated under dehydration conditions. A closer look at the mechanism revealed that the initially formed activated species by the iron precursor and silane led to the hydrosilylation of amide followed by rapid exchange to a bis(silyl) amide and bis(silyl)imidate, and final elimination of siloxane generated nitrile (Scheme 3.19). The siloxane and the monohydrolysis product were confirmed by ^{13}C, ^{29}Si NMR and by gas chromatography-mass spectrometry (GC-MS) analysis. TBAF-catalyzed dehydration also followed the same mechanistic pathway.

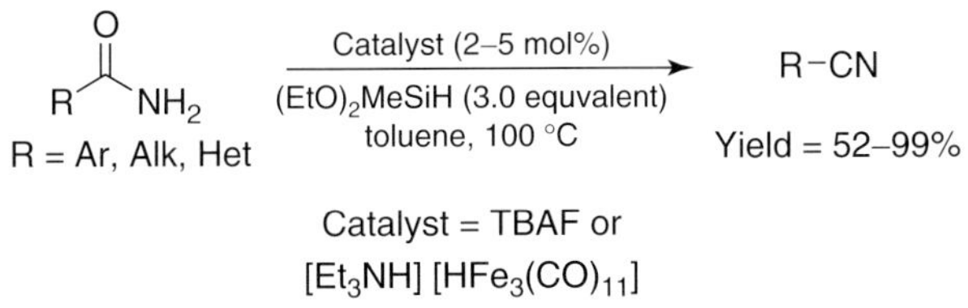

R–C(=O)–NH_2 —[Catalyst (2–5 mol%); $(EtO)_2MeSiH$ (3.0 equvalent), toluene, 100 °C]→ R–CN

R = Ar, Alk, Het

Yield = 52–99%

Catalyst = TBAF or $[Et_3NH]$ $[HFe_3(CO)_{11}]$

Scheme 3.17 Dehydration of primary amides.

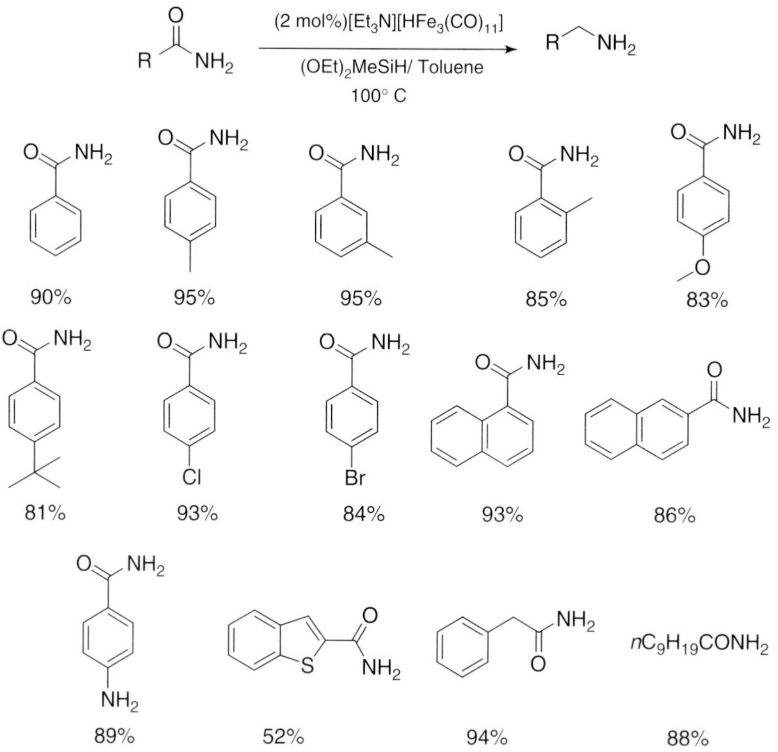

Scheme 3.18 Iron catalysed dehydration of different primary amides.

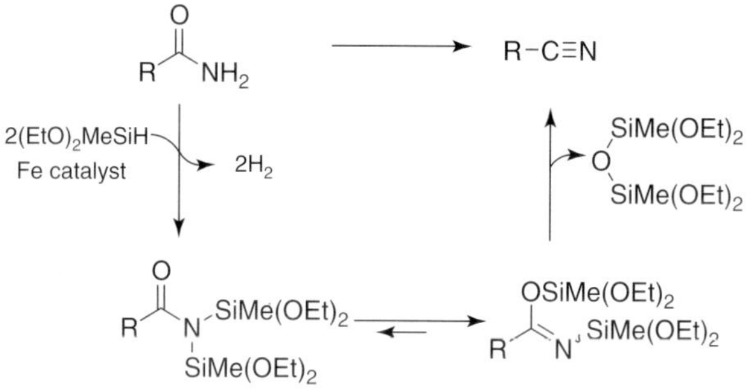

Scheme 3.19 Proposed mechanism for dehydration of primary amides.

3.5
Reduction of Imides

Isoindolinones represent an attractive target for organic synthesis and are a valuable scaffold in medicinal chemistry due to their widespread and diverse biological activities. In spite of a number of known stoichiometric or catalytic methods for the reduction of imides [19], catalytic direct synthesis of isoindolinone is always preferable.

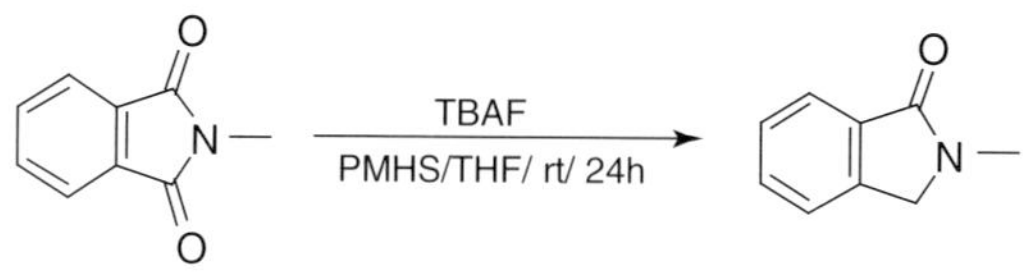

Scheme 3.20 TBAF-catalyzed mono-reduction of pthalimide.

In this respect, 5 mol% of TBAF and PMHS forms an excellent catalytic system for direct mono-reduction of pthalimide derivatives (Scheme 3.20) [20]. Other sources such as CsF, KF, KOtBu, and TBAF hydrate were less active, whereas tetra-*n*-butylammonium bromide TBAB and polymeric-supported fluoride were completely inactive. Once the optimal conditions were identified, the scope and limitations of the fluoride-catalyzed reduction of imides with PMHS were explored. A variety of *N*-substituted pthalimides were reduced smoothly to the corresponding isoindolinone, and in none of the cases did we observe the product with further reduction of isoindolinone moiety. Gratifyingly, a broad range of functional groups including halide, ether, acetal, epoxide, terminal, and internal alkyne were well tolerated by our catalyst (Scheme 3.21). Notably, a higher temperature was needed in case of substrates with substitution in the benzene ring and also non-aromatic imides. It is also noteworthy that the reaction can easily be scaled up to the 100 mmol scale.

Next, our investigation for the complete reduction of pthalimide showed that a combination of iron catalyst and TBAF could bring the reaction to complete conversion, although TBAF alone was unsuccessful to the further

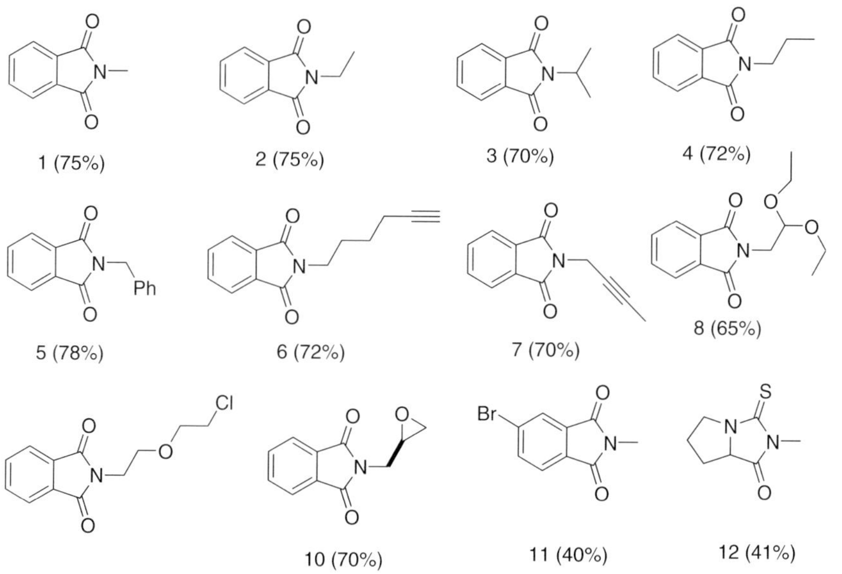

Scheme 3.21 Reduction of imides.

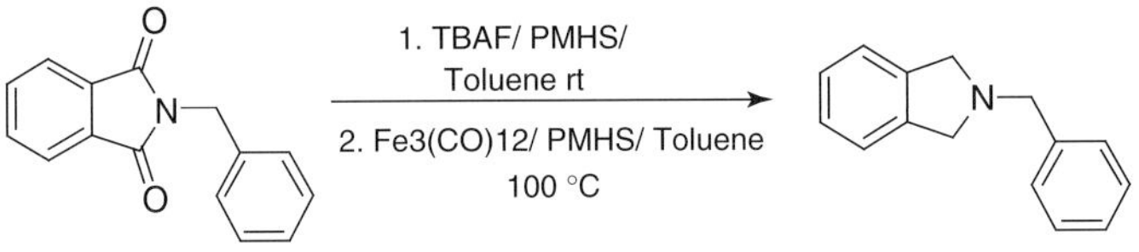

1. TBAF/ PMHS/ Toluene rt
2. Fe3(CO)12/ PMHS/ Toluene 100 °C

Scheme 3.22 Synthesis of isoindoline.

reduction of isoindolinone. Later, our effort on the one-pot synthesis of isoindoline derivative from *N*-benzyl phthalimide using dual catalysts gave 55% isolated yield (Scheme 3.22). Notable features of our protocol are the chemoselectivity, operational simplicity, safety, and mild reaction conditions.

3.6
Conclusion

In summary, we tried to focus here our recent successes in the field of catalytic reduction of amides. For a long time, asymmetric reduction was the prime concern to the academics and industries, but there are still more challenges in reduction of amides or other carboxylic acid derivatives. Notably, still no suitable method for hydrogenation is available. So, for this interim period, hydrosilylation might be an alternative which can selectively reduce different carboxylic acid derivatives. We believe that our new protocols will be successful for the synthesis of various natural products without using protecting and deprotecting groups and enhance the powerful application of green chemistry.

Acknowledgment

This work was funded by the state of Mecklenburg–Western Pomerania, the BMBF, the DFG, and the Leibniz society. I would like to especially thank Professor Dr. Matthias Beller for not only his immeasurable help and patience during the work but also for being a wonderful guide from both a scientific and human point of view. I am immensely thankful to Dr. Kathrin Junge, Dr. Daniele Addis, Dr. Naveen Kumar Mangu, and Shaolin Zhou for their unstinted support in the scientific part. Thanks are also due to Dr. W. Baumann, Dr. C. Fischer, Bianca Wendt, S. Buchholz, S. Schareina, A. Krammer, A. Koch, K. Mevius, and S. Rossmeisl (all at the Leibniz-Institut für Katalyse e.V. an der Universität Rostock) for the excellent analytical and technical support.

References

1. (a) Andersson, P.G. and Munslow, I.J. (2008) *Modern Reduction Methods*, Wiley-VCH Verlag GmbH, New York; (b) de Vries, J.G. and Elsevier, C.J. (eds) (2007) *Handbook of Homogeneous Hydrogenation*, Wiley-VCH Verlag GmbH, Weinheim; (c) Nishimura, S. (ed.) (2001) *Handbook of Heterogeneous*

Hydrogenation for Organic Synthesis, Wiley-Inter-Science, New York; (d) Rylander, P.N. (1979) *Catalytic Hydroenation in Organic Syntheses*, Academic Press; (e) Beller, M. and Bolm, C. (2004) *Transition Metals for Organic Synthesis*, vol. 1 & 2, Wiley-VCH Verlag GmbH, Weinheim.

2. (a) Seyden-Penne, J. (1997) *Reductions by the Alumino and Borohydride in organic Synthesis*, 2nd edn, Wiley-VCH Verlag GmbH, New York; (b) Gribble, G.W. (1998) *Chem. Soc. Rev.*, **27**, 395; (c) Ito, M. and Ikariya, T. (2007) *Chem. Commun.*, 5134; (d) Clarke, M.L., Daz-Valenzuela, M.B., and Slawin, A.M.Z. (2007) *Organometallics*, **26**, 16; (e) Nunez Magro, A.A., Eastham, G.R., and Cole-Hamilton, D.J. (2007) *Chem. Commun.*, 3154; (f) Saudan, L.A., Saudan, C.M., Debieux, C., and Wyss, P. (2007) *Angew. Chem. Int. Ed.*, **46**, 7473.
3. (a) Ito, M. and Ikariya, T. (2007) *Chem. Commun.*, 5134; (b) Clarke, M.L., Daz-Valenzuela, M.B., and Slawin, A.M.Z. (2007) *Organometallics*, **26**, 16; (c) Nunez Magro, A.A., Eastham, G.R., and Cole-Hamilton, D.J. (2007) *Chem. Commun.*, 3154; (d) Saudan, L.A., Saudan, C.M., Debieux, C., and Wyss, P. (2007) *Angew. Chem. Int. Ed.*, **46**, 7473.
4. (a) Marciniec, B., Gulinsky, J., Urbaniak, W., and Kornetka, Z.W. (1992) in *Comprehensive Handbook on Hydrosilylation* (ed. B. Marciniec), Pergamon Press, Oxford; (b) Ojima, I. (1989) in *The Chemistry of Organic Silicon Compounds*, vol. **1** (eds S. Patai and Z. Rapport), John Wiley & Sons, Ltd, Chichester; (c) Brook, M.A. (2000) *Silicon in Organic, Organometallic and Polymer Chemistry*, John Wiley & Sons, Inc., New York; (d) Pukhnarevich, V.B., Lukevics, E., Kopylova, L.T., and Voronkov, M.G. (1992) in *Perspectives of Hydrosilylation* (ed. E., Lukevics), Riga, Latvia; (e) Marciniec, B. (2005) *Coord. Chem. Rev.*, **249**, 2374.
5. (a) Hanada, S., Motoyama, Y., and Nagashima, H. (2008) *Eur. J. Org. Chem.*, **24**, 4097; (b) Zhou, S., Junge, K., Addis, D., Das, S., and Beller, M. (2009) *Org. Lett.*, **11**, 2461.
6. Kuwano, R., Takahashi, M., and Ito, Y. (1998) *Tetrahedron Lett.*, **39**, 1017.
7. Igarashi, M. and Fuchikami, T. (2001) *Tetrahedron Lett.*, **42**, 1945.
8. (a) Motoyama, Y., Itonaga, C., Ishida, T., Takasaki, M., and Nagashima, H. (2005) *Org. Synth.*, **82**, 188; (b) Motoyama, Y., Mitsui, K., Ishida, T., and Nagashima, H. (2005) *J. Am. Chem. Soc.*, **127**, 13150; (c) Hanada, S., Motoyama, Y., and Nagashima, H. (2006) *Tetrahedron Lett.*, **47**, 6173; (d) Sasakuma, H., Motoyama, Y., and Nagashima, H. (2007) *Chem. Commun.*, 4916; (e) Hanada, S., Ishida, T., Motoyama, Y., and Nagashima, H. (2007) *J. Org. Chem.*, **72**, 7551; (f) Hanada, S., Motoyama, Y., and Nagashima, H. (2008) *Eur. J. Org. Chem.*, **24**, 4097; (g) Hanada, S., Tsutsumi, E., Motoyama, Y., and Nagashima, H. (2009) *J. Am. Chem. Soc.*, **131**, 15032; (h) Sunada, Y., Kawakami, H., Imaoka, T., Motoyama, Y., and Nagashima, H. (2009) *Angew. Chem. Int. Ed.*, **48**, 9511.
9. Fernandes, A.C. and Romão, C.C. (2007) *J. Mol. Catal.*, **272**, 60.
10. Sakai, N., Fuhji, K., and Konakahara, T. (2008) *Tetrahedron Lett.*, **49**, 6873.
11. (a) Markò, L. and Palagyi, J. (1983) *Transit. Met. Chem.*, **8**, 207; (b) Jothimony, K., Vancheesan, S., and Kuriacose, J.C. (1985) *J. Mol. Catal.*, **32**, 11; (c) Jothimony, K., Vancheesan, S., and Kuriacose, J.C. (1989) *J. Mol. Catal.*, **52**, 301; (d) Bianchini, C., Farnetti, E., Graziani, M., Peruzzini, M., and Polo, A. (1993) *Organometallics*, **12**, 3753; (e) Bart, S.C., Lobkovsky, E., and Chirik, P.J. (2004) *J. Am. Chem. Soc.*, **126**, 13794; (f) Bart, S.C., Hawrelak, E.J., Lobkovsky, E., and Chirik, P.J. (2005) *Organometallics*, **24**, 5518; (g) Casey, C.P. and Guan, H. (2007) *J. Am. Chem. Soc.*, **129**, 5816; (h) Mimoun, H., De Saint Laumer, J.Y., Giannini, L., Scopelliti, R., and Floriani, C. (1999) *J. Am. Chem. Soc.*, **121**, 6158; (i) Inagaki, T., Yamada, Y., Phong, L., Furuta, A., Ito, J., and Nishiyama, H. (2009) *Synlett*, 253.
12. Zhou, S., Junge, K., Addis, D., Das, S., and Beller, M. (2009) *Angew. Chem. Int. Ed.*, **48**, 9507.

13. (a) Das, S., Addis, D., Zhou, S., Junge, K., and Beller, M. (2010) *J. Am. Chem. Soc.*, **132**, 1770; (b) Das, S., Zhou, S., Addis, D., Enthaler, S., Junge, K., and Beller, M. (2010) *Top. Catal.*, **53**, 979.

14. Buchwald, S.L. (1993) *Chem. Eng. News*, **71** (13), 2.

15. Das, S., Addis, D., Junge, K., Beller, M. (2011) *Chem. Eur. J.* **43**, 12186.

16. Barbe, G. and Charette, A.B. (2008) *J. Am. Chem. Soc.*, **130**, 18.

17. (a) Reisner, D.B. and Horning, E.C. (1963) *Org. Synth. Coll.*, **IV**, 144; (b) Rickborn, B. and Jensen, F.R. (1962) *J. Org. Chem.*, **27**, 4608; (c) Krynitsky, J.A. and Carhart, H.W. (1963) *Org. Synth. Coll.*, **IV**, 436; (d) Lehnert, W. (1971) *Tetrahedron Lett.*, **19**, 1501; (e) Ellzey, S.E., Mack, C.H., and Connick, W.J. (1967) *J. Org. Chem.*, **32**, 846; (f) Kuo, C.-W., Zhu, J.-L., Wu, J.-D., Chu, C.-M., Yao, C.-F., and Shia, K.-S. (2007) *Chem. Commun.*, 301; (g) Narsaiah, A.V. and Nagaiah, K. (2004) *Adv. Synth. Catal.*, **324**, 1271; (h) Mander, L.N. and McLachlan, M.M. (2003) *J. Am. Chem. Soc.*, **125**, 2400.

18. (a) Zhou, S., Addis, D., Junge, K., Das, S., and Beller, M. (2009) *Chem. Commun.*, 4883; (b) Zhou, S., Junge, K., Addis, D., Das, S., and Beller, M. (2009) *Org. Lett.*, **11**, 2461.

19. (a)Kuminobu, Y., Tokunaga, Y., Kawata, A., and Taka, K. (2006) *J. Am. Chem. Soc.*, **128**, 202; (b) Klump, D.A., Zhang, Y., O'Conor, M.J., Esteves, P.M., and Almeid, L.S. (2007) *Org. Lett.*, **9**, 3085; (c) Orito, K., Miyazawa, M., Nakamura, T., Horibata, A., Ushito, H., Nagashaki, H., Yuguchi, M., Yamashita, S., Yamazaki, T., and Tokuda, M. (2006) *J. Org. Chem.*, **71**, 5951; (d) Kobyashi, K., Hase, M., Hashimoto, K., Fujita, S., Tanmatsu, M., Morikawa, O., and Konishi, H. (2006) *Synthesis*, **15**, 2493; (e) Huang, X. and Xu, J. (2009) *J. Org. Chem.*, **74**, 8859; (f) Norman, M.H., Minick, D.J., and Rigdon, G.C. (1996) *J. Med. Chem.*, **39**, 149; (g) Cacchi, S., Fabrizi, G., and Moro, L. (1998) *Synlett*, 741; (h) Feng, S., Panetta, C.A., and Graves, D.E. (2001) *J. Org. Chem.*, **66**, 612; (i) Patton, D. E., Drago, R. S., (1993) *J. Chem. Soc. Perkin Trans.*, 1611; (j) Aoun, R., Renaud, J.L., Dixneuf, P.H., and Bruneau, C. (2005) *Angew. Chem. Int. Ed.*, **117**, 2057.

20. Das, S., Addis, D., Knöpke, L. R., Bentrup, U., Junge, K., Brückner, A., Beller, M., (2011) *Angew. Chem. Int. Ed.*, **50**, 9180.

4
Ionic Ozonides – From Simple Inorganic Salts to Supramolecular Building Blocks

Hanne Nuss and Martin Jansen

4.1
The Forgotten Oxygen Anion

In 1868, Wurtz observed the appearance of a yellow-brown color when exposing potassium hydroxide to ozone [1]. The color vanished immediately when the ozone flow was stopped, and initiated great efforts in investigating its origin. In the following years, controversial discussions started on the actual composition of this new substance. Baeyer and Villiger proposed a "potassium tetroxyde", K_2O_4, in analogy to the results of their experiments concerning higher potassium and rubidium oxides [2]. Bach, on the other hand, described the orange product as an acidic salt, "KHO_4", of "ozonic acid" [3]. These views persisted until 1908, when they were disproved by Manchot and Kampschulte. Their experiments showed that the conductivity of a solution of ozone in water is caused by the presence of nitric oxides and not by the dissociated "H_2O_4" [4]. Traube was the first to clearly distinguish between the intensely red-colored, metastable compounds, which he regarded as adhesion compounds of hydroxide ions and molecular oxygen, and their decomposition products, the yellow "alkali tetroxydes", nowadays known as superoxides ($M^+O_2^-$) [5, 6]. The controversy lasted until 1949, when Kazarnovskii and Nikol'shii suggested the correct formula for the first time and introduced the expression "alkali ozonide" for this type of compounds [7]. In the 1960s and 1970s, the potential application of ionic ozonides for regenerating breathing air in aeronautics was investigated [8, 9], but was dropped because of the high reactivity and difficult handling of salts containing the 19-valence-electron radical anion.

The homolog superoxide anion O_2^- was not only investigated in relationship to superoxide dismutase enzymes, but has also become a convenient oxidizing agent in organic synthesis [10–13]. In contrast, the ozonide anion O_3^- seemed to have been somehow forgotten, most probably due to the lack of a synthesis procedure for preparing pure ionic ozonides in reasonable amounts.

This chapter describes part of the investigations on ionic ozonides that have been performed over the last decades, focusing on synthesis and structure–property relationships. It is based, among others, on recent review articles, which give a

New Strategies in Chemical Synthesis and Catalysis, First Edition. Edited by Bruno Pignataro.

much more detailed report on this topic and to which the interested reader is directed [14–16].

4.2
The Synthesis of Ionic Ozonides

Remarkably, the formation of ozonides from alkali-metal superoxides and ozone and their decomposition into AO_2 and O_2 are exothermic reactions. Hence, alkali ozonides readily form following Eq. (4.3), but are unstable with respect to decay along a reaction channel according to Eq. (4.6). Supposing that the reactions involved are at (or close to) equilibrium, the heat balance can be derived from reported data [17, 18]. Mechanistically, one would expect the formation of an ozonide to proceed via the decomposition of ozone to dioxygen and monooxygen on the surface of the superoxide and eventually a reaction of monooxygen and O_2^- to form ozonide O_3^-. The relevant standard reaction enthalpies are given in Eqs. (4.1)–(4.3).

$$O_3 \longrightarrow O_2 + O \qquad \Delta H_R^0 = 107\ \text{kJ mol}^{-1} \tag{4.1}$$

$$KO_2 + O \longrightarrow KO_3 \qquad \Delta H_R^0 = -226\ \text{kJ mol}^{-1} \tag{4.2}$$

$$KO_2 + O_3 \longrightarrow KO_3 + O_2 \qquad \Delta H_R^0 = -119\ \text{kJ mol}^{-1} \tag{4.3}$$

The formation of potassium ozonide in the way described is clearly exothermic. Obviously, there is also a competing reaction in which monoxygen recombines to dioxygen ($\Delta H_R^0 = -249$ kJ mol^{-1}).

$$KO_3 \longrightarrow KO_2 + O \qquad \Delta H_R^0 = 226\ \text{kJ mol}^{-1} \tag{4.4}$$

$$O \longrightarrow 0.5O_2 \qquad \Delta H_R^0 = -249\ \text{kJ mol}^{-1} \tag{4.5}$$

$$KO_3 \longrightarrow KO_2 + 0.5O_2 \qquad \Delta H_R^0 = -23\ \text{kJ mol}^{-1} \tag{4.6}$$

Consequently, it is possible to define clear conditions for successful synthesis: on one hand, a sufficiently high temperature is required to activate the ozone cleavage and the transport of the monooxygen through the solid; on the other, it must not be so high as to set off the autocatalytic decay of the ozonide. This entails a well-controlled dissipation of the reaction heat, and the necessary and crucial control of heat flows and temperatures. A dedicated glass apparatus (Figure 4.1) was developed in order to satisfy the special requirements [17]. This setup permits the ozonization of a superoxide placed on a glass frit (b) and the subsequent separation of the ozonide from unreacted superoxide by extraction with liquid ammonia, without the need to open the glass apparatus during the process. In particular, such a setup provides for self-regulating temperature adjustment. If the ozonization process is started at room temperature and the reaction heat is directed away toward the cooling jacket (f), a temperature gradient results from the outer (cold) to the inner (warm) part of the sample, with a self-sustaining optimal reaction temperature contained in a cylindrical zone. In the course of the reaction

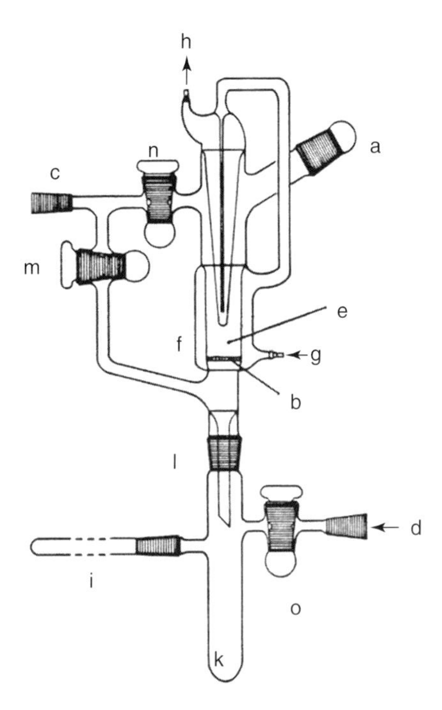

Figure 4.1 Glass apparatus for the sequential procedure of all steps of the synthesis of alkali-metal ozonides, (1) ozonization of alkali-metal superoxides, (2) extraction of the ozonide from the raw product with liquid ammonia, and (3) crystallization and isolation of the pure ozonide. (Taken from Ref. [16] with permission of Wiley-VCH.)

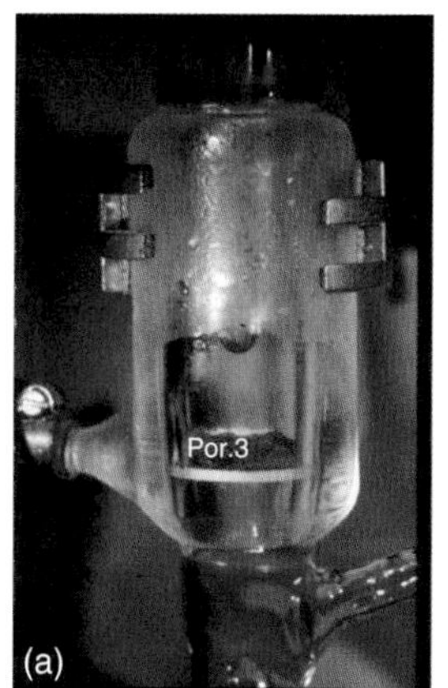

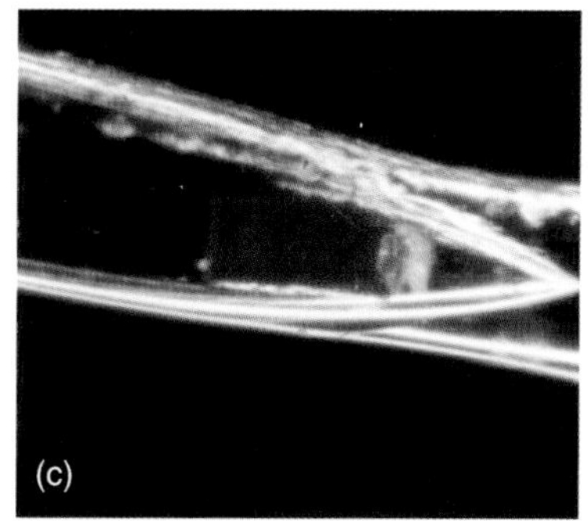

Figure 4.2 (a) Intermediate state of the ozonization of CsO_2 (red, ozonide; yellow, superoxide). (b) Solution of CsO_3 in liquid NH_3. (c) Single crystal of $(NMe_4)O_3$. (Taken from Ref. [16] with permission of Wiley-VCH.)

process, this zone slowly (over several hours) moves from the outside to the inside. The photograph in Figure 4.2(a) shows such an intermediate state.

Extraction with and recrystallization from liquid ammonia (Figures 4.2b,c) have yielded pure, coarsely crystalline ozonides of the heavy alkali-metals. Pure potassium, rubidium, and cesium ozonides have been synthesized via this method for the first time, which act as starting materials for all further synthetic efforts [17–21].

However, not even the lighter homolog NaO_3 can be synthesized this way anymore. Complications arise when ammonia is removed, which needs to be done in order to achieve crystallization. Because of the greater ionic strength of Na^+ compared to its heavier homolog, the coordinating NH_3 is more proton active, and via the protonation of O_3^- its decay is induced. If, however, dimethylamine is added to the solution of NaO_3 in liquid NH_3, this complication can be circumvented and microcrystalline sodium ozonide can be isolated in pure phase [22, 23]. Stable solutions of lithium, strontium, and barium ozonides can be prepared by ion exchange (IE). The corresponding solid ozonides, however, can only be extracted in the form of ammoniates [24, 25].

As already mentioned, KO_3, RbO_3, and CsO_3 built up the pool of starting materials for the synthesis of further ionic ozonides via two different methods

Table 4.1 Summary of ionic ozonides synthesized and characterized so far: OZ = ozonization of the respective superoxide, IE = ion exchange in liquid ammonia, MT = metathesis (starting from superoxide or fluoride), CO = coordination in liquid ammonia (adapted from Ref. [16] with permission of Wiley-VCH).

Ozonide	Method of synthesis
KO_3, RbO_3 [17], α-CsO_3 [20, 21], β-CsO_3 [20]	OZ
$LiO_3 \cdot 5\ NH_3$ [24], $[Li_8(OH)_6](O_3)_2 \cdot 8\ (CH_3)_2NH$ [28], NaO_3 [22, 23]	IE
$Cs_2Ba(O_3)_4 \cdot 2\ NH_3$ [29], $Sr(O_3)_2 \cdot 9\ NH_3$ [25]	IE
$(NMe_4)O_3$, $(NEt_4)O_3$ [27]	MT
$(NMe_3Ph)O_3$, $(NBzMe_3)O_3$, $(NBz^nBu_3)O_3$ [26], $(N^nBu_4)O_3$ [30], $(N^nPr_4)O_3$ [31]	IE
$(NC_8H_{16})O_3 \cdot 1.5\ NH_3$ [32],$[(Ph_3P)_2N)]O_3 \cdot (CH_3)_2NH$ [33]	IE
$[((Me_2N)_3PN)_4P]O_3$, $[((Me_2N)_3PN)_4P]O_3 \cdot (CH_3)_2NH$ [33]	MT
$(PMe_4)O_3$, $(PEt_4)O_3$, $(AsMe_4)O_3$ [34]	IE
$(Me_3N(CH_2)_3NMe_3)(O_3)_2 \cdot 3\ NH_3$, $(Me_3N(CH_2)_3NMe_3)(O_3)_2$, $(Me_3N(CH_2)_4NMe_3)(O_3)_2$, $(Me_3N(CH_2)_5NMe_3)(O_3)_2$, $(Me_3N(CH_2)_6NMe_3)(O_3)_2$ [35]	IE
[1,3-Bis(trimethylammonium)]benzene diozonide, [1,4-bis-(trimethylammonium)]benzene diozonide [36], [1,4-bis(trimethyl ammonium)]benzene diozonide ammoniate [37]	IE
$Li[2.1.1]O_3$, $Na[2.2.2]O_3$ [38]	IE + CO
$K[2.2.2]O_3$, $Rb[2.2.2]O_3$, $Cs[2.2.2]O_3$, $K[Benzo\text{-}18c6]O_3$, $Rb[15c5]_2O_3$, $Cs[15c5]_2O_3$ [38]	CO
$K[18c6]O_3 \cdot 2\ NH_3$ [39], $Rb[18c6]O_3 \cdot NH_3$ [40], $Cs[18c6]O_3 \cdot 8\ NH_3$ [39]	CO
$Cs_5[12c4]_2(O_3)_5$ [41]	CO
$K[12c4]_2O_3 \cdot 1.5\ NH_3$; $Rb[12c4]_2 \cdot 1.5\ NH_3$ [42]	CO

Abbreviations: Me, methyl; Et, ethyl; Pr, propyl; Bu, butyl; Bz, benzyl; Ph, phenyl; [2.1.1], [2.1.1]-cryptand; 12c4, [12]crown-4; 15c5, [15]crown-5; 18c6, [18]crown-6.

based on metathesis or IE, both in liquid ammonia [20, 26]. The first method, which relies on the usually high solubility of ozonides and the low solubility of a suitable complementary salt in ammonia, is shown using $(NMe_4)O_3$ (with Me = methyl) as an example [20, 27].

$$(NMe_4)O_2 + CsO_3 \longrightarrow (NMe_4)O_3 + CsO_2 \downarrow \tag{4.7}$$

Cation exchange in liquid NH_3 is much more versatile (IE = ion exchange resin):

$$IE{-}H + NR_4OH \longrightarrow IE{-}NR_4 + H_2O \tag{4.8}$$

$$IE{-}NR_4 + CsO_3 \longrightarrow IE{-}Cs + NR_4O_3 \tag{4.9}$$

Therefore, sharp conditioning of the IE resin is important in order to prevent hydrolysis or reduction of the extremely sensitive ozonides. A significant number of ionic ozonides with solvatized or coordinated alkali and alkaline-earth metals as well as organic cations were synthesized and characterized with the range of methods described, see Table 4.1.

4.3
The Structural Variety of Ionic Ozonides

4.3.1
Simple Binary and Pseudo-Binary Ozonides

The structural chemistry of the ozonide ion O_3^- is remarkably diverse. Ozonide ions have a bent molecular structure, enabling different interaction modes not only toward the respective counter cations but also toward each other. Population analyses of experimentally as well as theoretically derived electron densities have

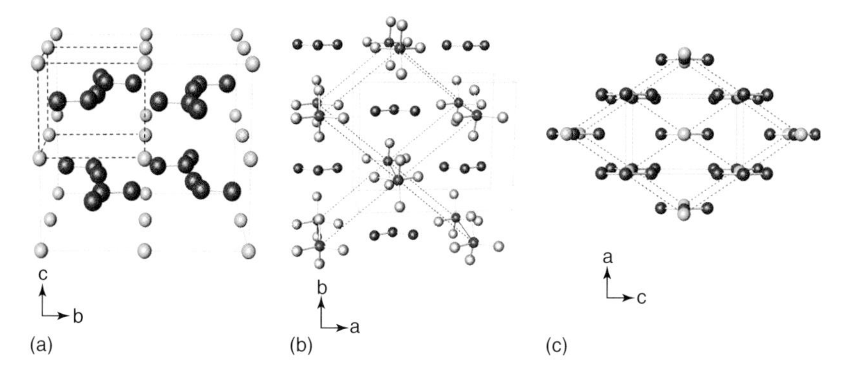

Figure 4.3 Perspective representation of the crystal structures of (a) KO_3, (b) $(NMe_4)O_3$, and (c) NaO_3 (alkali metal = green, oxygen = red, nitrogen = blue). Solid lines mark the crystallographic unit cell, and dashed lines the section analogous to the CsCl or NaCl type. (Taken from Ref. [16] with permission of Wiley-VCH.)

Table 4.2 Dimensions of the ozonide anion in ozonides with inorganic cation.

Ozonide	*d*(O–O)/(pm)	∠O–O–O (°)	Reference
NaO_3	135.3(3)	113.0(2)	[22, 23]
KO_3	134.6(2)	113.5(1)	[17]
RbO_3	134.1(6), 134.4(7)	113.7(5)	[17]
α-CsO_3	133.3(9), 133.2(8)	114.6(6)	[20, 21]
$Cs_2Ba(O_3)_4 \cdot 2\,NH_3$	133.7(3), 135.6(3)	114.2(2)	[29]
Best theoretical estimate	135 ± 3	114.5 ± 1.0	[44]

shown charges of +1.2 for the bridging and −1.0 for the terminal oxygen atoms [43], leading to a permanent dipole moment of this radical anion. In the binary alkali-metal ozonides, as well as in $(NMe_4)O_3$, which typically exhibit salt-like structures related to CsCl or NaCl types, this dipole moment often reflects in the closest intermolecular distance between adjacent ozonide ions. Most frequently, the negatively polarized terminal oxygen atom of one O_3^- ion approaches the positively charged central oxygen atom of its neighbor ("t-c interaction" as in KO_3, see Figure 4.3).

The molecular structure of the ozonide anion within these binary compounds changes only slightly with the cation radius. Table 4.2 shows the O–O bond lengths and O–O–O bond angles of the ozonide anion in alkali metal ozonides compared to the molecular dimensions in $Cs_2Ba(O_3)_4 \cdot 2\,NH_3$, and to the best theoretical estimate of the ozonide molecule in the gas phase [44]. The bond distances and angles are all in the same range as the theoretically determined values. At first sight, a trend is noticeable, namely a decrease in bond lengths and an increase in bond angles with increasing cation radius.

Such trends in the geometry of triatomic molecules are often discussed in terms of Walsh diagrams as shown in Figure 4.4 [45]. Obviously, the actual electron density within the antibonding $2b_1$ molecular orbital determines the geometry of the radical anion. In case of an ideal O_3^- with half-occupied highest occupied molecular orbital (HOMO), a decrease of the O–O–O angle leads to an energy gain and therefore to stabilization. At the same time, the O–O bond distances should be increased compared to primitive O_3 due to a reduction of bond order from 1.5 in ozone to 1.25 in the ozonide. On the other hand, if electron density is removed from the antibonding $2b_1$ orbital, widening of the bond angle as well as shortening of the bond lengths is expected.

Table 4.3 lists corresponding bond lengths and bond angles of ozonide compounds containing organic cations capable of forming hydrogen bonds. In comparison to the alkali-metal ozonides, significantly shorter bond lengths and wider bond angles are actually observed, indicating a certain removal of electron density from the $2b_1$ molecular orbital.

Applying this view to the alkali-metal ozonides at first leads to a contradiction. The smallest, most polarizing cation Na^+ forms ionic ozonides with a relatively

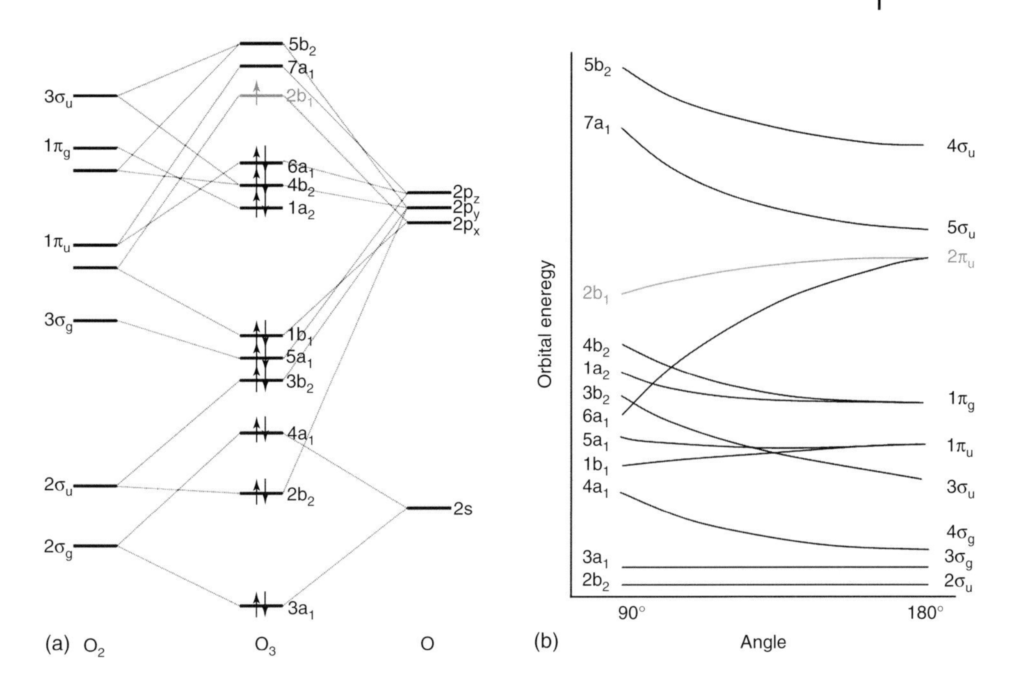

Figure 4.4 (a) Qualitative MO-scheme for O_3^- and (b) Walsh diagram for AB_2 molecules. (Adapted from the literature [45].)

Table 4.3 Dimensions of the ozonide anion of ozonides with organic cations.

Ozonide	***d*(O–O)/(pm)**	**∠O–O–O (°)**	**Reference**
$(NMe_4)O_3$	128.8(3)	119.6(4)	[27]
$(NMe_3CH_2C_6H_5)O_3$	128(1), 128.2(9), 128.2(9), 127.6(9)	118.9(7), 120.4(6)	[26]
$(NMe_3C_6H_5)O_3$	132.1(2), 130.8(2)	117.0(2)	[26]
$(N(C_8H_{16})O_3 \cdot 1.5\ NH_3$	131.5(10), 131.7(11)	114.4(4)	[32]
$(Me_3N(CH_2)_6NMe_3)(O_3)_2$	132.8(2), 131.9(2)	115.8(1)	[35]
$(Me_3N(CH_2)_3NMe_3)(O_3)_2 \cdot 3NH_3$	130.6(4), 130.9(4), 137.4(3), 129.5(3)	117.9(2), 114.7(2)	[35]
Solid ozone	127.2(0)	116.9(0)	[46]

long O–O length and a wide bond angle, whereas the cesium homolog, in which less electron density should be removed, exhibits a geometry closer to the ideal estimate. A closer look on the standard deviations reveals that this trend should be carefully discussed, especially because of the fact that the structure of sodium ozonide was determined only from powder diffraction data. The bonding situation in all known alkali-metal ozonides is described best to be equal and close to the

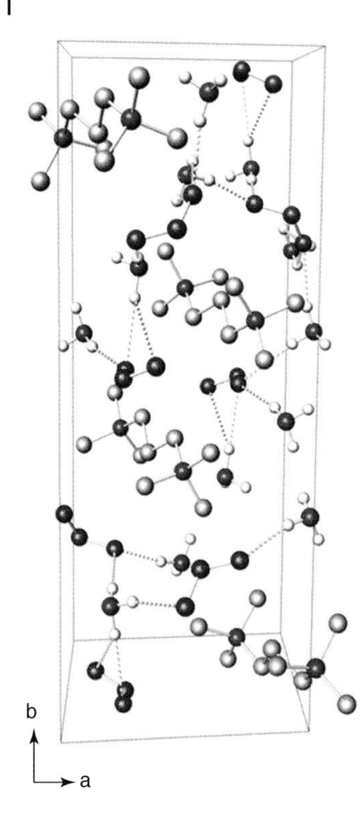

Figure 4.5 Crystal structure of $(Me_3N(CH_2)_3NMe_3)(O_3)_2 \cdot 3\ NH_3$ [35]; the H-bonds are indicated by dotted lines. (Taken from Ref. [16] with permission of Wiley-VCH.)

ideal value of an unperturbed O_3^- ion. Another situation is found in compounds containing hydrogen donors as organic cations, ligands, or even NH_3 solvent molecules within the crystal structure. Here, the ability of the ozonide ion to establish different bonding modes comes into play.

Figure 4.5 shows the unit cell of $[Me_3N(CH_2)_3NMe_3](O_3)_2 \cdot 3\ NH_3$ as an example for the incorporation of O_3^- into hydrogen-bonding networks with ammonia solvent molecules present in the crystal. The competition between hydrogen bond formation and metal coordination of O_3^-, however, is best illustrated by looking at ionic ozonides with complex cations. Figures 4.6 and 4.7 show the ozonide

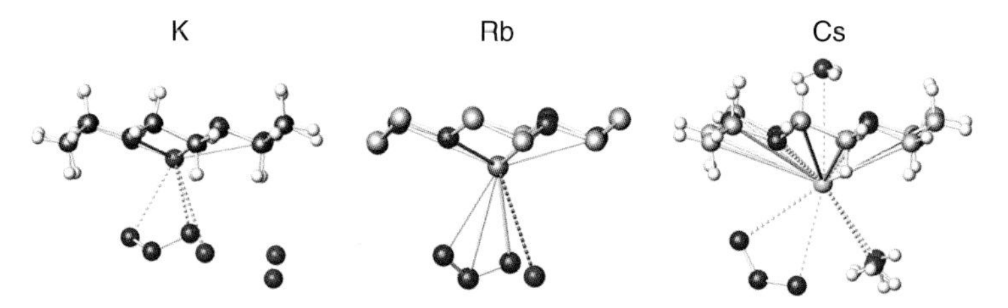

Figure 4.6 Series of the complex ozonides $M([18]crown\text{-}6)O_3 \cdot xNH_3$ with M = K ($x = 2$), Rb ($x = 1$), Cs ($x = 8$); the anion acts as bidentate ligand to the metal center [47].

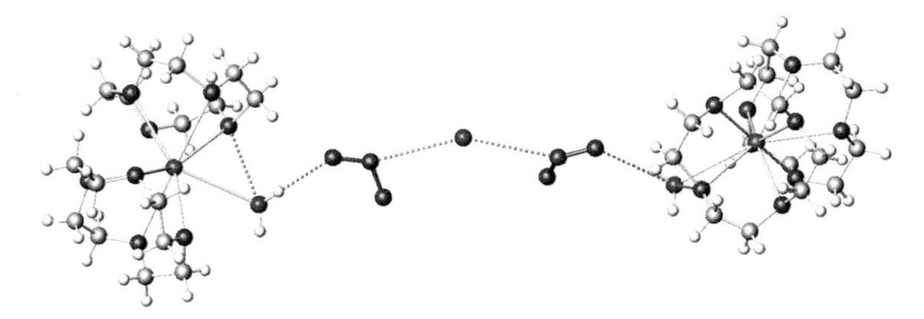

Figure 4.7 Cutout of the structure of $M([12]crown\text{-}4)O_3{\cdot}1.5\,NH_3$ with M = K, Rb; the anions are coordinated only by ammonia molecules via hydrogen bonds. There is no coordination to the metal center [47].

compounds $M([18]crown\text{-}6)O_3 \cdot x\ NH_3$ with M = K ($x = 2$), Rb ($x = 1$), Cs ($x = 8$) and $M([12]crown\text{-}4)_2O_3 \cdot 1.5\ NH_3$ with M = K, Rb, respectively [39, 42].

In the first series, the ozonide anion clearly points with its terminal oxygen atoms toward the alkali-metal, completing the coordination provided by the crown ether and additional ammonia molecules. In the second series, in which the metal center is shielded by two smaller crown ether molecules, the interaction of the molecular anion to the metal is suppressed and the O_3^- builds hydrogen bonds to solvent molecules instead.

4.3.2 $Cs_5([12]crown\text{-}4)_2(O_3)_5$ – from Simple Salts to Supramolecular Building Blocks

The structural flexibility of the ozonide anion, demonstrated above, has been utilized in combination with concepts of crystal engineering to build up entirely new structural motifs in order to study the resulting intermolecular interactions.

To strengthen intermolecular interactions, the distance between adjacent ozonide ions has to be reduced. This was attempted by utilizing organic bisammonium cations exhibiting short alkane spacer chains between the two positive charges to attract the anions as in $[Me_3N(CH_2)_3NMe_3](O_3)_2 \cdot 3\ NH_3$. As discussed above, such organic cations also facilitate the incorporation of solvent molecules into the crystal structure, which then leads to the undesirable reseparation of O_3^- ions.

In another approach, polar and nonpolar constituents are combined, which tend to occupy separate volumes in the crystal and might lead to a structural confinement, forcing ozonide anions to approach each other more closely, a concept often employed in crystal engineering [48, 49].

For cesium ozonide, this goal has been achieved by adding a smaller crown ether [12]crown-4, which coordinates only weakly on one side to the Cs^+ cations, because if size mismatch. The crystal structure of $Cs_5([12]crown\text{-}4)_2(O_3)_5$, see Figure 4.8, actually exhibits a one-dimensional host–guest arrangement and is the first example of an ionic ozonide compound showing a supramolecular building block [41]. The structure can be regarded as a distorted tetragonal packing of $^1_\infty\{Cs_8(O_3)_{10}\}^{2-}$ rods that run along [100] and are separated by $[Cs([12]crown\text{-}4)_2]^+$ units.

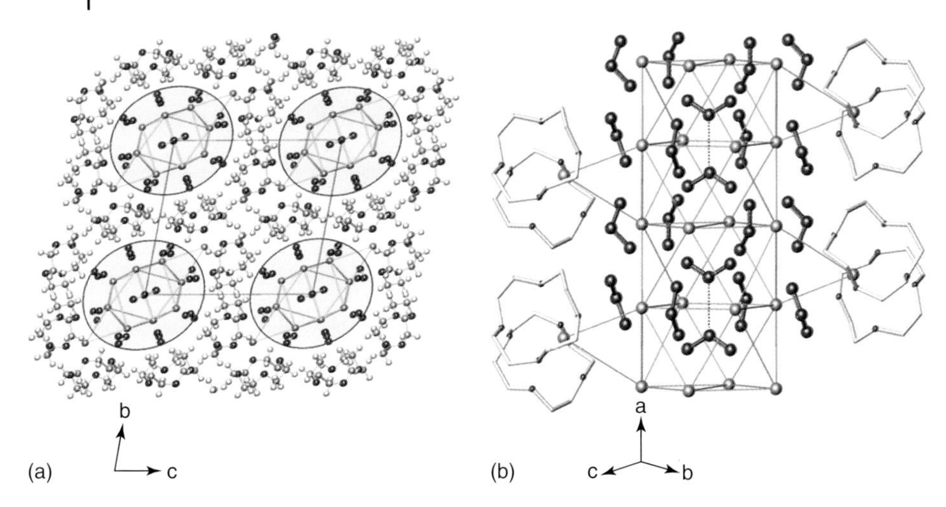

Figure 4.8 (a) View of the structure of $Cs_2([12]crown\text{-}4)_2(O_3)_5$, showing the distorted tetragonal packing of ${}^{1}_{\infty}\{Cs_8(O_3)_{10}\}^{2-}$ rods (indicated by red circles). (b) Side view of a ${}^{1}_{\infty}\{Cs_8(O_3)_{10}\}^{2-}$ rod and surrounding $[Cs([12]crown\text{-}4)_2]^+$ units. (Adapted from Ref. [41] with permission of Wiley-VCH.)

The rods have a diameter of approximately 1 nm and have a symmetry corresponding to the rod group $p\bar{1}$ (no. 2). The average Cs···Cs distance within the stacks of tetragonal antiprisms is 456 pm. Each antiprism of Cs atoms is centered by one and surrounded by four other crystallographically independent ozonide ions. The eightfold coordination environment of the central ozonide ion is reminiscent of the situation in α-CsO_3 and β-CsO_3. The anions outside the antiprisms are more distorted than the central ozonide ion, which has nearly perfect C_{2v} symmetry.

The structural restrictions imposed by this packing arrangement lead furthermore to the shortest intermolecular O···O distance ever observed between ozonide ions in the solid state. The distance between the central oxygen atoms of two adjacent anions within the stacks of antiprisms is 275.3(8) pm (indicated by the red dotted lines in Figure 4.8b). As a comparison, until then the shortest intermolecular O···O contact between oxygen atoms of adjacent ozonide anions of 291.6(1) pm was found in $Cs_2Ba(O_3)_4 \cdot 2\,NH_3$ [29]. Apart from this rather short intermolecular distance, the next contacts of 347–368 pm occur between the terminal oxygen atoms of adjacent ozonide anions outside of the stacks of antiprisms.

As mentioned above, the arrangement of adjacent ozonide ions in binary alkali-metal ozonides is mainly determined by dipole–dipole interactions between the positively charged central oxygen atom with one of the negatively charged terminal ones. In $Cs_5([12]crown\text{-}4)_2(O_3)_5$, the structural confinement of the ${}^{1}_{\infty}\{Cs_8(O_3)_{10}\}^{2-}$ substructure by the nonpolar shell of the $[Cs([12]crown\text{-}4)_2]^+$ units forces the inner radical anions to approach each other via their positively charged central oxygen atoms, which is a unique situation in ionic ozonide chemistry.

4.4 Magnetic Properties

The structural diversity of ionic ozonides might be reflected in variations of their physical properties. The radical nature of the ozonide ion moreover permits the use of magnetism as a probe for changes in the intermolecular interactions of adjacent anions caused by structural modifications.

Table 4.4 summarizes magnetic data of selected binary or pseudo-binary ozonides.

Simple ozonides in which the closest distance between two radical anions is found between a terminal and a central oxygen atom usually exhibit paramagnetic behavior with a tendency to form antiferromagnetic ordering at low temperatures. Magnetic moments derived from Curie–Weiss fits of the high-temperature region of magnetic data give values around the spin-only value of 1.73 μ_B.

The only exception is NaO_3. Sodium ozonide, as explained above, exhibits a remarkable solid state structure, in which the closest $O_3^- \cdots O_3^-$ distance is found between terminal O atoms of adjacent molecular anions, forming a chain-like arrangement. This unique structural feature manifests itself also in a significant deviation of the observed magnetic moment of 1.45 μ_B from the spin-only value and a remarkably high Weiss constant θ of −246.2 K, indicating strong exchange coupling. The actual reason for the decrease, however, is still not completely understood [23].

The incorporation of crown ether molecules into ozonide structures can lead to two different effects. In the series of complex ozonides of the formula $M([18]crown\text{-}6)O_3 \cdot x\ NH_3$ ($M = K$ ($x = 2$), Rb ($x = 1$), Cs ($x = 8$)), the complexation of the cation combined with the uptake of ammonia molecules and the interaction of the O_3^- with the metal lead to a larger separation of the radical anions. Figure 4.9 exemplarily shows a cutout of the crystal structure of $K([18]crown\text{-}6)O_3 \cdot 2NH_3$. Indicated in red are the shortest intermolecular contacts found to lie between 614 and 650 pm.

This separation leads to an almost ideal Curie behavior, with the magnetic moment determined to be 1.72 μ_B and the Weiss constant θ to be zero in the first approximation ($\theta = -0.12$ K) (Figure 4.10) [39].

Table 4.4 Magnetic data of selected ozonides [23, 50, 51].

Ozonide	μ/μ_B (250 K)	θ (K)	T_N (K)	$g_{isotrop}$ (Powder)	$g_x; g_y; g_z$ (Single crystal)	$g_x; g_y / g_z$ (Powder)
NaO_3	1.45	–	–	–	–	–
KO_3	1.74	−34	20	–	–	2.004; 2.014
RbO_3	1.80	−23	17	–	2.007; 2.013; 2.009	–
CsO_3	1.74	−10	6	2.006	–	–
$(NMe_4)O_3$	1.70	−5	<4	2.011	2.007; 2.017; 2.009	–
$(NEt_4)O_3$	1.73	≈ 0	<4	2.011	–	–

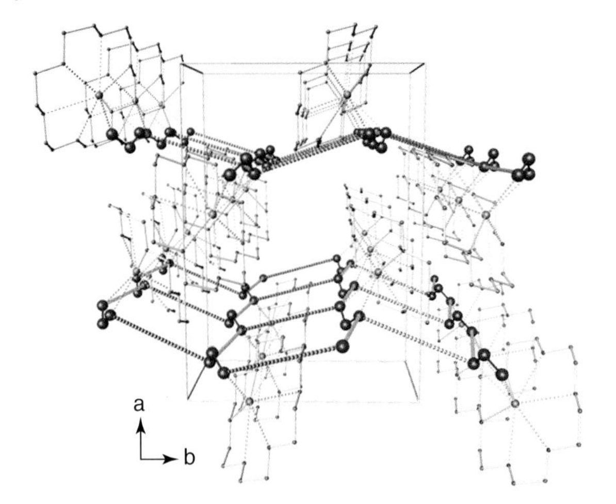

Figure 4.9 K([18]crown-6)O_3· 2 NH_3: view along [001]. Orange dotted lines indicate intermolecular O· · ·O distance shorter than 650 pm, and ammonia molecules are omitted for clarity [47].

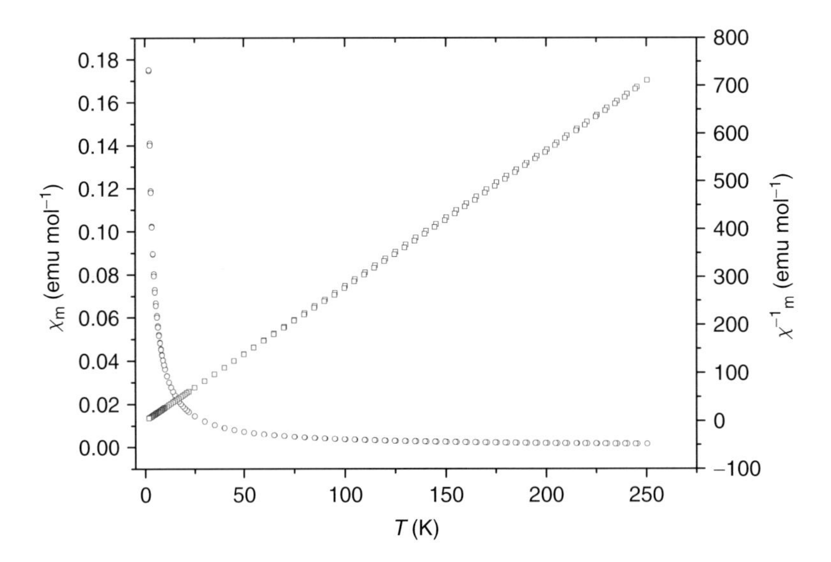

Figure 4.10 Temperature dependence of the molar susceptibility (○) and its inverse (□) of K([18]crown-6)O_3 ·2NH_3. (Taken from Ref. [39] with permission of Wiley-VCH.)

In Cs_5([12]crown-4)$_2$(O_3)$_5$, however, the crown ether molecules lead to the already discussed structural confinement, forcing ozonide anions to approach each other via their central oxygen atom to the shortest distance ever observed in the solid state.

Figure 4.11 shows the temperature dependence of the magnetic susceptibility and its inverse at 1 T.

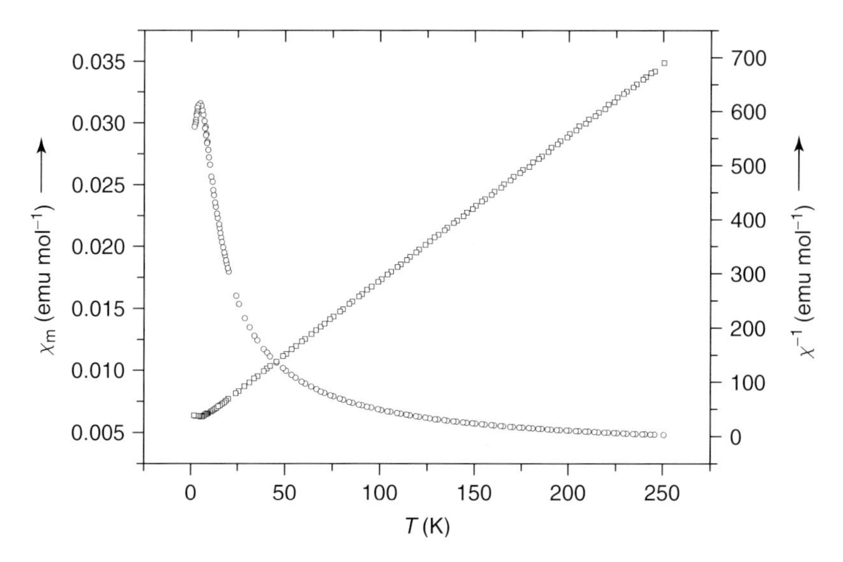

Figure 4.11 Temperature dependence of the molar susceptibility (○) and its inverse (□) of $Cs_5([12]crown\text{-}4)_2(O_3)_5$. (Taken from Ref. [41] with permission of Wiley-VCH.)

A significant interaction between unpaired electrons, imposed by the short inter-anion distance, would result in spin compensation and, thus, to a magnetic moment that is lower than the spin-only value. The magnetic behavior found for $Cs_5([12]crown\text{-}4)_2(O_3)_5$ can be described above approximately 20 K by the Curie–Weiss law $\chi = C/(T - \theta) + \chi_0$, with a temperature-independent contribution of $\chi_0 = 1.08(4) \times 10^{-3}$ emu mol^{-1}. The measured magnetic moment is 1.72 μ_B, as expected for a single unpaired electron. Antiferromagnetic ordering occurs below the Néel temperature of $T_N = 4.5$ K (the Weiss parameter is $\theta = -5.1(4)$ K).

Apparently, the interaction between the unpaired electrons is still weak. On the other hand, since superconducting quantum interference device (SQUID) measurements detect only the mean values of magnetization, from which the contribution of the interactions between the ozonide ions inside and outside the stacks of antiprisms cannot be distinguished, further experiments such as electron paramagnetic resonance (EPR) spectroscopy are needed.

Using EPR spectroscopy, a line shape analysis would allow the correlation of the peak-to-peak line width, ΔH_{pp}, with the actual structural environment of one spin-carrying species expressed via the second moment $\langle H^2 \rangle$ of the signal, the so-called method of moments [50, 52, 53]. By applying this method to the situation of the ozonide anions inside and outside the stacks of cesium antiprisms, it is possible to simulate the expected EPR spectrum, which is shown in Figure 4.12.

Obviously, the two different species contribute with significantly different line widths: a contribution exhibiting a relatively narrow line width for the outer ozonide anions, and a very broad signal, most probably caused by strong dipole–dipole

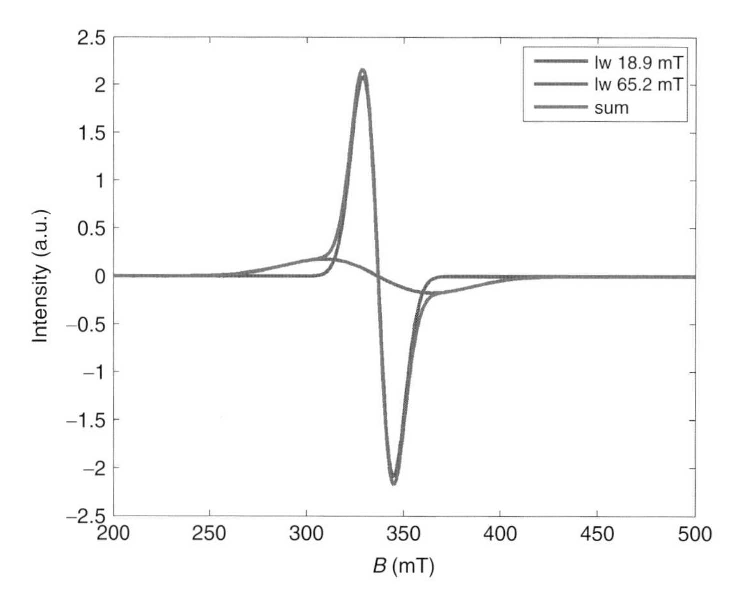

Figure 4.12 Simulated EPR spectrum using the method of moments to determine the line width of the inner (green) and the outer (blue) ozonide ions. The red curve is the sum of both and therefore represents the expected experimental EPR signal [47].

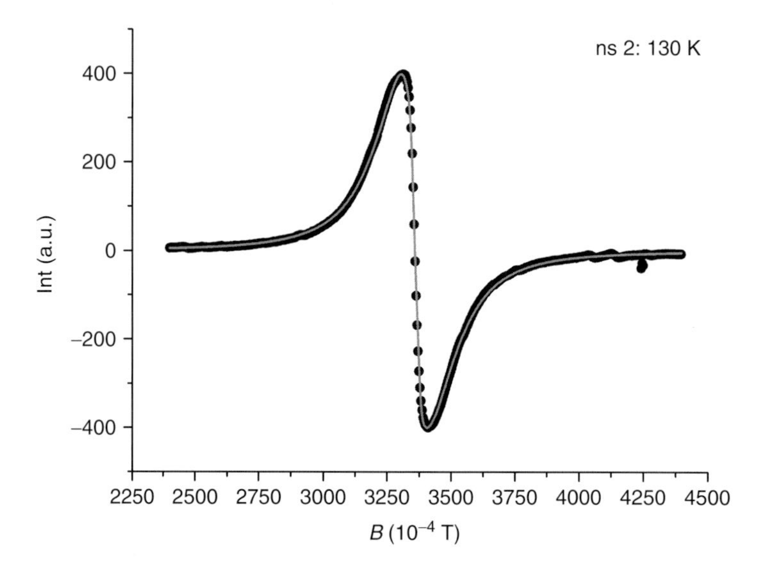

Figure 4.13 EPR spectrum of $Cs_5([12]crown\text{-}4)_2(O_3)_5$ recorded at 130 K: orange line: fitting curve under the assumption of two spin-carrying species [47].

interactions, for the inner lying ions is expected. The sum curve, therefore, is influenced only at its tails by the broad signal, which will hamper the possibility of differentiation.

EPR spectra recorded at different temperatures with an X-band EPR spectrometer actually showed a very broad signal (Figure 4.13). The curve fit improves significantly when assuming two instead of one spin-carrying species; however, since only one single line is fitted and the number of fit parameters is increased, caution has to be exercised when discussing two distinguishable types of radical anions.

4.5 Conclusions and Perspectives

The ozonide radical anion is a remarkable species with three atomic oxygens, exhibiting a variety of interesting chemical and physical properties such as high polarizability and p-electron paramagnetism. Its dipole moment and bent molecular shape lead to a diversity of crystal structures exhibiting various ways of metal-anion coordination and to different interaction modes between adjacent ozonide ions. Furthermore, structural confinement, for example via incorporation of big, nonpolar ligands, such as crown ether molecules, can force these radical species to approach each other to remarkably short intermolecular distances. Investigations on the magnetic properties of ozonide compounds are still at an early stage, but this field offers attractive options to get deeper insight into cooperative phenomena of p-electron paramagnets, a topic nowadays still rarely addressed [54–56].

Another aspect that has not been discussed so far is the improved solubility of ionic ozonides in organic solvents, as achieved by complexing the counter cations with crown ether ligands [38, 47]. The ozonide compounds $M([18]crown\text{-}6)O_3{\cdot}xNH_3$, for example, have been shown to form solutions in methylene chloride and are stable at $-70\ ^{\circ}C$ for several weeks. Hence, new possibilities have been opened up to explore the chemical reactivity of the O_3^- ion with respect to several organic and organometallic substrates.

In this chapter, several possible strategies to alter structural features have been discussed. Thereby, the overall goal to achieve is the modification of intermolecular interactions between these radical anions in order to get a better understanding and control over their chemical behavior. Maybe, one day it will be possible to learn a lesson from isoelectronic species such as the trisulfide anion S_3^- or the dithionite equilibrium, $S_2O_4^{2-} \rightarrow 2\ SO_2^-$, and a dimerization of two ozonide anions toward a new oxygen species can be realized.

Although the ozonide anion seemed to have been forgotten since the 1970s, there is still a lot of chemistry to be explored of this molecular oxygen anion and we hope to have motivated the reader to immerse into this fascinating field of molecular inorganic chemistry.

References

1. Wurtz, C.A. (1886) Dictionnaire de chimie pure et appliquée, II, Hachette, Paris.
2. Baeyer, A. and Villiger, V. (1902) *Ber. Dtsch. Chem. Ges.*, **35**, 3038.
3. Bach, A. (1902) *Ber. Dtsch. Chem. Ges.*, **35**, 342.
4. Manchot, W. and Kampschulte, W. (1908) *Ber. Dtsch. Chem. Ges.*, **41**, 471.
5. Traube, W. (1912) *Ber. Dtsch. Chem. Ges.*, **45**, 2201.
6. Traube, W. (1916) *Ber. Dtsch. Chem. Ges.*, **49**, 1670.
7. Kazarnovskii, I.A. and Nikol'skii, G.P. (1949) *Dokl. Akad. Nauk*, **64**, 69.
8. Petrocelli, A.W. and Capotosto, A. (1965) *Sci. Tech. Aerospace Rept.*, **3**, 1127.
9. Kacmarek, A. (1964) *US Govt. Res. Rept.*, **39**, 1.
10. Valentine, J.S. and Curtis, A.B. (1975) *J. Am. Chem. Soc.*, **97**, 224.
11. Scully, F.E. and Davis, R.C. (1978) *J. Org. Chem.*, **43**, 1467.
12. Kim, Y.H., Lim, S.C., and Kim, K.S. (1993) *Pure Appl. Chem.*, **65**, 661.
13. Singh, K.N. (2007) *Synth. Commun.*, **37**, 2651.
14. Volnov, I.I. (1966) *Peroxides, Superoxides, and Ozonides of Alkali and Alkaline Earth Metals*, Plenum Press, New York, p. 125.
15. Hesse, W., Jansen, M., and Schnick, W. (1989) *Progr. Solid State Chem.*, **19**, 47.
16. Jansen, M. and Nuss, H. (2007) *Z. Anorg. Allg. Chem.*, **633**, 1307.
17. Schnick, W. and Jansen, M. (1986) *Z. Anorg. Allg. Chem.*, **532**, 37.
18. Schnick, W. and Jansen, M. (1987) *Rev. Chim. Miner.*, **24**, 446.
19. Schnick, W. and Jansen, M. (1985) *Angew. Chem. Int. Ed. Engl.*, **24**, 54.
20. Jansen, M. and Hesse, W. (1988) *Z. Anorg. Allg. Chem.*, **560**, 47.
21. Jansen, M. and Assenmacher, W. (1991) *Z. Kristallogr.*, **194**, 315.
22. Klein, W., Armbruster, K., and Jansen, M. (1998) *Chem. Commun.*, p. 707.
23. Klein, W. and Jansen, M. (2000) *Z. Anorg. Allg. Chem.*, **626**, 136.
24. Klein, W. and Jansen, M. (1999) *Z. Naturforsch.*, **54b**, 287.
25. Klein, W. and Jansen, M. (2005) *Z. Naturforsch.*, **60b**, 426.
26. Korber, N. and Jansen, M. (1992) *Chem. Ber.*, **125**, 1383.
27. Hesse, W. and Jansen, M. (1991) *Inorg. Chem.*, **30**, 4380.
28. Klein, W. (2000) Über Ozonide der Alkali- und Erdalkalimetalle. PhD thesis, Universität Bonn.
29. Klein, W. and Jansen, M. (2000) *Z. Anorg. Allg. Chem.*, **626**, 947.
30. Dietzel, P.D.C. and Jansen, M. (2006) *Z. Anorg. Allg. Chem.*, **632**, 2006.
31. Jansen, M. and Seyeda, H. (1997) *Z. Kristallogr. NCS*, **212**, 229.
32. Klein, W. (1995) Neue ionische Ozonide. Diploma thesis, Universität Bonn.
33. Klein, W. and Jansen, M. (2001) *Z. Naturforsch.*, **56b**, 287.
34. Assenmacher, W. and Jansen, M. (1995) *Z. Anorg. Allg. Chem.*, **621**, 431.
35. Seyeda, H., Armbruster, K., and Jansen, M. (1996) *Chem. Ber.*, **129**, 997.
36. Seyeda, H. (1997) Ozonide und Hyperoxide bisquaternärer Ammoniumkationen. PhD thesis, Universität Bonn.
37. Seyeda, H. and Jansen, M. (1997) *Z. Kristallogr. NCS*, **212**, 233.
38. Korber, N. (1992) Neue präparative Zugänge zu ionischen Ozoniden. PhD thesis, Universität Bonn.
39. Nuss, H., Nuss, J., and Jansen, M. (2008) *Z. Anorg. Allg. Chem.*, **634**, 1291.
40. Korber, N. and Jansen, M. (1990) *J. Chem. Soc., Chem. Commun.*, 1654.
41. Nuss, H. and Jansen, M. (2006) *Angew. Chem. Int. Ed.*, **45**, 7969.
42. Nuss, H. and Jansen, M. (2009) *Z. Naturforsch.*, **64b**, 1325.
43. Kellersohn, T., Korber, N., and Jansen, M. (1993) *J. Am. Chem. Soc.*, **115**, 11–254.
44. Koch, W., Frenking, G., Steffen, G., Reinen, D., Jansen, M., and Assenmacher, W. (1993) *J. Chem. Phys.*, **99**, 1271.
45. Walsh, A.D. (1953) *J. Chem. Soc.*, 2266.
46. Marx, R. and Ibberson, R.M. (2001) *Solid State Sci.*, **3**, 195.

47. Nuss, H. (2007) Chemie metastabiler Anionen – Synthese und Charakterisierung neuer Auride und Ozonide. PhD thesis, Universität Stuttgart.
48. Müller, A., Reuter, H., and Dillinger, S. (1995) *Angew. Chem. Int. Ed. Engl.*, **34**, 2328.
49. Fyfe, M.C.T. and Stoddart, J.F. (1997) *Acc. Chem. Res.*, **30**, 393.
50. Steffen, G., Hesse, W., Jansen, M., and Reinen, D. (1991) *Inorg. Chem.*, **30**, 1923.
51. Lueken, H., Deussen, M., Jansen, M., Hesse, W., and Schnick, W. (1987) *Z. Anorg. Allg. Chem.*, **553**, 179.
52. Abragam, A. (1996) *Principles of Nuclear Magnetism*, Oxford University Press, Oxford, New York.
53. Poole C.P. Jr. (1996) *Electron Spin Resonance - A Comprehensive Treatise on Experimental Techniques*, Dover Publications, Mineola, New York.
54. Winterlik, J., Fecher, G.H., Felser, C., Muhle, C., and Jansen, M. (2007) *J. Am. Chem. Soc.*, **129**, 6990.
55. Winterlik, J., Fecher, G.H., Jenkins, C.A., Medvedev, S., Felser, C., Kuebler, J., Muhle, C., Doll, K., Jansen, M., Palasyuk, T., Trojan, I., Eremets, M.I., and Emmerling, F. (2009) *Phys. Rev. B*, **79**, 214–410.
56. Winterlik, J., Fecher, G.H., Jenkins, C.A., Felser, C., Muhle, C., Doll, K., Jansen, M., Sandratskii, L.M., and Kuebler, J. (2009) *Phys. Rev. Lett.*, **102**, 016–401.

5
Chemistry and Biological Properties of Amidinoureas: Strategies for the Synthesis of Original Bioactive Hit Compounds

Daniele Castagnolo

5.1
Amidinoureas: an Introduction

Ureas and guanidines are two important classes of organic compounds which are largely used in organic synthesis as well as in medicinal chemistry and have found application in different fields of research [1]. For instance, urea and guanidine functional groups can be often found as part of several drugs and organic catalysts [2].

The initiator of the urea family, the urea (**1**), also called *carbamide*, is the diamide of carbonic acid and its formula is H_2NCONH_2. The molecule has two amine ($-NH_2$) groups joined by a carbonyl (C=O) functional group. The urea molecule is planar, and the substance is colorless and crystalline and melts at 132.7 °C (271 °F) and decomposes before boiling. In solid urea, the oxygen center is engaged in two N–H···O hydrogen bonds. The carbon in urea is described as sp^2-hybridized, the C–N bonds have significant double bond character, and the carbonyl oxygen is basic compared to formaldehyde. Urea's high aqueous solubility reflects its ability to engage in extensive hydrogen bonding with water molecules. When urea is dissolved in water, it is neither acidic nor alkaline. Urea is the chief nitrogenous end product of the metabolic breakdown of proteins in all mammals and some fishes and has an important role in the metabolism of nitrogen-containing compounds. Urea occurs not only in the urine of all mammals but also in their blood, bile, milk, and perspiration. In the course of the breakdown of proteins, amino groups (NH_2) are removed from the amino acids that partly comprise proteins. These amino groups are converted to ammonia (NH_3), which is toxic to the body and thus must be converted to urea by the liver. The urea then passes to the kidneys and is eventually excreted in the urine. Urea in concentrations up to 10 M is a powerful protein denaturant, as it disrupts the noncovalent bonds in the proteins. Urea and its synthetic derivatives have important uses as fertilizers and feed supplements, as well as a starting material for the manufacture of plastics and drugs.

The simpler compound of the guanidine family is guanidine (**2**), also called *carbamidine*, a strongly alkaline and water-soluble compound structurally related

New Strategies in Chemical Synthesis and Catalysis, First Edition. Edited by Bruno Pignataro.

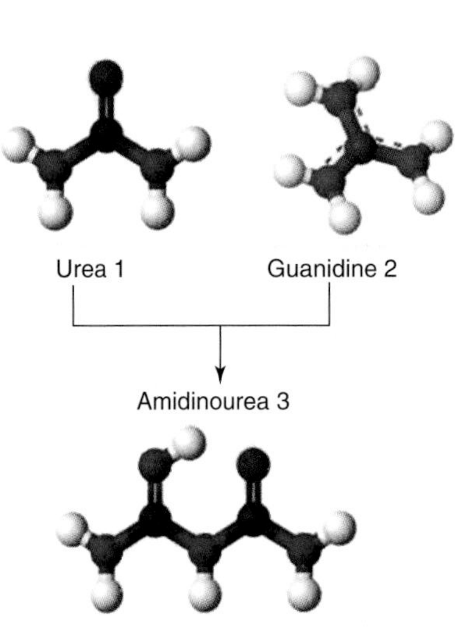

Figure 5.1 Amidinourea as the fusion product of a guanidine and a urea.

to urea. The molecule was first synthesized in 1861 by the oxidative degradation of an aromatic natural product, guanine, isolated from Peruvian guano. The general formula of guanidine is $NHC(NH_2)_2$. It contains a C=NH functional group in place of the urea's C=O carbonyl group. Guanidine is protonated in physiological conditions. The guanidine conjugate acid is called *guanidinium cation*, and possesses a pKa value in water of 12.5. The chief characteristic of guanidines is their high degree of basicity due to proton resonance. Guanidine is formed by the oxidation of guanine in urine as a normal product of protein metabolism in the body. Because of the strong N=C bond, guanidine and its modified derivatives are widely used in industry for the manufacture of plastics, resins, rubber chemicals, explosives (nitroguanidines), and photochemicals, and in medicinal chemistry as fungicides and disinfectants. Guanidines also have biotechnological application in protein separation and purification, and as protein denaturants.

However, despite their structural similarity, ureas and guanidines possess different chemical, physical, and biological properties. In addition to them, a third class of strictly related compounds exists, namely the amidinoureas. Amidinourea (**3**), also named *guanylurea,* and its derivatives could be considered as the fusion products between ureas and guanidines, possessing an NH group shared by these two compounds. As shown in Figure 5.1, amidinourea could be considered as the sum of a guanidine and a urea. An amidinourea contains the features of both guanidine and urea, but at the end it is a new and different molecule endowed with its own chemical and physical properties.

Amidinoureas have received less attention compared to their related compounds ureas and guanidines. However, the number of research articles on amidinourea derivatives is increasing and an even larger number of chemists have been attracted by the use of this uncommon and original functional group in different fields of research. The great interest in the chemistry of amidinoureas has led in recent years to the discovery of novel and different strategies for the synthesis of these compounds.

5.2 Amidinoureas in Chemistry

Amidinoureas are finding increasing interest and applications in different fields of chemistry. The structurally most simple amidinourea, namely the guanylurea (**3**), is a commercially available compound and it is sold as the sulfate or phosphate salt. However, other amidinourea salts have been also prepared, which exhibit in some cases interesting properties. As an example, amidinourea hydrogen phosphite and amidinourea tartrate have found applications in optics [3]. Amidinourea hydrochloride was used to study its electronic properties and chemical structure. The crystal structure and a complete vibrational analysis of amidinourea hydrochloride by Fourier transform infrared (FTIR) and Raman spectra showed a pronounced π-electron delocalization on the cation and the presence of strong intra and intermolecular hydrogen-bonding interactions between amino groups belonging to the guanidine moiety and the carbonyl oxygen of the ureic group. This arrangement gives rise to a polymer-like structure, in which the guanylurea cation chains are laterally hydrogen bonded by Cl^- anions and water molecules [4]. Amidinourea nitrate and amidinourea perchlorate show good thermal stabilities above 200 °C and interesting perspectives in the development of new environmentally friendly explosive formulations [5]. The amidinourea dinitramide, also known as *FOX-12*, is an energetic material with low sensitivity and good potential for use as a propellant or insensitive munition explosive [6]. Single crystals of amidinourea sulfate hydrate have been also obtained from a neutral aqueous solution containing amidinourea sulfate. By analogy with other molecular amidinourea salts, it has been demonstrated that amidinourea sulfate hydrate builds up an array of mutually linked chains of cations and anions, with the crystal packing being largely controlled by an extensive H-bonding network [7].

Substituted amidinoureas have found many applications in medicinal chemistry such as in the treatment of the irritable bowel syndrome, gastrointestinal, spasmolytic, and cardiovascular disorders, and parasitic infections [8]. Amidinoureas (**4**) and (**5**) were found to possess good antimalarial activity [9], while compound **6** showed high bactericidal activity. Finally, morpholino derivatives **7** proved to be endowed with high fungicidal properties [10] (Figure 5.2).

The naturally occurring antibiotics TAN-1057A–D (**8**–**11**) isolated from the bacteria Flexibacter sp. PK-74 and PK-176 contain an amidinourea structural feature (Figure 5.2).

These compounds showed potent antibacterial activity against methicillin-resistant *Staphylococcus aureus* (MRSA). TAN-1057A–D displayed better activity against gram-positive bacteria than against gram-negative bacteria. Moreover, TAN-1057 did not show any cross resistance with methicillin, erythromycin, and gentamycin [11]. TAN-1057's are dipeptides consisting of β-homoarginine and a heterocyclic amidinourea derivative of 2,3-diaminopropionic acid. TAN-1057A can be hydrolyzed in both acidic and basic media, affording the acyclic amidinourea (**12**) with attendant racemization of the *R*-amino acid stereogenic center. However, the acyclic compound **12** can be converted back into the cyclic form in a two-step

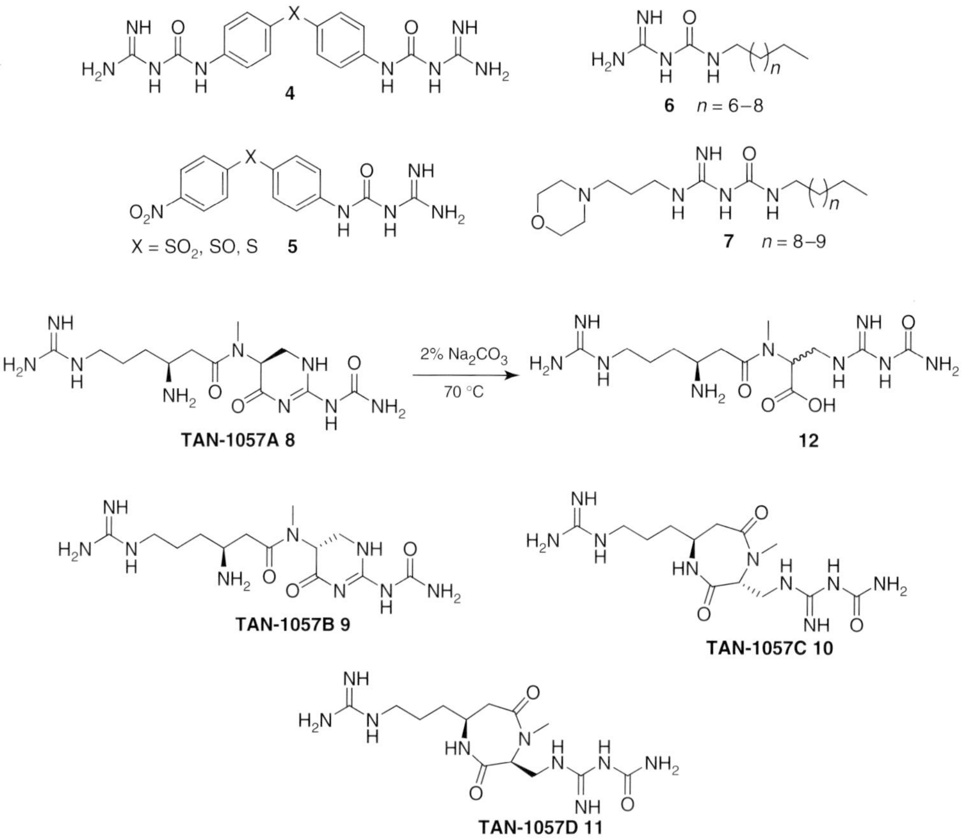

Figure 5.2 Amidinoureas endowed with antimicrobial activities and TAN-1057A–D antibiotics.

synthetic sequence, resulting in a diastereomeric mixture (1 : 1) of TAN-1057A and B [12].

Amidinoureas **13–16**, structurally related to TAN-1057, and **17–18** have been extensively studied and used for the construction of a variety of self-assembling systems. Each amidinourea possesses one (or two) acceptor (A) and two donor (D) hydrogen-bonding codes, which can lead to hydrogen-bond-mediated self-assembly and, as a result, to the synthesis of H-bonded dimers. These examples show the potential of amidinoureas in the formation of noncovalent interactions with different molecules. This trait makes amidinoureas versatile compounds both in organic synthesis as noncovalent organocatalysts and in medicinal chemistry as drug-like compounds able to interact with proteins through hydrogen-bond interactions [13] (Figure 5.3).

The amidinourea functional group can also be considered as a modification of the guanidine moiety. Many biologically active compounds contain a guanidine function in their backbone and the synthesis of modified derivatives of guanidine

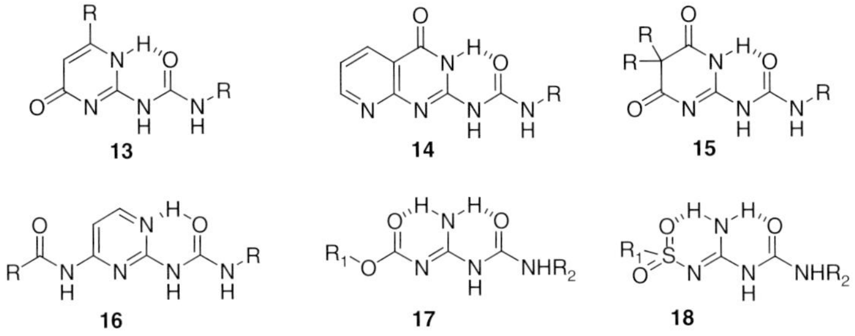

Figure 5.3 Amidinoureas used in the construction of self-assembling systems.

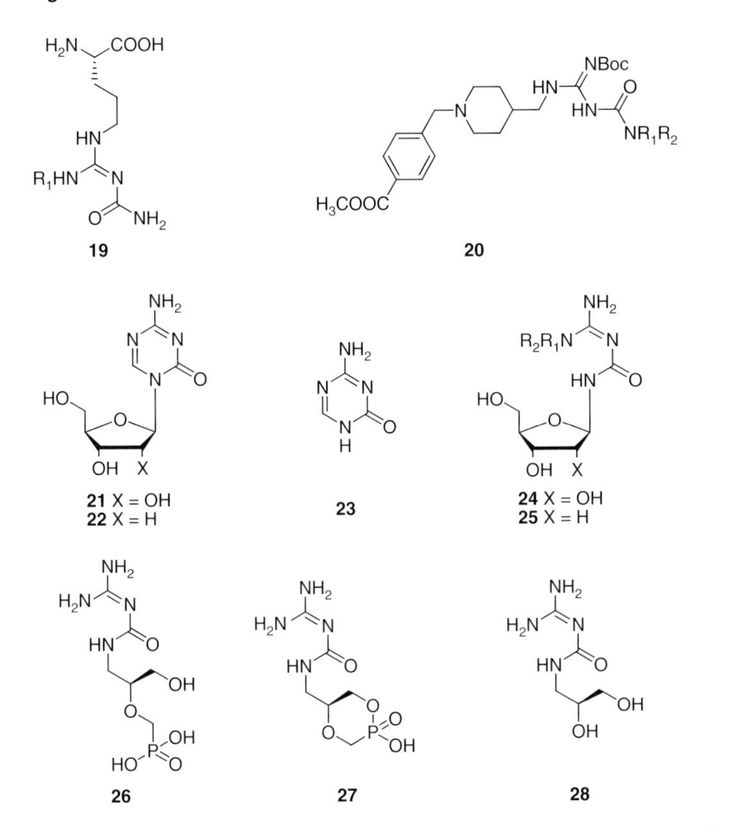

Figure 5.4 Amidinourea as modification of guanidine groups and 5-azacitydine analogs.

compounds is often sought by chemists with the aim to discover new molecules endowed with improved activity. Amidinourea functional group might in some cases be an effective alternative and a bioisosteric group of the guanidine moieties. In Figure 5.4, an arginine derivative (**19**) and amidinoureas (**20**) are shown [14].

In this context, the work made by Czech researchers on the synthesis of a series of 5-azacytidine analogs is noteworthy. 5-Azacytidine (**21**) is an analog of the DNA

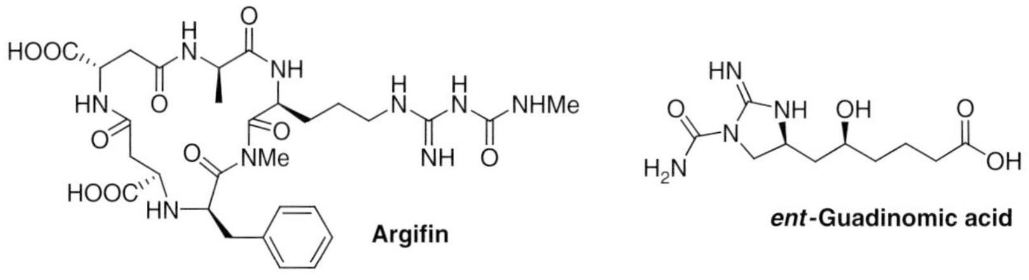

Figure 5.5 Natural amidinoureas.

and RNA nucleoside cytidine, and it is used as chemotherapeutic agent mainly in the treatment of myelodysplastic syndrome (MDS). 2′-Deoxy-5-azacytidine (**22**) also showed clinical activity against metastatic lung cancer and hormone-independent prostate cancer. The 5-azacytidine (**21**) is phosphorylated and incorporated in the RNA and, after deoxygenation, also in the DNA, while **22** is incorporated directly into the DNA. The base 5-azacytosine (**23**) can interact with DNA nucleobases through hydrogen-bonding interactions, leading to the formation of a 5-azacytosine-containing DNA, which is a potent inhibitor of the enzyme DNA-methyltranferase. The inhibition process is due to a covalent interaction of the thiol group in the active site of the enzyme with the labile 5-azacytosine ring. Analogs of **21** and **22** were then synthesized with the aim of finding improved compounds. The 5-azacytosine base was replaced by the amidinourea moiety leading to compounds **24** and **25**, which showed the same ability of **21** to form hydrogen-bonding interactions with nucleobases [15]. Similarly, modifications of the sugar moiety led to the discovery of derivatives **26–28** as potential antiviral compounds [16].

Finally, natural compounds containing the amidinourea structural motif and endowed with biological properties have been described [17] (Figure 5.5). The peptide Argifin was isolated from *Gliocladium* fungal strains and proved to inhibit chitinase B with a K_i of 20 nM. The solid-phase total synthesis of Argifin was first published by Eggleston [18] and followed almost at the same time by Sunazuka [19]. The *ent*-guadinomic acid is another interesting amidinourea natural product which has been isolated from *Streptomyces* sp. (K01-0509 B) [20] and whose total synthesis has been recently described [21].

5.3 Synthetic Strategies for the Preparation of Amidinoureas

Only a few methods for the synthesis of amidinoureas have been described so far. These approaches, despite being apparently easy, in some cases turned out to be inefficient or required numerous steps. Hence, the synthesis of amidinoureas still remains complicated and the development of novel synthetic strategies is highly desirable. The existing methods for the synthesis of amidinoureas are reported below.

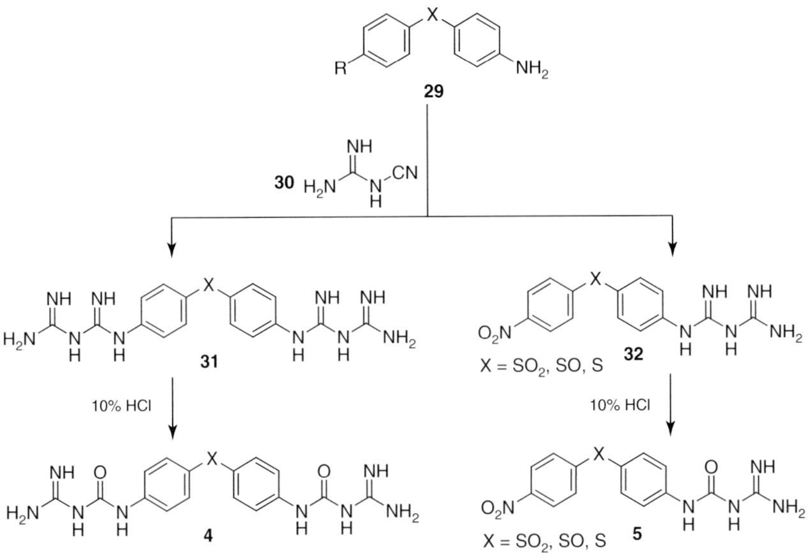

Scheme 5.1 Amidinoureas through the hydrolysis of biguanides.

5.3.1 Hydrolysis of Biguanides

One of the first examples of amidinourea synthesis was described in 1969 by Serafin *et al.* during their studies on antimalarial 4,4′-diaminodiphenylsulfone (DDS) analogs. Amidinoureas (**4**) and (**5**) were synthesized in two steps from the common starting compound **29**. This latter compound was converted into the biguanides (**31**) and (**32**) through reaction with cyanoguanidine (**30**). Hydrolysis of biguanides with diluted hydrochloridic acid led to amidinoureas in variable yields (33–75%) after recrystallization from appropriate solvents (Scheme 5.1) [9].

5.3.2 Reaction of Guanidines with Isocyanates

In 1980, Tilley *et al.* described the synthesis of a series of amidinoureas in the effort to obtain 1,2,4-oxadiazoles and 1,2,4-triazoles. In this context, the authors synthesized a series of phenylamidinoureas (**35**), which were in turn cyclized affording oxadiazoles (**37**) and triazoles (**38**). The substituted amidinoureas (**35**) were obtained through the reaction of a guanidine (**33**) with the equimolar amount of the appropriate isocyanate (**34**) as reported in Scheme 5.2. The nucleophilic nitrogen of guanidine (**33**) reacts with the isocyanate affording the desired **35**. Amidinoureas (**35**) were obtained at −20 °C in 30–68% yields. The low yield of the hydrolysis reaction has to be ascribed in some cases to the formation of acylated side products whose characterization proved difficult and which were removed through fractional crystallization. The more the guanidine (**33**) and the isocyanate

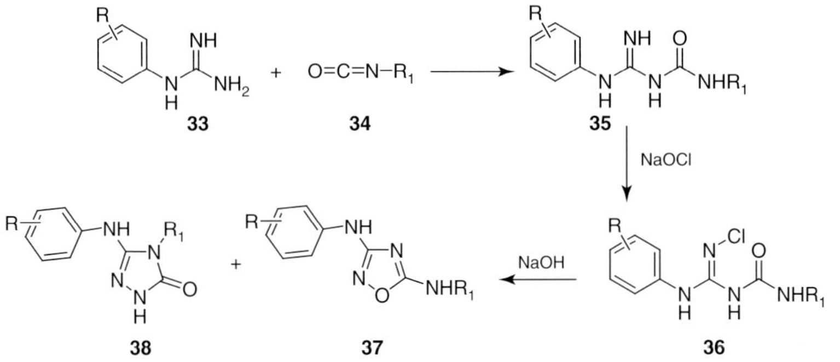

Scheme 5.2 Synthesis of phenylamidinoureas and transformation into heterocyclic compounds.

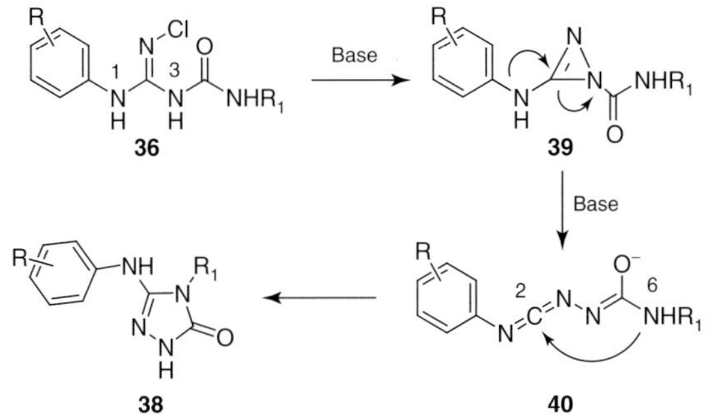

Scheme 5.3 Rearrangement of N-chloro-phenylamidinoureas.

(34) were allowed to react, the more side products were formed. Hence, short reaction periods were required to get amidinoureas in acceptable amounts. Finally, amidinoureas were converted in a two-step sequence into 1,2,4-oxadiazole (**37**). In some cases, the byproducts **38** were isolated owing to the rearrangement of chloro intermediates **36**.

The formation of triazolones (**38**) was rationalized by the mechanism proposed in Scheme 5.3. An intramolecular displacement of chloride by the nitrogen N3 led to the tricyclic intermediate diazirine (**39**). Further rearrangement catalyzed by a base afforded carbodiimide (**40**), which in turn was cyclized by the attack of nitrogen N6 on the electrophilic carbodiimide carbon C2, affording triazole (**38**). The rearrangement of **36** seems to be determined by the substituents on the phenyl ring of the starting amidinoureas (**35**). It has been shown, in fact, that the electron-withdrawing groups on the aromatic ring of the amidinourea (**35**) favor the formation of 1,2,4-oxadiazole [22].

It is noteworthy that hydrogenation of 5-amino-3-amino-1,2,4-oxadiazoles (**37**) with Pd/C gets back the starting amidinoureas (**35**). Oxadiazole (**37**) can be synthesized not only from amidinoureas (**35**) but also from thioureas (**41**) through reaction with NH_2OH at 50 °C. Thioureas (**41**) are easily accessible from commercially

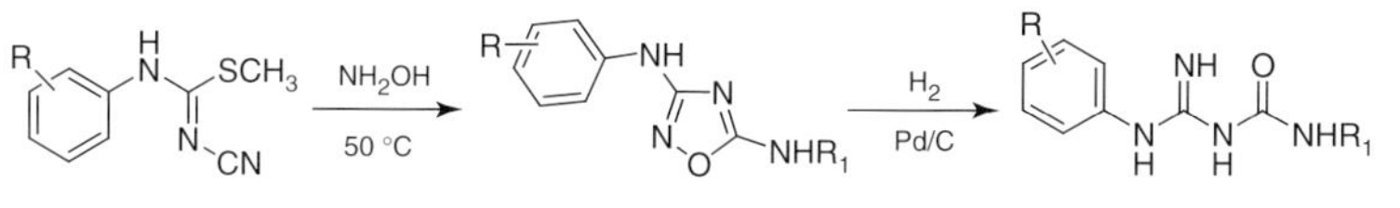

Scheme 5.4 Hydrogenation of 5-amino-3-amino-1,2,4-oxadiazoles.

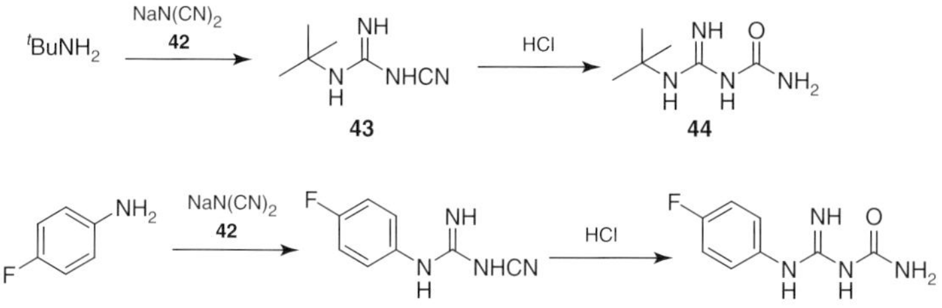

Scheme 5.5 Hydrolysis of cyanoguanidines.

available phenyl isothiocyanates. Hence, hydrogenation of **37** represents a further and valid alternative to synthesize the desired amidinoureas (**35**) [23] (Scheme 5.4).

5.3.3 Hydrolysis of Cyanoguanidines

It is well known that a nitrile can be hydrolyzed under acidic conditions affording the corresponding amide. As a consequence, it might be expected that the acidic hydrolysis of a guanidine bearing a cyano group could lead to the formation of an amidinourea. Actually, the hydrolysis of cyanoguanidines, described in Scheme 5.5, is one of the most used methods for the synthesis of amidinoureas. Cyanoguanidines such as **43** and **45** can be obtained through the reaction of anilines or amines with sodium dicyanoimide (**42**) [22, 23].

However, although this approach might appear at first sight quite easy, the hydrolysis of cyanoguanidines might present some problems and generally has to be carried out under stringent conditions requiring the use of strong acids and/or long reaction times. In addition, the use of dicyanoimide might represent sometimes a limitation for this procedure since the synthesis of *N*-polysubstituted amidinoureas is not allowed or would require numerous synthetic steps. A clear example in this context is represented by the synthesis of amidinourea (**51**). The reaction of the ornithine derivative **47** with dicyanoimide (**42**) led in fact to the intermediate **48** bearing no substituents on the guanidine group. Hydrolysis of **48** with 4N HCl afforded unsubstituted amidinourea (**50**). Hence, for the synthesis of substituted derivative **51**, a different synthetic pathway was chosen. Compound **47** was first reacted with cyano derivative **52**, affording cyanocarbamate (**53**). This latter compound was then reacted with an appropriate amine, affording the desired intermediate **49**, which after acidic hydrolysis led to the substituted amidinourea (**51**) [14a] (Scheme 5.6).

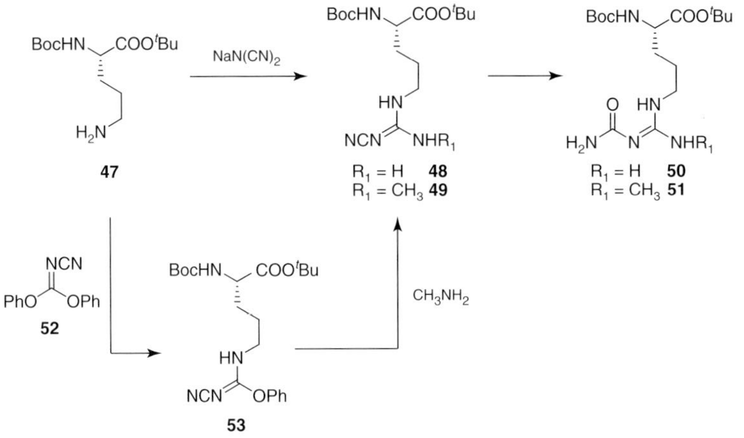

Scheme 5.6 N-Substituted amidinoureas through the hydrolysis of cyanoguanidines.

5.3.4
Reaction of Acyl-S-Methylisothiourea with Amines

In 1997 Williams and Yuan described the first total synthesis of the anti-MRSA peptide antibiotics TAN-1057A–D (**8–11**) [12b]. These natural compounds contain an amidinourea moiety whose guanidine portion is part of a six- or seven-membered ring. The authors noticed that none of the methods described earlier for the synthesis of amidinoureas was suitable for accessing the labile TAN-1057 amidinourea substructure. Hence a new and general synthetic strategy was developed to prepare the amidinoureas to be used for the total synthesis of **8–11**. As guanidines can be easily synthesized via reaction of amines with *S*-methylisothioureas or acyl-*S*-methylisothioureas such **54** and **55** in weak basic media [24], it was assumed that amidinoureas could be obtained similarly from (acyl)ureido-*S*-methylisothioureas such as **56**. These latter compounds were synthesized in quantitative yields through the reaction of Cbz-*S*-methyl-isothiourea (**55**) with Cbz-isocyanate. Reaction of Cbz-ureido-*N*′-Cbz-*S*-methylisothiourea (**56**) with different amines (**57a–d**) in the presence of an excess of Et_3N in DMF at room temperature afforded Cbz-proteced amidinoureas (**58a–d**) in high yields. Cleavage of the Cbz protecting groups was achieved by hydrogenation with H_2 at 60 psi in the presence of 20% $Pd(OH)_2$ and gave amidinoureas (**59a–d**) in excellent yields (Scheme 5.7). Table 5.1 shows the yields and chemical structures of the synthesized amidinoureas.

An interesting side reaction occurred when nBuNH$_2$ (**57d**) was reacted with **56**. As shown in Table 5.1 (entry 4) compound **58d** was obtained only in moderate yield (48%) unlike compounds **58a–c**. During the course of the reaction, a cyclic triazine side product **58'd** was obtained together with **58d** in comparable yield. The amount of triazine (**58'd**) increased when the reaction was allowed to proceed longer. The formation of **58'd** was assumed to be due to the nucleophilic attack of

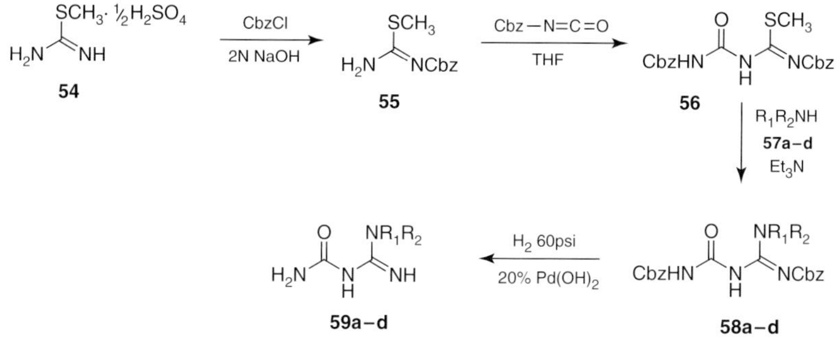

Scheme 5.7 Synthesis of amidinoureas from acyl-S-methylisothiourea.

Table 5.1 Yields and chemical structures of amidinoureas (**58**) and (**59**).

Entry		Amines		58 Yield (%)	59 Yield (%)
		R_1	R_2		
1	**57a**	$CH_2(CH_2)_3CH_2$		81	96
2	**57b**	$CH_2CH_2OCH_2CH_2$		75	97
3	**57c**	H	Cyclohexyl	99	93
4	**57d**	H	*n*Bu	48	99

nitrogen N1 on the electrophilic C=O of CBz protecting group and favored by the stability of the 1,3,5-triazine-2,4-dione six-membered ring. Amidinoureas (**58a–b**) formed from secondary amines were precluded from forming the corresponding side products as well as the sterically hindered cyclohexylamine derivative (**58c**) [25] (Scheme 5.8).

5.3.5
Reaction of Di-Boc-Guanidines with Amines

In the previous example, it was shown how nucleophilic nitrogens might react with electrophilic Cbz carbamate, leading to the formation of a new ureido group NH(C=O)NH and elimination of BnOH. Hence, as a general protected amine, $BocNHR_1$ might react with another amine R_2NH_2 affording a urea, so the reaction

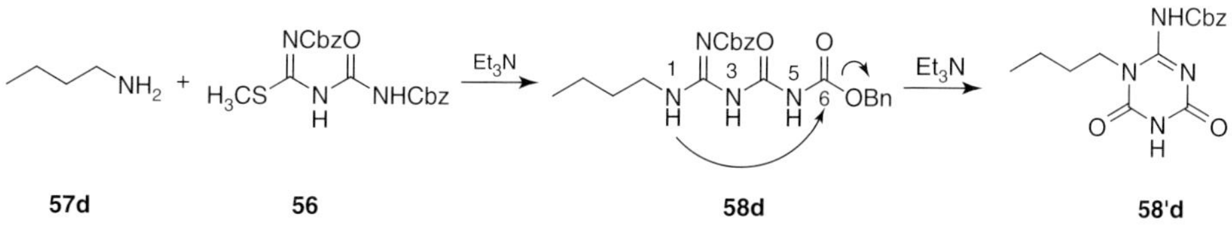

Scheme 5.8 Formation of triazine side product 58′d from amidinourea (58d).

of the same amine R_2NH_2 with a Boc-protected guanidine is expected to lead to the formation of an amidinourea. In this context, the development of a new method for the synthesis of amidinoureas, described by Rault and Miel in 1998, fits in. During the course of a work aimed at synthesizing guanidine derivatives endowed with biological activity, they found that the reaction of di-Boc-*S*-Me-isothiourea (**60**) with benzylamine (**61**) led under certain conditions (prolonged heating and use of excess of **61**) to the formation of the desired product **62** together with the amidinourea side product **63**. The reaction of di-Boc-guanidines (**62**) with different amines (**64**) was then explored, revealing a new synthetic method for the synthesis of amidinoureas. Starting from guanidines (**62**), a series of amidinoureas (**63b–i**) has been then synthesized in high yields. Even poorly nucleophilic amines such as aniline or the bulky diisopropylamine proved to be able to react with **62**, affording, respectively, the corresponding amidinoureas **63d** and **63f** (Scheme 5.9, Table 5.2) [13e, 26].

Despite the fact that the Boc-protecting group is resistant toward many nucleophilic reagents [27], in 1996 Lamothe *et al.* [28] demonstrated that Boc-protected amines could be converted into ureas in the presence of a strong base such BuLi or NaH via an isocyanate intermediate. A similar mechanism of reaction was hypothesized for the formation of amidinoureas from di-Boc-guanidines. However, since the reaction of **62** with amines (**64**) took place in absence of strong bases, it was assumed that amines (**64**) themselves might act both as bases to promote the formation of isocyanate (**65**) and as nucleophiles toward this latter intermediate. In fact, the N–H protons of a di-Boc-guanidine are expected to be much more acidic than those of a Boc-amine because of the possibility of delocalizing the resulting negative charge through the guanidine and Boc-functional groups [29]. Hence, a weaker base, such as an amine, is enough to deprotonate the di-Boc-guanidine, leading to the isocyanate intermediate **65**.

The plausible mechanism of reaction for the formation of amidinoureas (**63**) is outlined in Scheme 5.10.

More recently, Sanjayan *et al.* demonstrated that tri-Boc-guanidine (TBG) (**66**) might react with amines, affording directly, under certain conditions, amidinoureas (**69**) [30]. TBG is a well-known and versatile guanylating reagent for the synthesis

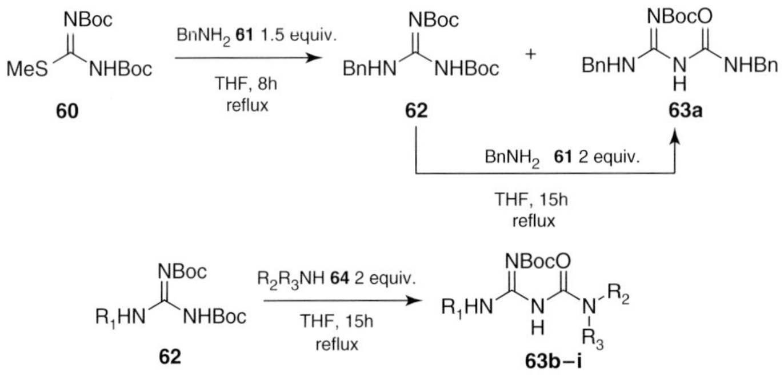

Scheme 5.9 Reaction of di-Boc-guanidines with amines.

Table 5.2 Structures and yields of amidinoureas **63a–i**.

Entry	Amidinourea	R1	R2	R3	Yield (%)
1	**63a**	Bn	Bn	H	82
2	**63b**	Bn	iPr	H	79
3	**63c**	Bn	Cyclohexyl	H	78
4	**63d**	Bn	Ph	H	65
5	**63e**	Bn	Et	Et	85
6	**63f**	Bn	iPr	iPr	73
7	**63g**	Cyclohexyl	Et	Et	82
8	**63h**	Cyclohexyl	Bn	H	77
9	**63i**	Ph	Et	Et	81

Scheme 5.10 Plausible mechanism for the formation of amidinoureas from di-Boc-guanidines.

of *N*-substituted guanidines by reaction with alcohols under Mitsunobu conditions [31]. The reaction of TBG (**66**) with different primary amines (**67**) has been widely investigated, showing the possibility to obtain selectively amidinoureas or *N*-alkylated-di-Boc-guanidines depending on the reaction conditions (Scheme 5.11). When TBG and primary amines are reacted in a 1 : 1 ratio in refluxing THF, both guanidines (**68**) and amidinoureas (**69**) were obtained in equimolar amounts. Increasing the amount of primary amine in refluxing conditions led to a dramatic increase in the yield of amidinoureas (**69**). As shown in Table 5.3 (entries 3, 6, 10, 12, and 14) amidinoureas (**69**) can be obtained as the only reaction products. It is plausible to assume that *N*-alkylated-guanidines (**68**) are formed first by the nucleophilic attack of the amines at the electrophilic quaternary carbon of TBG. Further reaction of **68** with the excess of amines afforded amidinoureas (**69**), which are presumably formed through an isocyanate intermediate **72** as shown on Scheme 5.11. The addition of the external base 1,8-diazabiciclo[5.4.0]undec-7-ene (DBU) to the reaction also furnished important information regarding the rate of formation of guanidines and amidinoureas. In the presence of DBU, the reaction of TBG with isobutylamine in 1 : 1 ratio proceeds with no significant differences in the yield of **68** (entry 7). Raising the temperature of the reaction also has only minor effects on the product distribution (entry 8). On the basis of these results, it is reasonable to suppose that the formation of guanidine (**68**) precedes amidinourea (**69**). In fact, even strong bases such as DBU fail to deprotonate TBG and to form

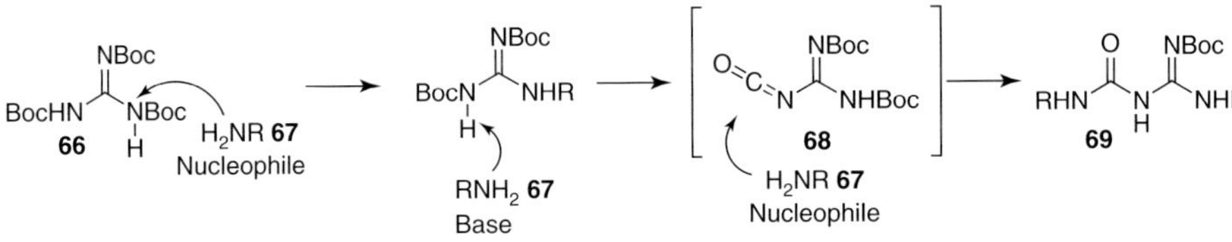

NBoc BocHN NBoc H 66 H2NR 67 Nucleophile → NBoc BocN NHR H RNH2 67 Base → [O=C=N NBoc NHBoc 68 H2NR 67 Nucleophile] → O NBoc RHN N NHR H 69

Scheme 5.11 Synthesis of amidinoureas from TBG.

Table 5.3 Structure and yields of guanidines (**68**) and amidinoureas (**69**).

Entry	Amine	TBG/ amine (equiv.)	Base	Solvent temperature (°C)	Guanidine (68) yield (%)	Amidinourea (69) yield (%)
1	$BnNH_2$	1 : 1	–	THF, rt	98	–
2	$BnNH_2$	1 : 1	–	THF, reflux	38	30
3	$BnNH_2$	1 : 4	–	THF, reflux	–	90
4	$iBuNH_2$	1 : 1	–	THF, rt	60	–
5	$iBuNH_2$	1 : 1	–	THF, reflux	50	22
6	$iBuNH_2$	1 : 3	–	THF, reflux	–	86
7	$iBuNH_2$	1 : 1	DBU	THF, rt	61	–
8	$iBuNH_2$	1 : 1	DBU	DMSO, 50	40	40
9	$DodecylNH_2$	1 : 1	–	THF, rt	97	–
10	$DodecylNH_2$	1 : 3	–	THF, reflux	–	91
11	$CyclohexylNH_2$	1 : 1	–	THF, rt	52	–
12	$CyclohexylNH_2$	1 : 3	–	THF, reflux	–	60
13	$iPrNH_2$	1 : 1	–	THF, rt	40	–
14	$iPrNH_2$	1 : 3	–	THF, reflux	–	52

an isocyanate intermediate such as **70**, which would have led directly to formation of amidinoureas (**71**). The different reactivities of TBG (**66**) and di-Boc-guanidine (**68**) must be due to the increased electrophilicity at the quaternary carbon of the guanidine moiety of TBG because of the additional Boc protecting group. The increased electrophilicity favors the nucleophilic attack of a first amine over the deprotonation of TBG with subsequent formation of guanidine (**68**). On the other hand, the second amine acts as a base deprotonating the poor electrophile **68** and thus forming isocyanate intermediate **72**. Finally, the third amine makes a nucleophilic attack on **72** leading to amidinourea (**69**).

5.4 Macrocyclic Amidinoureas

The intermolecular reaction of an amine with di-Boc-guanidine constitutes probably the most efficient among the previously described methods to synthesize amidinoureas. This reaction leads to linear amidinoureas. However, some molecules,

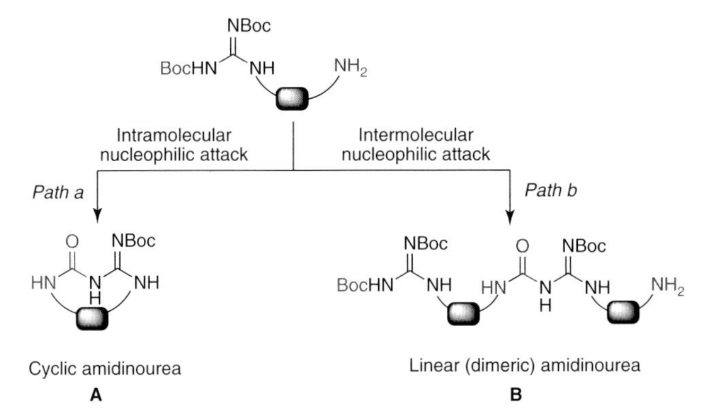

Scheme 5.12 Intramolecular versus intermolecular formation of amidinoureas.

namely, the guanylated (poly-) diamines, contain both a Boc-protected guanidine moiety and an amino moiety. It is likely that, under the same conditions as shown before, the amino group of these molecules would react with one Boc-protecting group affording an amidinourea. Nevertheless, in this case we might expect two different possibilities: (i) the amine reacts with the Boc-protecting groups of the same molecule leading in an intramolecular fashion to a cyclic amidinourea with general structure **A** (path *a*); and (ii) the amine reacts with the Boc-protecting group bound to another molecule affording a cross-coupling amidinourea product with general structure **B** (path *b*) (Scheme 5.12).

In this context, guanylated diamines and polyamines represent interesting substrates as potential precursors for the synthesis of cyclic and/or linear (dimeric) amidinoureas. An in-depth study on the transformation of guanylated polyamines into amidinoureas has been recently reported by Botta and coworkers. This section treats the chemistry of guanylated polyamines and their conversion into cyclic amidinoureas, the latter derivatives proving to be excellent bioactive lead compounds.

5.4.1 Guanylated Polyamines

Polyamines (PAs) are organic compounds having two or more amino groups. Some polyamines are biosynthesized by many organisms from bacteria to humans and play important roles in both eukaryotic and prokaryotic cells. The human PAs putrescine, spermidine, and spermine are fundamental for cell viability, having a multitude of in vivo functions, and their study is an expanding field of research.

When one or more amino groups of a PA are part of a guanidine function, the resulting compound is named a *guanylated polyamine*. Replacement of an amino group in a biologically active polyamine compound by the strongly basic guanidinium results often in a significant increase of its potency and/or selectivity. Guanylated polyamines show interesting biological properties or chemical behavior and have therefore found important applications in medicinal, bioorganic, and supramolecular chemistry, as well as in asymmetric synthesis, possessing in some cases biochemical and biophysical properties surpassing those of their parent polyamines.

Depending on the number of guanidino moieties, guanylated polyamines can be classified as mono-, bi-, and tri-guanylated polyamines. Similarly, a further classification can be done on the basis of the number of amino moieties. Hence, a diamine bearing one guanidino moiety can be named *monoguanylated diamine* while a triamine bearing two guanidino moieties will be classified as *biguanylated triamine*. An example of monoguanylated diamine is represented by the natural compound agmatine (**73**), known to be an *intermediate* in the polyamine metabolism of various bacteria, fungi, parasites, and marine fauna as well as in the metabolism of mammals where polyamines have been attributed an important function in cellular growth. Hirudonine (**74**), namely, the $N^1 - N^8$-biguanylated spermidine, and the Caldine derivative (**75**) represent, respectively, examples of biguanylated and monoguanylated triamines [32] (Figure 5.6).

Guanylated polyamines can be generally synthesized through the simple guanylation reaction of polyamines by the use of different guanylating agents. However, despite their simple chemical structure, syntheses could be sometimes problematic due the peculiar structures of some of these compounds. Guanylation of polyamine is generally carried out with *S*-Me-isothiourea (**54**) or *O*-Me-isourea (**76**), or with thioureas (**77**) in the presence of activating agents such as $HgCl_2$ or Mukayama's reagent **78**. From early 1990s, the use of di-Boc-*S*-Me-isothiourea (**60**) as a guanylating agent emerged as an efficient method for the synthesis of guanidines. The use of this reagent offers advantages, especially with regard to purification procedures. Guanidines and in particular guanylated polyamines containing both a guanidine and an amine moiety are highly polar compounds and are diffficult to handle and to purify, being sometimes also soluble in water. The use of di-Boc-isothioureas in

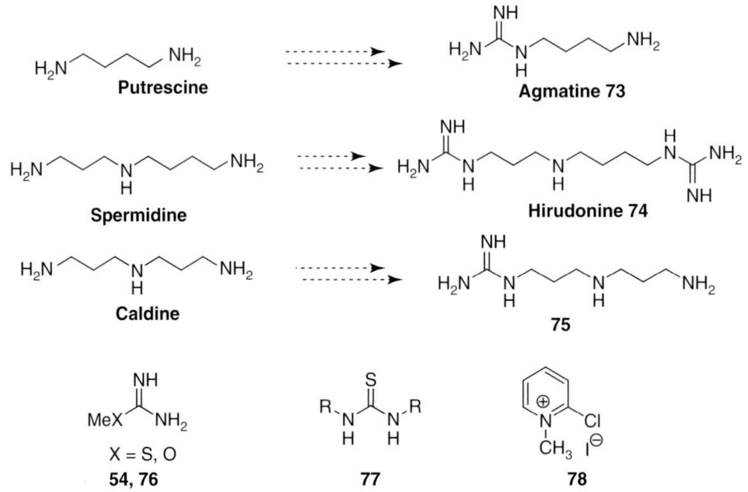

Figure 5.6 Guanylated polyamines and guanylating agents.

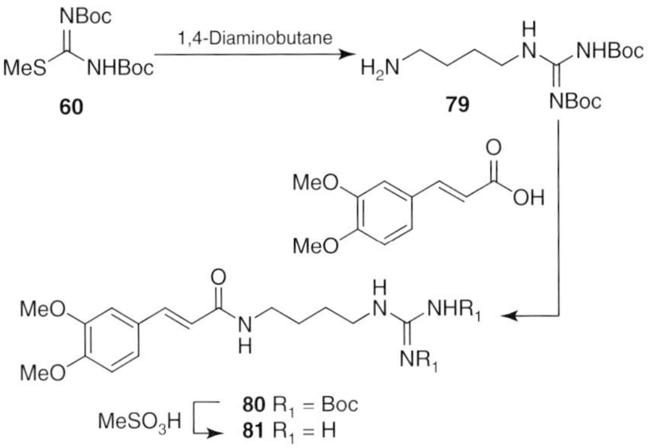

Scheme 5.13 Synthesis of guanylated diamine G3.

multistep syntheses allows the introduction of the guanidine moiety into molecules in the early stages of the synthetic pathway without purification problems. The Boc-protecting group can then be removed in the final step of the synthesis. As an example, Botta *et al.* described the synthesis of the natural compound G5 (**81**) endowed with antihypertensive properties through the use of di-Boc-isothiourea as shown in Scheme 5.13 [33]. The intermediate **79** constitutes an ideal substrate to investigate the formation of amidinoureas from guanylated polyamines.

5.4.2 Conversion of Di-Boc-Guanylated Diamines into Amidinoureas

Starting from early 1990s, Botta and coworkers described the synthesis of several guanylated diamines and polyamines endowed with antimicrobial as well as polyamine oxidase (PAO) and nitric oxide synthase (NOS) inhibitorial activity [34]. It is exactly from these studies on guanylated polyamines that the discovery of macrocyclic amidinoureas took place. In the case in point, Botta *et al.* recently reported an in-depth study on the commercially available antifungal agent guazatine which was found to be a mixture of guanylated polyamines that were separated by HPLC [35]. Each component of the guazatine mixture showed peculiar antifungal properties and was chosen as a template for the synthesis of analogs with the aim of improving its biological activity. During the synthesis of a series of guanylated diamines and triamines related to guazatine components, the authors encountered difficulties while trying to guanylate the compound **82**. It was observed that when **82** was reacted with crotyl-di-Boc-*S*-Me-isothiourea at room temperature, the expected biguanylated diamine (**83**) was formed only in poor yield. On the other hand, raising the temperature of the reaction led to the formation of a side product in 40% yield, together with traces of the desired **83**. Simple heating of guanylated amine (**82**) in refluxing THF also afforded the same side product (Scheme 5.14).

On the basis of data in the literature, it was hypothesized that the side compound formed in the course of the previous reaction could be an amidinourea. As

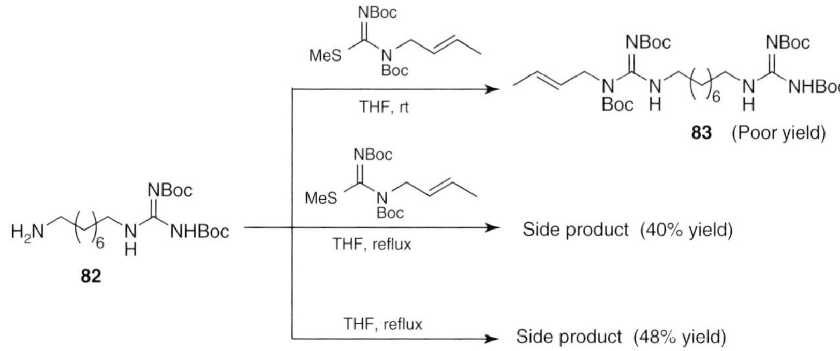

NBoc
MeS N Boc
THF, rt
NBoc NBoc
N N N NHBoc
Boc H 6 H
83 (Poor yield)

NBoc
MeS N Boc
NBoc
H_2N 6 N NHBoc
H
82
THF, reflux
Side product (40% yield)

THF, reflux
Side product (48% yield)

Scheme 5.14 Discovery of amidinoureas.

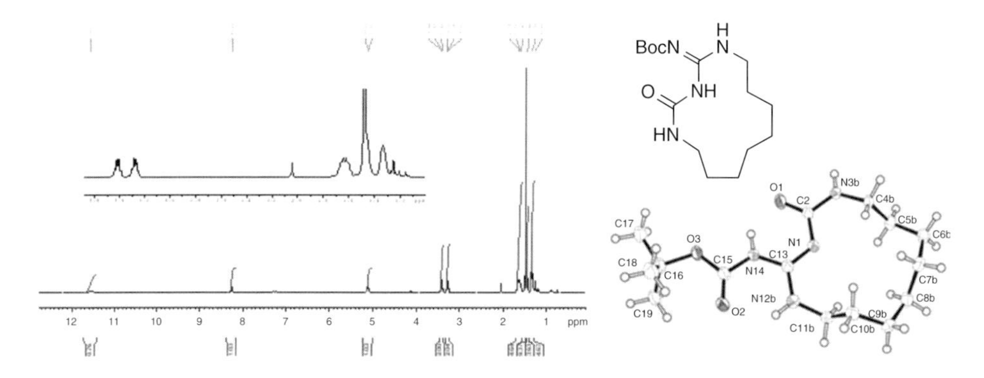

Figure 5.7 ^{1}H-NMR and crystal structure of amidinourea (**84**).

previously shown, two potential amidinoureas could have been formed from guanylated diamine (**82**), due to either an intramolecular or an intermolecular nucleophilic attack of the primary amine on the Boc-protecting group. The low molecular weight (313, M + 1, revealed by HPLC-MS analysis) together with NMR spectral analysis results suggested that the side compound was a 13-membered ring amidinourea (**84**). Finally, definitive confirmation of the chemical structure of **84** came from X-ray analysis (Figure 5.7).

A second product was also formed in 8% yield from heating of **82** in THF. Surprisingly, this compound did not turn out to be the expected linear cross-coupled amidinourea product but the cross-coupled dimerized macrocyclic compound bis-amidinourea (**85**). The proposed mechanism for the formation of cyclic amidinourea (**84**) is shown in Scheme 5.15. In analogy with the Rault and Miel mechanism, the formation of **84** was supposed to proceed via an isocyanate intermediate **86**, formed from the Boc-protecting group by an E1cB mechanism with the primary amine. Hence, the amino group of **82** works both as the conjugate base in the formation of the isocyanate intermediate and as the nucleophilic agent. The formation of the bis-amidinourea macrocycle (**85**) is supposed to be due to the double reciprocal cross attack of two isocyanate intermediates [36].

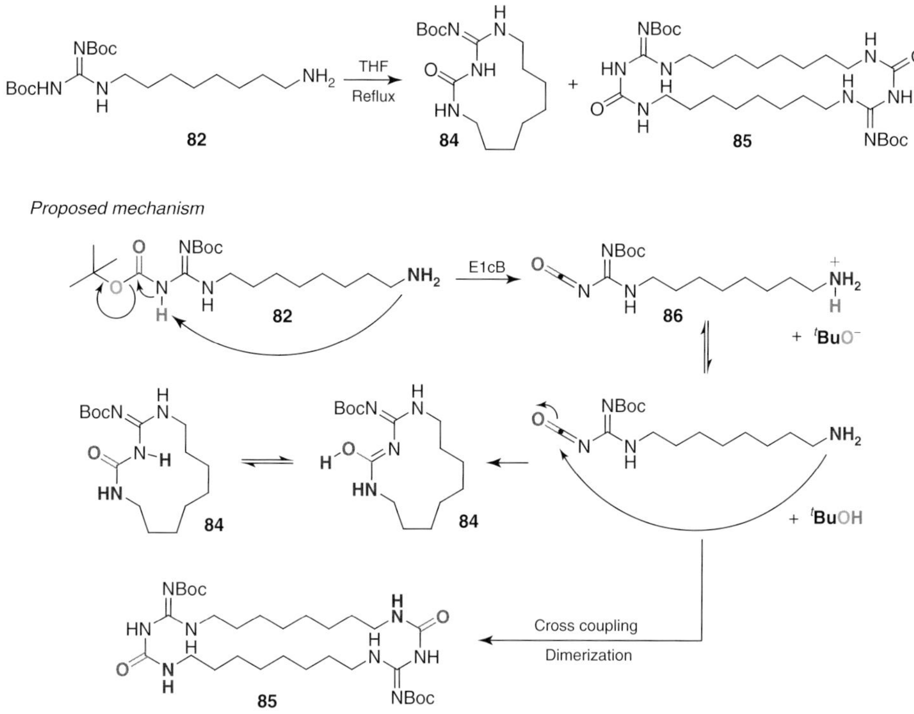

Scheme 5.15 Proposed mechanism for the formation of cyclic amidinoureas.

5.4.3 Synthesis of Cyclic Amidinoureas

Macrocyclic amidinoureas represent original scaffolds that could be potentially manipulated and used as a template for the synthesis of biologically active compounds. The design and synthesis of macrocyclic structures with an appropriate ring size and in a predictable function is an important research topic and still remains an attractive challenge in organic synthesis. For instance, macrocyclic peptides play a significant role in biology since they can interact with a large protein surface disrupting protein–protein interaction, which is a new, important topic in medicinal chemistry. Less common natural macrocyclic polyamine lactams and macrocyclic lactams containing the biogenetic base spermine as well as novel macrocyclic urea compounds are of great interest in view of their broad activity. In this context, amidinourea functional group could be considered as a bioisosteric group of the common urea, lactam, amide, and guanidine moieties, smoothening the way to the design and discovery of novel bioactive compounds.

However, the synthesis of cyclic amidinoureas encountered some difficulties. An in-depth investigation on the synthesis of cyclic amidinoureas has been described recently. A series of guanylated diamines precursors **88** bearing alkyl chains of different lengths was first prepared in order to investigate and generalize

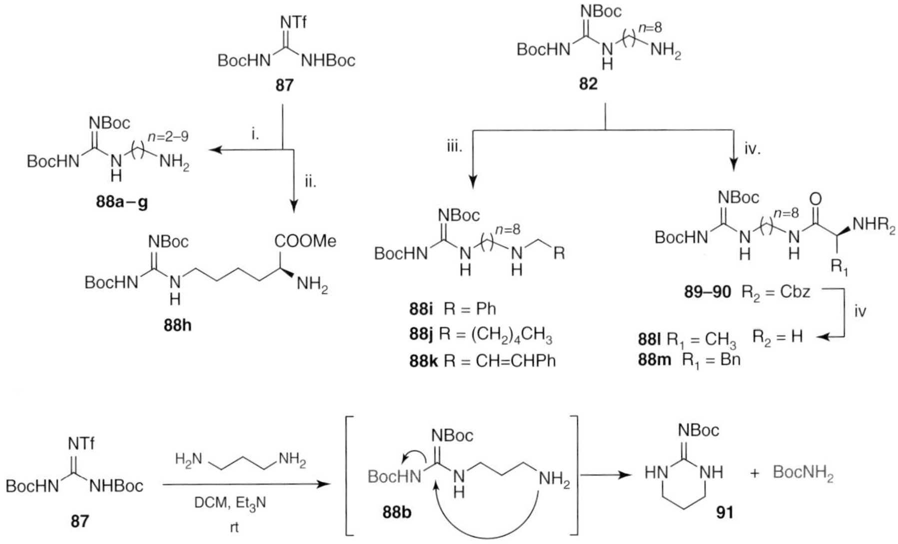

Scheme 5.16 Synthesis of cyclization precursors.

their tendency to form cyclic amidinoureas. Compounds **88a–h** were synthesized through guanylation reaction of the appropriate diamines with the triflic guanylating agent **87**. The advantage resulting from the use of **87** instead of thiourea (**60**) lies in the possibility to carry out guanylation reaction at room temperature overcoming possible side reactions at this stage. Guanyalted diamines (**88i–k**) containing a secondary amino moiety were also synthesized through reductive amination of **82** with different aldehydes in order to evaluate the influence of steric factors in the course of the cyclization reaction. Finally, enantiomerically pure precursors (**88l–m**) were also synthesized (Scheme 5.16).

The guanylation reaction between 1,3-diaminopropane with **87** for the synthesis of precursor **88b** is noteworthy. This reaction led to the formation of compound **91**, which was isolated in 94% yield as the only product while no traces of **88b** were formed. It is possible that in the early stage of the reaction compound **88b** was formed. However, as **88b** was being formed, its free amino group immediately collapsed on the quaternary electrophilic carbon atom, affording in this way the more stable six-membered cyclic guanidine (**91**). The greater stability of **91** compared to **88b** represents the driving force for this reaction [36].

Precursors **88** were refluxed in THF to evaluate their tendency to afford cyclic amidinoureas (**92**). Results (Table 5.4) showed how heating guanylated diamines with shorter alkylic chains in refluxing THF led to the formation of bis-amidinoureas (**94**) as the sole or main product. On the other hand, guanylated amines with longer alkylic chains afforded amidinoureas (**92**) as major products, while dimeric compounds **94** were not formed at all or obtained just in low yields. These simple experiments clearly show that a linear correlation exists between the length of the chain of **88** and the formation of **92** and/or **94**.

Table 5.4 Structures and yields of cyclic amidinoureas.

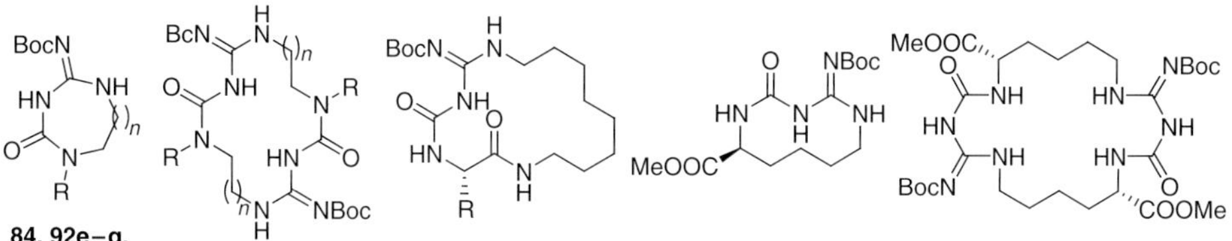

84, 92e–g, 92i–k **94c–f, 94h–j** **92l–m** **92h** **94h**

Entry	Precursor	n	Amidinourea (yield %)	Bis-amidinourea (yield %)
1	**88a**	2	**92a** (n.d.)	**94a** (n.d.)
2	**88c**	4	**92c** (n.d.)	**94c** (20)
3	**88d**	5	**92d** (n.d.)	**94d** (15)
4	**88e**	6	**92e** (6)	**94e** (12)
5	**88f**	7	**92f** (30)	**94f** (9)
6	**82**	8	**84** (48)	**85** (8)
7	**88g**	9	**92g** (51)	**94g** (n.d.)
8	**88h**	5	**92h** (n.d.)	**94h** (13)
9	**88i**	8	**92i** (31)	**94i** (6)
10	**88j**	8	**92j** (39)	**94j** (4)
11	**88k**	8	**92k** (32)	**94k** (traces)
12	**88l**	8	**92l** (44)	**94l** (n.d.)
13	**88m**	8	**92m** (42)	**94m** (n.d.)

When the length of the chain of **88** is $n = 4–6$, dimers **94c–e** and **94h** were isolated as the major (or only) products (entries 2–4 and 8). On the contrary, when the chain length is $n = 9$, no traces of the dimeric compound **94i** were detected, and **92i** was the only product. A plausible explanation of the behavior of the different guanylated diamines in refluxing THF might be found in the planar and electronic nature of the amidinourea functional group. In fact, the amidinourea moiety being planar, the macrocyclization is conformationally hindered when the length of the chain is $n = 4, 5$. The amino group is too far away to reach the isocyanate moiety on the same molecule, and thus it prefers to cross-react with the isocyanate moiety of a second molecule, leading to the formation of dimers **94**. On the other hand, when $n = 6$, aminoguanidine (**88**) has greater conformational flexibility, and the amino group can reach the isocyanate moiety easily, affording also macrocyles **92**. Hence, the geometric and stereoelectronic constraints on the transition state for the ring closure play a fundamental role in the formation of dimers **94** over the cycles **92** or vice versa. This could explain, for instance, why the 18-membered ring **94c** was formed instead of the 9-membered ring **92c**, whose rate of ring closure would be expected to be faster. Finally, the formation of the cyclic guanidine (**96**) when **88a**

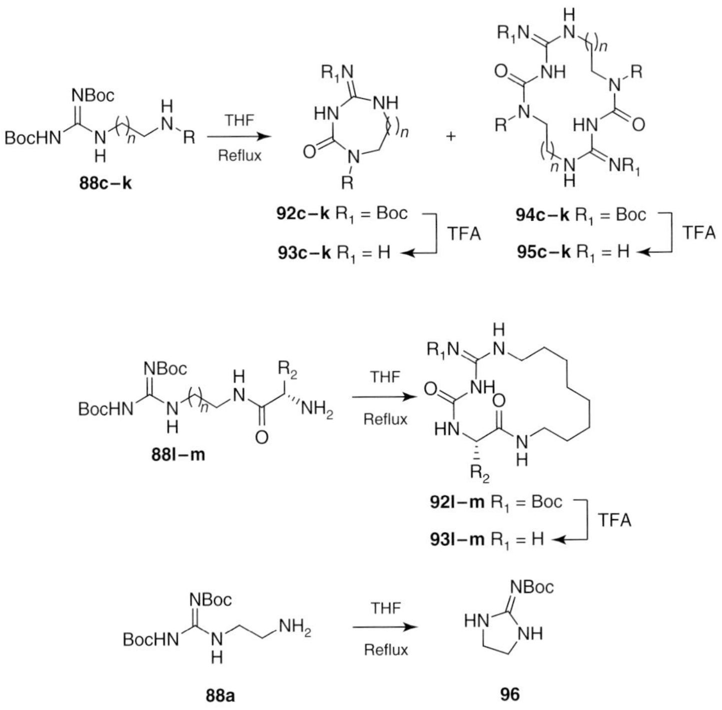

Scheme 5.17 Macrocyclization of Boc-guanylated diamines.

was refluxed in THF is noteworthy. Compound **96** was formed after only 2 h and isolated as the only product of the reaction in 92% yield (Scheme 5.17). No traces of the desired amidinourea (**92a**) were detected (entry 1). Also in this case, the formation of the five-membered ring **96** was favored because of its greater stability compared to the expected but more constrained seven-membered ring **92a**.

5.4.4
Synthesis of Macrocyclic Amidinoureas from Di-Boc-Monoguanylated Triamines

As di-Boc-guanylated diamines are converted into macrocyclic amidinoureas, the same behavior might be expected by di-Boc-monoguanylated triamines. A general monoguanylated triamine such **97** (Scheme 5.18) contains both a primary and a secondary amino group. Hence, assuming that di-Boc-monoguanylated triamines could be converted into cyclic amidinoureas through the same mechanism of reaction shown earlier via an isocyanate intermediate, two possible cyclic products might be expected. Nucleophilic attack of the primary amino group of **97** on the isocyanate would lead to cyclic amidinourea (**98**), while nucleophilic attack of the secondary amine would lead to **99**. In principle, the secondary amine is expected to react faster because of its higher nucleophilicity. However, steric hindrance could be a limitation and favor the nucleophilic attack of the primary amine [37].

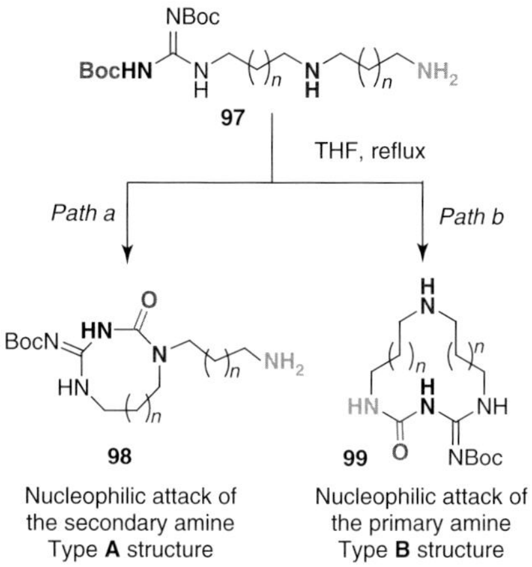

Scheme 5.18 Amidinoureas from di-Boc-monoguanylated triamines.

In this context, Botta *et al.* recently described the synthesis of cyclic amidinoureas from di-Boc-monoguanylated triamine (**101**) in the course of their studies on the antifungal agent guazatine. Compound **101** was obtained from triamine (**100**) in nearly quantitative yield by guanylation with triflic guanidine (**87**) carried out at room temperature. In order to get biguanyalted derivatives **103**, a second guanylation reaction was carried out on **101**. Botta and coworkers found that the guanylation of **101** with substituted *N*-alkyl-di-Boc-*S*-Me-isothioureas (**102**) in refluxing THF in some cases did not afford the expected biguanylted triamines products **103** and that side compounds were often isolated as major products from the reaction mixtures. These side products, resulting from a one-pot, two-step guanylation–cyclization reaction, were assumed to be, on the basis of spectral data and in connection with previous studies, guanylated cyclic amidinoureas. However, the formed guanylated amidinoureas might have either a type **A** or a type **B** structure. Which is the right structure of the formed amidinoureas? Since in some cases linear biguanylated triamines (**103**) were isolated in poor yields or as main product, the authors reasoned that the guanylation step would occur first at the less hindered primary amine and that the cyclization step would start as soon as the guanylation step was completed. Thus, the resulting cyclized products were assumed to be the type **A** amidinoureas (**104**) [38] (Scheme 5.19).

However, in order to confirm these assumptions, a synthetic approach to obtain unambigously amidinoureas (**104**) was also planned by the authors. The synthetic strategy planned was based on the synthesis of the key intermediate monguanylated triamine (**112**) with the primary amino group protected as benzyloxycarbamate. The linear triamine backbone was synthesized through reductive amination between guanylated diamine (**82**) and aldehyde (**111**). This latter compound was in turn obtained in a four-step synthetic sequence starting from aminocaprilic acid (**107**). Triamine (**112**) was then heated in refluxing THF, affording cyclic amidinourea (**113**). Being the primary amino group protected as carbamate, only the secondary amino group can act as nucleophile in the course of cyclization reaction, leading unambiguously to the type **A** 13-membered amidinuorea (**113**). Subsequent

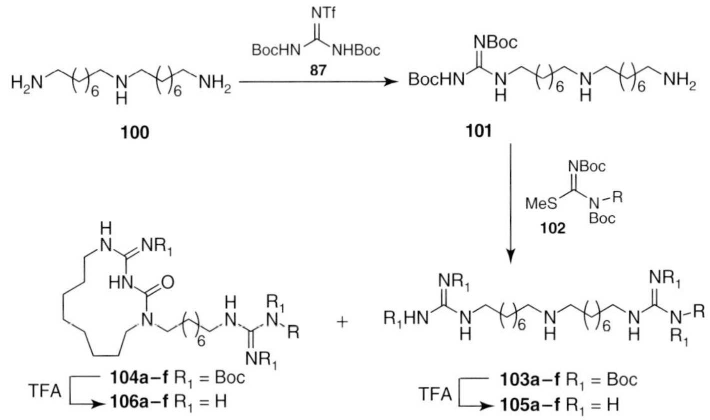

Scheme 5.19 Macrocyclic amidinoureas.

Table 5.5 Ratios of linear guanidines/cyclic amidinoureas obtained from **101**.

R	Linear guanidine	Macrocyles
Benzyl	–	100
Propargyl	–	100
Cyclopropylmethyl	100	–
Buten-2-yl	1	9
Isobutenyl	1	8
Prenyl	1	3

hydrogenolysis, followed by guanylation of the resulting primary amine, led to compound **104d**. Spectral data of this latter compound were compared with those of amidinourea obtained from **101** through the guanylation/cyclization reaction (Table 5.5), confirming in this way the right assignment of the chemical structure [39] (Scheme 5.20).

The plausible mechanism of the one-pot, two-step guanylation/cyclization reaction of **101** is reported in Scheme 5.21, and, according to previous hypotheses, was also assumed to occur through an isocyanate intermediate.

5.4.5
Biological Properties of Cyclic Amidinoureas

Macrocyclic amidinoureas were found to possess high antifungal activity against different *Candica* and *Aspergillus* spp. The pathogenic species of *Candida* derive their importance from the severity of their infections and from their ability to develop resistance against a variety of antifungal agents. On the other hand, also invasive filamentous fungal infections, such as aspergillosis, represent a major problem

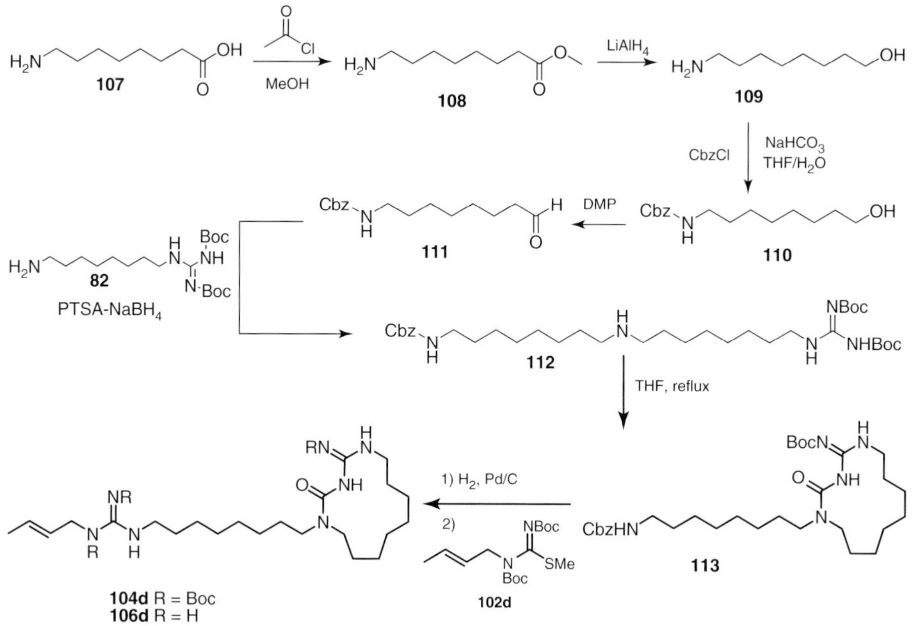

Scheme 5.20 Synthesis of cyclic amidinourea (**104d**) through the convergent approach.

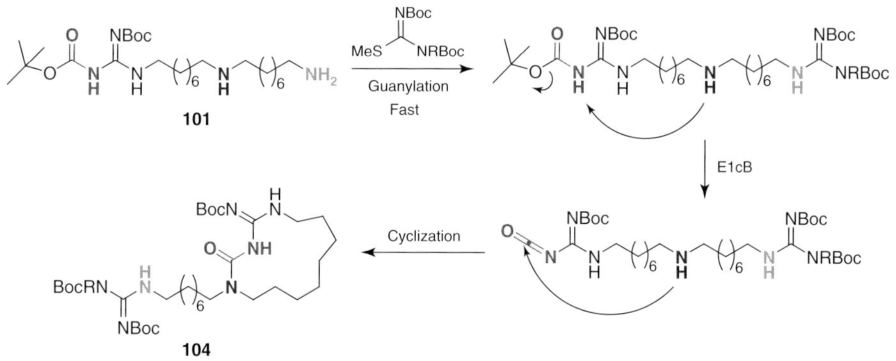

Scheme 5.21 Mechanism of guanylation/cyclization reaction of **101**.

for certain groups of patients. Clinically, candidosis and aspergillosis account for 80–90% of systemic fungal infections in immunocompromised patients. While many drugs have proven to be effective against *Candida* infections, aspergillosis still remains very hard to overcome. However, most active drugs against candidosis such as fluconazole often suffer from severe drug resistance and are not active against invasive aspergillosis. For these reasons, chemists focused their efforts in the synthesis of conceptually new antifungal, and more generally antimicrobial, agents having innovative chemical structures and with a mechanism of action different from that of common drugs.

Table 5.6 Biological activity of macrocyclic amidinoureas **106a–f**.

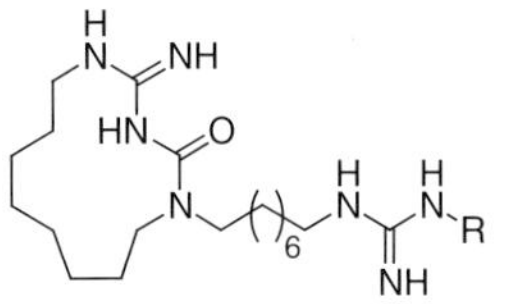

106a R = benzyl, 106b R = propargyl, **106d** R = buten-2-yl, **106e** R = isobutenyl, **106f** R = prenyl

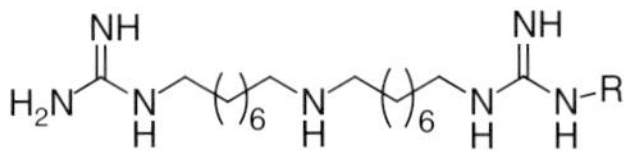

105c R = cyclopropylmethyl; **105f** R = prenyl

***Candida* strains**	**Antifungal activity (MIC, μM)[a]**								
	G[b]	**105c[c]**	**105f[c]**	**106a**	**106b**	**106d**	**106e**	**106f**	**F[d]**
C. albicans ATCC 60193	40	20	80	2.5	2.5	2.5	20	40	0.8
C. albicans 4T	20	10	80	2.5	80	1.25	40	20	209
C. albicans 53T	40	10	80	2.5	5	2.5	40	20	418
C. albicans 15T	80	20	40	5	2.5	1.25	20	20	209
C. krusei ATCC 14243	20	5	40	20	80	5	40	10	209
C. krusei 193T	20	10	20	10	40	5	80	20	418
C. parapsilosis ATCC 34136	10	80	>80	80	40	5	>80	>80	6.5
C. parapsilosis 64E	20	5	>80	20	40	5	>80	>80	32
C. parapsilosis 81E	20	20	40	20	80	5	>80	40	13
C. glabrata 70E	40	20	80	40	80	20	80	80	209
C. tropicalis 86E	10	5	20	2.5	5	1.25	40	20	52

[a] MIC values were determined at 24 h both visually and spectrophotometrically.
[b] G is the guazatine mixture.
[c] Compounds **105d** and **105e** were isolated in amounts not sufficient for assays.
[d] F is fluconazole.

	Antifungal activity (MIC, μM)[a]					
	A. niger	***A. niger***	***A. fumigatus***	***A. versicolor***	***A. versicolor***	***A. flavus***
106d	18	18	36	18	4	36
AMB	0.27	0.135	0.27	0.54	0.135	2.1
ITRA	0.35	0.35	0.04	0.02	0.35	0.04

[a] MIC values were determined at 48 h.
AMB: amfotericin B.
ITRA: itraconazole.

In this context, macrocyclic amidinoureas appear as promising molecules because of their original backbone, and their antifungal activity seem to be due to a mechanism different from that of common azoles and other antifungal drugs. The potential of these classes of compounds is shown in Table 5.6. Amidinoureas (**106a–f**) showed antifungal activity against *Candida* species comparable to or higher than that of fluconazole. In addition, one of the cyclic compounds, namely amidinourea (**106d**), was shown also to be active against *Aspergillus* species with 4–18 μM minimum inhibitory concentration (MIC) values. It is noteworthy that cyclic amidinoureas proved to be more active than the corresponding linear biguanylated triamines (**105**) and also the antifungal agent guazatine [40]. Finally, in vitro cytotoxicity of **106d** checked in human lymphocytes at 25 and 37 °C with concentrations between 0 and 50 μM has been also reported. The therapeutic indices (TI50) related to the various *Candida* species and evaluated on the basis of the 50% viability and MIC_{50} were between 38 and 2.4 μM at 25 °C and between 30 and 1.9 μM at 37 °C [38].

Hence, amidinoureas (**106**) showed a broad spectrum antifungal activity, low cytotoxicity, and high solubility in aqueous media because of the ability of the amidinourea group to be both an acceptor and a donor of H-bonds. All these properties make these bioactive compounds excellent drug candidates. Moreover, the uncommon and novel chemical structure of macrocyclic amidinoureas make them promising and unique in the field of antifungal and, more generally, antimicrobial agents.

5.5 Perspectives

Amidinoureas constitute a class of underexplored compounds possessing several properties both from chemical and biological points of view. Unlike ureas and guanidines which have been largely documented in chemistry, amidinoureas still remain "mysterious" compounds with a hidden power. Their biological potential has been already documented in a few patent applications which showed amidinoureas as possessing strong activities in particular as antimicrobial agents. In addition, the strict correlation of amidinourea with urea and/or guanidine groups makes it the candidate of choice in the synthesis of analogs of the two latter classes of compounds. Amidinoureas could be potential bioisosters of ureas and guanidines as well as lactams, amides, and in general carbonyl- and nitrogen-containing functional groups. In particular, both linear and cyclic amidinoureas could constitute interesting bioisosters in the synthesis of peptidomimetics. The ability of amidinoureas to be both an acceptor and a donor of H-bonds makes them capable of interacting with biologic systems such proteins or DNA and consequently to be potentially interesting bioactive hit compounds. The ability to form noncovalent bond interactions could be also important in the synthesis of supramolecular systems to be applied in different fields of research, such as medicine, biology, and materials science.

Despite their apparently simple chemical structures, amidinoureas show difficulties in their synthesis. As shown throughout this chapter, only a few methods exist for the synthesis of compounds containing the amidinourea functional group, most of them being sometimes ineffective or requiring several synthetic steps or harsh conditions. Hence, also from a synthetic point of view, the development of new methodologies for the synthesis of amidinoureas is highly desiderable. For instance, the development of novel protecting-group-free synthetic pathways would allow the reduction of the number of steps during a synthetic sequence. In this context, the synthesis of macrocyclic amidinoureas from Boc-protected guanylated polyamines represents a goal, since it allows the formation of amidinourea directly from the protecting group under mild conditions. In this way, protecting groups can not only be considered as such, but also as constituents of target compounds, leading to synthetic strategies with higher atom economy.

Finally, as ureas and thioureas found in the last decade a wide range of applications as organic catalysts, the same might be expected of the amidinoureas. Their ability in forming noncovalent bond interactions with organic compounds could lead to the development of new organocatalysts. Moreover, amidinoureas, owing to their chemical structure, could complex different kind of metals, leading in this way to organometallic catalysts or metal ligands.

In conclusion, the race for amidinoureas has just started, and their full potential has not yet been realized. The development of further strategies in the synthesis of amidinoureas would allow a deeper knowledge of these attractive compounds, offering interesting perspectives in different fields of chemistry.

Acknowledgments

At the end of this work, I would like to acknowledge Prof. Maurizio Botta who supervised the research on macrocyclic amidinoureas. I am deeply indebted to him for helpful and stimulating discussions and encouragement. Dr. Francesco Raffi and Dr. Marco Radi are gratefully acknowledged for helpful discussions and precious suggestions. University of Siena is finally acknowledged for partial financial support.

References

1. (a) Matsuda, K. (1994) *Med. Res. Rev.*, **14**, 271–292; (b) Berlinck, R.G.S. (1996) *Nat. Prod. Rep.*, **13**, 377–421; (c) Collins, J.L., Shearer, B.G., Oplinger, J.A., Lee, S., Garvey, E.P., Salter, M., Duffy, C., Burnette, T.C., and Furfine, E.S. (1998) *J. Med. Chem.*, **41**, 2858–2871; (d) Mori, A., Cohen, B.D., and Lowenthal, A. (1983) *Historical, Biological, Biochemical, and Clinical Aspects of the Naturally Occurring Guanidino Compounds*, Plenum, New York.
2. (a) Pihko, P.M. (2009) *Hydrogen Bonding in Organic Synthesis*, Wiley-VCH Verlag GmbH & Co; (b) Sohtome, Y., Tanatani, A., Hashimoto, Y., and Nagasawa, K. (2004) *Chem. Pharm. Bull.*, **52**, 477–480; (c) Kimmel, K.L., Robak, M.T., and Ellman, J.A. (2009) *J. Am. Chem. Soc.*, **131**, 8754–8755; (d) Cao,

X.-Y., Zheng, J.-C., Li, Y.-X., Shu, Z.-C., Sun, X.-L., Wang, B.-Q., and Tang, Y. (2010) *Tetrahedron*, **66**, 9703–9707.
3. Fridrichová, M., Němec, I., Císařová, I., and Nìmec, P. (2010) *Cryst. Eng. Comm.*, 7, 2054–2056.
4. Scoponi, M., Polo, E., Bertolasi, V., Carassiti, V., and Bertelli, G. (1991) *J. Chem. Soc., Perkin Trans. 2*, 1619–1624.
5. Klapotke, T.M. and Sabate, C.M. (2008) *Heteroatom Chem.*, **19**, 301–306.
6. Östmark, H., Bemm, U., Bergman, H., and Langlet, A. (2002) *Thermochim. Acta*, **384**, 253–259.
7. Lotsch, B.V. and Schnick, W. (2005) *Z. Anorg. Allg. Chem.*, **631**, 2967–2969.
8. (a) Yelnosky, J. and Ghulam, M.N. (1987) US Patent 4,701,457; (b) Studt, W.L. Zimmerman, H.K., and Dodson, S.A. (1986) Can. CA Patent 1 210 394 A1; (c) Goday, E. and Puigdellivol, L.P. (1986) Span. ES Patent 550 020 A1; (d) Diamond, J. and Douglas, G.H. (1982) US Patent 4,353,842.
9. Serafin, B., Urbanski, T., and Warburst, D.C. (1969) *J. Med. Chem.*, **2**, 336–337.
10. Cutler, R.A. and Schalit, S. (1976) US Patent 3,988,370.
11. Katayama, N., Fukusumi, S., Funabashi, Y., Iwahi, T., and Ono, H. (1993) *J. Antibiot.*, **46**, 606.
12. (a) Funabashi, Y., Tsubotani, S., Koyama, K., Katayama, N., and Harada, S. (1993) *Tetrahedron*, **49**, 13; (b) Yuan, C. and Williams, R.M. (1997) *J. Am. Chem. Soc.*, **119**, 11777–11784.
13. (a) Beijer, F.H., Sijbesma, R.P., Kooijman, H., Spek, A.L., and Meijer, E.W. (1998) *J. Am. Chem. Soc.*, **120**, 6761–6769; (b) Beijer, F.H., Kooijman, H., Spek, A.L., Sijbesma, R.P., and Meijer, E.W. (1998) *Angew. Chem. Int. Ed.*, **37**, 75–78; (c) Corbin, P.S. and Zimmerman, S.C. (1998) *J. Am. Chem. Soc.*, **120**, 9710–9711; (d) Baruah, P.K., Gonnade, R., Phalgune, U.D., and Sanjayan, G.J. (2005) *J. Org. Chem.*, **70**, 6461–6467; (e) Prabhakaran, P., Puranik, V.G., and Sanjayan, G.J. (2005) *J. Org. Chem.*, **70**, 10067–10072.
14. (a) Wagenaar, F.L. and Kerwin, J.F. Jr. (1993) *J. Org. Chem.*, **58**, 4331–4338; (b) Sun, C.-M. and Shey, J.-Y. (1999) *J. Comb. Chem.*, **1**, 361–363.
15. Piskala, A., Hanna, N.B., Masojidkova, M., Otmar, M., Fiedler, P., and Ubik, K. (2003) *Collect. Czech. Chem. Commun.*, **69**, 711–743.
16. (a) Rogstad, D.K., Herring, J.L., Theruvathu, J.A., Burdzy, A., Perry, C.C., Neidigh, J.W., and Sowers, L.C. (2009) *Chem. Res. Toxicol.*, **22**, 1194–1204; (b) Krecmerova, M., Masojidkova, M., and Holy, A. (2010) *Bioorg. Med. Chem.*, **18**, 387–395; (c) Dracinsky, M., Krecmerova, M., and Holy, A. (2008) *Bioorg. Med. Chem.*, **16**, 6778–6782.
17. Berlinck, R.G.S., Burtoloso, A.C.B., Trinidade-Silva, A.E., Romminger, S., Morais, R.P., Bandeira, K., and Mizuno, C.M. (2010) *Nat. Prod. Rep.*, **27**, 1871–1907.
18. Dixon, M.J., Nathubhai, A., Andersen, O.A., van Aalten, D.M.F., and Eggleston, I.M. (2009) *Org. Biomol. Chem.*, **7**, 259.
19. Sunazuka, T., Sugawara, A., Iguchi, K., Hirose, T., Nagai, K., Noguchi, Y., Saito, Y., Yamamoto, T., Ui, H., Gouda, H., Shiomi, K., Watanabe, T., and Omura, S. (2009) *Bioorg. Med. Chem.*, **17**, 2751.
20. Iwatsuki, M., Uchida, R., Yoshijima, H., Ui, H., Shiomi, K., Matsumoto, A., Takahashi, Y., Abe, A., Tomoda, H., and Omura, S. (2008) *J. Antibiot.*, **61**, 222.
21. Kim, H., Kim, M.-Y., and Tae, J. (2009) *Synlett*, 2949.
22. Tilley, J.W., Ramuz, H., Levitan, P., and Blount, J.F. (1980) *Helv. Chim. Acta*, **63**, 841–859.
23. Tilley, J.W. and Blount, J.F. (1980) *Helv. Chim. Acta*, **63**, 832–840.
24. Katritzky, A.R. and Rogovoy, B.V. (2005) *ARKIVOC*, **iv**, 49–87.
25. Yuan, C. and Williams, R.M. (1996) *Tetrahedron Lett.*, **37**, 1945–1948.
26. Miel, H. and Rault, S. (1998) *Tetrahedron Lett.*, **39**, 1565–1568.
27. Myers, A.G. and Yoon, T. (1995) *Tetrahedron Lett.*, **36**, 9429–9432.
28. Lamothe, M., Perez, M., Colovray-Gotteland, V., and Halazy, S. (1996) *Synlett*, 507–508.
29. Ko, S.Y., Lerpiniere, J., and Christophi, M.A. (1995) *Synlett*, 815–816.
30. Prabhakaran, P. and Sanjayan, G.J. (2007) *Tetrahedron Lett.*, **48**, 1725–1727.

31. Feichtinger, K., Sings, H.L., Baker, T.J., Matthews, K., and Goodman, M. (1998) *J. Org. Chem.*, **63**, 8432–8439.
32. Castagnolo, D., Schenone, S., and Botta, M. (2011) *Chem. Rev.*, **111**, 5247–5300.
33. Carmignani, M., Volpe, A.R., Botta, B., Espinal, R., De Bonevaux, S.C., De Luca, C., Botta, M., Corelli, F., Tafi, A., Sacco, R., and Delle Monache, G. (2001) *J. Med. Chem.*, **44**, 2950–2958.
34. (a) Manetti, F., Cona, A., Angeli, L., Mugnaini, C., Raffi, F., Capone, C., Dreassi, E., Zizzari, A.T., Tisi, A., Federico, R., and Botta, M. (2009) *J. Med. Chem.*, **52**, 4774–4785; (b) Corelli, F., Federico, R., Cona, A., Venturini, G., Schenone, S., and Botta, M. (2002) *Med. Chem. Res.*, **11**, 309–321.
35. (a) Dreassi, E., Zizzari, A.T., D'Arezzo, S., Visca, P., and Botta, M. (2007) *J. Pharm. Biomed. Anal.*, **43**, 1499–1506; (b) Dreassi, E., Zizzari, A.T., Zanfini, A., Corbini, G., and Botta, M. (2007) *J. Agric. Food Chem.*, **55**, 6850–6856.
36. Castagnolo, D., Raffi, F., Giorgi, G., and Botta, M. (2009) *Eur. J. Org. Chem.*, **3**, 334–337.
37. (a) Xu, D., Prasad, K., Repic, O., and Blacklock, T.J. (1995) *Tetrahedron Lett.*, **36**, 7357–7360; (b) Feichtinger, K., Zapf, C., Sings, H.L., and Goodman, M. (1998) *J. Org. Chem.*, **63**, 3804–3805; (c) Mc Arnett, E., Miller, J.G., and Day, A.R. (1950) *J. Am. Chem. Soc.*, **72**, 5635; (d) Mc Arnett, E., Miller, J.G., and Day, A.R. (1951) *J. Am. Chem. Soc.*, **73**, 5393.
38. Manetti, F., Castagnolo, D., Raffi, F., Zizzari, A.T., Rajamaki, S., D'Arezzo, S., Visca, P., Cona, A., Fracasso, M.E., Doria, D., Posteraro, B., Sanguinetti, M., Fadda, G., and Botta, M. (2009) *J. Med. Chem.*, **52**, 7376–7379.
39. (a) Raffi, F., Corelli, F., and Botta, M. (2007) *Synthesis*, **19**, 3013–1016; (b) Raffi, F. (2008) Synthesis and biological evaluation of a new class of macrocyclic compounds. PhD. Thesis, University of Siena.
40. Botta, M., Raffi, F., and Visca, F. (2009) WIPO Patent Application WO/2009/113033.

Part II
Catalysis

New Strategies in Chemical Synthesis and Catalysis, First Edition. Edited by Bruno Pignataro.

6
DNA Catalysts for Synthetic Applications in Biomolecular Chemistry

Claudia Höbartner and P.I. Pradeepkumar

Abbreviations

A	adenine, adenosine
G	guanine, guanosine
C	cytosine, cytidine
T	thymine, thymidine
U	uracil, uridine
R	purine nucleoside (A, G)
Y	pyrimidine nucleoside (C, U, T)
D	A or G or U, but not C
M	A or C
W	A or U/T
N	any nucleotide
rN	any ribonucleotide
br. site	branch site
p	phosphate
ppp	5′-triphosphate
App	5′-adenylate
>p	2′,3′-cyclic phosphate

6.1
Introduction

The nucleic acids DNA and RNA are fundamentally important biopolymers in all types of organisms. DNA is best known for its famous double helical structure and its unique biological role in the storage of genetic information. RNA, on the other hand, plays diverse roles in biology, ranging from the transfer of genetic information, to RNA maturation, and regulation of gene expression [1]. Additionally, RNA is a biocatalyst of vital importance. RNA enzymes are known as *ribozymes* and were first discovered in the early 1980s [2]. Natural ribozymes catalyze essential RNA splicing reactions. In addition, the arguably most important

New Strategies in Chemical Synthesis and Catalysis, First Edition. Edited by Bruno Pignataro.

reaction in a living cell – the formation of peptide bonds during protein biosynthesis – is catalyzed by RNA at the core of the ribosome [3]. In the laboratory, RNA catalysts have been generated for other types of reactions beyond their natural parallels [4]. The diverse functions of RNA depend on the formation of complex three-dimensional structures containing various secondary structure elements and tertiary interactions [5].

Besides the classical double helical form of natural double-stranded DNA, single-stranded DNA can also adopt complex structures which are the basis for sophisticated functions, such as catalysis. Single-stranded DNA indeed has the ability to catalyze chemical reactions with impressive rate enhancements, as was first reported in 1994 [6]. DNA catalysts are also named deoxyribozymes or DNA enzymes. Deoxyribozymes have not been found in nature, but are generated in the laboratory by *in vitro* selection from random sequence DNA pools (see Section 6.2). From the practical perspective, DNA is preferable over RNA to be used as an in vitro research tool, because of its higher chemical stability and much lower cost. Indeed, DNA has attracted great interest in the chemical community over the past 15 years, not only for its catalytic potential, but also because it provides many desirable properties for other areas of research [7]. In the emerging area of DNA-based asymmetric catalysis, double-stranded DNA is heavily investigated as an effective chiral ligand in transition-metal-catalyzed reactions [8]. The most prominent property of DNA, the ability to encode information in the nucleotide sequence, has been explored for reaction discovery by DNA-templated synthesis [9]. In the area of nanotechnology, DNA is appreciated as a construction material owing to the highly programmable nature of this remarkable biopolymer [10].

This chapter focuses on deoxyribozymes as only one partial aspect of DNA's applications in current chemical research. A brief overview on the scope of reactions catalyzed by deoxyribozymes (Section 6.3) is followed by selected examples of DNA catalysts that are of synthetic interest for the preparation of biomolecules that are challenging to access by other methods (Sections 6.4–6.7). The major focus is on DNA enzymes that enable the ligation of nucleic acid fragments in different topologies (Sections 6.5 and 6.6). Limited available information on mechanistic studies on DNA catalysis is briefly summarized (Section 6.8) and the potential for future applications is discussed.

6.2
In vitro Selection of Deoxyribozymes

In vitro selection, an iterative process of selection and amplification, has the potential to mimic Darwinian evolution in the laboratory [11]. Starting from a random pool of DNA sequences, the cyclic enrichment process is experimentally repeated until individual DNA sequences with the desired properties (i.e., catalytic activity) dominate the population (Figure 6.1). The random starting library is generated by solid-phase DNA synthesis using a mixture of all four standard deoxyribonucleotide

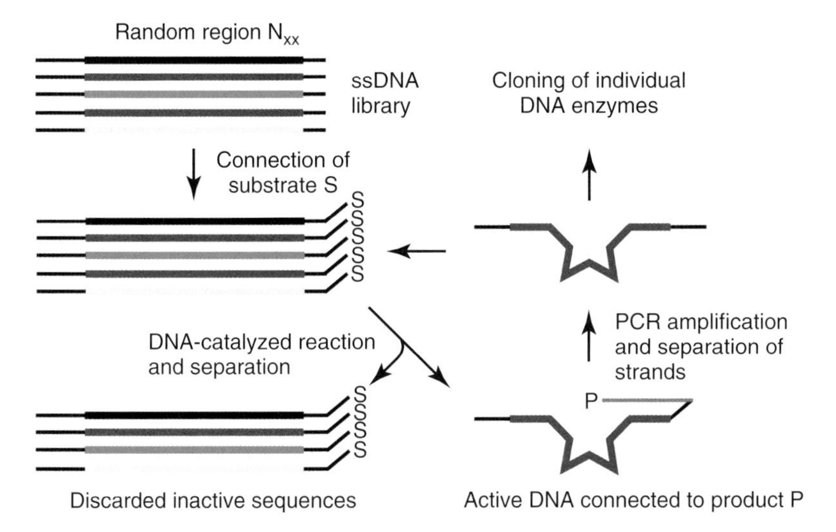

Figure 6.1 *In vitro* selection of catalytic DNA starting from a random pool of single-stranded DNA (ssDNA). The substrate is covalently connected to the DNA library. Upon incubation under selection conditions, the active DNA sequences form product P. Separation is achieved, for example, via gel electrophoresis or affinity chromatography (in case P contains biotin or another affinity label). Active DNA sequences are amplified by PCR and strands are separated (e.g., via gel electrophoresis based on different lengths achieved by a non-extendable primer). Substrate S is again connected and selection is continued for another round. This cycle is repeated until the desired activity is obtained. Individual deoxyribozymes are then cloned, sequenced, and further characterized.

phosphoramidites, which results in equal probability of each of the four nucleotides at every position of the random region. Usually, random regions are 25–70 nucleotides long (in exceptional cases more than 200 positions were randomized) [4]. For a commonly used N_{40} pool, $4^{40} = 10^{24}$ sequences are theoretically possible. In a practical *in vitro* selection experiment, only $10^{14}-10^{16}$ molecules are examined. Despite the low coverage of sequence space per experiment, diverse active DNA sequences have been found, indicating that a large number of DNA sequences are catalytically active.

The *in vitro* selection of catalytic DNA sequences is conceptually simple (Figure 6.1). It requires the substrate S being covalently bound to the DNA pool. The substrate library is then incubated under desired reaction conditions that enable active DNA candidates to generate the product P, which is connected to the DNA pool. In this step, a "selectable function" is generated that allows the active DNA sequences to be separated from the vast majority of inactive variants. Depending on the activity sought, this selectable function is either a shortened or an elongated "tail" of the nucleic acid that allows for separation by gel electrophoresis. Alternatively, an affinity label can be gained or lost during the selection step. A commonly used label is biotin, as was also employed in the first reported *in vitro* selection of a catalytic DNA in 1994 [6].

This first DNA catalyst was selected for cleavage of a ribonucleotide linkage in the oligonucleotide substrate (see Section 6.4 for more details on RNA-cleaving deoxyribozymes). Upon successful cleavage, the active fraction of the DNA library was released from the streptavidin matrix to which it was connected via the biotin label.

In the amplification step, the active candidate molecules are regenerated by polymerase chain reaction (PCR) using appropriate primers. As PCR generates double-stranded DNA, an important step after the amplification is the separation of the complementary strands into single strands. This is again achieved by gel electrophoresis or affinity separation (depending on the primer used; one primer either contains a spacer unit that prevents polymerase extension and therefore leads to the generation of different lengths of the amplified DNA strands; or one primer carries a biotin label that allows strand separation under denaturing conditions using a streptavidin matrix). Since the separation of active from inactive sequences cannot be achieved with 100% efficiency in a single step, a number of inactive variants will remain in the amplified fraction. By repeating the selection and amplification steps for several rounds of *in vitro* selection, the active DNA molecules are enriched, and can then be isolated and identified by cloning and sequencing using standard molecular biology techniques. The following section briefly summarizes the different types of chemical reactions for which DNA catalysts have been generated by *in vitro* selection.

6.3 Scope of DNA-Catalyzed Reactions

The first reported deoxyribozyme catalyzed the site-specific cleavage of a ribonucleotide phosphodiester bond in an oligonucleotide substrate [6]. In a typical setup for this and other DNA enzymes, the substrate is bound via Watson–Crick base-pairing and the catalytic reaction, assisted by divalent metal ion cofactors, generates two nucleic acid fragments, terminating in a 2′,3′-cyclic phosphate and a free 5′-hydroxyl end. This type of DNA-catalyzed reaction can be generalized as "cleavage reaction," where two products are generated out of one substrate (possibly assisted by an additional reagent) (Figure 6.2a). The second class of DNA-catalyzed reactions generally involves formation of a new covalent bond between two substrates, also referred to as "*ligation reactions*" (e.g., by displacement of a leaving group L, among other possibilities; Figure 6.2b). The majority of known DNA catalysts use DNA and RNA oligonucleotides as substrates and catalyze various types of phosphodiester transformations that fall into these two classes of cleavage and ligation reactions. In addition, several other deoxyribozymes have been described that accept non-nucleosides as substrates and catalyze small-molecule transformations, such as the Diels–Alder reaction between anthracene and maleimide. Table 6.1 summarizes known DNA-catalyzed reactions and specifies the substrates involved.

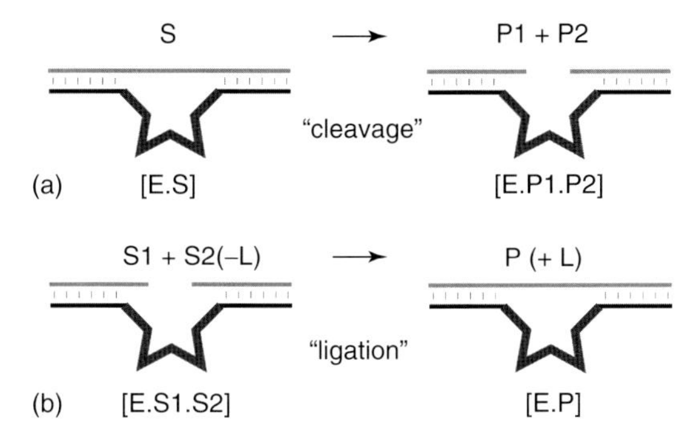

Figure 6.2 General depiction of DNA-catalyzed cleavage and ligation reactions of substrates bound via Watson–Crick base-pairing. S, S1, S2 = substrates; L = leaving group; P, P1, P2 = products; E = DNA enzyme (bold line = enzyme core region). Enzyme–substrate and enzyme–product complexes are indicated. See Table 6.1 for specific examples of both types of DNA-catalyzed reactions.

6.4 Synthetic Applications of RNA-Cleaving Deoxyribozymes

RNA-cleaving deoxyribozymes represent the best studied and most widespread class of DNA catalysts [35]. The DNA-mediated, site-specific cleavage reaction is depicted in detail in Figure 6.3a and consists of the nucleophilic attack of the 2′-OH group on the neighboring phosphodiester group. Two well-known examples are the 10–23 and 8–17 deoxyribozymes (Figure 6.3b). The catalytic core of these deoxyribozymes consists of only 13–15 nucleotides. In the presence of Mg^{2+} or Mn^{2+}, these single-stranded DNA molecules enable highly efficient cleavage of all-RNA substrates. The DNA enzymes are targeted to their cleavage sites by Watson–Crick base-pairing to the substrate; 10–23 prefers cleavage sites between purines and pyrimidines, and the prototypical 8–17 cleaves between adenosine and guanosine. A variety of derivatives of these DNA catalysts have been developed for all 16 possible di-ribonucleotide junctions embedded in DNA substrates [36, 37].

RNA-cleaving deoxyribozymes have found many applications in basic and applied research, which have been reviewed elsewhere [38, 39]. Briefly, potential areas for utilization include DNA enzymes as components of biosensor, for example, when combined with fluorescently labeled substrates. The potential application as therapeutic agents for the specific cleavage of viral, bacterial, cardiovascular-, or cancer-related mRNA targets has also been investigated [40]. In addition, 10–23 deoxyribozymes have been useful reagents for the analysis and identification of natural RNA modifications [41].

In the context of synthetic applications of deoxyribozymes, the utilization of RNA-cleaving DNA catalysts as in vitro endoribonucleases for the preparation of homogeneous RNA products is particularly important [38, 42]. Micura and coworkers recently reported the application of the 10–23 deoxyribozyme as a critical player in their strategy for the semi-synthesis of nonhydrolyzable peptidyl-tRNAs that contain all natural tRNA modifications (Figure 6.4) [43, 44]. The natural tRNA

Table 6.1 Scope of DNA-catalyzed reactions.[a]

Reaction	Substrates	Original pool size (nt)	Required (most common) cofactor	First report (year)	References
Cleavage reactions					
RNA cleavage	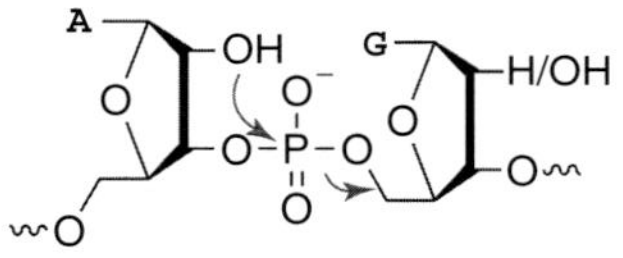				
Embedded rN in DNA	A, OH, O, O^-, O–P–O, O, G, H/OH, O, O	50	Pb^{2+}	1994	[6]
All-RNA substrate		50	Mg^{2+}, Mn^{2+}	1997	[12]
Phosphoramidate cleavage	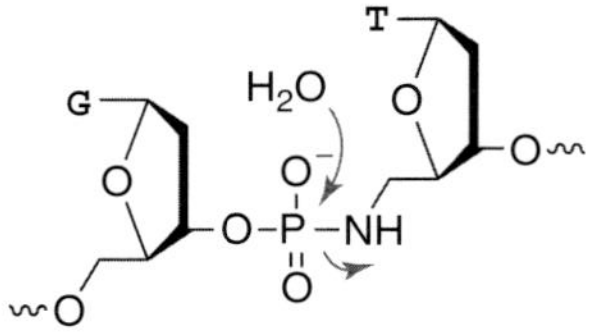 G, O, H_2O, O^-, O–P–NH, O, T, O, O, O	72	Mg^{2+}	1997	[13]
DNA deglycosylation	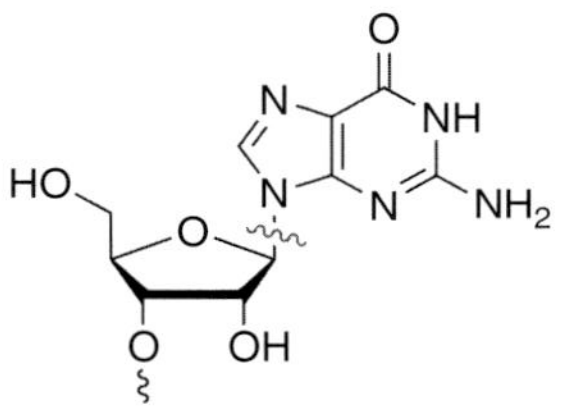				
Internal guanosine	O, N, NH, N, N, NH_2, HO, O, O, OH	85	Ca^{2+}	2000	[14]
5′-Terminal guanosine		70	IO_4^-	2007	[15]

Thymidine dimer photoreversion	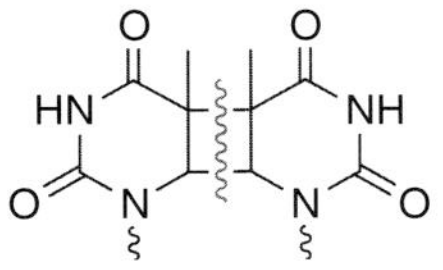	40		2004	[16]
DNA cleavage					
Oxidative	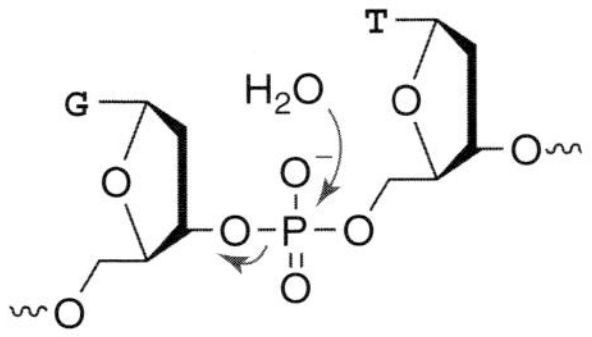	50	Cu^{2+}	1996	[17, 18]
Hydrolytic		40	$Mn^{2+} + Zn^{2+}$	2009	[19]
"Ligation" reactions					
DNA phosphorylation	ATP + 5′-OH-DNA	70	$Ca^{2+}Mn^{2+}$	1999	[20]
DNA adenylation	ATP + 5′-p-DNA	70	$Mn^{2+}Cu^{2+}$	2000	[21]
DNA ligation	3′-p-Imidazolide-DNA + 5′-OH-DNA	116	Zn^{2+}	1995	[22]
	3′-OH-DNA + 5′-App-DNA	150	Mn^{2+}	2004	[23]
RNA ligation (see Sections 6.5 and 6.6)					
2′,5′ Linear	2′,3′-cp-RNA + 5′-OH-RNA	40	Mg^{2+}	2003	[24]
3′,5′-Linear	2′,3′-cp-RNA + 5′-OH-RNA	40	Zn^{2+}	2005	[25]
	3′-OH-RNA + 5′-ppp-RNA	40	Mg^{2+}	2005	[26]
2′,5′-Branched	Internal RNA 2′-OH + 5′-ppp-RNA	40	Mg^{2+}	2004	[27]
NTP ligation	Internal RNA 2′-OH + GTP		Mg^{2+}	2007	[28]

(continued overleaf)

Table 6.1 (*Continued*)

Reaction	Substrates	Originate pool size (nt)	Required (most common) cofactor	First report (year)	References
RNA–DNA ligation	Internal RNA 2′-OH + 5′-App-DNA	22	Mg^{2+}	2007	[29]
Nucleopeptide ligation (see Section 6.7)	Tyrosine-OH + 5′-ppp-RNA	40	Mg^{2+},Mn^{2+}	2008	[30]
Diels–Alder reaction		36	Ca^{2+}	2008	[31]
Other reactions					
Porphyrin metalation	N-Methylmesoporphyrin IX + Cu^{2+}(Zn^{2+})	228	K^+	1996	[32, 33]
Peroxidase reaction	Hemin (Fe(III)protoporphyrin IX) + H_2O_2		K^+	1998	[34]

p = phosphate, cp = cyclic phosphate, ppp = triphosphate, ATP = adenosine triphosphate, App = adenylate, GTP = guanosine triphosphate, NTP = nucleoside triphosphate (see Figure 6.7 for structure).

[a]Only references in which the activity was first reported are given. Therefore this is not a comprehensive listing (reports on other cofactors or incorporation of modified nucleotides have not been considered here).

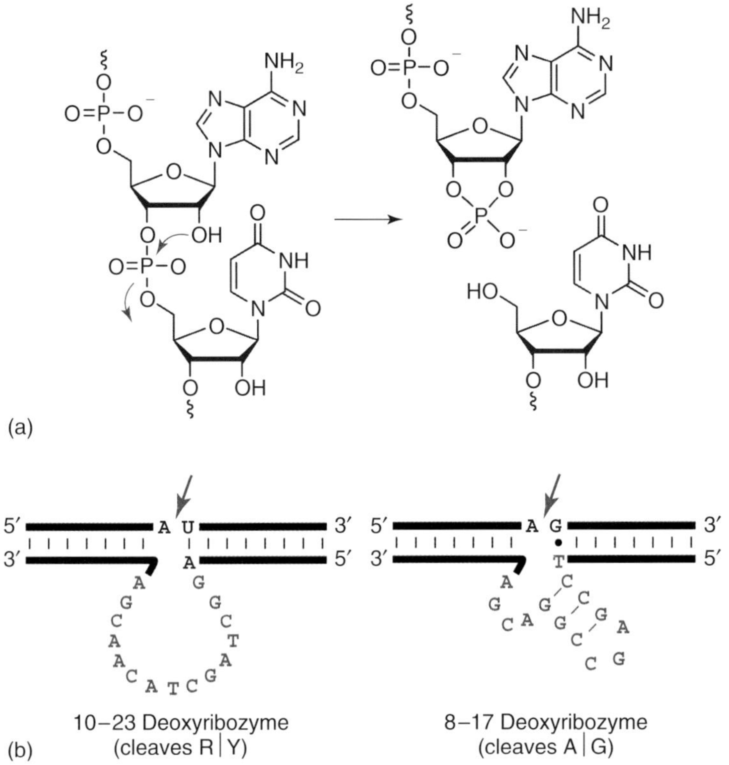

(a)

5′ A U 3′ 5′ A G 3′
3′ A 5′ 3′ T 5′

10–23 Deoxyribozyme
(cleaves R|Y)

8–17 Deoxyribozyme
(cleaves A|G)

(b)

Figure 6.3 (a) DNA-catalyzed RNA cleavage. Nucleophilic attack of the 2′-OH group onto the adjacent phosphodiester bond leads to generation of a 2′,3′-cyclic phosphate and a free 5′-OH fragment. (b) Two prominent examples of RNA-cleaving deoxyribozymes, 10–23 and 8–17, that have short catalytic core sequences and that select their targets via specific Watson–Crick base-pairing. The preferred cleavage sites are indicated.

was cleaved at a defined position in the TΨC-loop by the action of 10–23, thus generating a fully modified truncated tRNA fragment as the key intermediate in the semisynthetic strategy. After generation of a free 2′,3′-diol terminus by enzymatic dephosphorylation, the tRNA fragment was ligated to a synthetic RNA fragment that carried a specific peptide fragment attached to the 3′-end via a nonhydrolyzable amide bond [43–45]. This efficient approach generates important products for biophysical and structural investigations of ribosomal mechanisms.

6.5 DNA-Catalyzed Linear Ligation of RNA

A productive venture in deoxyribozyme research was the development of RNA ligase catalysts that enable the formation of a native 3′,5′-phosphodiester linkage between two RNA termini. Such DNA catalysts could be practical alternatives to protein

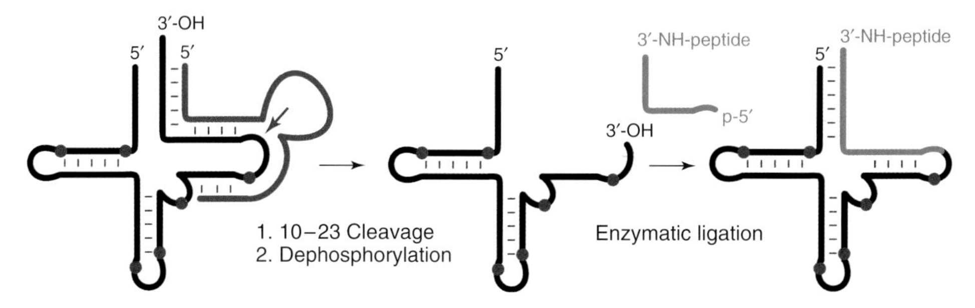

Figure 6.4 Application of RNA-cleaving deoxyribozyme for the synthesis of nonhydrolyzable peptidyl-tRNAs containing all natural nucleobase modifications (indicated as red spheres). Starting from a natural tRNA, the 3′-terminal domain is released by the action of the site-specific 10–23 deoxyribozyme (blue line) that binds to the tRNA substrate and cleaves the phosphodiester bond between two specific loop nucleotides (indicated by blue arrow). After dephosphorylation of the generated 2′,3′-cyclic phosphate, the new 3′-OH terminus of the truncated tRNA can be ligated to a synthetic peptidyl-RNA fragment to generate the full-length peptidyl-tRNA [43].

enzyme-mediated ligation of RNA substrates with T4 DNA or T4 RNA ligase, and facilitate the access to large chemically modified RNAs for structure–function studies.

Silverman and coworkers developed a series of DNA catalysts for this purpose. The initial selection strategy employed RNA substrates with 2′,3′-cyclic phosphate and 5′-hydroxyl termini (Figure 6.5a, reaction 1) in the presence of Mg^{2+} as divalent metal ion cofactor [24]. The desired ring-opening reaction with the 2′-oxygen acting as leaving group is conceptually based on the reversal of the DNA-catalyzed RNA cleavage reaction (compare Figure 6.3a). Several active DNA catalysts were obtained in the initial selection efforts. However, all isolated DNA enzymes generated non-native 2′,5′ rather than the desired 3′,5′ phosphodiester linkages. In this case, the cyclic phosphate was opened with the 3′-oxygen rather than the 2′-oxygen acting as the leaving group (reaction 2 in Figure 6.5a). Several alterations in the selection setup and metal ion conditions resulted in distinct classes of RNA ligase catalysts that synthesize native 3′,5′ linkages by this approach [25]. Interestingly, Zn^{2+} instead of Mg^{2+} was required to achieve the desired regioselectivity. The most recently reported deoxyribozymes that catalyze Zn^{2+}-dependent 3′,5′-RNA ligation via opening of cyclic phosphates, for example, 7CX6 (Figure 6.5c) demonstrate reasonable output of 50–60% ligation product with k_{obs} of up to 0.03 min^{-1} ($t_{1/2}$ of 20–60 min, see Table 6.2) [46]. The ligation yield is considerably below 100%, which is likely due to the inherent reversibility of the ligation reaction. Catalysis of the reverse cleavage reaction has also been demonstrated for Mg^{2+}-dependent RNA ligases that form 2′,5′ linkages [47]. Another limitation of all reported DNA catalysts that use 2′,3′-cyclic phosphates as substrates is their nontrivial sequence requirements near the ligation junction, which comprises at least four nucleotides.

The second class of RNA-ligating deoxyribozymes that create linear 3′,5′ RNA linkages was obtained from unrelated *in vitro* selection experiments

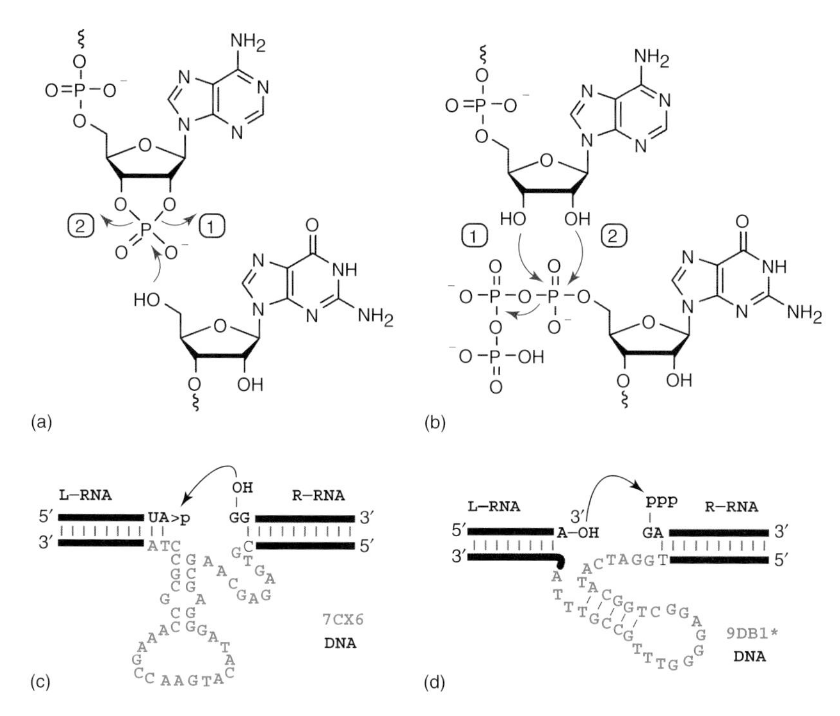

Figure 6.5 Substrate combinations and deoxyribozymes for linear RNA ligation reactions. (a) 5′-OH nucleophile with 2′,3′-cyclic phosphate electrophile. (b) 3′-Terminal ribose as nucleophile and 5′-triphosphate as electrophile. (c, d) Selected deoxyribozymes that catalyze the formation of native 3′,5′ phosphodiester linkages. 7CX6 [46] uses substrate combination (a), and 9DB1 [26] uses substrate combination (b). (*9DB1 was originally reported with a 40 nt core region, but was recently minimized to the depicted shortened version [48] which displays similar catalytic activity.)

that used 5′-triphosphorylated RNA as electrophilic substrate. Unmodified 2′,3′-diol-terminated RNA was used as the nucleophile to displace pyrophosphate as the leaving group (Figure 6.5b). This essentially irreversible ligation reaction was the basis for the selection of DNA catalysts in the presence of Mg^{2+} or Zn^{2+} as metal ion cofactors. To generate deoxyribozymes that use the terminal 3′-OH as the nucleophile (as opposed to the terminal or any internal 2′-OH), a strict selection pressure was applied that allowed only desired RNA products (covalently linked to their DNA catalysts) to survive the selection step. In one setup, the ligation junction was part of a double helical arrangement that prefers the 3′,5′-connection because of the higher stability of the formed RNA:DNA hybrid duplex [49]. The resulting 8AY12 deoxyribozyme indeed is able to create native linkages. However, it requires eight fixed nucleotides around the ligation junction, which prevents general applicability. In a second set of efforts, the selection pressure was imposed by selective cleavage of 3′,5′-RNA linkages by the RNA-cleaving 8–17 deoxyribozyme. The resulting 9DB1 and 7DE5 DNA catalysts are the most generally applicable DNA-based RNA ligases known today. The

Table 6.2 A collection of selected deoxyribozymes for linear ligation of RNA.

Name	Conditions	Ligation junction	k_{obs} (min^{-1})	Generality/sequence requirement	References
2′,3′-Cyclic phosphate and 5′-OH					
9A12	40 mM Mg^{2+}, pH 9.0	2′,5′	0.07	* (Not examined)	[24]
12BB2	1 mM Zn^{2+}, pH 7.5	2′,5′	0.08	* (Not examined)	[25]
12BB5	1 mM Zn^{2+}, pH 7.5	3′,5′	0.02	* (Not examined)	[25]
7CX6	1 mM Zn^{2+}, pH 7.5	3′,5′	0.03	UA|GG	[46]
3′-OH and 5′-ppp					
8AY13	160 mM Mg^{2+}, pH 9.0	3′,5′	0.03	UAUA|GGAA	[49]
9DB1	40 mM Mg^{2+}, pH 9.0	3′,5′	0.06	D|RA	[26]
7DE5	1 mM Zn^{2+}, pH 7.5	3′,5′	0.02	A|R	[26]

ligation reactions achieve practically useful ligation yields of 50–70% after 2–3 h of reaction time (Table 6.2) [26]. Even though 9DB1 and 7DE5 allow reasonable flexibility of substrate sequences, a collection of deoxyribozymes will be needed for the ligation of all possible RNA ligation junction sequences.

6.6
DNA-Catalyzed Synthesis of 2′,5′-Branched Nucleic Acids

The term "*branched nucleic acids*" refers to different types of nucleic acid structures where multiple (three or more) strands emerge from defined branch points. For double-stranded nucleic acids, the (noncovalent) branch point can be the intersection of multiple helices, as for example, formed during natural DNA replication (replication fork) and recombination (Holliday junction) [50]. Nucleic acid nanotechnology is an emerging area where noncovalently branched DNA and RNA have been attracting increasing interest [51, 52]. Noncovalently branched RNA conformations are found as part of complex tertiary structures of natural functional RNAs, such as ribozymes and riboswitches [53]. In the single-stranded form, branched nucleic acids contain a branch point to which individual strands are covalently connected. For branched RNA, the most prominent example is the lariat RNA formed during RNA splicing [54]. At the core of this structure, a branch site adenosine nucleoside (as part of a normal 3′,5′- phosphodiester-linked RNA strand) contains an additional covalent connection via a 2′,5′-phosphodiester bond to the ribose moiety (Figure 6.6a). In the case of lariat RNA, this structure originates from a single piece of RNA (i.e., from an intramolecular branching reaction). The analogous architecture that originates from two individual RNA substrates is named 2′,5′-branched RNA. Generally, the two oligonucleotides forming this 2′,5′-branched nucleic acid structure, the vertically drawn scaffold strand (also named primary strand, target strand, or L strand) and the horizontally drawn

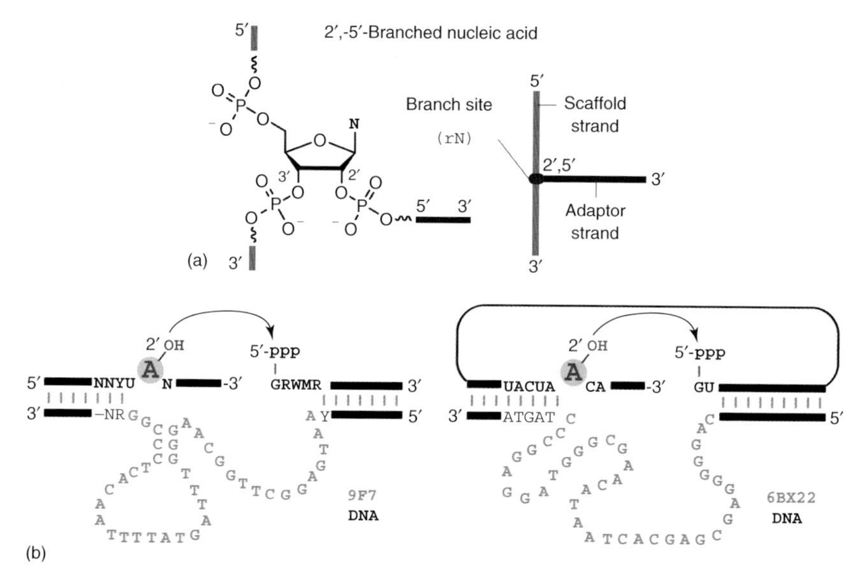

Figure 6.6 (a) 2′,5′-Branched nucleic acids contain a core ribonucleotide, at which the adaptor strand is connected to the scaffold strand via a 2′,5′-phosphodiester bond. (b) The 9F7 [55] and 6BX22 [56] deoxyribozymes synthesize branched or lariat RNA.

adaptor strand (also named secondary strand, tagging strand, or R strand), can be composed of RNA or DNA. This results in four possible combinations, with the scaffold and adaptor strands being DNA or RNA (see Figure 6.7). In the case of 2′,5′-branched DNA, only the branch site nucleotide is a ribonucleotide that contains a 2′-OH group.

Covalently branched nucleic acids have previously been prepared by multistep chemical syntheses. For branched RNA, the synthesis is very challenging, and only small constructs of a few nucleotides can be obtained [57, 58]. The bottom-up chemical synthesis of biologically relevant lariat RNA has not been achieved by solid-phase synthesis. For branched DNA, several strategies involving sophisticated protecting group strategies have been reported. Most approaches involve connection of the DNA strands via non-nucleosidic linker units [59–61], or the secondary strands are attached via the amino groups of the nucleobase at the branch site, such as branch point cytidines [62, 63].

Deoxyribozymes offer a unique opportunity for the synthesis of covalently 2′,5′-branched nucleic acids, for all four combinations of DNA and RNA scaffold and adaptor strands. The different combinations have been explored at different levels of detail. Prototypical deoxyribozymes for all four combinations are depicted in Figure 6.7. Table 6.3 summarizes typical properties of branch-forming deoxyribozymes. Below, we highlight selected achievements and potential applications for DNA-catalyzed synthesis of branched nucleic acids.

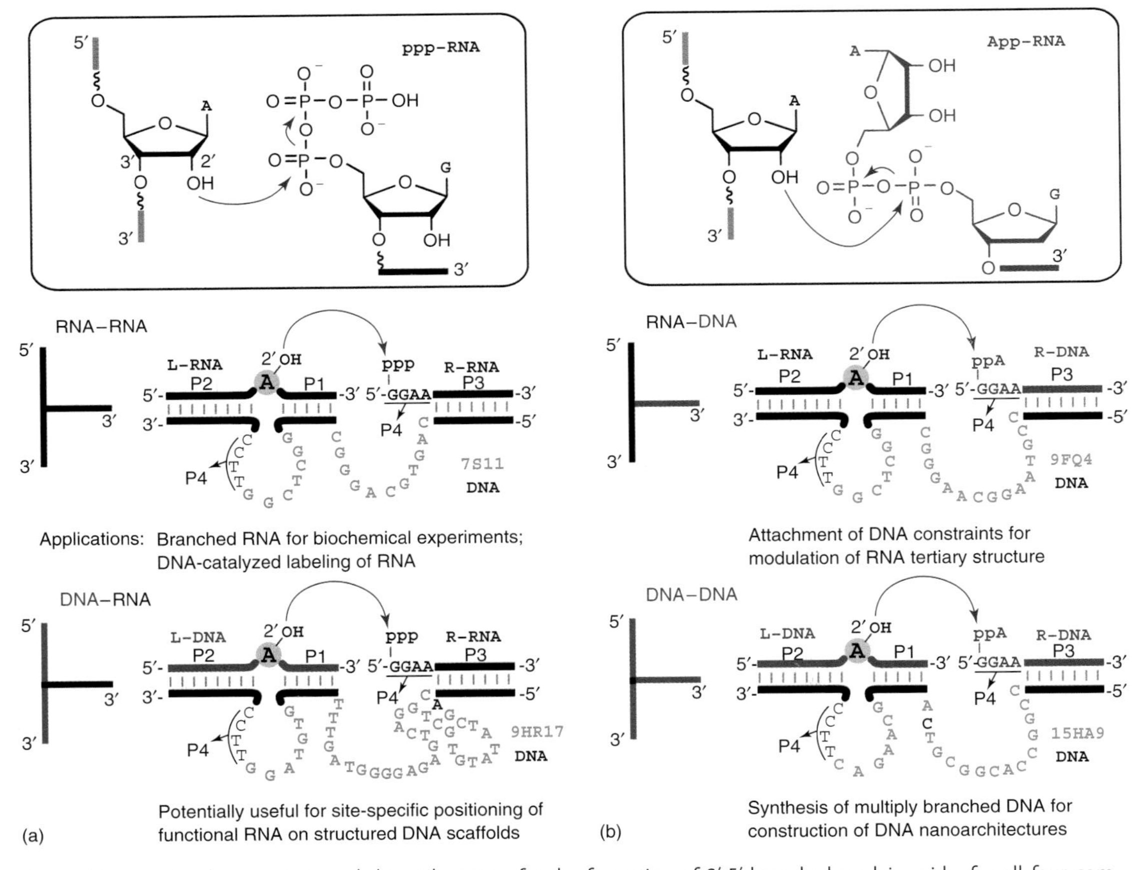

Figure 6.7 Depicted are prototypical deoxyribozymes for the formation of 2′,5′-branched nucleic acids, for all four combinations of DNA and RNA for L- and R-substrates. For RNA–RNA and DNA–DNA branching reactions, more general variants have been reported recently [64, 65].

Table 6.3 A collection of deoxyribozymes for synthesis of 2′,5′-branched nucleic acids.

Name	Metal ion, pH	Scaffold/ target strand	Accessory/ tagging strand	k_{obs} (min^{-1})	Generality/ sequence requirement	References
9F7	20 mM Mn^{2+}, pH 7.5 or	RNA	RNA	0.81	YUAN(br. site)|GRMWR	[55]
	80 mM Mg^{2+}, pH 7.5	RNA	RNA	0.0043	(as above)	[55]
6BX22	20 mM Mn^{2+}, pH 7.5	RNA	RNA	0.08	A > U > C(br. site)|G (lariat)	[56]
6CE8	20 mM Mn^{2+}, pH 7.5	RNA	RNA	0.01	U > C ~ A >G(br. site)|GA	[66]
7S11	40 mM Mg^{2+}, pH 9.0 or	RNA	RNA	0.56	A(br. site)|R	[27]
	20 mM Mn^{2+}, pH 7.5	RNA	RNA	0.5	A(br. site)|R	[67]
10DM24	40 mM Mg^{2+}, pH 9.0	RNA	RNA	0.26	A > G > U >C(br. site)|R	[64]
9FQ4	40 mM Mg^{2+}, pH 9.0	RNA	DNA	not reported	A(br. site)|R	[29]
15HA9	40 mM Mg^{2+}, pH 9.0	DNA	DNA	0.0015	r(U > A ≫ G > C)(br. site)|dN	[68]
8LV13	20 mM Mn^{2+}, pH 7.5	DNA	DNA	0.14	dGrNdC(br. site)|d(GGAAGA)	[65]
9HR17	40 mM Mg^{2+}, pH 9.0	DNA	RNA	0.01	(Not enough data)	[30]

6.6.1
2′,5′-Branched RNA

Deoxyribozymes that synthesize 2′,5′-branched RNA were the first class of DNA catalysts described for the synthesis of covalently branched nucleic acids. Such deoxyribozymes were discovered serendipitously during *in vitro* selection experiments that aimed at the selection of RNA ligase catalysts. The substrates were a 2′,3′-hydroxyl-terminated RNA fragment (the L-RNA) and a 5′-triphosphorylated RNA substrate (the R-RNA). Interestingly, the initial set of experiments resulted in an additional class of deoxyribozymes, which specifically use an internal 2′-OH group as a nucleophile instead of the terminal hydroxyl group [55]. Silverman and coworkers realized the great potential offered by these new deoxyribozymes, and undertook further selection efforts that targeted the synthesis of branched and lariat RNAs. Two examples are depicted in Figure 6.6b. The 9F7 deoxyribozyme is one of the fastest RNA-ligating deoxyribozymes known today and achieves excellent ligation yield with k_{obs} of 0.8 min^{-1} in the presence of 20 mM Mn^{2+} [69]. With Mg^{2+} as the metal ion cofactor, the ligation reaction is about 200-fold slower (same pH, four times higher divalent ion concentration). However, the sequence requirement around the ligation junction is considerably large. The 6BX22 deoxyribozyme requires Mn^{2+} as obligatory cofactor for the intramolecular ligation of the branch site 2′-OH to the 5′-ppp of a single RNA substrate. This DNA catalyst has been successfully used for the synthesis of biologically relevant lariat RNAs of various lengths ranging from 50 to 300 loop nucleotides [56]. The sequence requirements in the RNA substrates around the ligation junction have been carefully characterized, but the specific nucleotides and potential interactions of the DNA sequences have not been investigated for any of these practically useful DNA catalysts. Nevertheless, the extremely high regioselectivities of the DNA-catalyzed reactions enable the efficient synthesis of products that would be very challenging or impossible to prepare by alternative methods. DNA-catalyzed synthesis of branched and lariat RNAs has already enabled biochemical experiments to address biologically relevant questions in the context of group II intron splicing [70] and retrotransposition [71]. A better understanding of the minimal sequence requirements in the catalytic core region of the DNA catalysts is expected to contribute to the development of improved reagents with finely tuned properties for the synthesis of complex branched RNA targets.

A second class of RNA-ligating deoxyribozymes is exemplified by 7S11 [27] in Figure 6.7a. This and related DNA catalysts form a three-helix junction (3HJ) structure with their RNA substrates [67]. The branch site is selected by Watson–Crick base-pairing of the DNA enzyme to the flanking sequences in the L-RNA substrate, forming paired regions P1 and P2, that is, the branch site nucleotide is the only non-base-paired (bulged) nucleotide in the L-RNA substrate. The R-RNA substrate forms P3 with the R-binding arm of the DNA. The first four nucleotides of the R-RNA substrate hybridize to the complementary region in the DNA loop, thus forming P4, which draws the nucleophile and the electrophile into close proximity. The ligation reactions by 7S11 and the related deoxyribozyme 10DM24 [64] are highly efficient with k_{obs} values up to of 0.6 min^{-1} in the presence of Mg^{2+} or Mn^{2+} as metal ion cofactors.

The current knowledge of the ligation mechanism is extremely limited. RNA-ligating DNA catalysts have not been characterized in great detail by structural, biochemical, or biophysical methods. Nevertheless, the 10DM24 deoxyribozyme has found interesting applications for deoxyribozyme-catalyzed labeling of RNA [72] and for the ligation of single nucleotides to the RNA branch site [28]. The concept of DNA-catalyzed site-specific modification of RNA with single nucleotides is now further explored with chemically modified nucleotide analogs. This enables the installation of unique functional groups for the conjugation of spectroscopic probes to RNA (L. Büttner and C. Höbartner, unpublished results). The general architecture of 3HJ-forming deoxyribozymes also formed the basis for the development of all other combinations of 2′,5′-branch-forming DNA catalysts (see below).

6.6.2
2′,5′-Branched Nucleic Acids Containing RNA as Scaffold and DNA as Adaptor

The combination of RNA as the scaffold strand with DNA as adaptor strand can be achieved with deoxyribozymes that use 5′-adenylated DNA as the activated R-substrate (Figure 6.7b). In this case, adenosine 5′-phosphate is released upon formation of the new phosphodiester bond in the branching reaction. In contrast to 5′-triphosphorylated RNA, which is easily obtained by *in vitro* transcription, the 5′-activated DNA substrates are more difficult to prepare, as there is no direct enzymatic method available for the installation of 5′-triphosphate groups on DNA. The chemical synthesis of 5′-triphosphorylated oligonucleotides is experimentally challenging, although optimized methods have been reported recently [73–75]. The preparation of 5′-adenylated oligonculeotide substrates can be achieved by enzymatic methods that use T4 DNA or T4 RNA ligase and ATP. The mechanism of enzymatic ligation by these enzymes involves activation of the 5′-phosphorylated donor substrate by adenylation. In the absence of any competent acceptor substrate, this activated intermediate can be isolated [73]. Chemical methods for the synthesis of 5′-adenylated substrates have also been reported [74], as adenylated adaptor oligonucleotides are gaining increasing importance for deep sequencing protocols [75].

All reported *in vitro* selection experiments for deoxyribozymes with activated DNA as R-substrate for the synthesis of branched nucleic acids used enzymatically generated 5′-adenylated DNA as substrate. In combination with RNA as the L-substrate, this resulted in the identification of 9FQ4 [29] as Mg^{2+}-dependent DNA catalyst that enables the efficient attachment of DNA adaptor strands at defined positions in an RNA substrate.

The 9FQ4 deoxyribozyme has been used to attach DNA oligonucleotides to specific adenosine positions in functional RNA molecules. The DNA pieces served as efficient constraints for the modulation of catalytic activities of the hammerhead [29] and the group I intron ribozymes [76]. The concept of using DNA constraints for the modulation of RNA tertiary structures had been demonstrated earlier by the attachment of DNA constraints by chemical means [77, 78]. The application of deoxyribozymes for this purpose facilitates the experimental approach because the target RNA can be provided as a single large substrate, and tedious RNA ligation steps can be avoided. With the help of appropriate disruptor oligonucleotides that break RNA secondary structures, the accessibility of RNA sequences presenting the branch site nucleotides can be achieved.

6.6.3 2′,5′-Branched DNA

The second implementation of 5′-adenylated DNA substrates in DNA-catalyzed synthesis of 2′,5′-branched nucleic acids enables the preparation of covalently branched DNA (Figure 6.7c). In this setup, the scaffold strand is a chemically modified DNA that provides single internal ribonucleotides as branch site nucleotides. The first deoxyribozyme that synthesized branched DNA was 15HA9 with reported k_{obs} of only 1.5×10^{-3} min^{-1} [68]. Recently, improved variants with much faster ligation kinetics were reported that use Mn^{2+} as metal ion cofactor instead of Mg^{2+}. The 8LV13 [65] deoxyribozyme (Table 6.3) originated from a selection experiment with 40 randomized positions in the DNA (as opposed to only 22 randomized positions in the HA selection). The contribution of metal ion and loop size to the efficiency of DNA-catalyzed DNA branching reaction remains to be determined. Nevertheless, the available catalysts enable efficient syntheses and, in contrast to RNA-branching deoxyribozymes, do not show pronounced preference for branch site adenosine. It is expected that the 8LV13 and related DNA enzymes will find applications for the synthesis of multiply branched DNAs that are potentially interesting building blocks for complex DNA nanoarchitectures.

6.6.4
2′,5′-Branched Nucleic Acids Containing DNA as Scaffold and RNA as "Adaptor"

The fourth combination of branched DNA–RNA in which DNA is used as the scaffold strand to which RNA is attached via a 2′,5′-phosphodiester bond is so far the least investigated combination. The 9HR17 deoxyribozyme originates from an N_{40} pool and has not been optimized in any way [30]. (It is fair to say that this was not the intention of the study in which 9HR17 was identified.) Nevertheless, this activity should be mentioned here for completeness. In addition, we would like to put forward the idea of utilizing such deoxyribozymes for the site-specific positioning of functional RNA molecules on DNA scaffolds. This could be an interesting application for the future design of functionalized DNA architectures.

6.7
DNA-Catalyzed Synthesis of Nucleopeptide Conjugates

The formation of a new phosphodiester bond between the phenolic OH group of a tyrosine side chain and the 5′-terminus of an RNA oligonucleotide was the first example that demonstrated the capability of DNA to covalently modify amino acid side chains (Figure 6.8) [30]. The Tyr1 deoxyribozyme used a tyrosine amino acid embedded in a DNA oligonucleotide, which provided binding arms for the DNA catalyst. This design enabled the favorable positioning of the reactive nucleophile at the center of a 3HJ architecture, which is similar to the RNA-branch-forming deoxyribozymes described in Section 6.6. The *in vitro* selection started from a random pool of 40 nucleotides that were split into two single-stranded loops (of 7 and 33 nucleotides) by the fixed DNA sequences of the substrate-binding arms in

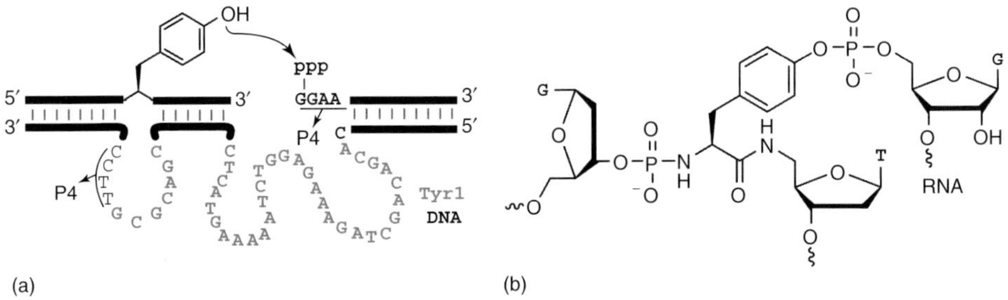

Figure 6.8 DNA-catalyzed nucleopeptide synthesis. (a) Schematic of the Tyr1 deoxyribozyme in complex with its substrates [30]. (b) Chemical structure of the product showing the product phosphodiester linkage of the Tyr phenolic OH group to the 5′-end of RNA.

the paired regions P1, P2, and P3. The Tyr1 deoxyribozyme was found to catalyze the tyrosine-RNA ligation with k_{obs} of 3.6 h^{-1} in the presence of 20 mM Mn^{2+} at pH 7.5 and 37 °C.

The second generation of deoxyribozymes catalyzing nucleopeptide bond formation use tripeptides instead of single amino acids at the center of the 3HJ architecture [79]. New *in vitro* selection experiments with tyrosine and serine at the central position of the tripeptide yielded DNA catalysts TyrB1 and SerB1 which catalyze tyrosine-RNA ligation with k_{obs} of 0.30 and 0.24 h^{-1}, respectively, in the presence of 20 mM Mn^{2+} and 40 mM Mg^{2+} at pH 7.5 and 37 °C. Interestingly, the Tyr1 and TyrB1 deoxyribozymes show much higher substrate selectivity for tyrosine over serine (350- and 600-fold) compared to the SerB1 deoxyribozyme, which ligates serine only 12 times faster than tyrosine. Although the substrate selectivity of SerB1 and some related serine-ligating deoxyribozyme is still suboptimal, the results for tyrosine and serine incorporation demonstrate the general ability of DNA to catalyze reactions of amino acid side chains. Future challenges in this area are the expansion of DNA catalysis to other amino acids such as threonine or lysine and to train deoxyribozymes to catalyze reactions of larger peptide and protein substrates.

6.8 Mechanistic Aspects of DNA Catalysis

The catalytic strategies employed by DNA have not yet been characterized in great detail. Although the number of different functional groups in DNA is limited by the small number of four types of deoxyribonucleotides, remarkable catalytic ability has been demonstrated with rate enhancements up to 10^{12} over the uncatalyzed background reaction [19, 80]. Most deoxyribozymes depend on the presence of divalent metal ions as "cofactors" for catalytic activity. The exact binding sites have not been identified, but it is plausible that the negatively charged phosphate backbone as well as the nucleobases (via nitrogen and oxygen groups) provide coordination sites for divalent metal ions, which are likely involved in the formation of complex tertiary structures. Additionally, coordinated metal ions could participate in the catalytic mechanisms by acting as Lewis acids. A properly positioned metal hydroxide could function as a general base during DNA catalysis. However, the detailed catalytic mechanism of any deoxyribozyme-mediated reaction is not yet known. Moreover, and despite considerable efforts, the three-dimensional structure of any deoxyribozyme in an active conformation could not yet be determined at atomic resolution [81]. Nevertheless, the best characterized class of DNA catalysts comprises the RNA-cleaving deoxyribozymes, in particular 8–17 and 10–23 (see Figure 6.3, for detailed reviews see [35, 82]). Comprehensive mutation and nucleobase modification analyses have been performed [83, 84], and nucleotide deletion mutants [85, 86] and phosphorothioate derivatives [87] have been investigated for their catalytic behavior. The global folding of RNA-cleaving deoxyribozymes has been studied by fluorescence spectroscopy, in ensemble [88, 89] and single-molecule experiments

[90]. These studies revealed that deoxyribozymes can use different metal ions to exploit various modes of folding and substrate activation. Other spectroscopic studies investigated metal ion binding by Tb^{3+} luminescence [91] or circular dichroism (CD) spectroscopy [92]. The analysis of electron hole flow patterns in the catalytic core of the 8–17 deoxyribozymes explored additional aspects of DNA folding [93]. Local interactions between substrate and deoxyribozyme nucleotides have been investigated by photocrosslinking experiments [94, 95].

Other classes of deoxyribozymes have not yet been investigated in so much detail. The first step toward understanding DNA folding and catalysis is the identification of catalytically essential nucleotides and their functional groups that define the deoxyribozyme active sites. We have recently reported a combinatorial approach to mutation interference analysis (CoMA) [48] that serves as a general tool for the

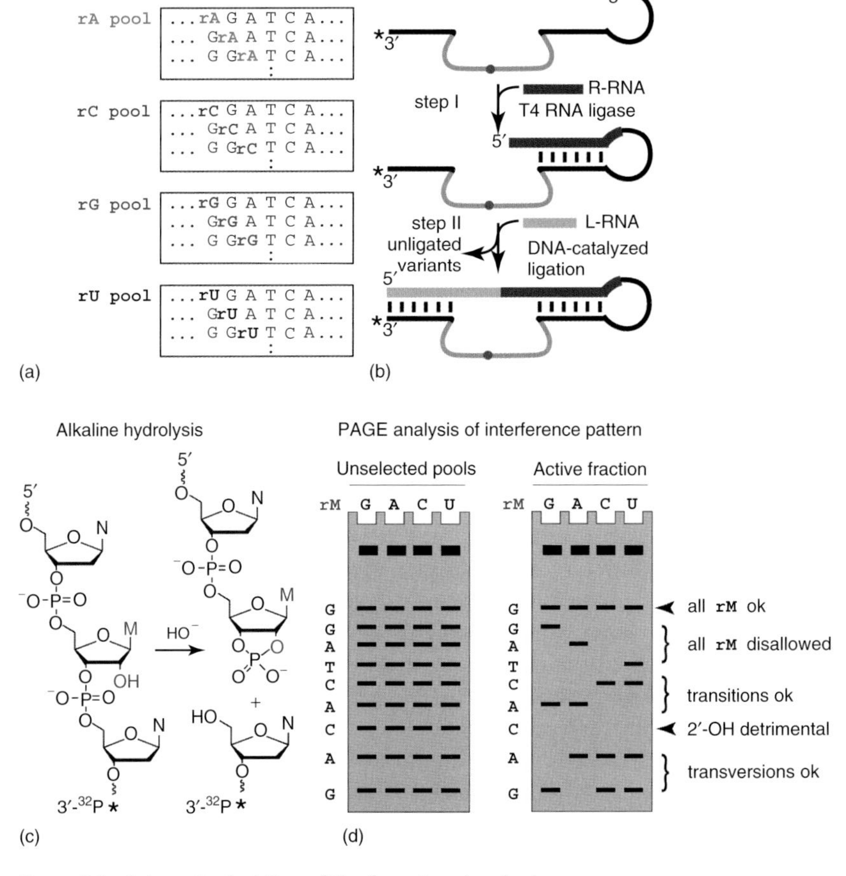

Figure 6.9 Schematic depiction of the four steps involved in combinatorial mutation interference analysis (CoMA).

characterization of single-stranded DNA, that is, it is not limited to deoxyribozymes but can be applied to any single-stranded DNA that has a selectable function, such as binding small or large molecules (i.e., aptamers). This method involves solid-phase synthesis of mutant DNA libraries, in which the nucleobase mutations (or functional group modifications) are encoded by ribose 2′-OH groups that serve as chemical tags for analysis. This strategy allows direct mapping of the sites of mutation interference by specific cleavage of the phosphodiester backbone at the site of mutation. The workflow consists of four steps: (a) solid-phase synthesis of 2′-OH encoded libraries, (b) separation of active from inactive mutants, (c) specific backbone cleavage by alkaline hydrolysis, and (d) analysis of interference pattern by denaturing polyacrylamide gel electrophoresis (Figure 6.9). Hydrolysis bands that are missing in the active fraction of the deoxyribozyme library (compared to the unselected library) indicate nucleotide positions that do not tolerate any mutation and are therefore essential for catalytic activity.

The design of our method for combinatorial mutation analysis of DNA is conceptually related to nucleotide analog interference mapping (NAIM) that is applied for studying functional RNA [96]. NAIM is based on the enzymatic preparation of modified RNA using phosphorothioate-tagged triphosphates that are statistically incorporated by *in vitro* transcription. In sharp contrast, our approach for combinatorial mutation interference analysis is entirely based on chemical solid-phase synthesis, which renders this method completely independent of template-dependent polymerase enzymes and allows us to address mismatch

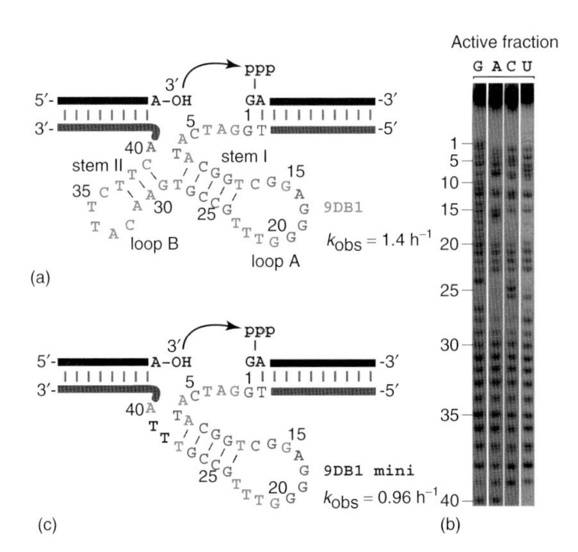

Figure 6.10 Combinatorial mutation interference analysis guided minimization of 9DB1 deoxyribozyme. (a) original 9DB1 with substrates; essential nucleotides are colored red, mutable positions are colored green; (b) interference pattern of active fraction of CoMA pools, analyzed by denaturing PAGE; and (c) minimized 9DB1 with substrates.

mutations that cannot be incorporated enzymatically. Moreover, we can investigate functional group changes that interfere with Watson–Crick base-pairing (e.g., m^1G or m^3U, Wachowius and Höbartner, unpublished results).

Combinatorial mutation interference analysis was applied to a variety of RNA-ligating deoxyribozymes, and was shown to be an extremely efficient tool to facilitate minimization of the deoxyribozyme sequence to the essential core nucleotides. As example, the results for the RNA-ligating 9DB1 deoxyribozyme are depicted in Figure 6.10. The original sequence is drawn on top in the form of a predicted secondary structure (by mfold [97]). The interference pattern of the active fraction suggests that a stretch of nucleotides between position 29 and 41 is not required for catalysis, since all mutations arc tolerated in this region. This analysis suggested the truncated version of the deoxyribozyme, denoted 9DB1 mini [48]. The core sequence has been shortened by nine nucleotides, without considerably affecting the ligation activity. Similarly, we have shown that the RNA-branch-forming deoxyribozymes 9F7 and 6BX22 can be shortened to about 60–70% of their original length (F. Javadi-Zarnaghi and C. Höbartner, unpublished results, see Figure 6.6 for original deoxyribozyme sequences). The contributions of individual functional groups of the remaining guanine-rich sequences are currently investigated by a combination of chemical and biochemical methods.

6.9
Conclusions and Outlook

DNA catalysts can considerably facilitate synthetic access to complex nucleic acid structures that are difficult to obtain by other approaches. In addition, cleaving and ligating deoxyribozymes have found diverse applications in various areas of research that are beyond the scope of this chapter, but have been summarized in recent reviews [7, 39, 98]. Although the catalytic mechanisms of DNA catalysis are not yet well understood, DNA enzymes serve as useful tools in biomolecular synthetic chemistry. In particular, we expect that RNA-ligating deoxyribozymes will find a broader range of applications, as they are further developed into convenient tools for the synthesis of chemically modified RNA.

References

1. Gesteland, R.F., Cech, T.R., and Atkins, J.F. (2006) *The RNA World*, 3rd edn, CSHL Press.
2. Kruger, K., Grabowski, P.J., Zaug, A.J., Sands, J., Gottschling, D.E., and Cech, T.R. (1982) Self-splicing RNA: autoexcision and autocyclization of the ribosomal RNA intervening sequence of Tetrahymena. *Cell*, **31**, 147–157.
3. Nissen, P., Hansen, J., Ban, N., Moore, P.B., and Steitz, T.A. (2000) The structural basis of ribosome activity in peptide bond synthesis. *Science*, **289**, 920–930.
4. Silverman, S.K. (2009) in *Wiley Encyclopedia of Chemical Biology*, vol. 3 (ed. T.P. Begley), John Wiley & Sons, Inc. pp. 450–466.

5. Cruz, J.A. and Westhof, E. (2009) The dynamic landscapes of RNA architecture. *Cell*, **136**, 604–609.
6. Breaker, R.R. and Joyce, G.F. (1994) A DNA enzyme that cleaves RNA. *Chem. Biol.*, **1**, 223–229.
7. Silverman, S.K. (2010) DNA as a versatile chemical component for catalysis, encoding, and stereocontrol. *Angew. Chem. Int. Ed.*, **49**, 7180–7201.
8. Boersma, A.J., Megens, R.P., Feringa, B.L., and Roelfes, G. (2010) DNA-based asymmetric catalysis. *Chem. Soc. Rev.*, **39**, 2083–2092.
9. Kanan, M.W., Rozenman, M.M., Sakurai, K., Snyder, T.M., and Liu, D.R. (2004) Reaction discovery enabled by DNA-templated synthesis and in vitro selection. *Nature*, **431**, 545–549.
10. Zhang, D.Y. and Seelig, G. (2011) Dynamic DNA nanotechnology using strand-displacement reactions. *Nat. Chem.*, **3**, 103–113.
11. Joyce, G.F. (2004) Directed evolution of nucleic acid enzymes. *Annu. Rev. Biochem.*, **73**, 791–836.
12. Santoro, S.W. and Joyce, G.F. (1997) A general purpose RNA-cleaving DNA enzyme. *Proc. Natl. Acad. Sci. U.S.A.*, **94**, 4262–4266.
13. Burmeister, J., von Kiedrowski, G., and Ellington, A.D. (1997) Cofactor-assisted self-cleavage in DNA Libraries with a 3′-5′ Phosphoramidate bond. *Angew. Chem. Int. Ed.*, **36**, 1321–1324.
14. Sheppard, T.L., Ordoukhanian, P., and Joyce, G.F. (2000) A DNA enzyme with *N*-glycosylase activity. *Proc. Natl. Acad. Sci. U.S.A.*, **97**, 7802–7807.
15. Höbartner, C., Pradeepkumar, P.I., and Silverman, S.K. (2007) Site-selective depurination by a periodate-dependent deoxyribozyme. *Chem. Commun.*, 2255–2257.
16. Chinnapen, D.J. and Sen, D. (2004) A deoxyribozyme that harnesses light to repair thymine dimers in DNA. *Proc. Natl. Acad. Sci. U.S.A.*, **101**, 65–69.
17. Carmi, N., Shultz, L.A., and Breaker, R.R. (1996) *In vitro* selection of self-cleaving DNAs. *Chem. Biol.*, **3**, 1039–1046.
18. Carmi, N., Balkhi, S.R., and Breaker, R.R. (1998) Cleaving DNA with DNA. *Proc. Natl. Acad. Sci. U.S.A.*, **95**, 2233–2237.
19. Chandra, M., Sachdeva, A., and Silverman, S.K. (2009) DNA-catalyzed sequence-specific hydrolysis of DNA. *Nat. Chem. Biol.*, **5**, 718–720.
20. Li, Y. and Breaker, R.R. (1999) Phosphorylating DNA with DNA. *Proc. Natl. Acad. Sci. U.S.A.*, **96**, 2746–2751.
21. Li, Y., Liu, Y., and Breaker, R.R. (2000) Capping DNA with DNA. *Biochemistry*, **39**, 3106–3114.
22. Cuenoud, B. and Szostak, J.W. (1995) A DNA metalloenzyme with DNA ligase activity. *Nature*, **375**, 611–614.
23. Sreedhara, A., Li, Y., and Breaker, R.R. (2004) Ligating DNA with DNA. *J. Am. Chem. Soc.*, **126**, 3454–3460.
24. Flynn-Charlebois, A., Wang, Y., Prior, T.K., Rashid, I., Hoadley, K.A., Coppins, R.L., Wolf, A.C., and Silverman, S.K. (2003) Deoxyribozymes with 2′-5′ RNA ligase activity. *J. Am. Chem. Soc.*, **125**, 2444–2454.
25. Hoadley, K.A., Purtha, W.E., Wolf, A.C., Flynn-Charlebois, A., and Silverman, S.K. (2005) Zn2+-dependent deoxyribozymes that form natural and unnatural RNA linkages. *Biochemistry*, **44**, 9217–9231.
26. Purtha, W.E., Coppins, R.L., Smalley, M.K., and Silverman, S.K. (2005) General deoxyribozyme-catalyzed synthesis of native 3′-5′ RNA linkages. *J. Am. Chem. Soc.*, **127**, 13124–13125.
27. Coppins, R.L. and Silverman, S.K. (2004) A DNA enzyme that mimics the first step of RNA splicing. *Nat. Struct. Mol. Biol.*, **11**, 270–274.
28. Höbartner, C. and Silverman, S.K. (2007) Engineering a selective small-molecule substrate binding site into a deoxyribozyme. *Angew. Chem. Int. Ed.*, **46**, 7420–7424.
29. Zelin, E. and Silverman, S.K. (2007) Allosteric control of ribozyme catalysis by using DNA constraints. *ChemBioChem*, **8**, 1907–1911.
30. Pradeepkumar, P.I., Höbartner, C., Baum, D.A., and Silverman, S.K. (2008) DNA-catalyzed formation of nucleopeptide linkages. *Angew. Chem. Int. Ed.*, **47**, 1753–1757.

31. Chandra, M. and Silverman, S.K. (2008) DNA and RNA can be equally efficient catalysts for carbon-carbon bond formation. *J. Am. Chem. Soc.*, **130**, 2936–2937.
32. Li, Y. and Sen, D. (1996) A catalytic DNA for porphyrin metallation. *Nat. Struct. Biol.*, **3**, 743–747.
33. Li, Y. and Sen, D. (1997) Toward an efficient DNAzyme. *Biochemistry*, **36**, 5589–5599.
34. Travascio, P., Li, Y., and Sen, D. (1998) DNA-enhanced peroxidase activity of a DNA-aptamer-hemin complex. *Chem. Biol.*, **5**, 505–517.
35. Schlosser, K. and Li, Y. (2010) A versatile endoribonuclease mimic made of DNA: characteristics and applications of the 8-17 RNA-cleaving DNAzyme. *ChemBioChem*, **11**, 866–879.
36. Cruz, R.P.G., Withers, J.B., and Li, Y. (2004) Dinucleotide junction cleavage versatility of 8-17 deoxyribozyme. *Chem. Biol.*, **11**, 57–67.
37. Schlosser, K., Gu, J., Lam, J.C., and Li, Y. (2008) In vitro selection of small RNA-cleaving deoxyribozymes that cleave pyrimidine-pyrimidine junctions. *Nucleic Acids Res.*, **36**, 4768–4777.
38. Silverman, S.K. (2005) In vitro selection, characterization, and application of deoxyribozymes that cleave RNA. *Nucleic Acids Res.*, **33**, 6151–6163.
39. Schlosser, K. and Li, Y. (2009) Biologically inspired synthetic enzymes made from DNA. *Chem. Biol.*, **16**, 311–322.
40. Baum, D.A. and Silverman, S.K. (2008) Deoxyribozymes: useful DNA catalysts in vitro and in vivo. *Cell. Mol. Life. Sci.*, **65**, 2156–2174.
41. Meusburger, M., Hengesbach, M., and Helm, M. (2011) A post-labeling approach for the characterization and quantification of RNA modifications based on site-directed cleavage by DNAzymes. *Methods Mol. Biol.*, **718**, 259–270.
42. Pyle, A.M., Chu, V.T., Jankowsky, E., and Boudvillain, M. (2000) Using DNAzymes to cut, process, and map RNA molecules for structural studies or modification. *Methods Enzymol.*, **317**, 140–146.
43. Graber, D., Moroder, H., Steger, J., Trappl, K., Polacek, N., and Micura, R. (2010) Reliable semi-synthesis of hydrolysis-resistant 3′-peptidyl-tRNA conjugates containing genuine tRNA modifications. *Nucleic Acids Res.*, **38**, 6796–6802.
44. Steger, J., Graber, D., Moroder, H., Geiermann, A.S., Aigner, M., and Micura, R. (2010) Efficient access to nonhydrolyzable initiator tRNA based on the synthesis of 3′-azido-3′-deoxyadenosine RNA. *Angew. Chem. Int. Ed.*, **49**, 7470–7472.
45. Moroder, H., Steger, J., Graber, D., Fauster, K., Trappl, K., Marquez, V., Polacek, N., Wilson, D.N., and Micura, R. (2009) Non-hydrolyzable RNA-peptide conjugates: a powerful advance in the synthesis of mimics for 3′-peptidyl tRNA termini. *Angew. Chem. Int. Ed.*, **48**, 4056–4060.
46. Kost, D.M., Gerdt, J.P., Pradeepkumar, P.I., and Silverman, S.K. (2008) Controlling the direction of site-selectivity and regioselectivity in RNA ligation by Zn2+-dependent deoxyribozymes that use 2′,3′-cyclic phosphate RNA substrates. *Org. Biomol. Chem.*, **6**, 4391–4398.
47. Flynn-Charlebois, A., Prior, T.K., Hoadley, K.A., and Silverman, S.K. (2003) In vitro evolution of an RNA-cleaving DNA enzyme into an RNA ligase switches the selectivity from 3′-5′ to 2′-5′. *J. Am. Chem. Soc.*, **125**, 5346–5350.
48. Wachowius, F., Javadi-Zarnaghi, F., and Höbartner, C. (2010) Combinatorial mutation interference analysis reveals functional nucleotides required for DNA catalysis. *Angew. Chem. Int. Ed.*, **49**, 8504–8508.
49. Coppins, R.L. and Silverman, S.K. (2004) Rational modification of a selection strategy leads to deoxyribozymes that create native 3′-5′ RNA linkages. *J. Am. Chem. Soc.*, **126**, 16426–16432.
50. Lilley, D.M. (2008) Analysis of branched nucleic acid structure using comparative gel electrophoresis. *Q. Rev. Biophys.*, **41**, 1–39.
51. Seeman, N.C. (2010) Nanomaterials based on DNA. *Annu. Rev. Biochem.*, **79**, 65–87.

52. Guo, P. (2010) The emerging field of RNA nanotechnology. *Nat. Nanotechnol.*, **5**, 833–842.
53. Serganov, A. and Patel, D.J. (2007) Ribozymes, riboswitches and beyond: regulation of gene expression without proteins. *Nat. Rev. Genet.*, **8**, 776–790.
54. Padgett, R.A., Konarska, M.M., Grabowski, P.J., Hardy, S.F., and Sharp, P.A. (1984) Lariat RNA's as intermediates and products in the splicing of messenger RNA precursors. *Science*, **225**, 898–903.
55. Wang, Y. and Silverman, S.K. (2003) Deoxyribozymes that synthesize branched and lariat RNA. *J. Am. Chem. Soc.*, **125**, 6880–6881.
56. Wang, Y. and Silverman, S.K. (2005) Efficient one-step synthesis of biologically related lariat RNAs by a deoxyribozyme. *Angew. Chem. Int. Ed.*, **44**, 5863–5866.
57. Mitra, D. and Damha, M.J. (2007) A novel approach to the synthesis of DNA and RNA lariats. *J. Org. Chem.*, **72**, 9491–9500.
58. Reese, C.B. and Song, Q. (1999) A new approach to the synthesis of branched and branched cyclic oligoribonucleotides. *Nucleic Acids Res.*, **27**, 2672–2681.
59. Eckardt, L.H., Naumann, K., Pankau, W.M., Rein, M., Schweitzer, M., Windhab, N., and von Kiedrowski, G. (2002) DNA nanotechnology: chemical copying of connectivity. *Nature*, **420**, 286.
60. Yang, H. and Sleiman, H.F. (2008) Templated synthesis of highly stable, electroactive, and dynamic metal-DNA branched junctions. *Angew. Chem. Int. Ed.*, **47**, 2443–2446.
61. Utagawa, E., Ohkubo, A., Sekine, M., and Seio, K. (2007) Synthesis of branched oligonucleotides with three different sequences using an oxidatively removable tritylthio group. *J. Org. Chem.*, **72**, 8259–8266.
62. Horn, T. and Urdea, M.S. (1989) Forks and combs and DNA: the synthesis of branched oligodeoxyribonucleotides. *Nucleic Acids Res.*, **17**, 6959–6967.
63. Chandra, M., Keller, S., Gloeckner, C., Bornemann, B., and Marx, A. (2007) New branched DNA constructs. *Chemistry*, **13**, 3558–3564.
64. Zelin, E., Wang, Y., and Silverman, S.K. (2006) Adenosine is inherently favored as the branch-site RNA nucleotide in a structural context that resembles natural RNA splicing. *Biochemistry*, **45**, 2767–2771.
65. Lee, C.S., Mui, T.P., and Silverman, S.K. (2011) Improved deoxyribozymes for synthesis of covalently branched DNA and RNA. *Nucleic Acids Res.*, **39**, 269–279.
66. Pratico, E.D., Wang, Y., and Silverman, S.K. (2005) A deoxyribozyme that synthesizes 2′,5′-branched RNA with any branch-site nucleotide. *Nucleic Acids Res.*, **33**, 3503–3512.
67. Coppins, R.L. and Silverman, S.K. (2005) A deoxyribozyme that forms a three-helix-junction complex with its RNA substrates and has general RNA branch-forming activity. *J. Am. Chem. Soc.*, **127**, 2900–2907.
68. Mui, T.P. and Silverman, S.K. (2008) Convergent and general one-step DNA-catalyzed synthesis of multiply branched DNA. *Org. Lett.*, **10**, 4417–4420.
69. Wang, Y. and Silverman, S.K. (2003) Characterization of deoxyribozymes that synthesize branched RNA. *Biochemistry*, **42**, 15252–15263.
70. Wang, Y. and Silverman, S.K. (2006) Experimental tests of two proofreading mechanisms for 5′-splice site selection. *ACS Chem. Biol.*, **1**, 316–324.
71. Pratico, E.D. and Silverman, S.K. (2007) Ty1 reverse transcriptase does not read through the proposed 2′,5′-branched retrotransposition intermediate in vitro. *RNA*, **13**, 1528–1536.
72. Baum, D.A. and Silverman, S.K. (2007) Deoxyribozyme-catalyzed labeling of RNA. *Angew. Chem. Int. Ed.*, **46**, 3502–3504.
73. Patel, M.P., Baum, D.A., and Silverman, S.K. (2008) Improvement of DNA adenylation using T4 DNA ligase with a template strand and a strategically mismatched acceptor strand. *Bioorg. Chem.*, **36**, 46–56.
74. Dai, Q., Saikia, M., Li, N.S., Pan, T., and Piccirilli, J.A. (2009) Efficient chemical synthesis of AppDNA by adenylation of

immobilized DNA-5′-monophosphate. *Org. Lett.*, **11**, 1067–1070.
75. Hafner, M., Landgraf, P., Ludwig, J., Rice, A., Ojo, T., Lin, C., Holoch, D., Lim, C., and Tuschl, T. (2008) Identification of microRNAs and other small regulatory RNAs using cDNA library sequencing. *Methods*, **44**, 3–12.
76. Zelin, E. and Silverman, S.K. (2009) Efficient control of group I intron ribozyme catalysis by DNA constraints. *Chem. Commun.*, **45**, 767–769.
77. Miduturu, C.V. and Silverman, S.K. (2005) DNA constraints allow rational control of macromolecular conformation. *J. Am. Chem. Soc.*, **127**, 10144–10145.
78. Miduturu, C.V. and Silverman, S.K. (2006) Modulation of DNA constraints that control macromolecular folding. *Angew. Chem. Int. Ed.*, **45**, 1918–1921.
79. Sachdeva, A. and Silverman, S.K. (2010) DNA-catalyzed serine side chain reactivity and selectivity. *Chem. Commun.*, **46**, 2215–2217.
80. Xiao, Y., Allen, E.C., and Silverman, S.K. (2011) Merely two mutations switch a DNA-hydrolyzing deoxyribozyme from heterobimetallic (Zn^{2+} /Mn^{2+}) to monometallic (Zn^{2+}-only) behavior. *Chem. Commun.*, **47**, 1749–1751.
81. Nowakowski, J., Shim, P.J., Prasad, G.S., Stout, C.D., and Joyce, G.F. (1999) Crystal structure of an 82-nucleotide RNA-DNA complex formed by the 10-23 DNA enzyme. *Nat. Struct. Biol.*, **6**, 151–156.
82. Peracchi, A. (2005) DNA catalysis: potential, limitations, open questions. *ChemBioChem*, **6**, 1316–1322.
83. Zaborowska, Z., Furste, J.P., Erdmann, V.A., and Kurreck, J. (2002) Sequence requirements in the catalytic core of the "10-23" DNA enzyme. *J. Biol. Chem.*, **277**, 40617–40622.
84. Peracchi, A., Bonaccio, M., and Clerici, M. (2005) A mutational analysis of the 8-17 deoxyribozyme core. *J. Mol. Biol.*, **352**, 783–794.
85. Zaborowska, Z., Schubert, S., Kurreck, J., and Erdmann, V.A. (2005) Deletion analysis in the catalytic region of the 10-23 DNA enzyme. *FEBS Lett.*, **579**, 554–558.
86. Wang, B., Cao, L., Chiuman, W., Li, Y., and Xi, Z. (2010) Probing the function of nucleotides in the catalytic cores of the 8-17 and 10-23 DNAzymes by abasic nucleotide and C3 spacer substitutions. *Biochemistry*, **49**, 7553–7562.
87. Nawrot, B., Widera, K., Wojcik, M., Rebowska, B., Nowak, G., and Stec, W.J. (2007) Mapping of the functional phosphate groups in the catalytic core of deoxyribozyme 10-23. *FEBS J.*, **274**, 1062–1072.
88. Liu, J. and Lu, Y. (2002) FRET study of a trifluorophore-labeled DNAzyme. *J. Am. Chem. Soc.*, **124**, 15208–15216.
89. Kim, H.K., Liu, J., Li, J., Nagraj, N., Li, M., Pavot, C.M., and Lu, Y. (2007) Metal-dependent global folding and activity of the 8-17 DNAzyme studied by fluorescence resonance energy transfer. *J. Am. Chem. Soc.*, **129**, 6896–6902.
90. Kim, H.K., Rasnik, I., Liu, J., Ha, T., and Lu, Y. (2007) Dissecting metal ion-dependent folding and catalysis of a single DNAzyme. *Nat. Chem. Biol.*, **3**, 763–768.
91. Kim, H.K., Li, J., Nagraj, N., and Lu, Y. (2008) Probing metal binding in the 8-17 DNAzyme by TbIII luminescence spectroscopy. *Chemistry*, **14**, 8696–8703.
92. Mazumdar, D., Nagraj, N., Kim, H.K., Meng, X., Brown, A.K., Sun, Q., Li, W., and Lu, Y. (2009) Activity, folding and Z-DNA formation of the 8-17 DNAzyme in the presence of monovalent ions. *J. Am. Chem. Soc.*, **131**, 5506–5515.
93. Leung, E.K. and Sen, D. (2007) Electron hole flow patterns through the RNA-cleaving 8-17 deoxyribozyme yield unusual information about its structure and folding. *Chem. Biol.*, **14**, 41–51.
94. Liu, Y. and Sen, D. (2008) A contact photo-cross-linking investigation of the active site of the 8-17 deoxyribozyme. *J. Mol. Biol.*, **381**, 845–859.
95. Liu, Y. and Sen, D. (2010) Local rather than global folding enables the lead-dependent activity of the 8-17 deoxyribozyme: evidence from contact photo-crosslinking. *J. Mol. Biol.*, **395**, 234–241.

96. Ryder, S.P. and Strobel, S.A. (1999) Nucleotide analog interference mapping. *Methods*, **18**, 38–50.

97. Zuker, M. (2003) Mfold web server for nucleic acid folding and hybridization prediction. *Nucleic Acids Res.*, **31**, 3406–3415.

98. Teller, C. and Willner, I. (2010) Functional nucleic acid nanostructures and DNA machines. *Curr. Opin. Biotechnol.*, **21**, 376–391.

7
Iron-Catalyzed Csp^3–H Oxidation with H_2O_2: Converting a Radical Reaction into a Selective and Efficient Synthetic Tool

Laura Gómez

7.1
Introduction and Scope

Unactivated Csp^3–H bonds are ubiquitous in organic molecules; however, they are not usually seen as functionalities to build up molecular complexity for obtaining value-added products. Conversion of alkyl C–H into C–O bonds is a thermodynamically highly favorable transformation, but generally a large kinetic barrier is associated with the C–H bond cleavage event, which is required before or during functionalization. In order to overcome this problem, highly reactive oxidants are needed, making the reaction more challenging. Most organic molecules contain many different types of C–H bonds, and usually there is very little difference in reactivity between various C–H bonds. Moreover, other a priori more reactive functionalities are present in the same molecule. Therefore, developing transformations that regioselectively functionalize a single C–H bond within a complex structure remains a challenge in this field. Furthermore, the ability to stop functionalization at the required oxidation state is difficult, as over-oxidation of functionalized products is often highly thermodynamically favored.

There is great interest in developing methodologies for selective oxidation of unactivated Csp^3–H bonds, since the scientific and technological impact will be very important. These reactions will have application in the functionalization of petroleum-derived products and in organic synthesis for the preparation of biologically and pharmaceutically relevant compounds. The design of straighter synthetic strategies toward complex organic molecules will be allowed by limiting the protection–deprotection stages, and eventually the efficient and fast preparation of natural products will be possible. Currently, only a few methods for such type of transformations are available, and their selectivity and scope are quite limited [1–5].

Nature is the source of inspiration for many chemists. The selectivity, efficiency, and mild conditions of the reactions that take place in the active site of metallo-oxygenases are nowadays unattainable by artificial systems. The aim of this chapter is to review the development of bioinspired iron catalysts for unactivated

New Strategies in Chemical Synthesis and Catalysis, First Edition. Edited by Bruno Pignataro.

Csp^3–H oxidation over the last 20 years, and analyze the important features for the design of efficient and selective catalysts.

7.2 Environmentally Benign C–H Oxidation

As already stated, the inertness of non-activated Csp^3–H bonds implies the use of powerful oxidation agents. Only a few methods are available, and many of these are unselective, yielding a complicated mixture of products (such as Gif processes) [3]. Others make use of stoichiometric amount of oxidants that are difficult to handle (*in situ* preparation is required, such as dioxiranes) [6] and/or generate unacceptable toxic waste (such as permanganate) [7, 8].

From the ecological and economic points of view, today there is considerable pressure to employ environmentally benign reaction conditions. While significant advances have been made, important challenges remain. These include substrate conversion with high selectivity and the use of inexpensive and environmentally friendly metal-based catalysts in low loadings. Moreover, most oxidants (such as iodosylbenzene (PhIO), NaOCl, organic peracids, etc.) have the disadvantage that, in addition to the oxidized products, stoichiometric amounts of waste products are also formed. Therefore, there is special interest on the use of environmentally friendly oxidants ("green" oxidants) to minimize the amount of waste [9]. Among the environmentally benign oxidants, aside from air and O_2, the most efficient are peroxides. In particular, hydrogen peroxide is a suitable "green" oxidant since the only side product after an oxidation reaction is water; it is relatively cheap and easier to handle than dioxygen.

7.3 Inspiration from Nature

Many enzymes have been found to activate O_2 to oxidize stereospecifically and efficiently isolated C–H bonds. Among them, iron-containing enzymes stand out. Its high availability and various attainable oxidation states have likely led to the evolutionary selection of iron in many life processes. Iron proteins implicated in O_2 activation are present in nature in a wide range of configurations. They can be classified into three different types depending on the structure of their active site [10, 11]:

1) **Heme proteins**: Cytochrome P450 (Cyt P450) family is the prototypical example. These enzymes contain a thiolate-ligated porphyrinic iron center in their active site [12–15].
2) **Mononuclear non-heme systems**: Rieske dioxygenases are the most versatile group of this class. In the oxygenase component of their active site, it is found an iron center coordinated by the common structural motive 2-his-1-carboxylate facial triad [10, 11, 16–18].

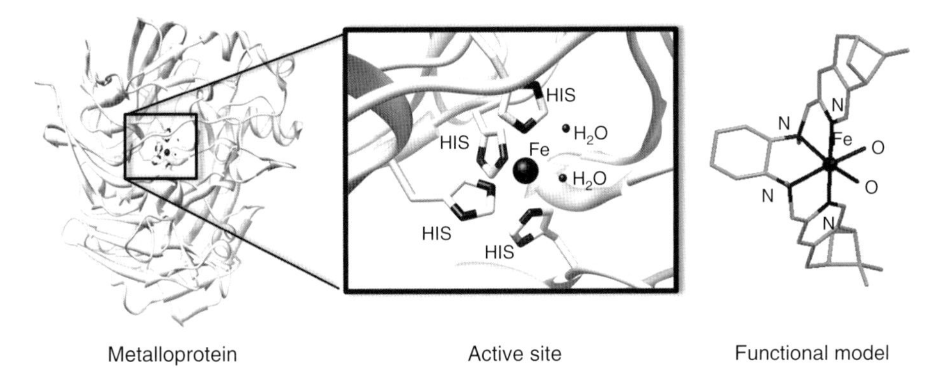

Figure 7.1 Bioinorganic chemistry gaining inspiration from metal active site of enzymes to design structural or functional model compounds (2BIW.pdb).

3) **Dinuclear non-heme enzymes**: The most challenging substrate for alkane oxidation, namely, methane, is oxidized to methanol by soluble methane monooxygenase [19–21].

The direct use of oxidative enzymes for synthetic purposes presents several difficulties in terms of purification, handling, and stability. Therefore, designing and studying low molecular weight synthetic complexes that can reproduce structural, spectroscopic, and/or chemical properties of an enzyme (Figure 7.1) is an interesting strategy for avoiding the use of enzymes. Ideally, these complexes should be able to perform reactions with comparable selectivity and efficiency to that performed by enzymes.

7.4 Mechanistic Considerations

Features of the first coordination sphere of the active site of oxidative iron enzymes have been a source of inspiration for catalyst design (Figure 7.1). The simplest approach is the combination of iron salts with H_2O_2. In this "bioinspired" approach, the combination of O_2 and $2e^-$ is replaced by peroxides, which can be understood as a $2e^-$ reduced version of O_2. From the ecological and economic point of view, this represents an attractive choice. However, radical pathways are usually involved and inefficient, and nonselective reactions occur (Scheme 7.1).

Oxidative reactions by free diffusing radical pathways have found applications in the oxidation of raw materials coming from natural gas or oil. Nevertheless, the low efficiency and selectivity of these reactions prevent their use in the synthesis of complex molecules. Furthermore, molecules of relevance in pharmaceutical industry and organic chemistry contain stereocenters. Even though the enantioselective oxidation of C–H bonds is still a challenge, the stereoretention observed in selected metal-centered reactions could open the door for obtaining enantiopure molecules.

$$Fe^{2+} + H_2O_2 \rightarrow Fe^{3+} + HO^- + \cdot OH$$

$$Fe^{3+} + H_2O_2 \rightarrow Fe^{2+} + H^+ + \cdot OOH$$

$$R\text{-}H + \cdot OH \rightarrow R\cdot + H_2O$$

Scheme 7.1 Generation of free diffusing radicals by reaction between iron ions and H_2O_2.

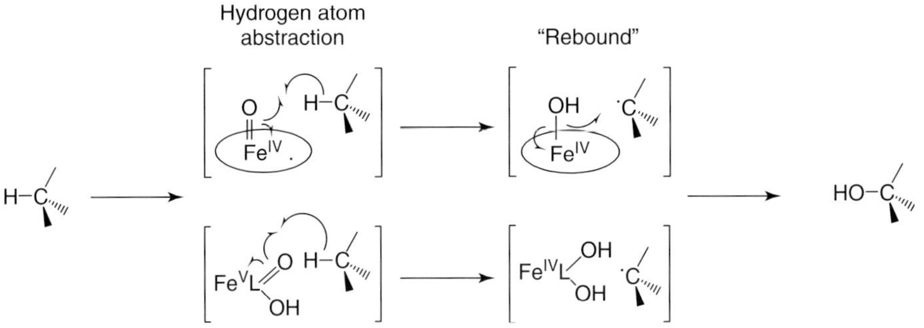

Hydrogen atom abstraction

"Rebound"

Scheme 7.2 Stereospecific hydroxylation mechanism proposed for heme catalysts (up) [24] and non-heme catalysts (down) [23].

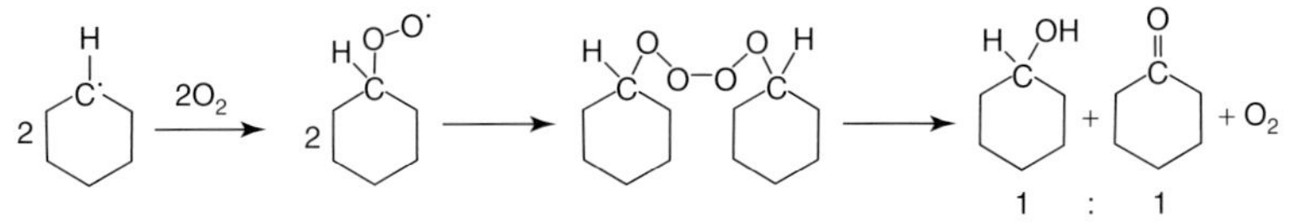

Scheme 7.3 Formation of equimolar amounts of alcohol and ketone product by O_2 trapping of cyclohexyl radicals following a Russell-type termination [26].

It has been suggested that high-valent iron-oxo species are involved in the catalytic cycle of both heme and non-heme iron enzymes and synthetic models that mediate stereospecific monohydroxylation reactions. Such species are highly reactive and have been proposed to abstract a hydrogen atom from the substrate in the rate-determining step (Scheme 7.2). Hydrogen atom abstraction by the iron-oxo species results in the formation of a short-lived carbon-centered radical and an iron hydroxide unit. Fast hydroxyl ligand transfer (rebound) forms the hydroxylated product [22, 23].

Several tests have been developed to determine the species responsible for the oxidation [25]. A very simple one is the oxidation of cyclohexane. At low conversion, cyclohexanol is the major product obtained by metal-based oxidants. When free diffusing radicals are formed, they can be trapped by O_2, leading to the formation of equal amounts of cyclohexanol and cyclohexanone (Scheme 7.3) [26].

In the oxidation of tertiary (3°) C–H bonds, high retention of configuration is observed when high-valent oxo species are responsible for the oxidation

cis-1,2-DMCH [Ox] 1R,2R and 1S,2S 1R,2S and 1S,2R

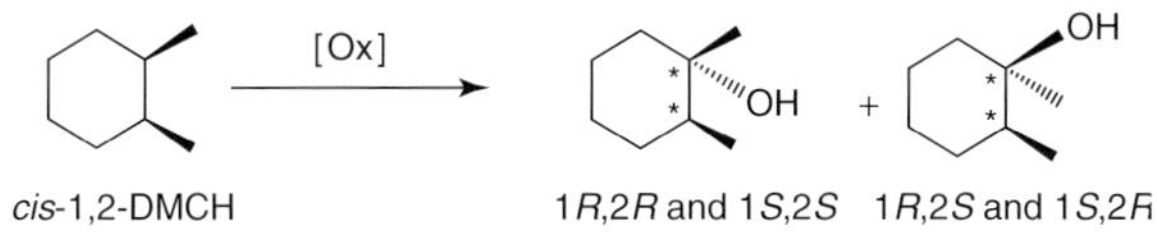

Scheme 7.4 Oxidation of *cis*-1,2-dimethylcyclohexane (*cis*-1,2-DMCH).

reaction. Instead, when free radicals are involved in the reaction, epimerization of tertiary C–H bonds occurs. The prototypical example is the oxidation of *cis*-1,2-dimethylcyclohexane (*cis*-1,2-DMCH) (Scheme 7.4).

Isotopic labeling experiments, by performing oxidation reactions in the presence of $H_2{}^{18}O$, constitute a very informative mechanistic tool to further address this scenario. Water exchange can occur at $LFe^{V}(O)(OH)$ and $Por^{\cdot+}Fe^{IV}(O)(OH)$ units by means of an oxo–hydroxo tautomerism [27, 28]. ^{18}O incorporation into products in stereospecific transformations constitutes an indirect evidence for the implication of such high-valent metal-oxo species in these reactions.

7.5 Bioinspired C–H Oxidation Catalysts

7.5.1 Porphyrinic Catalysts

The active site of Cyt P450 monooxygenases has been and still is a source of inspiration for catalyst design. In this context, porphyrins have been widely studied as chemical models for many years and extensive reviews have been published [13, 24, 29, 30]. For this reason, in this chapter only the design parameters extracted from the heme-oxygenase field that can be applied in the non-heme catalysis field for obtaining more efficient and selective catalysts are discussed.

Groves and coworkers were the first to demonstrate Cyt P450-like activity in a model system using iron PhIO as the oxidant (Scheme 7.5) [31]. This first-generation, however, was prone to oxidative decomposition and therefore synthetic applications were hampered by rapid catalyst deactivation.

This problem was overcome by attaching alkyl or halogen substituents at the ortho, meta, or para positions of the phenyl groups located at the macrocycle meso positions, in order to (i) provide steric effects to avoid the formation of catalytically inactive oxo complexes and/or (ii) enhance the electrophilicity of the metal-oxo

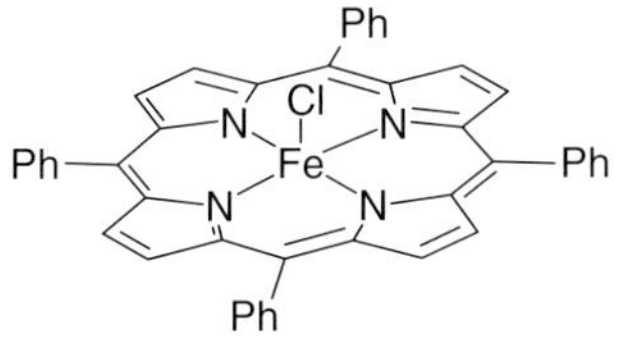

Scheme 7.5 First porphyrinic synthetic complex as catalyst for alkane hydroxylation developed by Grobes [31].

entity by electron-withdrawing substituents on phenyl rings. In a third generation, β-pyrrole positions were halogenated generating considerable electronic activation of the catalyst and severe saddling of the macrocyclic structure [32], favoring monomeric porphyrin complexes with regard to formation of oxo-bridged dimers. Exceptional catalyst stabilities and efficiencies in hydroxylation reactions were achieved.

A variety of oxidants such as iodosylarenes, alkylhydroperoxides, peracids, and hypochlorite have been commonly employed. However, H_2O_2 has been scarcely used [30, 33].

7.5.2 Non-porphyrinic Mononuclear Iron Catalysts

In order to obtain selective and efficient catalysts proceeding through metal-based oxidants, accurate design of the ligand is necessary, which is not obvious. An attractive feature of non-heme iron catalysts is that ligand modification and catalyst tuning are relatively straightforward compared to porphyrin-type ligands. However, from the wide variety of iron complexes synthesized, only a few examples presented high efficiency and selectivity when combined with H_2O_2 and a catalytic behavior that clearly distinguished from radical-type chemistry.

The first non-heme iron complex capable of performing stereospecific alkane hydroxylation with H_2O_2 was reported by Que and coworkers in 1997 ($[Fe(TPA)(CH_3CN)_2]^{2+}$, Table 7.1) [34]. The selectivity for cyclohexanol (A) in front of cyclohexanone (K) ($TN_{A+K}(A/K) = 3.2(5)$, TN = turnover number) in the oxidation of cyclohexane together with the high stereoretention of configuration in the oxidation of *cis*-1,2-DMCH ($RC > 99$) and a kinetic isotope effect (KIE) value of 3.5 in the oxidation of cyclohexane pointed toward metal-based oxidants rather than free diffusing radical pathways. Since then, several groups have tried to design catalysts that overcome this precedent, but only a few have been successful. From these studies, some general rules for the design of efficient stereospecific alkane oxidation iron catalysts have been extracted.

The TPA ligand has been modified in several ways. Que and coworkers have shown that introduction of methyl groups in the 3 and 5 positions of the pyridine rings increased both efficiency and selectivity for alcohol product [23]. In contrast, introduction of an electron-withdrawing group such as methyl carboxylate in the 5 position of more than one pyridine decreased significantly the efficiency, but not the selectivity ($TN_{A+K}(A/K) = 2.3$ (19) for 5-$(COOMe)_2$-TPA, Table 7.1). These results indicate that the donating ability of the ligand is a key factor in the stability of the catalyst. On the other hand, α-substitution in more than one pyridine ring of the ligand lowered the efficiency levels and displayed reactivity patterns indicating the involvement of hydroxyl radicals in the reaction ($TN_{A+K}(A/K) = 2.9(2)$, 1.4 (1), and 1.7 (2) for 6-Me_2-TPA, 6-Me_3-TPA, and BQPA ligands, respectively; see Table 7.1) [23]. Pyridyl α-substitution introduces steric effects that prevent the pyridyl groups from approaching the iron center, thereby increasing the chances of complex decomposition.

Table 7.1 Oxidation of alkanes with H_2O_2 catalyzed by Fe^{II} complexes.[a]

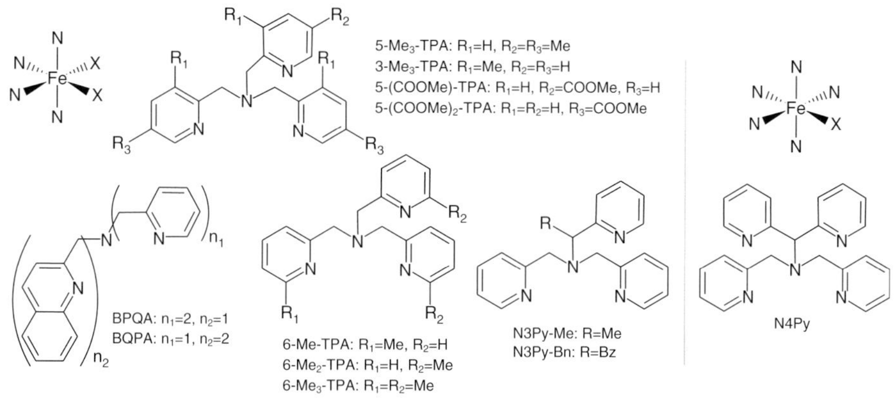

5-Me_3-TPA: R_1=H, R_2=R_3=Me
3-Me_3-TPA: R_1=Me, R_2=R_3=H
5-(COOMe)-TPA: R_1=H, R_2=COOMe, R_3=H
5-$(COOMe)_2$-TPA: R_1=R_2=H, R_3=COOMe

BPQA: n_1=2, n_2=1
BQPA: n_1=1, n_2=2

6-Me-TPA: R_1=Me, R_2=H
6-Me_2-TPA: R_1=H, R_2=Me
6-Me_3-TPA: R_1=R_2=Me

N3Py-Me: R=Me
N3Py-Bn: R=Bz

N4Py

Ligand	Cyclohexane	*cis*-1,2-DMCH	References
	TN_{A+K} (A/K)[b]	RC[c]	
TPA	3.2 (5)	100	[23]
5-Me_3-TPA	4.0 (9)	100	[23]
3-Me_3-TPA	4.5 (14)	100	[23]
6-Me-TPA	4.0 (7)	85	[23]
6-Me_2-TPA	2.9 (2)	64	[23]
6-Me_3-TPA	1.4 (1)	54	[23]
BPQA	5.8 (10)	89	[23]
BQPA	1.7 (2)	74	[23]
5-(COOMe)-TPA	4.0 (5)	100	[23]
5-$(COOMe)_2$-TPA	2.3 (19)	100	[23]
N3Py-Me	9.6 (4.6)[d]	73	[35]
N3PyBn	9.5 (5.8)[d]	85	[35]
N4Py	4.3 (1.4)[e]	27	[36]

[a] Catalyst:H_2O_2:Substrate:1 : 10 : 1000 in CH_3CN.
[b] A: cyclohexanol, K: cyclohexanone. A/K = mol A/mol K.
[c] Percentage of retention of configuration in the oxidation of the tertiary C–H bonds of *cis*-1,2-DMCH.
[d] 50 equiv. of H_2O_2.
[e] 100 equiv. of H_2O_2.

Further studies on tripodal tetradentate ligands were carried out more recently by Britovsek's group. They synthesized a series of complexes containing ligands where the pyridyl donors of TPA were successively replaced by dimethylamine donors to give *iso*-BPMEN, Me_4-BENPA, and Me_6-TREN ligands (Table 7.2) [37]. The efficiency in the oxidation of cyclohexane varies substantially for the different catalysts. As in the case of TPA, *iso*-BPMEN ($TN_{A+K}(A/K)$ = 3.2 (6.8)) showed good

Table 7.2 Oxidation of alkanes with H_2O_2 catalyzed by Fe^{II} complexes containing tripodal ligands.[a]

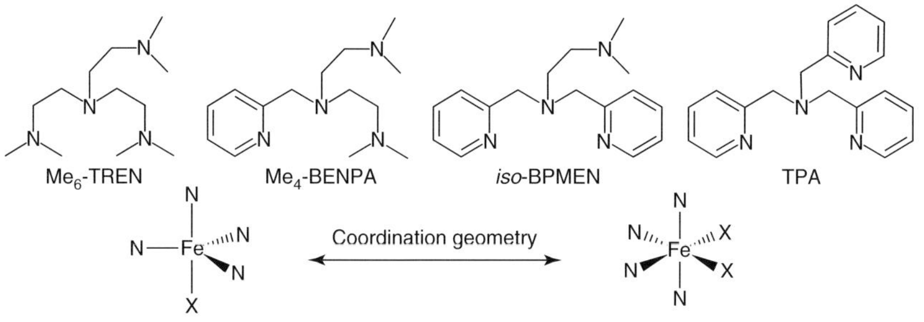

Ligand	Cyclohexane	References
	TN_{A+K} (A/K)[b]	
Me_6-TREN	0.3 (1.3)	[37]
Me_4-BENPA	0.3 (3.7)	[37]
iso-BPMEN	3.2 (6.8)	[37]

[a]Catalyst:H_2O_2:Substrate:1 : 10 : 1000 in CH_3CN.
[b]A: cyclohexanol, K: cyclohexanone. A/K = mol A/mol K.

efficiency and selectivity, indicating metal-based oxidation (Table 7.2). Instead, Me_4-BENPA and Me_6-TREN complexes, containing more amine than pyridine donors, appear to be less stable and degrade quickly under reaction conditions ($TN_{A+K}(A/K)$ = 0.3 (3.7) and 0.3 (1.3), respectively; see Table 7.2). Two main reasons are proposed for this behavior. First, pyridine donors increase the ligand field, which stabilizes the intermediates responsible for metal-based oxidation, and appear to be essential for high catalytic activity and selectivity in these systems. Second, the coordination geometry of the complexes changes from six-coordinate to five-coordinate as the number of methyl pyridyl arms decreases. Five-coordinative complexes lack the availability of two cis labile coordination sites at the metal center, which is thought to be important for catalytic activity. This is in agreement with the implication of hydroxyl radicals in the oxidation reactions catalyzed by the complex $[Fe(N4Py)(CH_3CN)_2]^{2+}$ containing a pentadentate ligand (Table 7.1) [36]. Also in this case, only one labile site is available for the coordination of an exogenous ligand.

Menage and coworkers provided further evidences of the need for two cis labile sites for coordination with exogenous ligands (oxidant and/or substrate) [38]. In this study, it was shown that weakly coordinating triflate anions or acetonitrile molecules in combination with noncoordination anions led to metal-based pathways, while

Table 7.3 Oxidation of alkanes with H_2O_2 catalyzed by BPMEN-related Fe^{II} complexes.[a]

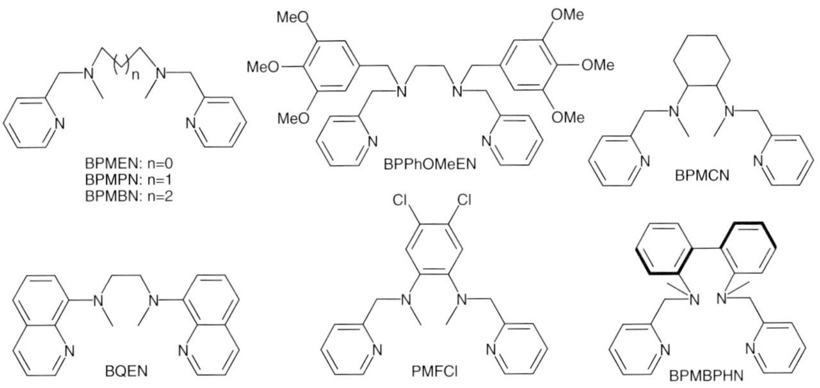

BPMEN: n=0
BPMPN: n=1
BPMBN: n=2
BPPhOMeEN
BPMCN
BQEN
PMFCl
BPMBPHN

Entry	Ligand	Cyclohexane	*cis*-1,2-DMCH	References
		$TN_{A+K}(A/K)$[b]	RC[c]	
1	BPMEN	6.3 (8)	96	[23]
2	BPPhOMeEN[d]	0.6 (1)	0	[38]
3	BPPhOMeEN[e]	3.8 (8)	82	[38]
4	BPMPN	4.2 (1.4)	–	[39]
5	BPMBN	0 (0)	–	[39]
6	α-BPMCN	5.9 (9)	>99	[40]
7	β-BPMCN	1.9 (0.9)	68	[40]
8	BPMBPH	0.14 (1.9)	–	[41]
9	PMFCl	6.2 (2.4)	–	[39]
10	BQEN	5.1 (5)	–	[42]

[a] Catalyst:H_2O_2:cyclohexane:1 : 10 : 1000.
[b] A: cyclohexanol, K: cyclohexanone. A/K = mol A/mol K.
[c] Percentage of retention of configuration in the oxidation of the tertiary C–H bonds of *cis*-1,2-DMCH.
[d] [$FeLCl_2$] complex.
[e] [$FeL(CH_3CN)_2(ClO_4)_2$] complex.

strongly binding ligands such as chloride led to Fenton-type reactions (Table 7.3, entries 2 and 3).

Tetradentate N-donor linear ligands containing two aliphatic amines and two pyridines have shown very good performance in alkane oxidation. For many years, the iron(II) complex of the BPMEN ligand was the most efficient and selective non-heme catalyst (Table 7.3). The effect of the structure and atom donor nature of the backbone was studied for the linear BPMEN and BQEN families (Table 7.3) [39–43]. Rigidity of the ligand backbone appears to be an important factor, since too flexible backbones lead to a mixture of coordination modes, leading to less active catalysts (Table 7.3). A nice example of the dramatic differences in the

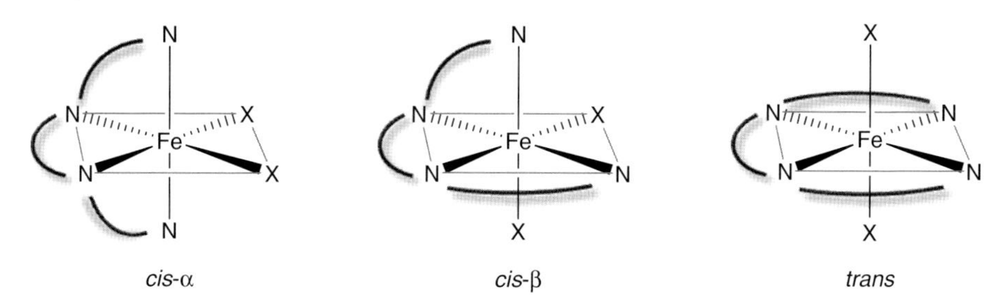

Scheme 7.6 Three different topologies that can be adopted by lineal tetradentate ligands.

catalytic outcome of an iron complex depending on its coordination structure is the comparison between the *cis-α* and *cis-β* topological isomeric forms of $[Fe(CF_3SO_3)_2(BPMCN)]$ (Table 7.3, entries 6 and 7, and Scheme 7.6). The former is an efficient catalyst for stereospecific C-H hydroxylation, while the latter gives rise to carbon-centered diffusing radicals [40].

A different structural approach for designing tetradentate ligands was reported by Company *et al.* The authors functionalized the triazacyclononane (TACN) ring with a picolyl arm. This PyTACN platform presents multiple possibilities for ligand tuning. The type of substitution on the N atoms of the triazamacrocycle and on the pyridine ring can be easily modified, converting this scaffold in a unique and versatile platform for studying structure–activity relations in catalytic oxidation reactions. A series of iron(II) complexes were prepared and studied accordingly (Table 7.4). Contradicting the aforementioned rule that 2 is the magic number of pyridine rings necessary for efficient catalytic oxidation reactions, these TACN-derived complexes showed excellent efficiency in the conversion of H_2O_2 into oxidized products (up to 76%). $[Fe(CF_3SO_3)_2(^{Me,H}PyTACN)]$ and $[Fe(CF_3SO_3)_2(^{Me,Me}PyTACN)]$, containing a single pyridine moiety, bypassed state-of-the art oxidations catalyzed by TPA and BPMEN complexes. Moreover, complexes $[Fe(CF_3SO_3)_2(^{Me,H}PyTACN)]$ and $[Fe(CF_3SO_3)_2(^{Me,Me}PyTACN)]$ exhibit remarkable large A/K ratios (12.3 and 10.2, respectively) in the oxidation of cyclohexane. On the other hand, complexes $[Fe(CF_3SO_3)_2(^{iPr,H}PyTACN)]$ and $[Fe(CF_3SO_3)_2(^{iPr,Me}PyTACN)]$ exhibit significantly lower yields, very likely due to the weak tertiary C–H bond of the isopropyl group that renders the complexes susceptible to self-oxidation. This is in agreement with the need for oxidatively robust ligands.

Despite the high efficiencies exhibited by these systems, their application in synthetic organic chemistry is hampered by low substrate conversion. So far, only two catalysts have been described as capable of sustaining selective C-H oxidation reactions in synthetic yields: the $[Fe(PDP)(CH_3CN)_2](SbF_6)_2$ reported by White and Chen in 2007 [45–47] and the $[Fe(CF_3SO_3)_2((S,S)\text{-MCPP})]$ [48] reported by Costas and coworkers in 2009 (Scheme 7.7). These complexes are capable of performing efficient alkane oxidation in a preparative way (limiting substrate conditions) and show unprecedented predictable selectivity, making these systems a useful tool for synthetic chemistry. The use of acetic acid as additive and the structure of the ligand are key to their success. The PDP ligand contains a very rigid bis-pyrrolidine

Table 7.4 Cyclohexane oxidation catalyzed by different Fe^{II} complexes containing PyTACN-type of ligands [44].

Complex	TN_{A+K} $(A/K)^a$
$[Fe(CF_3SO_3)_2(^{iPr,H}PyTACN)]$	2.4 (3.6)
$[Fe(CF_3SO_3)_2(^{iPr,Me}PyTACN)]$	0.8 (2.7)
$[Fe(CF_3SO_3)_2(^{Me,H}PyTACN)]$	6.5 (12.3)
$[Fe(CF_3SO_3)_2(^{Me,Me}PyTACN)]$	7.6 (10.2)

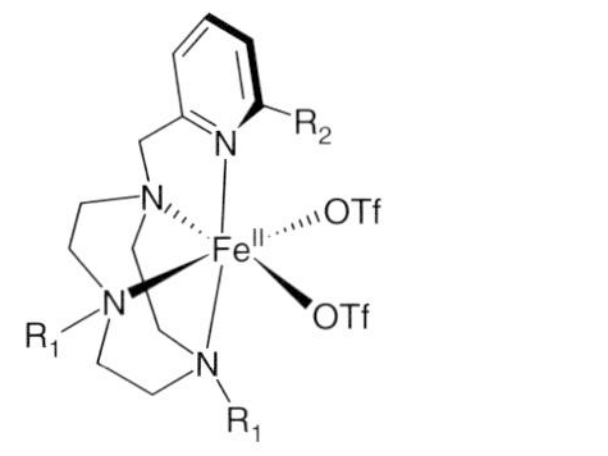

$[Fe(CF_3SO_3)_2(^{iPr,H}PyTACN)]$, R_1=*i*Pr, R_2=H
$[Fe(CF_3SO_3)_2(^{iPr,Me}PyTACN)]$, R_1=*i*Pr, R_2=Me
$[Fe(CF_3SO_3)_2(^{Me,H}PyTACN)]$, R_1=Me, R_2=H
$[Fe(CF_3SO_3)_2(^{Me,Me}PyTACN)]$, R_1=Me, R_2=Me

^{a}Catalyst:H_2O_2:cyclohexane:1 : 10 : 1000.
A: cyclohexanol, K: cyclohexanone. A/K = mol A/mol K.

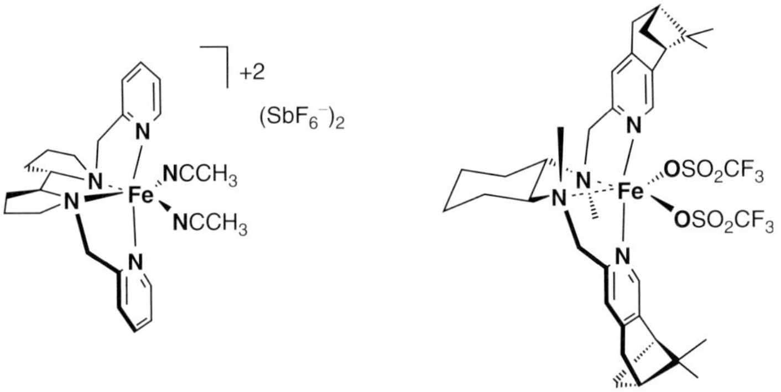

Scheme 7.7 Complexes (a) $[Fe((S,S)\text{-PDP})(CH_3CN)_2](SbF_6)_2$ and (b) $[Fe(CF_3SO_3)_2((S,S,R)\text{-MCPP})]$.

backbone, and the MCPP ligand incorporates a bulky pinene ring fused to the 4 and 5 positions of the pyridine rings of the BPMCN ligand.

Both systems are capable of oxidizing several substrates in moderate to good yields. As a general trend, tertiary C–H bonds are preferentially oxidized over secondary (2°) C–H bonds. Moreover, the oxidation of 3° C–H bonds is attained with high retention of configuration, indicating that reactions are metal-centered.

A particularly interesting example is the different selectivities observed in the oxidation of *cis*- and *trans*-4-methoxycyclohexyl pivalate when the MCPP complex is employed as catalyst (Table 7.5). *cis*-4-Methylcyclohexyl pivalate is selectively hydroxylated at the 3° C–H bond in 59% yield, while *trans*-4-methoxycyclohexyl

Table 7.5 Oxidation of *cis* and *trans*-4-methylcyclohexyl pivalate [47, 48].

Substrate	Products	Catalyst (loading)	Yield[a] (%)	RC[b]
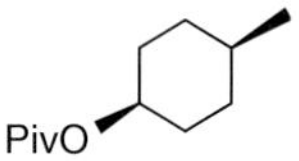 PivO	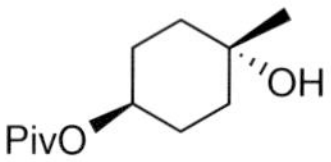 PivO, OH	$[Fe(BPMEN)(CH_3CN)_2](SbF_6)_2$ (5%)	26	–
		$[Fe((S,S)\text{-}PDP)(CH_3CN)_2](SbF_6)_2$ (15%)	60	99
		$[Fe(CF_3SO_3)_2((S,S,R)\text{-}MCPP)]$ (3%)	59	99
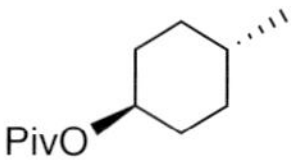 PivO	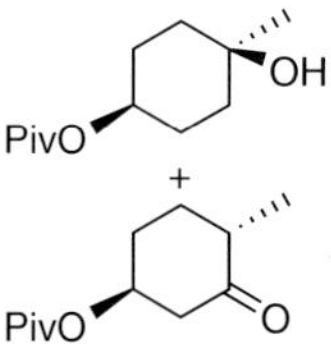 PivO, OH + PivO, O	$[Fe(CF_3SO_3)_2((MCPP)])$ (2%)	39	99
			16	–

[a]Isolated yield.
[b]Percentage of retention of configuration.

pivalate is oxidized to a mixture of 3° alcohol and ketone, resulting from oxidation at 2° positions. This observation can be reasoned by the fact that equatorial C–H bonds react more rapidly than those oriented axially. The rate enhancement in these reactions is attributed to a release of strain in the 1,3-diaxial interactions, as was recently described by Baran, Eschenmoser, and coworkers [49] and also observed by White and coworkers [46] in the oxidation of *trans* and *cis*-1,2-DMCH.

Site selectivity directed by electronic parameters was observed with both catalysts in the oxidation of substrates containing multiple C–H bonds (Table 7.6). Oxidation of 2,6-dimethyloctane (S1H) afforded a 1 : 1 mixture of tertiary alcohol products (O1H and O2H), while functionalization of S1Br and S1OAc occurred selectively at the distal C–H bond. In all cases examined, hydroxylation occurred preferentially at the most electron-rich tertiary C–H bond, despite the fact that secondary C–H bonds have a significant statistical advantage.

Electronic and/or steric factors play a role in discriminating between different C–H bonds, thus making site-selective oxidation predictable. The role of steric factors was established by performing the oxidation of (−)-acetoxy-*p*-menthane (S2, Table 7.7). This substrate presents two tertiary C–H bonds at the same distance to an acetate group, and they are electronically equivalent [45], but C1 is less sterically hindered (Table 7.7). Both catalysts afforded the C1-hydroxylated alcohol as the major product, O3, with good selectivity.

It is important to notice that, in all the aforementioned cases, the MCPP complex achieve comparable efficiencies using at least five times less catalyst loadings than the PDP analog; demonstrating that $[Fe(CF_3SO_3)_2((S,S,R)\text{-}MCPP)]$ is more robust under the harsh oxidizing conditions (see below).

Table 7.6 Electronic discrimination in the oxidation of substrates with multiple tertiary C–H bonds.

S1X [Ox] CH_3CN O1X + O2X

Catalyst Cat: H_2O_2: Subs: AcOH	X	Yield (%)	O1X/O2X
[Fe((*S*,*S*)-PDP)(CH_3CN)$_2$](SbF_6)$_2$	H	48	1/1
15 : 360 : 100 : 150 [45]	Br	39	9/1
	OAc	43	5/1
[Fe(CF_3SO_3)$_2$((*S*,*S*,*R*)-MCPP)]	H	35	1/1
2 : 240 : 100 : 100 [48]	Br	48	7/1
	OAc	46	5/1

Table 7.7 Regioselective oxidation of (−)-acetoxy-*p*-menthane.

(-)-Acetoxy-p-menthane, S2 [OX] CH_3CN O3 + O4

Catalyst	Cat: H_2O_2:S2:AcOH	Yield (%)	O3/O4
[Fe((*S*,*S*)-PDP)(CH_3CN)$_2$](SbF_6)$_2$ [47]	15 : 300 : 100 : 50	54	11/1
[Fe(CF_3SO_3)$_2$((*S*,*S*,*R*)-MCPP)] [48]	2 : 240 : 100 : 100	62	17/1

Synthetic applications were demonstrated by White in the oxidation of biologically relevant substrates. The aforementioned selectivity patterns allowed prediction of the obtained product. An interesting example is the oxidation of (+)-artemisin [47]. Only C_{10}-H, the most electron rich of the five 3° C–H bonds along its tetracyclic skeleton, was oxidized to the corresponding tertiary alcohol (Scheme 7.8). The presence of endoperoxide and ester moieties at α or β positions at the other 3° C–H bonds, prevented them from being oxidized. Furthermore, the oxidized compound was obtained in diasteriomerically pure form.

Although not necessary, directing groups can be used to obtain interesting products. In the oxidation of 16β-tetrahydrogibberellate diacetate, the presence of a terminal carboxylate moiety directs the oxidation to a secondary position to form a unique lactone product (Scheme 7.9) [47].

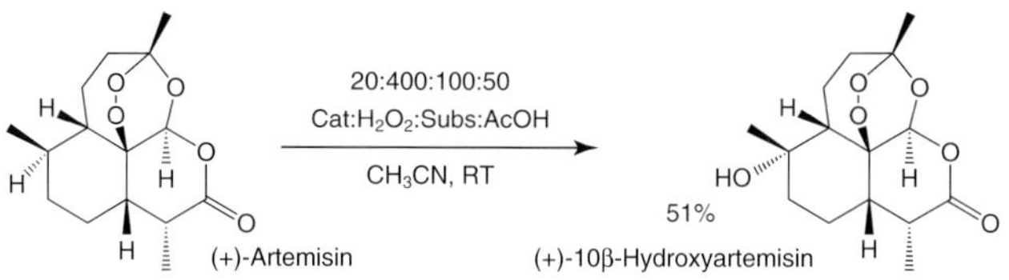

20:400:100:50
$Cat:H_2O_2:Subs:AcOH$
CH_3CN, RT
(+)-Artemisin
51%
(+)-10β-Hydroxyartemisin

Scheme 7.8 Stereospecific oxidation of (+)-artemisin by complex $[Fe((S,S)\text{-PDP})(CH_3CN)_2](SbF_6)_2$ [47].

25:500:100
$Cat:H_2O_2:Subs$
CH_3CN, RT
16β-Tetrahydrogibberellate diacetate
51%

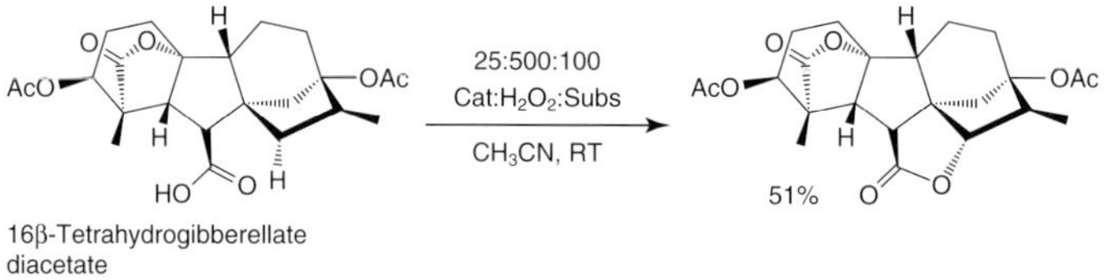

Scheme 7.9 Carboxylate-directed lactone formation using complex $[Fe((S,S)\text{-PDP})(CH_3CN)_2](SbF_6)_2$ [47].

15:360:100:150
$Cat:H_2O_2:Subs:AcOH$
CH_3CN, RT
(-)-Ambroxide
(+)-Sclareolide
80%
25:500:100:50
$Cat:H_2O_2:Subs:AcOH$
CH_3CN, RT
(+)-2-Oxo-sclareolide
46%
32%
(+)-3-Oxo-sclareolide

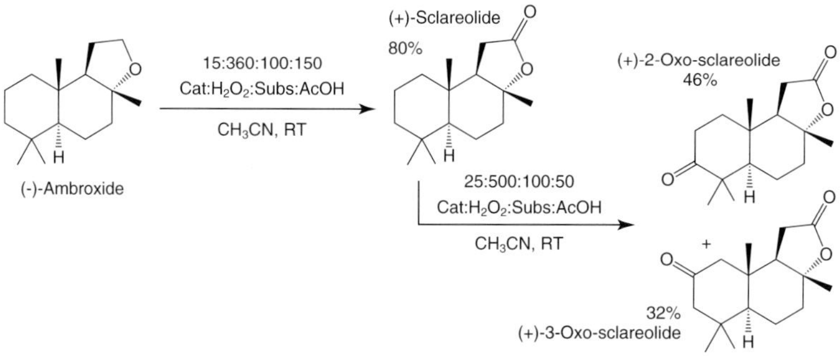

Scheme 7.10 Sequential oxidation of (−)-ambroxide to (+)-sclareolide and (+)-2- and (+)-3-oxo-ambroxide by catalyst $[Fe((R,R)\text{-PDP})(CH_3CN)_2](SbF_6)_2$.

Another nice example is the sequential oxidation of (+)-ambroxide (Scheme 7.10) [46]. In the first step, the methylene group next to the ester moiety is selectively oxidized as a result of the electronic activation produced by the aforementioned group. In a subsequent oxidation, the newly formed lactone deactivates this part of the molecule, shifting the preference of oxidation toward the opposite part of the molecule and obtaining a mixture of two ketone products.

Although these reactions entail great advances in synthetic planning, the $[Fe((S,S)\text{-PDP})(CH_3CN)_2](SbF_6)_2$ loadings required remain an important drawback. In most cases, less than five TNs are obtained (sub-stoichiometric reaction). The most plausible reason is the decomposition of the catalysts [47].

As mentioned earlier, catalyst $[Fe(CF_3SO_3)_2((S,S,R)\text{-MCPP})]$ showed comparable yields to $[Fe((S,S)\text{-PDP})(CH_3CN)_2](SbF_6)_2$ using much less catalyst loadings. The profound effect that the bulky pinene groups have in the efficiency and stability of

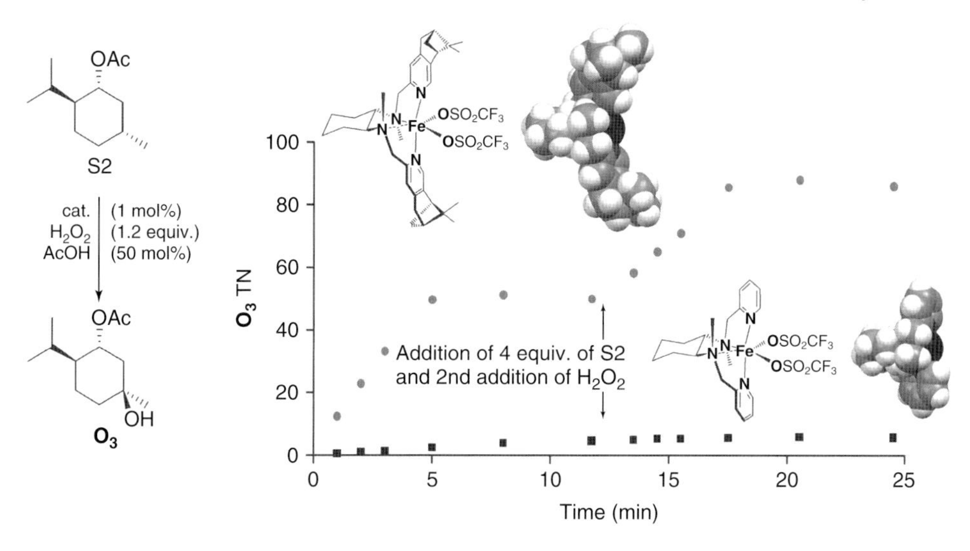

Figure 7.2 O_3 formation in the hydroxylation of S2 versus time in the two-step addition of H_2O_2 in presence of large excess of substrate, showing the relative stability of MCPP and BPMCN iron(II) catalysts [48].

the catalyst is evidenced by performing a time-profile analysis of the oxidation of S2 (see Figure 7.2) [48]. This analysis shows that the BPMCN complex (lacking the pinene rings) has a much lower catalytic activity than the MCPP complex during all H_2O_2 addition. Moreover, a second addition of H_2O_2 reassumes the catalytic activity of [Fe(CF_3SO_3)$_2$((*S*,*S*,*R*)-MCPP)] but not that of [Fe(CF_3SO_3)$_2$(BPMCN)].

Moreover, electrospray ionization mass spectrometry (ESI-MS) analyses (Figure 7.3) show that monomeric $[Fe^{III}(CF_3SO_3)(OR)(L)]^+$ (OR = OH, OAc) species (L = MCPP, $m/z = 734.4$ and 776.4, respectively; L = BPMCN, $m/z = 546.3$ and 588.3, respectively) are rapidly formed upon reaction of [Fe(CF_3SO_3)$_2$((*S*,*S*,*R*)-MCPP)] and [Fe(CF_3SO_3)$_2$(BPMCN)] with H_2O_2 in presence of AcOH and S2 [48]. These species are presumed to be the precursors of the mononuclear high-valent iron-oxo species responsible for catalytic activity [23, 44, 50, 51]. Most remarkably, such species still remain in the final reaction solutions of Fe(CF_3SO_3)$_2$((*S*,*S*,*R*)-MCPP)], whereas they have rapidly disappeared during H_2O_2 addition when [Fe(CF_3SO_3)$_2$(BPMCN)] is used. The authors conclude that the pinene rings play a key role in stabilizing such mononuclear species, most likely via steric encumbrance.

7.6
Perspectives

Spectacular advances have been achieved in the Csp^3-H oxidation at non-heme iron sites during the last two decades. Combination of iron complexes with H_2O_2

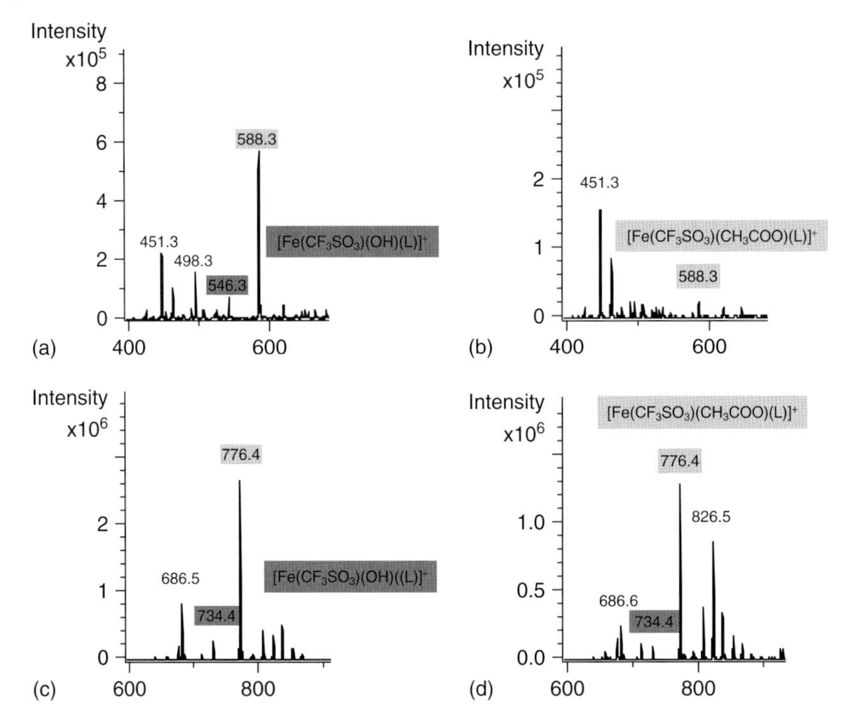

Figure 7.3 ESI-MS spectrum obtained in the catalytic oxidation of S2 by $[Fe(CF_3SO_3)_2(BPMCN)]$ (a) at time = 3 min during H_2O_2 addition and (b) after H_2O_2 addition. ESI-MS spectrum obtained in the catalytic oxidation of S2 by $[Fe(CF_3SO_3)_2((S,S,R)\text{-MCPP})]$ (c) at time = 3 min during H_2O_2 addition and (d) after H_2O_2 addition [48].

have evolved from free diffusing radical transformations with poor selectivity and synthetic value, to exquisitely site-selective steoreospecific C-H hydroxylation of complex organic molecules with high potential applications in synthetic chemistry. In some cases, it has already bypassed that attained by traditional oxidation reagents. The bases are currently set, but the huge potential of iron-based catalytic technologies for selective oxidation is waiting to be released.

References

1. Dyker, G. (2005) *Handbook of C-H Transformations*, Vol. 1–2, Wiley-VCH Verlag GmbH, Weinheim.
2. Shilov, A.E. and Shul'pin, G.B. (2000) *Activation and Catalytic Reactions of Saturated Hydrocarbons in the Presence of Metal Complexes*, Springer-Verlag, Boston.
3. Barton, D.H.R. and Doller, D. (1992) *Acc. Chem. Res.*, **25**, 504.
4. Sheldon, R.A. and Kochi, J.A. (1981) *Metal-Catalyzed Oxidations of Organic Compounds*, Academic Press, New York.
5. Newhouse, T. and Baran, P.S. (2011) *Angew. Chem. Int. Ed.* **50**, 3362.

6. Curci, R., d'Accolti, L., and Fusco, C. (2006) *Acc. Chem. Res.*, **39**, 1.
7. Arndt, D. (1981) *Manganese Compounds as Oxidizing Agents in Organic Chemistry*, Open Court La Salle, New York.
8. March, J. (1985) *Advanced Organic Chemistry*, 3rd edn, Wiley-Interscience, New York.
9. Sheldon, R.A. (2000) in *Biomimetic Oxidations Catalyzed by Transition Metal Complexes* (ed. B. Meunier), Imperial College Press, London, pp. 613.
10. Costas, M., Mehn, M.P., Jensen, M.P., and Que, L. Jr. (2004) *Chem. Rev.*, **104**, 939.
11. Bruijnincx, P.C.A., van Koten, G., and Klein Gebbink, R.J.M. (2008) *Chem. Soc. Rev.*, **37**, 2716.
12. Meunier, B. and Bernadou, J. (2000) *Struct. Bonding*, **97**, 1.
13. Meunier, B., de Visser, S.P., and Shaik, S. (2004) *Chem. Rev.*, **104**, 3947.
14. de Montellano, P.O. (2005) *Cytochrome P450: Structure, Mechanism, and Biochemistry*, 3rd edn, Springler Edition, New York.
15. Schlichting, I., Berendzen, J., Chu, K., Stock, A.M., Maves, S.A., Benson, D.E., Sweet, R.M., Ringe, D., Petsko, G.A., and Sligar, S.G. (2000) *Science*, **287**, 1615.
16. Tshuva, E.Y. and Lippard, S.J. (2004) *Chem. Rev.*, **104**, 987.
17. Abu-Omar, M.M., Loaiza, A., and Hontzeas, N. (2005) *Chem. Rev.*, **105**, 2227.
18. Solomon, E.I., Brunold, T.C., Davis, M.I., Kemsley, J.N., Lee, S.-K., Lehnert, N., Neese, F., Skulan, A.J., Yang, Y.-S., and Zhou, J. (2000) *Chem. Rev.*, **100**, 235.
19. Merkx, M., Kopp, D.A., Sazinsky, M.H., Blazyk, J.L., Merkx, J., and Lippard, S.J. (2001) *Angew. Chem. Int. Ed.*, **40**, 2782.
20. Feig, A.L. and Lippard, S.J. (1994) *Chem. Rev.*, **94**, 759.
21. Wallar, B.J. and Lipscomb, J.D. (1996) *Chem. Rev.*, **96**, 2625.
22. Groves, J.T. and McCluskey, G.A. (1976) *J. Am. Chem. Soc.*, **98**, 859.
23. Chen, K. and Que, L. Jr. (2001) *J. Am. Chem. Soc.*, **123**, 6327.
24. Ortiz de Montellano, P.R. (2010) *Chem. Rev.*, **110**, 932.
25. Costas, M., Chen, K., and Que, L. Jr. (2000) *Coord. Chem. Rev.*, **200–202**, 517.
26. Russell, G.A. (1957) *J. Am. Chem. Soc.*, **79**, 3871.
27. Bernadou, J. and Meunier, B. (1998) *Chem. Commun.*, 2167.
28. Lee, K.A. and Nam, W. (1997) *J. Am. Chem. Soc.*, **119**, 1916.
29. Mansuy, D. (1993) *Coord. Chem. Rev.*, **125**, 129.
30. Meunier, B. (1992) *Chem. Rev.*, **92**, 1411.
31. Groves, J.T., Nemo, T.E., and Myers, R.S. (1979) *J. Am. Chem. Soc.*, **101**, 1032.
32. Grinstaff, M.W., Hill, M.G., Labinger, J.A., and Gray, H.B. (1994) *Science*, **264**, 1311.
33. Katsuki, T. (1995) *Coord. Chem. Rev.*, **143**, 189.
34. Kim, C., Chen, K., Kim, J., and Que, L. Jr. (1997) *J. Am. Chem. Soc.*, **119**, 5964.
35. Klopstra, M., Roelfes, G., Hage, R., Kellogg, B.H., and Feringa, B.L. (2004) *Eur. J. Inorg. Chem.*, **4**, 846.
36. Roelfes, G., Lubben, M., Hage, R., Que, L., and Feringa, B.L. Jr. (2000) *Chem. Eur. J.*, **6**, 2152.
37. Britovsek, G.J.P., England, J., and White, A.J.P. (2005) *Inorg. Chem.*, **44**, 8125.
38. Mekmouche, Y., Ménage, S., Toia-Duboc, C., Fontecave, M., Galey, J.-B., Lebrun, C., and Pecaut, J. (2001) *Angew. Chem. Int. Ed.*, **40**, 949.
39. England, J., Davies, C.R., Banaru, M., White, A.J.P., and Britovsek, G.J.P. (2008) *Adv. Synth. Catal.*, **350**, 883.
40. Costas, M. and Que, L. Jr. (2002) *Angew Chem. Int. Ed.*, **41**, 2179.
41. Britovsek, G.J.P., England, J., and White, A.J.P. (2006) *Dalton Trans.*, **11**, 1399.
42. England, J., Britovsek, G.J.P., Rabadia, N., and White, A.J.P. (2007) *Inorg. Chem.*, **46**, 3752.
43. England, J., Gondhia, R., Bigorra-Lopez, L., Petersen, A.R., White, A.J.P., and Britovsek, G.J.P. (2009) *Dalton Trans.*, **27**, 5319.
44. Company, A., Gómez, L., Fontrodona, X., Ribas, X., and Costas, M. (2008) *Chem. Eur. J.*, **14**, 5727.
45. Chen, M.S. and White, M.C. (2007) *Science*, **318**, 783.

46. Chen, M.S. and White, M.C. (2010) *Science*, **327**, 566.
47. Vermeulen, N.A., Chen, M.S., and White, M.C. (2009) *Tetrahedron*, **65**, 3078.
48. Gómez, L., Garcia-Bosch, I., Company, A., Benet-Buchholz, J., Polo, A., Sala, X., Ribas, X., and Costas, M. (2009) *Angew. Chem. Int. Ed.*, **48**, 5720.
49. Chen, K., Eschenmoser, A., and Baran, P.S. (2009) *Angew. Chem. Int. Ed.*, **48**, 9705.
50. Company, A., Gómez, L., Güell, M., Ribas, X., Luis, J.M., Que, L., and Costas, M. Jr. (2007) *J. Am Chem. Soc.*, **129**, 15766.
51. Mas-Balleste, R. and Que, L. Jr. (2007) *J. Am. Chem. Soc.*, **129**, 15964.

8
Hydrogen Bonds as an Alternative Activation

Eugenia Marqués-López and Raquel P. Herrera

8.1
Introduction

In the last decade, a remarkable expansion of a new field has been accomplished, namely, *asymmetric organocatalysis* [1]. This new discipline has appeared as a useful, complementary, and alternative strategy to the earlier and more broadly explored metal and enzymatic catalyses. In this context, an impressive number of highly enantioselective approaches have been successfully and efficiently developed for the most common organic reactions, and the synthesis of new complex products has been achieved [2].

The immense number of organocatalytic processes could be classified into four large groups depending on the nature of catalyst activation, as previously proposed by Dalko and Moisan [3]: (i) reactions via covalent activation complexes, such as those performed with chiral secondary or primary amines [4]; (ii) reactions via noncovalent activation transition states, using hydrogen-bond catalysis [5]; (iii) enantioselective phase-transfer reactions by chiral quaternary ammonium salts [6]; and (iv) asymmetric transformations in a chiral cavity [7]. Among them, and within the second group, catalysts acting by hydrogen-bond interactions represent a significant contribution to this large area, receiving special attention in the last decade [5]. The main organocatalytic structures covering this large group are thiourea/urea derivatives [8], guanidines [9], Trans-α,α'-(Dimethyl-1,3-dioxolane-4,5-diyl)bis(diphenylmethanol) (TADDOL) [10] or 1,1′-Binaphthalene-2,2′-diol (BINOL) analogs [11], "chiral proton catalysts" [12], and chiral phosphoric acid derivatives [13]. Some of these representative structures are illustrated in Figure 8.1.

8.1.1
Chiral Thiourea/Urea Organocatalysts

Intensive efforts have been made in studying the behavior of (thio)urea structures as suitable catalysts in a large number of efficient processes. In this respect, the center of inspiration for the subsequent reactions concerning this area was the work

New Strategies in Chemical Synthesis and Catalysis, First Edition. Edited by Bruno Pignataro.

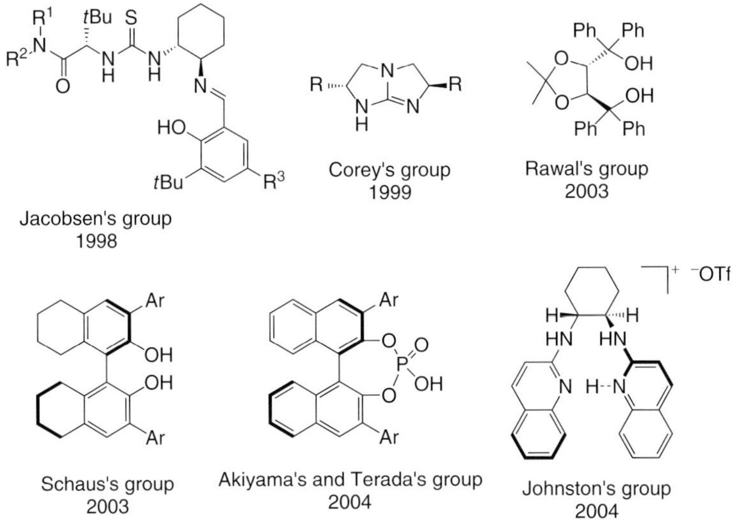

Jacobsen's group 1998

Corey's group 1999

Rawal's group 2003

Schaus's group 2003

Akiyama's and Terada's group 2004

Johnston's group 2004

Figure 8.1 Representative organocatalysts acting as hydrogen-bond donors.

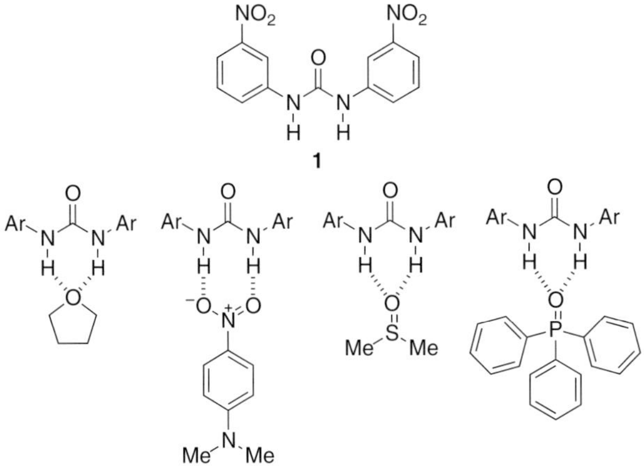

Figure 8.2 Co-crystallized functional groups using urea **1**. Model of bidentate activation.

reported by Etter and coworkers in which aryl urea **1** was able to form co-crystal complexes with a variety of hydrogen-bond acceptors (Figure 8.2) [14].

Nevertheless, the first application of this kind of bidentate activation employing such structures was reported by Curran and coworkers, using the urea catalyst **2**. They observed that the presence of urea **2** modified the rate and the stereochemical course of allylation reactions employing cyclic sulfinyl radicals, as depicted in Scheme 8.1 [15]. Moreover, urea **2** was also able to accelerate a Claisen rearrangement of electron-rich allyl vinyl ethers [16]. Both studies became key works in this growing area and they were a main reference for the following research works using (thio)urea catalysts.

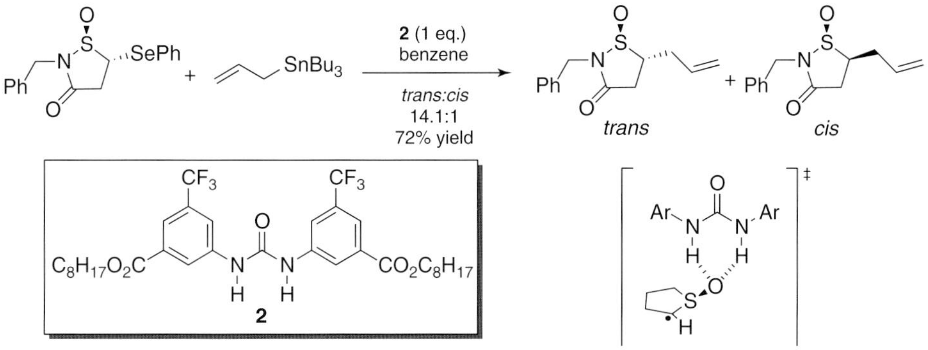

Scheme 8.1 Allylation reaction promoted by urea **2**.

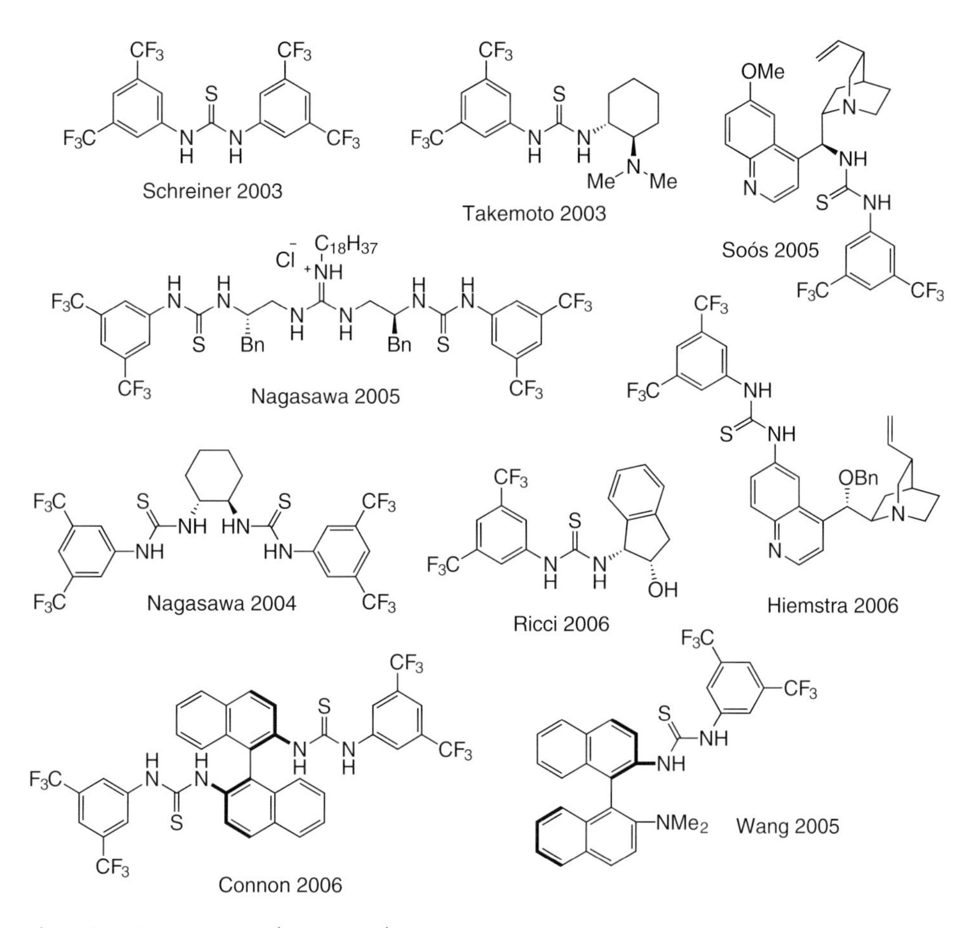

Figure 8.3 Representative thiourea catalysts.

However, the first examples using hydrogen-bond catalysis promoted by bidentate motifs could be attributed to Hine *et al.* [17] and Kelly's group [18], although in these cases biphenylenediol derivatives were used. Following these original examples, a large number of chiral thioureas and several ureas have been developed and effectively applied in a number of catalytic systems (Figure 8.3) [8].

This chapter focuses on our own contribution to the emerging area of chiral thiourea catalysis. Our work will be illustrated and discussed by comparison with pioneering models concerning different explored reactions. Diverse developed strategies for the preparation of interesting building blocks will be highlighted.

8.2 Thiourea Catalysts

8.2.1 Friedel–Crafts Alkylation Reaction

Friedel–Crafts alkylation reaction has become one of the most powerful carbon–carbon bond-forming processes in organic synthesis, mainly by means of metal catalysis [19], and less frequently explored by organocatalytic procedures [20]. Among the most significant Friedel–Crafts reactions, those concerning the Michael addition of indoles to electron-deficient alkenes have attracted special attention [20a]. Indole is considered a privileged structure and it is a frequent motif in a number of natural products (Figure 8.4) [21]; for this reason, the development of new enantioselective methodologies for its chiral synthesis is always an important achievement.

In this context, we developed a pioneering example of organocatalyzed Friedel–Crafts alkylation reaction between nitroolefins and aromatic and heteroaromatic systems [22]. Until then, only metal-catalyzed examples of this reactions using Lewis acids such as $Yb(OTf)_3 \cdot 3H_2O$ [23], $Sc(OTf)_3$ [24], or $Bi(OTf)_3$ [25] had been reported. We revealed, for the first time, the useful application of organocatalysis in this process using neutral hydrogen-bond donors, namely, urea- and thiourea-type catalysts **3** and **4** respectively (Scheme 8.2), in a comparative study where all examples were tested in toluene and solventless conditions affording very good yields.

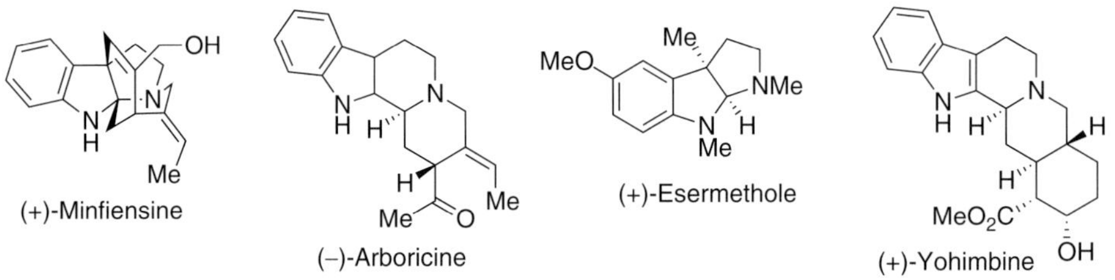

Figure 8.4 Indole motifs in natural products.

NO2
R1
R5
Yield up to 86%
R5
NO2
3 or 4 (10 mol%), r.t.
Toluene or
solventless
R2
N
R3
R4
CF3
X
F3C
N
H
X = O(3), S(4)
Yield up
to quantitative
R1 = Me, Ph
R2 = Et, Me, -CH2CH2-
R3 = H, -CH=CH-
R4 = H, OMe, -CH=CH-
R5 = Ph, C5H11

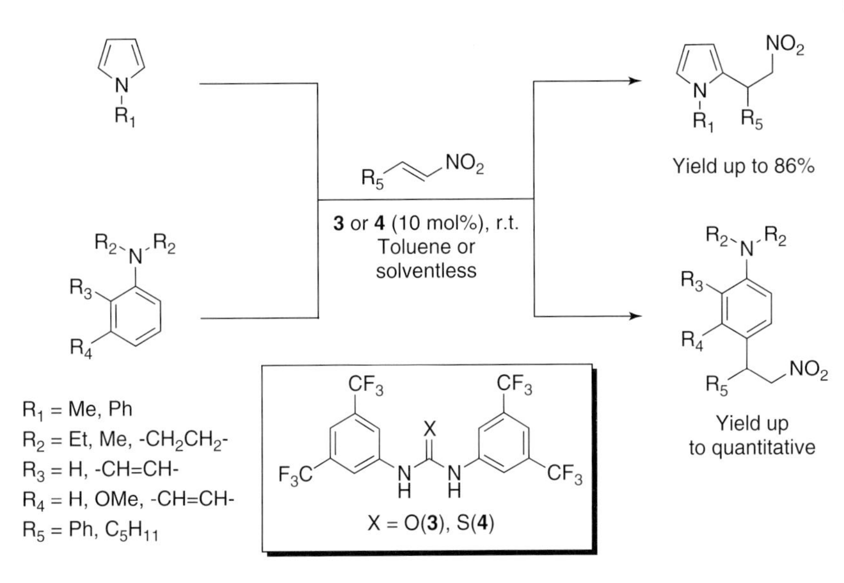

Scheme 8.2 Organocatalyzed Friedel–Crafts alkylation reaction of aromatic and heteroaromatic systems.

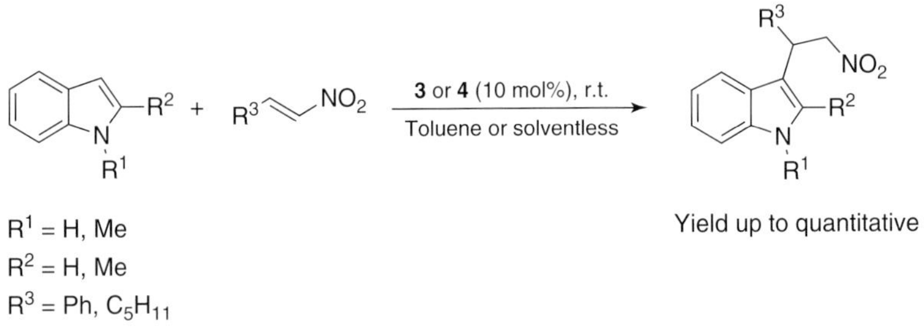

R2
+
R3
NO2
3 or 4 (10 mol%), r.t.
Toluene or solventless
N
R1
Yield up to quantitative
R1 = H, Me
R2 = H, Me
R3 = Ph, C5H11

Scheme 8.3 Friedel–Crafts alkylation reaction of indoles with nitroalkenes.

We also turned our attention to the Friedel–Crafts alkylation of indoles (Scheme 8.3), because only a practical and efficient $InBr_3$-catalyzed addition of indoles to nitroalkenes in aqueous media had been reported previously [26].

The outcomes using thiourea organocatalyst **4** were superior in all cases to those obtained employing urea **3**. This tendency was in agreement with the greater hydrogen-bond donor ability of thiourea derivatives supported by the enhanced differences in acidities (pK_{HA} thiourea = 21.0; pK_{HA} urea = 26.9) [27]. Furthermore, the lower tendency of sulfur atom to make self-association between the N–H group of one molecule and the thiocarbonyl group of another would justify the different results achieved with these catalysts [28].

In order to explain the mode of activation, a similar double hydrogen-bond interaction between the (thio)urea-type derivatives with the nitro group was proposed, in analogy to the hypothesis reported earlier for this functional group (Figure 8.5) [14].

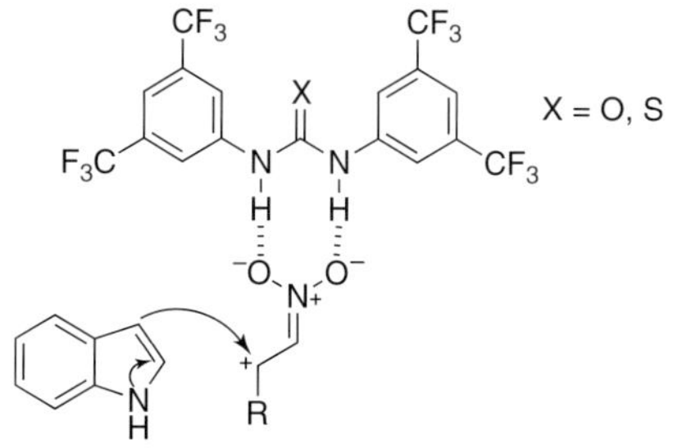

Figure 8.5 Mechanistic hypothesis.

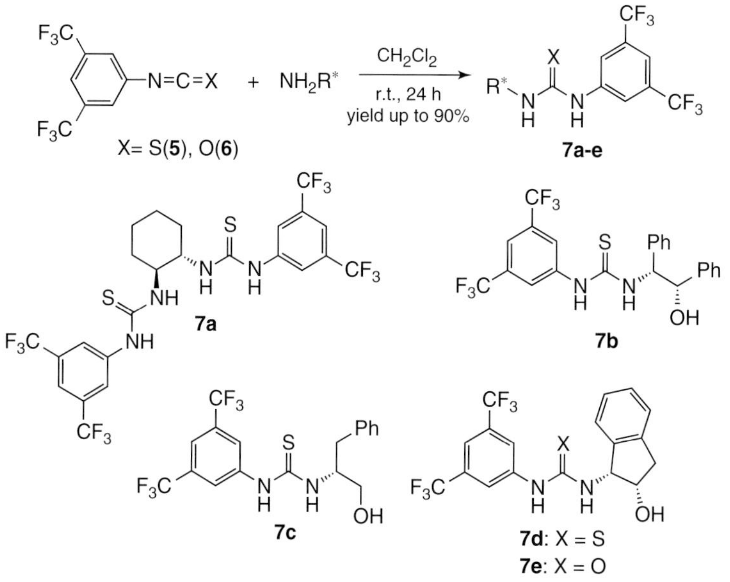

Scheme 8.4 Synthesis of chiral (thio)ureas **7a–e**.

Even more interesting was the original development of the enantioselective version of the latter Friedel–Crafts alkylation reaction [29]. For this purpose, the synthesis of simple and easily accessible chiral (thio)urea catalysts was required. Initially, we prepared and screened catalysts **7a–e** (Scheme 8.4), which were easily obtained in a one-step reaction and in very good yields from the coupling reaction between 3,5-bis(trifluoromethyl)phenyl iso(thio)cyanate **5** or **6** and the corresponding chiral amines or amino alcohols, all commercially available.

In all cases, an increased reactivity of *trans-β*-nitrostyrene against indole was achieved with every synthesized catalyst. However, only thiourea catalyst **7d** exhibited promising results in terms of both reactivity and enantioselectivity. After exploring different parameters and under the optimized reaction conditions, we extended our method to diverse indoles and nitroalkenes, affording final adducts with very good yields and high enantioselectivities. The scope of the reaction is summarized in Scheme 8.5.

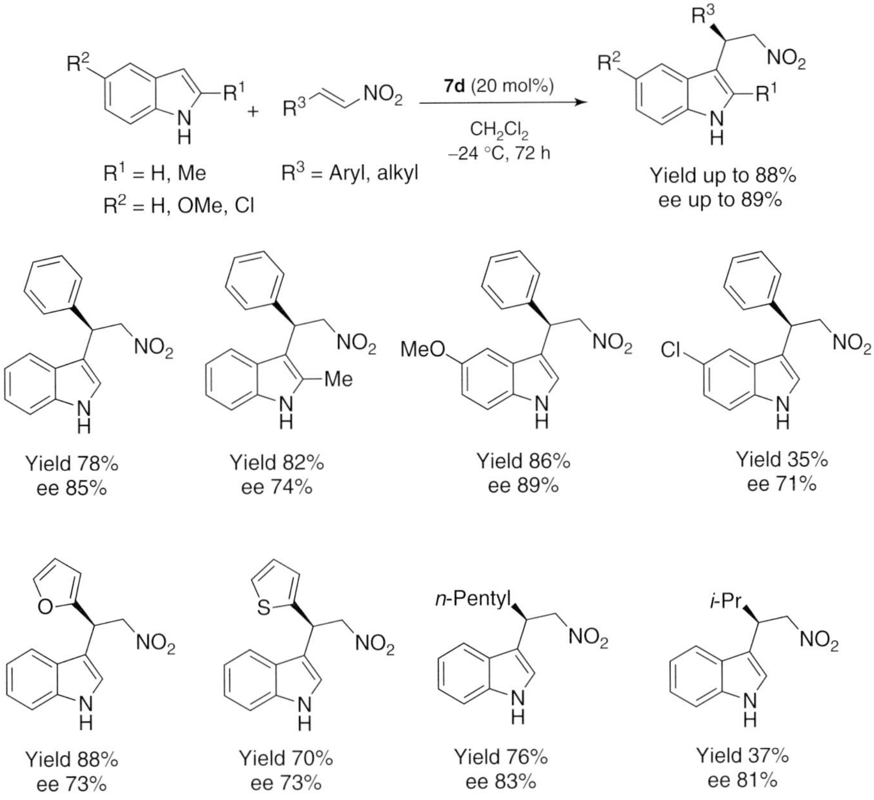

Scheme 8.5 Thiourea-catalyzed Friedel–Crafts alkylation reaction.

In order to demonstrate the synthetically valuable aspect of this process, the optically active adduct **8** was transformed into important compounds of biological interest such as tryptamine **9** [30] and 1,2,3,4-tetrahydro-β-carboline **11** [31] in very good yield and without racemization, as represented in Scheme 8.6. Moreover, derivatization of compound **8** in the corresponding *N*-tosylated tryptamine **10** allowed the assignment of the absolute configuration of the final products.

In an attempt to understand the possible mode of action of novel catalyst **7d** in the activation complex that led to the observed enantioinduction in final products, we synthesized catalyst **7f**, which lacked the alcoholic function, and **7g**, in which the hydroxy group was protected by a trimethylsilyl group. Both catalysts showed poorer results in comparison with **7d**, not only with regard to the enantioselectivity but also in terms of the catalyst activity, as shown in Figure 8.6.

This fact, in addition to the results obtained with different indoles in which the NH in the structure seemed to be also crucial, made us to envision that our catalyst could act as a bifunctional catalyst [8f, 32]. Our proposal was that the thiourea would activate the nitro group through the N–H atoms, as previously reported, and the hydroxyl function would interact with the indolic proton by means of a weak

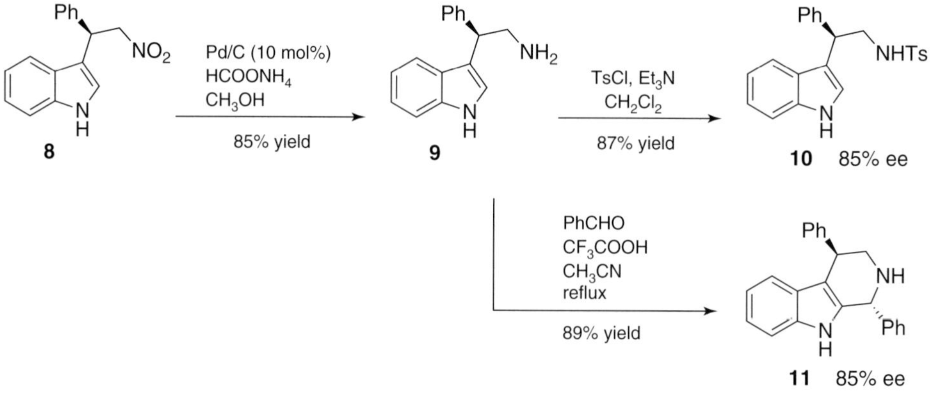

Scheme 8.6 Synthesis of tryptamine **9** and 1,2,3,4-tetrahydro-β-carboline **11**.

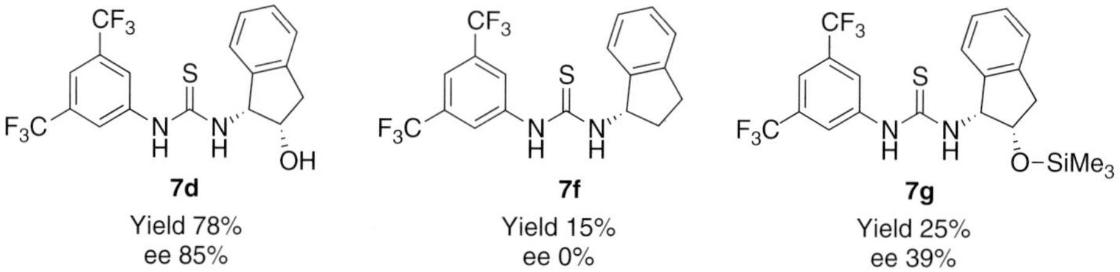

Figure 8.6 Additional mechanistic proofs.

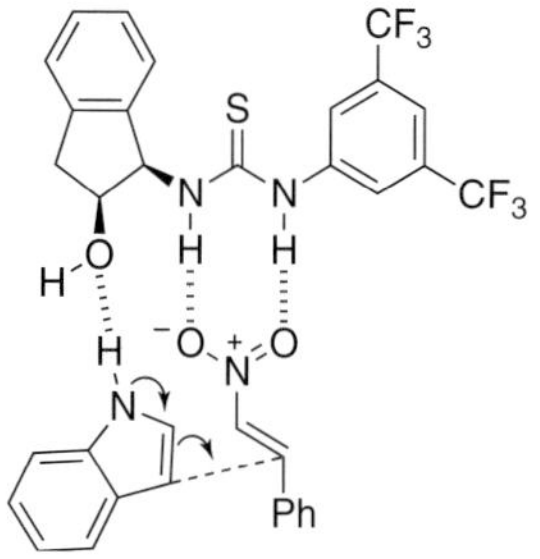

Figure 8.7 Mechanistic proposal.

hydrogen bond, approaching the nucleophile on the *Si* face of the nitroolefin as depicted in Figure 8.7. This attack would afford the absolute configuration observed in the final products.

The rigidity of the aminoindanol skeleton in catalyst **7d** seemed to be also crucial for the success of the process compared with the major flexibility expected by catalysts **7b** and **7c** in the transition state, in which the OH group was also present, but afforded poorer results overall in terms of enantioselectivities.

Concurrently with our original work, Jørgensen and coworkers [33] and more recently Connon's [34], Seidel's [35], and Akiyama's [36] groups have developed attractive and efficient catalytic systems, namely, bis-sulfonamide **12**, bis-thiourea

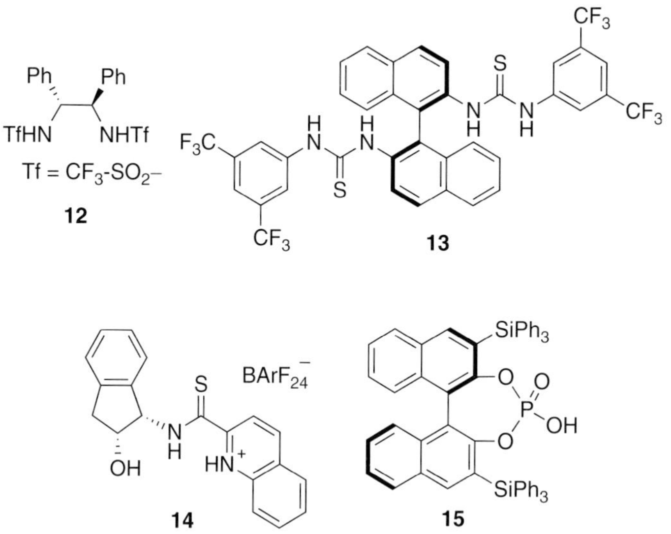

Figure 8.8 Efficient catalysts for activating the Friedel–Crafts alkylation reaction.

13, quinolinium thioamide catalyst **14**, and phosphoric acid silyl-derivative **15** respectively, for their applications as promoters in the Friedel–Crafts alkylation reaction (Figure 8.8).

Chiral bis-arylthiourea **13** was employed in the study of the asymmetric organocatalytic Friedel–Crafts addition between *N*-methylindole and nitroolefins, affording low enantioselectivities and good yields after long reaction time. However, with the other catalysts, non-*N*-substituted indoles were efficient nucleophiles against a range of nitroolefins, being essential the presence of the N-H moiety in the indole ring to achieve high yields and enantioselectivities. Interestingly, in all cases a hydrogen-donor catalyst was required to promote the reaction.

In spite of the importance of this process, it is still rarely investigated and documented; therefore the search for more effective catalysts is still open. Additional investigations are expected to take place in this area for the preparation of attractive asymmetric indole structures.

8.2.2
Michael Addition Reactions

Among the most significant and common reactions developed so far in organocatalysis, those concerning conjugate Michael additions of a nucleophile to an electron-deficient olefin are the most relevant ones in this expanding area [37]. This strategy allows the stereoselective formation of carbon-carbon [38] or carbon-heteroatom bonds [39] and has been successfully applied in a great number of reactions. Its synthetic potential has been demonstrated by application in the total synthesis of molecules of high complexity [2, 40].

8.2.2.1 Michael Addition Reaction of *N,N*-Dialkylhydrazones to Nitroalkenes

Suitable but less explored nucleophiles are the *N,N*-dialkylhydrazones, which may exhibit an ambiphilic behavior: on the one hand demonstrating an aza-enamine character and reacting with electrophiles such as activated Michael acceptors under the appropriate conditions [41], and on the other acting as imine surrogates against nucleophiles. However, this aspect has been less considered [42], contrary to the well-known electrophilic character of *N*-acylhydrazones [43]. As aza-enamines, formaldehyde *N,N*-dialkylhydrazones have been extensively used as synthetic equivalents of formyl and cyanide anions [44]. However, and despite their diverse reactivity and synthetic versatility, these privileged species have been less considered in organocatalysis [45].

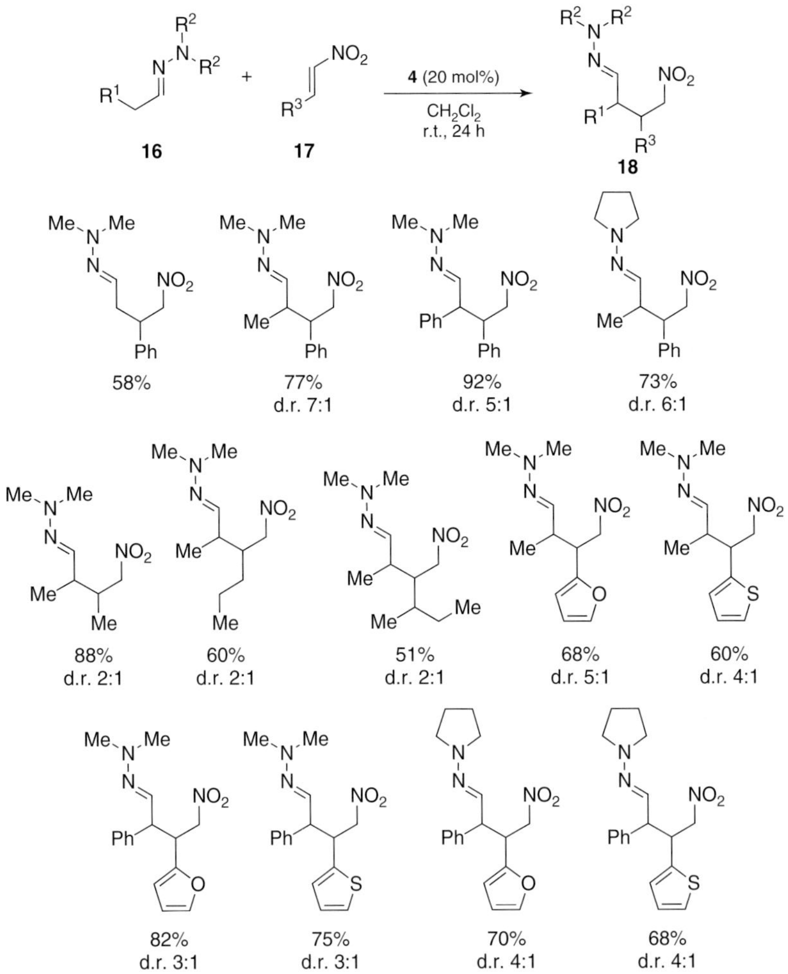

16 + 17 → 18: **4** (20 mol%), CH_2Cl_2, r.t., 24 h

58%; 77%, d.r. 7:1; 92%, d.r. 5:1; 73%, d.r. 6:1

88%, d.r. 2:1; 60%, d.r. 2:1; 51%, d.r. 2:1; 68%, d.r. 5:1; 60%, d.r. 4:1

82%, d.r. 3:1; 75%, d.r. 3:1; 70%, d.r. 4:1; 68%, d.r. 4:1

Scheme 8.7 Michael addition reaction of *N,N*-dialkylhydrazones **16** to nitroalkenes **17**.

In this context, another developed application of thiourea (4) is the activation of the nucleophilic addition of *N,N*-dialkylhydrazones to diverse nitroolefins [46]. Formaldehyde *N,N*-dialkylhydrazones have been extensively investigated, in comparison with hydrazones synthesized with other aldehydes, which have been rarely considered because of their inherent lower reactivity [42b]. We started a new work in the field of conjugate additions using less reactive aliphatic aldehyde hydrazones and nitroalkenes catalyzed by thiourea **4** and in absence of base, as depicted in Scheme 8.7.

Some remarkable aspects of this process should be underlined: first, we were glad to observe that thiourea **4** promoted the reaction affording higher yields at the same reaction time as the uncatalyzed reaction. Surprisingly, instead of the formation of the expected products from the attack by the azomethine carbon atom of the hydrazone, as had been observed with formaldehyde hydrazone derivatives, we obtained the products resulting from the nucleophilic attack at the α-carbon of the hydrazone. This reactivity under non-basic conditions was totally unprecedented and had not been observed previously in the literature for enolizable hydrazones since the use of a strong base was always required via enamine formation and subsequent trapping by electrophiles [47]. To explain this surprising reactivity, we suggested that the hydrazones **16** should be in equilibrium with its ene-hydrazine form **19**, which would furnish the observed adducts by attacking the activated nitroalkene, as represented in Scheme 8.8. Even when the presence of **19** could not be detected by NMR experiments because the equilibrium is shifted to the most favorable hydrazone form **16**, the small amount of **19** in the medium was enough to allow the reaction and to move the equilibrium toward the appropriate direction.

We have illustrated a novel example of reactivity for thiourea **4** promoting an unprecedented conjugate addition of enolizable hydrazones to nitroalkenes, reacting at the α-carbon in absence of base and affording γ-nitrohydrazones **18** in very good yields. This kind of reactivity is still open to study from an enantioselective point of view.

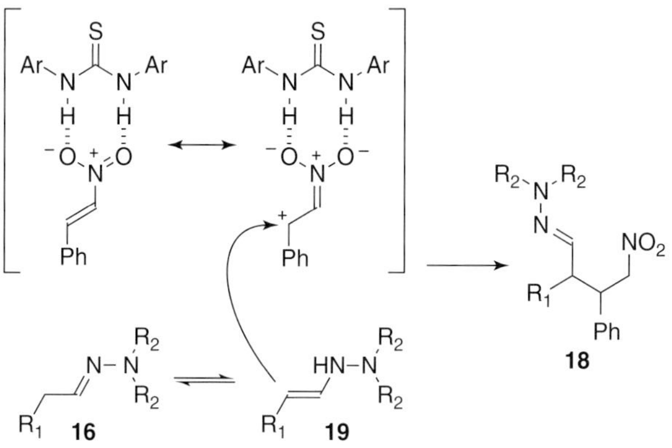

Scheme 8.8 Proposed mechanism via hydrazone-ene-hydrazine equilibrium.

8.2.2.2 Michael Addition Reaction of Formaldehyde *N,N*-Dialkylhydrazones to β,γ-Unsaturated α-Keto Esters

Among the different Michael addition approaches, the nucleophilic reactivity of formaldehyde dialkylhydrazones seems to be too low for reactions using weak electrophiles, such as α,β-unsaturated carbonyl compounds, and therefore, hydrazones have been rarely considered as formyl anion equivalents of synthetic value against these substrates [48]. Encouraged for the lack of catalytic examples and in order to contribute to this less explored field, we studied the viability of thiourea-catalyzed Michael addition reaction of hydrazone **20** to β,γ-unsaturated α-ketoesters **21** (Scheme 8.9) [49]. After testing different parameters such as chiral catalysts, solvents, the dilution, and temperature, we finally explored the scope of this new umpolung strategy with β,γ-unsaturated α-ketoester derivatives **21**, which resulted in the corresponding adducts **22a–f** in suitable yields and good enantioselectivities at −40 or −60 °C (Scheme 8.9). In this process, thiourea **7d** was once more identified as the most promising catalyst for promoting this Michael addition.

The synthetic importance of these substrates lies in the possibility of further cleavage of the hydrazone moiety for regenerating the carbonyl function under a

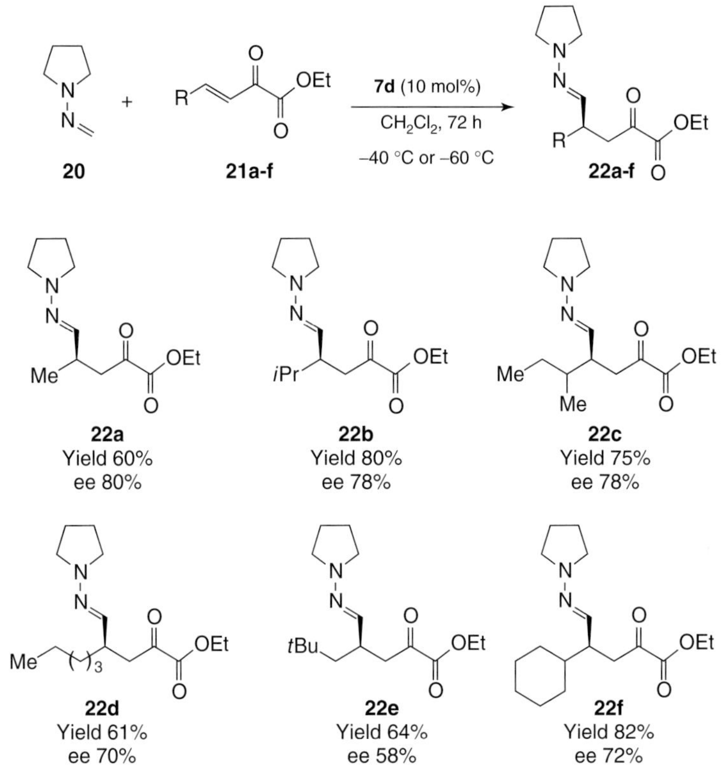

20 + 21a-f → **7d** (10 mol%), CH_2Cl_2, 72 h, −40 °C or −60 °C → 22a-f

22a Yield 60% ee 80%
22b Yield 80% ee 78%
22c Yield 75% ee 78%
22d Yield 61% ee 70%
22e Yield 64% ee 58%
22f Yield 82% ee 72%

Scheme 8.9 Organocatalytic Michael addition reaction using hydrazone pyrrolidine derivative **20**.

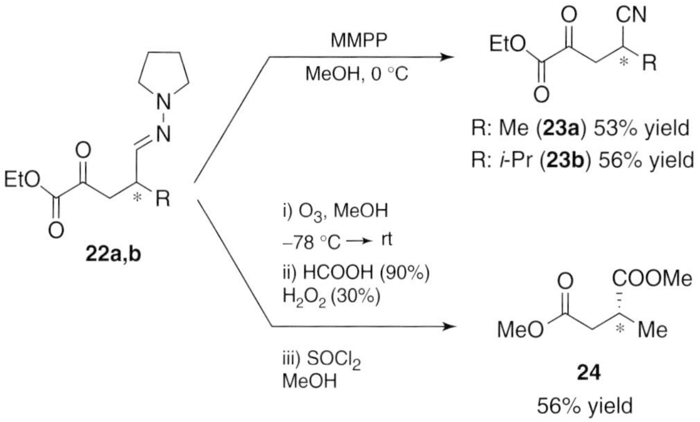

Scheme 8.10 Syntheses of target compounds.

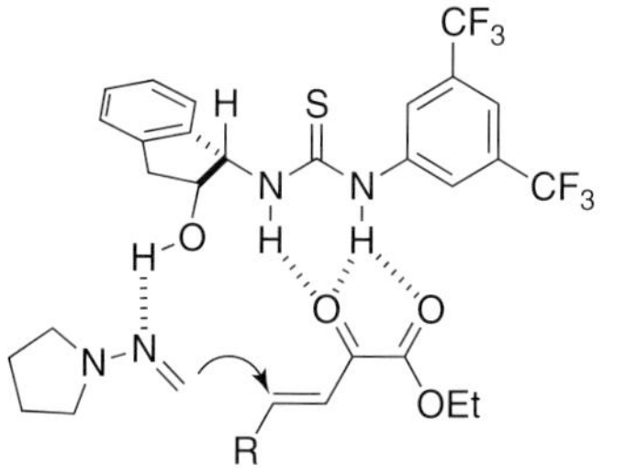

Figure 8.9 Possible bifunctional mode of action.

range of mild conditions, and the later transformation of final products into useful chiral building blocks. In this regard, the utility of adducts **22** was established by transforming the hydrazone group into some target compounds such as the corresponding nitriles **23a,b** by oxidative cleavage and into the succinate **24** resulting from oxidative decarboxylation of an unstable intermediate *in situ* generated under the specified conditions as described in Scheme 8.10. The latter allowed the assignment of the absolute configuration of product **22a** as *R*, and the same configuration was assumed for adducts **22b–f** by analogy.

In order to explain the absolute configuration obtained in the final products, we envisioned that catalyst **7d** would also act in a bifunctional way as previously shown in Figure 8.7 [29]. Therefore, whereas the two thiourea hydrogen atoms would activate the carbonyl group, the free alcoholic function would interact with the hydrazone with a weak hydrogen bond, directing the attack of the incoming nucleophile on the *Re* face of the β,γ-unsaturated α-ketoester, as illustrated in Figure 8.9.

This synthetically useful class of hydrazones, from formaldehyde *N,N*-dialkylhydrazones or from enolizable aldehydes, represents an interesting type of umpolung carbonyl reagent or imine surrogates, respectively. Moreover, their neutral character as electrophiles or nucleophiles makes them compatible

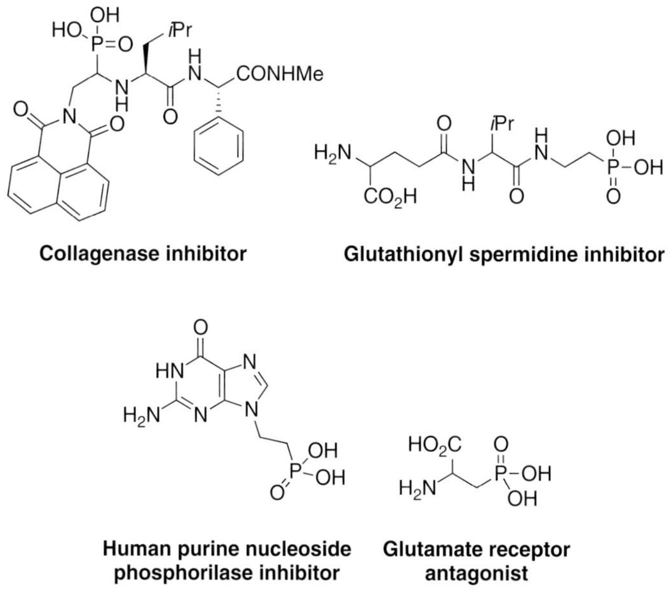

Collagenase inhibitor **Glutathionyl spermidine inhibitor**

Human purine nucleoside phosphorilase inhibitor **Glutamate receptor antagonist**

Figure 8.10 Biologically active β-amino phosphoric acid derivatives.

with many functional groups, being potential substrates in many catalytic approaches. Therefore, we expect that further efforts will be directed to the development of new applications based on the combination of organocatalysis and *N,N*-dialkylhydrazones, and a novel but promising area of research could grow up around these compounds.

8.2.2.3 Hydrophosphonylation Reaction of Nitroalkenes

Currently, the stereoselective synthesis of α- [50] and β-aminophosphonic [51] acids and derivatives has attracted particular interest because of their appealing biological activities as structural analogs to α- and β-amino acids, respectively, and because in general phosphorus compounds are important substrates in biochemical processes (Figure 8.10).

However, although great progress has been made in the preparation of enantioenriched α-aminophosphonates [50, 52], the strategies for the access to β-aminophosphonic acids have been less explored [51], and only a few examples based in organocatalytic protocols have been reported [53]. The Michael addition of phosphorus compounds to nitroalkenes [54, 55] has been applied as an effective tool for the straightforward synthesis of P-C bonds, which has become a convenient method for the construction of functionalized β-aminophosphonates. In this respect, only three interesting and versatile methods concerning the organocatalytic conjugate addition of diphenyl phosphite to nitroalkenes have been reported using chiral guanidine **25** [53a], quinine **26** [53b] and squaramide **27** [53c] (Figure 8.11). Therefore, the development of new catalytic methodologies to provide β-nitrophosphonates is still of notable importance.

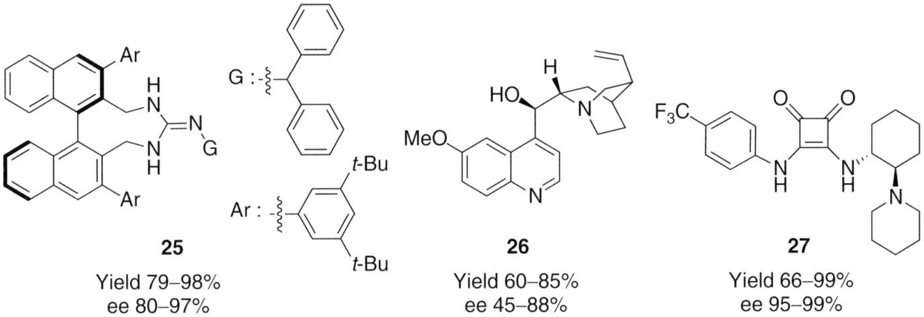

Figure 8.11 Efficient catalysts for the synthesis of β-nitrophosphonates.

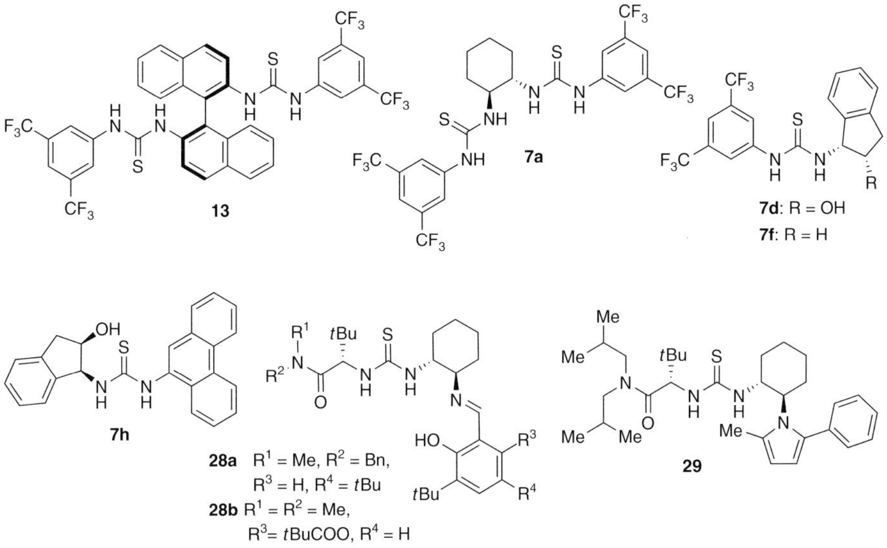

Figure 8.12 Thiourea catalysts tested.

Recently, we explored this process using chiral thiourea catalysts as a new viable route to obtain the desired β-aminophosphonate adducts [56]. Among all the thioureas tested in a preliminary study to obtain the most efficient catalyst (Figure 8.12), only **31** afforded promising results in terms of both reactivity and enantioselectivity (Scheme 8.11). It is noteworthy that, with all catalysts shown in Figure 8.12, the addition of an external base ($i\text{Pr}_2\text{EtN}$) was always required in order to achieve reactivity in the explored reaction, except when using thiourea **31** which already has a Brønsted base moiety in its structure.

After an exhaustive exploration of different parameters such as solvents, various dialkylphosphites, temperature, and reagent concentration, the best reaction conditions were accomplished with bifunctional thiourea **31** using CH_2Cl_2 and 10 mol% of catalyst at $-10\,^{\circ}\text{C}$, and we extended the applicability of our developed procedure to a variety of aromatic and aliphatic nitroalkenes (Scheme 8.11).

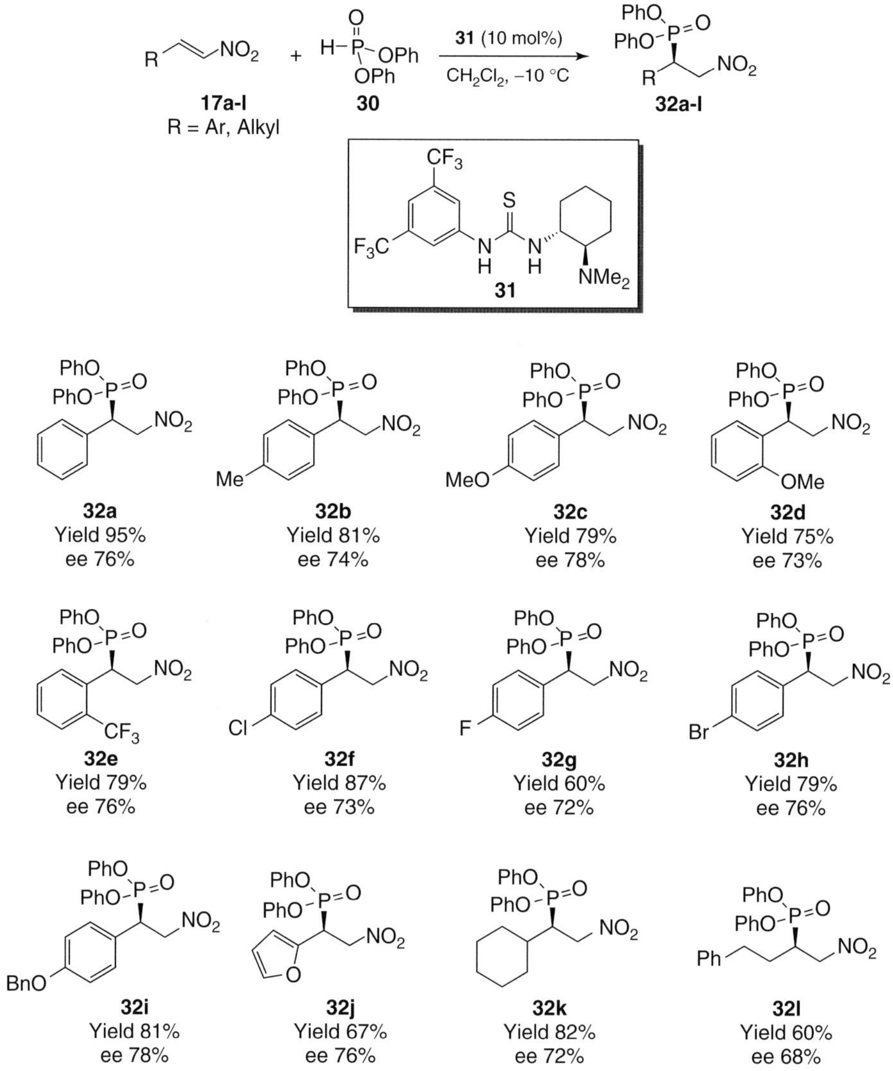

Scheme 8.11 Thiourea-catalyzed organocatalytic hydrophosphonylation of nitroalkenes.

Our moderate results are comparable to those previously reported by Wang and coworkers [53a], but in shorter reaction times because in our case it was not necessary to decrease the temperature to $-55\,^{\circ}$C. However, we could not improve upon the excellent results reported by Terada's [53b] and Rawal's [53c] groups. Nevertheless, the accessibility to the commercially available thiourea **31** supports the simplicity and viability of our methodology. The absolute configuration of the Michael adducts **32a–l** was determined by comparison between the optical rotation values with those previously reported in the literature for the same products [53].

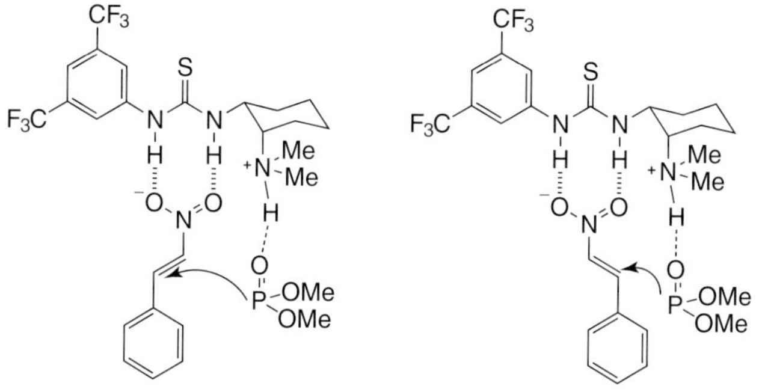

Figure 8.13 DFT calculations at B3LYP/6-31+G* level. Proposed transition-state models.

On the basis of the experimental results and comparing with other thioureas tested in our study, we envisioned that catalyst **31** could act following a bifunctional model as previously proposed in the literature for the same structure [57]. In order to give additional support to this hypothesis, we performed computational calculations using a density functional theory (DFT) model: B3LYP/6-31+G* [58]. We have studied different approaches, and the more stable **TS-1** was found to be in agreement with the experimental results, since it would afford the observed enantiomer in our final adducts (Figure 8.13).

Even though this field has only recently emerged, the growing interest for β-aminophosphonic acids showing biological activity or as useful intermediates for the preparation of peptidomimetic derivatives should stimulate further research and provide important achievements in this area.

8.2.3 Aza-Henry Reaction

The addition of nitroalkanes to C=N bond (aza-Henry or nitro-Mannich reaction) [59] is an important synthetic procedure that affords C–C bond formation, and represents a simple route to achieve two different vicinal nitrogenated functionalities. The resulting β-nitroamine could be further transformed into a variety of valuable building blocks and interesting biological compounds such as 1,2-diamines [60] or α-amino acids [61]. Much attention has been paid to the catalytic asymmetric version of this reaction over the years [59, 62], but the search of new, effective catalysts to promote highly selective processes is still continuing.

In this area, we focused our research on the study of modified Cinchona alkaloids [32, 63] in order to explore the capability of these structures to promote the addition of nitromethane to a variety of protected aromatic and heteroaromatic imines (Scheme 8.12) [64]. In a preliminary screening of a variety of catalysts (Figure 8.14),

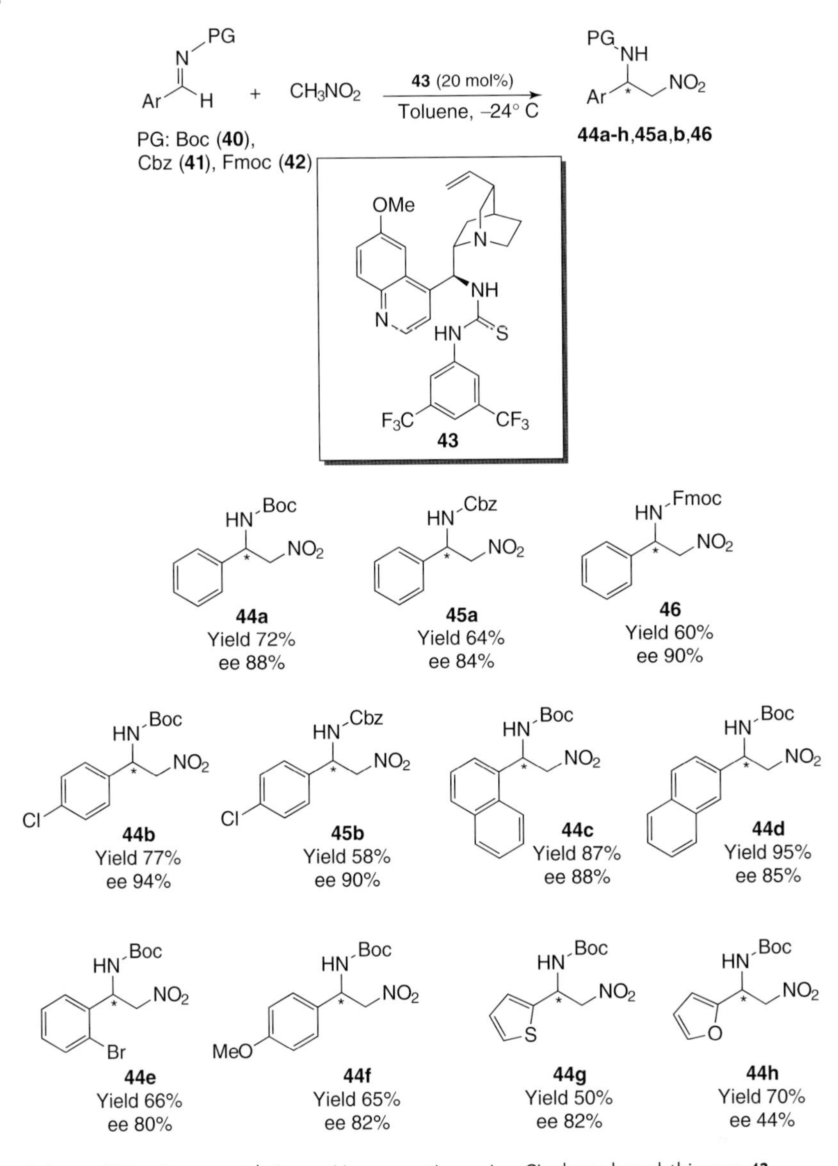

Scheme 8.12 Organocatalytic aza-Henry reaction using Cinchona-based thiourea **43**.

Cinchona-based thiourea organocatalyst **43** proved to be the most efficient one (Scheme 8.12).

The final adducts were isolated in moderate to good yields and high levels of enantioselectivity (up to 94%). Moreover, several protecting groups on the nitrogen atom were tested, including *N*-Boc, *N*-Cbz, and *N*-Fmoc groups, and although the best results were obtained with the *N*-Boc group, the high values achieved with the other two protecting groups confirmed the wide range of tolerance of this process to

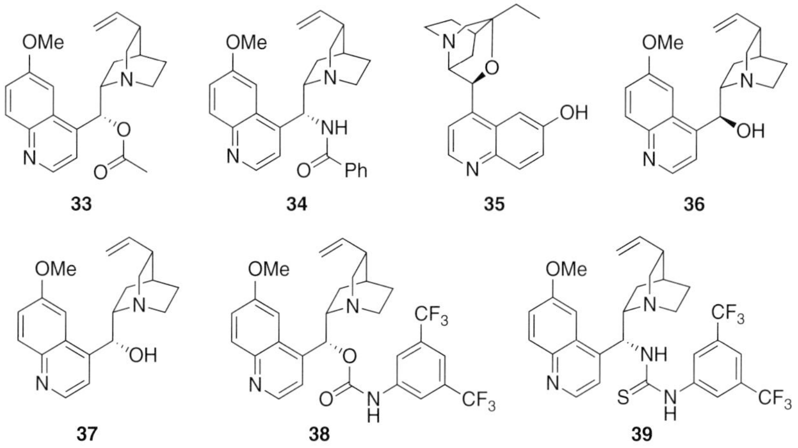

Figure 8.14 Representative Cinchona alkaloids screened in the aza-Henry reaction.

different *N*-carbamoyl groups and imines. Although a mechanistic explanation was not given, we cannot rule out a possible bifunctional mode of action by the catalyst activating the reaction through the thiourea moiety and with the quinuclidinic base function as previously suggested for this catalyst [65]. However, we observed that catalysts with the OH group of quinine derivative protected via acetylation or carbamate formation, as derivatives **33** and **38**, or replaced by a benzomido moiety as in **34**, afforded the formation of the desired adducts with very poor yields. This fact supports the importance of the presence of the thiourea moiety in the backbone of the catalyst.

The products resulting from these strategies are very important motifs in biologically active compounds and in medicinal chemistry, and for this reason this area of research has encouraged significant efforts in order to improve the previously published results; however, several drawbacks have not been overcome yet, and this will prompt further investigations in this field.

8.3 Conclusions

In this chapter, our small contribution to the huge field of organocatalysis in general and, more specifically, with chiral thioureas has been reported. Different thiourea-catalyzed processes have been illustrated, some of them original and pioneering in this area, which became a precedent for further explored reactions. Nevertheless, the exciting organocatalytic world is still expanding, particularly in the direction of development of new catalysts and more complex systems. We hear every day about the discovery of new and amazing approaches and efficient catalysts, and among them thiourea catalysts still represent an interesting research area. In the coming years, improved and valuable catalysts will be reported and more complex versions of these and other useful processes will emerge. In fact, the search for environmentally friendly catalysts is nowadays a challenge and one of the main goals in the scientific world, but its accomplishment will require further investigations.

Acknowledgments

We would like to thank the research groups where all these projects have been performed and reached: Prof. Ricci's group (Bologna, Italy, 2003–2006), Prof. Lassaletta and Prof. Fernández's group (Seville, Spain, 2006–2007), and Prof. Merino and Prof. Tejero's group (Zaragoza, Spain, 2008-present). We wish to thank the Spanish Ministry of Science and Innovation (MICINN, Madrid, Spain. Project CTQ2009-09028 and CTQ2010-19606), FEDER Program, and the Government of Aragon (Zaragoza, Spain. Project PI064/09 and Research Group E-10) for their financial support. E. M.-L. thanks CSIC for a JAE-Doc postdoctoral contract. R.P.H. thanks the ARAID Foundation for a permanent position.

References

1. (a) Pellissier, H. (ed.) (2010) *Recent Developments in Asymmetric Organocatalysis*, RSC Publishing, Cambridge; (b) Berkessel, A. and Groger, H. (2005) *Asymmetric Organocatalysis*, Wiley-VCH Verlag GmbH, Weinheim; (c) Dalko, P.I. (ed.) (2007) *Enantioselective Organocatalysis*, Wiley-VCH Verlag GmbH, Weinheim; (d) List, B. (ed.) (2010) *Asymmetric Organocatalysis*, Topics in Current Chemistry, Vol. 291, Springer, Heidelberg.
2. (a) de Figueiredo, R.M. and Christmann, M. (2007) Organocatalytic synthesis of drugs and bioactive natural products. *Eur. J. Org. Chem.*, 2575–2600; (b) Marqués-López, E., Herrera, R.P., and Christmann, M. (2010) Asymmetric organocatalysis in total synthesis – a trial by fire. *Nat. Prod. Rep.*, **27**, 1138–1167.
3. Dalko, P.I. and Mosan, L. (2004) In the golden age of organocatalysis. *Angew. Chem. Int. Ed.*, **43**, 5138–5175.
4. (a) Pihko, P.M., Majander, I., and Erkkilä, A. (2010) Enamine catalysis. *Top. Curr. Chem.*, **291**, 29–75; (b) Brazier, J.B. and Tomkinson, N.C.O. (2010) Secondary and primary amine catalysts for iminium catalysis. *Top. Curr. Chem.*, **291**, 281–347.
5. (a) Berkessel, A. (2008) Organocatalysis by hydrogen bonding networks, in *Organocatalysis*, vol. **2** (eds M.T. Reetz, B. List, H. Jaroch, and H. Weinmann), Springer-Verlag, Berlin, Heidelberg, pp. 281–297; (b) Pihko, P.M. (ed.) (2009) *Hydrogen Bonding in Organic Synthesis*, Wiley-VCH Verlag GmbH, Weinheim; (c) Kerstin, E.-E. and Berkessel, A. (2010) Noncovalent organocatalysis based on hydrogen bonding: elucidation of reaction paths by computational methods. *Top. Curr. Chem.*, **291**, 1–27.
6. (a) Ooi, T. and Maruoka, K. (2007) Recent advances in asymmetric phase-transfer catalysis. *Angew. Chem. Int. Ed.*, **46**, 4222–4266; (b) Hashimoto, T. and Maruoka, K. (2007) Recent development and application of chiral phase-transfer catalysts. *Chem. Rev.*, **107**, 5656–5682; (c) Maruoka, K. (ed.) (2008) *Asymmetric Phase Transfer Catalysis*, Wiley-VCH Verlag GmbH, Weinheim.
7. Engeldinger, E., Armspach, D., and Matt, D. (2003) Capped cyclodextrins. *Chem. Rev.*, **103**, 4147–4174.
8. (a) Schreiner, P.R. (2003) Metal-free organocatalysis through explicit hydrogen bonding interactions. *Chem. Soc. Rev.*, **32**, 289–296; (b) Takemoto, Y. (2005) Recognition and activation by

ureas and thioureas: stereoselective reactions using ureas and thioureas as hydrogen-bonding donors. *Org. Biomol. Chem.*, **3**, 4299–4306; (c) Connon, S.J. (2006) Organocatalysis mediated by (thio)urea derivatives. *Chem. Eur. J.*, **12**, 5418–5427; (d) Taylor, M.S. and Jacobsen, E.N. (2006) Asymmetric catalysis by chiral hydrogen-bond donors. *Angew. Chem. Int. Ed.*, **45**, 1520–1543; (e) Doyle, A.G. and Jacobsen, E.N. (2007) Small-molecules H-bond donors in asymmetric catalysis. *Chem. Rev.*, **107**, 5713–5743; (f) Miyabe, H. and Takemoto, Y. (2008) Discovery and Application of asymmetric reaction by multi-functional thioureas. *Bull. Chem. Soc. Jpn.*, **81**, 785–795; (g) Zhang, Z. and Schreiner, P.R. (2009) (Thio)urea organocatalysis – What can be learnt from anion recognition? *Chem. Soc. Rev.*, **38**, 1187–1198; (h) Marqués-López, E. and Herrera, R.P. (2009) El renacer de un Nuevo campo: la organocatálisis asimetrica. Tioureas como organocatalizadores. *An. Quím.*, **105**, 5–12.

9. (a) Ishikawa, T. and Kumamoto, T. (2006) Guanidines in organic synthesis. *Synthesis*, 737–752; (b) Leow, D. and Tan, C.-H. (2010) Catalytic reactions of chiral guanidines and guanidinium salts. *Synlett*, 1589–1605.
10. (a) Huang, Y., Unni, A.K., Thadani, A.N., and Rawal, V.H. (2003) Single enantiomers from a chiral-alcohol catalyst. *Nature*, **424**, 146–146; (b) Thadani, A.N., Stankovic, A.R., and Rawal, V.H. (2004) Enantioselective Diels-Alder reactions catalyzed by hydrogen bonding. *Proc. Natl. Acad. Sci. U.S.A.*, **101**, 5846–5850.
11. (a) McDougal, N.T. and Schaus, S.E. (2003) Asymmetric Morita-Baylis-Hillman reactions catalyzed by chiral Brønsted acids. *J. Am. Chem. Soc.*, **125**, 12094–12095; (b) McDougal, N.T., Trevellini, W.L., Rodgen, S.A., Kliman, L.T., and Schaus, S.E. (2004) The development of the asymmetric Morita-Baylis-Hillman reaction catalyzed by chiral Brønsted acids. *Adv. Synth. Catal.*, **346**, 1231–1240; (c) Takizawa, S., Horii, A., and Sasai, H. (2010) Acid-base organocatalysts for the aza-Morita-Baylis-Hilmann reaction of nitroalkenes. *Tetrahedron: Asymmetry*, **21**, 891–894.
12. (a) Nugent, B.M., Yoder, R.A., and Johnston, J.N. (2004) Chiral proton catalysis: a catalytic enantioselective direct aza-Henry reaction. *J. Am. Chem. Soc.*, **126**, 3418–3419; (b) Hess, A.S., Yoder, R.A., and Johnston, J.N. (2006) Chiral proton catalysis: pK_a determination for a BAM-HX Brønsted acid. *Synlett*, 147–149; (c) Singh, A., Yoder, R.A., Shen, B., and Johnston, J.N. (2007) Chiral proton catalysis: enantioselective Brønsted acid catalysed additions of nitroacetic acid derivatives as glycine equivalents. *J. Am. Chem. Soc.*, **129**, 3466–3467; (d) Shen, B. and Johnston, J.N. (2008) A formal enantioselective acetate Mannich reaction: the nitro functional group as a traceless agent for the activation and enantiocontrol in the synthesis of β-amino acids. *Org. Lett.*, **10**, 4397–4400.
13. (a) Connon, S.J. (2006) Chiral phosphoric acids: powerful organocatalysts for asymmetric addition reactions to imines. *Angew. Chem. Int. Ed.*, **45**, 3909–3912; (b) Akiyama, T. (2007) Stronger Brønsted acids. *Chem. Rev.*, **107**, 5744–5758; (c) Adair, G., Mukherjee, S., and List, B. (2008) TRIP-A powerful Brønsted acid catalyst for asymmetric synthesis. *Aldrichimica Acta*, **41**, 31–39; (d) Terada, M. (2008) Binaphthol-derived phosphoric acid as a versatile catalyst for enantioselective carbon-carbon bond forming reactions. *Chem. Commun.*, 4097–4112; (e) Terada, M. (2010) Chiral phosphoric acids as versatile catalysts for enantioselective transformation. *Synthesis*, 1929–1982; (f) Terada, M. (2010) Chiral phosphoric acids as versatile catalysts for enantioselective carbon-carbon forming reactions. *Bull. Chem. Soc. Jpn.*, **83**, 101–119.
14. (a) Etter, M.C. and Panunto, T.W. (1988) 1,3-Bis (*m*-nitrophenyl)urea: an exceptionally good complexing agent for proton acceptors. *J. Am. Chem. Soc.*, **110**, 5896–5897; (b) Etter, M.C., Urbañczyk-Lipkowska, Z., Zia-Ebrahimi, M., and Panunto, T.W.

(1990) Hydrogen-bond direct cocrystallization and molecular recognition properties of diarylureas. *J. Am. Chem. Soc.*, **112**, 8415–8426; (c) Kelly, T.R. and Kim, M.H. (1994) Relative binding affinity of carboxylate and its isosteres: nitro, phosphate, phosphonate, sulfonate, and γ-lactone. *J. Am. Chem. Soc.*, **116**, 7072–7080.

15. Curran, D.P. and Kuo, L.H. (1994) Altering the stereochemistry of allylation reactions of cyclic α-sulfinyl radicals with diarylureas. *J. Org. Chem.*, **59**, 3259–3261.
16. Curran, D.P. and Kuo, L.H. (1995) Acceleration of a dipolar Claisen rearrangement by hydrogen bonding to a soluble diaryl urea. *Tetrahedron Lett.*, **36**, 6647–6650.
17. Hine, J., Linden, S.-M., and Kanagasabapathy, V.M. (1985) 1,8-Biphenylenediol is a double-hydrogen-bonding catalyst for reaction of an epoxide with a nucleophile. *J. Am. Chem. Soc.*, **107**, 1083–1984.
18. Kelly, T.R., Meghani, P., and Ekkundi, V.S. (1990) Diels-Alder reactions: rate acceleration promoted by a biphenylenediol. *Tetrahedron Lett.*, **31**, 3381–3384.
19. (a) Bandini, M., Melloni, A., and Umani-Ronchi, A. (2004) New catalytic approaches in the stereoselective Friedel-Crafts alkylation reaction. *Angew. Chem. Int. Ed.*, **43**, 550–556; (b) Bandini, M., Melloni, A., Tommasi, S., and Umani-Ronchi, A. (2005) A journey across recent advances in catalytic and stereoselective alkylation of indoles. *Synlett*, 1199–1222; (c) Bandini, M. and Umani-Ronchi, A. (eds) (2009) *Catalytic Asymmetric Friedel-Crafts Alkylations*, Wiley-VCH Verlag GmbH, Weinheim.
20. (a) Marqués-López, E., Diez-Martinez, A., Merino, P., and Herrera, R.P. (2009) The role of the indole in important organocatalytic enantioselective Friedel-Crafts alkylation reactions. *Curr. Org. Chem.*, **13**, 1585–1609; (b) Terrasson, V., de Figueiredo, R.M., and Campagne, J.M. (2010) Organocatalyzed asymmetric Friedel-Crafts reactions. *Eur. J. Org. Chem.*, 2635–2655.
21. (a) Borschberg, H.-J. (2005) New strategies for the synthesis of monoterpene indole alkaloids. *Curr. Org. Chem.*, **9**, 1465–1491; (b) Cacchi, S. and Fabrizi, G. (2005) Synthesis and functionalization of indoles through palladium catalyzed reactions. *Chem. Rev.*, **105**, 2873–2920; (c) Poulsen, T.B. and Jørgensen, K.A. (2008) Catalytic asymmetric Friedel-Crafts alkylation reactions-copper showed the way. *Chem. Rev.*, **108**, 2903–2915.
22. Dessole, G., Herrera, R.P., and Ricci, A. (2004) H-bonding organocatalysed Friedel-Crafts alkylation of aromatic and heteroatomatic systems with nitroolefines. *Synlett*, 2374–2378.
23. Harrington, P.E. and Kerr, M.A. (1996) Reaction of indoles with electron deficient olefins catalyzed by $Yb(OTf)_3{\cdot}3H_2O$. *Synlett*, 1047–1048.
24. Komoto, I. and Kobayashi, S. (2002) 1-Dodecyloxy-4-perfluoroalkylbenzene as a novel efficient additive in aldol reactions and Friedel-Crafts alkylation in supercritical carbon dioxide. *Org. Lett.*, **4**, 1115–1118.
25. Alam, M.M., Varala, R., and Adapa, S.R. (2003) Conjugate addition of indoles and thiols with electron-deficient olefins catalyzed by $Bi(OTf)_3$. *Tetrahedron Lett.*, **44**, 5115–5119.
26. Bandini, M., Melchiorre, P., Melloni, A., and Umani-Ronchi, A. (2002) A practical indium tribromide catalysed addition of indoles to nitroalkenes in aqueous media. *Synthesis*, 1110–1114.
27. Bordwell, F.G., Algrim, D.J., and Harrelson, J.A. Jr. (1988) The relative ease of removing a proton, a hydrogen atom, or an electron from carboxamides versus thiocarboxamides. *J. Am. Chem. Soc.*, **110**, 5903–5904.
28. Scheerder, J., Engbersen, J.F.J., Casnati, A., Ungaro, R., and Reinhoudt, D.N. (1995) Complexation of halide anions and tricarboxylate anions by neutral urea-derivatized *p-tert*-butylcalix[6]arenes. *J. Org. Chem.*, **60**, 6448–6454.
29. Herrera, R.P., Sgarzani, V., Bernardi, L., and Ricci, A. (2005) Catalytic enantioselective Friedel-Crafts alkylation of indoles with nitroalkenes by using a simple thiourea organocatalyst. *Angew. Chem. Int. Ed.*, **44**, 6576–6579.

30. Freeman, S. and Alder, J.F. (2002) Arylethylamine psychotropic recreational drugs: a chemical perspective. *Eur. J. Med. Chem.*, **37**, 527–539.
31. Oh, S.J., Ha, H.-J., Chi, D.Y., and Lee, H.K. (2001) Serotonin receptor and transporter ligands-current status. *Curr. Med. Chem.*, **8**, 999–1034.
32. Connon, S.J. (2008) Asymmetric catalysis with bifunctional Cinchona alkaloid-based urea and thiourea organocatalysts. *Chem. Commun.*, 2499–2510.
33. Zhuang, W., Hazell, R.G., and Jørgensen, K.A. (2005) Enantioselective Friedel-Crafts type addition of indoles to nitro-olefins using a chiral hydrogen-bonding catalyst-synthesis of optically active tetrahydro- β-carbolines. *Org. Biomol. Chem.*, **3**, 2566–2571.
34. Fleming, E.M., McCabe, T., and Connon, S.J. (2006) Novel axially chiral bis-arylthiourea-based organocatalysts for asymmetric Friedel-Crafts type reactions. *Tetrahedron Lett.*, **47**, 7037–7042.
35. Ganesh, M. and Seidel, D. (2008) Catalytic enantioselective additions of indoles to nitroalkenes. *J. Am. Chem. Soc.*, **130**, 16464–16465.
36. Itoh, J., Fuchibe, K., and Akiyama, T. (2008) Chiral phosphoric acid catalyzed enantioselective Friedel-Crafts alkylation of indoles with nitroalkenes: cooperative effect of 3 Å molecular sieves. *Angew. Chem. Int. Ed.*, **47**, 4016–4018.
37. Roca-Lopez, D., Sadaba, D., Delso, I., Herrera, R.P., Tejero, T., and Merino, P. (2010) Asymmetric organocatalytic synthesis of γ-nitrocarbonyl compounds through Michael and domino reactions. *Tetrahedron: Asymmetry*, **21**, 2561–2601.
38. (a) Vicario, J.L., Badia, D., and Carrillo, L. (2007) Organocatalytic enantioselective Michael and hetero-Michael reactions. *Synthesis*, 2065–2092; (b) Tsogoeva, S.B. (2007) Recent advances in asymmetric organocatalytic 1,4-conjugate addition. *Eur. J. Org. Chem.*, 1701–1716; (c) Almaşi, D., Alonso, D.A., and Nájera, C. (2007) Organocatalytic asymmetric conjugate addition. *Tetrahedron: Asymmetry*, **18**, 299–365.
39. (a) Enders, D., Wang, C., and Liebich, J.X. (2009) Organocatalytic asymmetric aza-Michael additions. *Chem. Eur. J.*, **15**, 11058–11076; (b) Krishna, P.R., Sreeshailam, A., and Srinivas, R. (2009) Recent advances and applications in asymmetric aza-Michael addition chemistry. *Tetrahedron*, **65**, 9657–9572.
40. Guo, H.-C. and Ma, J.-A. (2006) Catalytic asymmetric tandem transformations triggered by conjugate additions. *Angew. Chem. Int. Ed.*, **45**, 354–366.
41. Fernández, R. and Lassaletta, J.M. (2000) Formaldehyde *N,N*-dialkylhydrazones as C-1 building-blocks in asymmetric synthesis. *Synlett*, 1228–1240.
42. (a) Marqués-López, E., Herrera, R.P., Fernández, R., and Lassaletta, J.M. (2008) Uncatalyzed Strecker-type reaction of *N,N*-dialkylhydrazones in pure water. *Eur. J. Org. Chem.*, 3457–3460; (b) Lazny, R. and Nodzewska, A. (2010) *N,N*-dialkylhydrazones in organic synthesis. From simple *N,N*-dimethylhydrazones to supported chiral auxiliaries. *Chem. Rev.*, **110**, 1386–1434.
43. Sugiura, M. and Kobayashi, S. (2005) *N*-Acylhydrazones as versatile electrophiles for the synthesis of nitrogen-containing compounds. *Angew. Chem. Int. Ed.*, **44**, 5176–5186.
44. Brehme, R., Enders, D., Fernández, R., and Lassaletta, J.M. (2007) Aldehyde *N,N*-dialkylhydrazones as neutral acyl anion equivalents: umpolung of the imine reactivity. *Eur. J. Org. Chem.*, 5629–5660.
45. (a) Dixon, D.J. and Tillman, A.L. (2005) Enantioselective Brønsted acid catalyzed addition reactions of methyleneaminopyrrolidine to imines. *Synlett*, 2635–2638; (b) Perdicchia, D. and Jørgensen, K.A. (2007) Asymmetric aza-Michael reactions catalyzed by Cinchona alkaloids. *J. Org. Chem.*, **72**, 3565–3568; (c) Rueping, M., Sugiono, E., Theissmann, T., Kuenkel, A., Köckritz, A., Pews-Davtyan, A., Nemati, M., and Beller, M. (2007) An enantioselective chiral Brønsted acid catalyzed imino-azaenamine reaction. *Org. Lett.*, **9**, 1065–1068; (d) Hashimoto, T., Hirose, M., and Maruoka, K. (2008) Asymmetric imino

aza-enamine reaction catalyzed by axially chiral dicarboxylic acid: use of arylaldehyde *N,N*-dialkylhydrazones as acyl anion equivalent. *J. Am. Chem. Soc.*, **130**, 7556–7557.

46. Pettersen, D., Herrera, R.P., Bernardi, L., Fini, F., Sgarzani, V., Fernández, R., Lassaletta, J.M., and Ricci, A. (2006) A broadened scope for the use of hydrazones as neutral nucleophiles in the presence of H-bonding organocatalysts. *Synlett*, 239–242.

47. (a) Enders, D., Wortmann, L., and Peters, R. (2000) Recovery of carbonyl compounds from *N,N*-dialkylhydrazones. *Acc. Chem. Res.*, **33**, 157–169; (b) Job, A., Janeck, C.F., Bettray, W., Peters, R., and Enders, D. (2002) The SAMP-/RAMP-hydrazone methodology in asymmetric synthesis. *Tetrahedron*, **58**, 2253–2329.

48. (a) Lassaletta, J.-M., Fernández, R., Martín-Zamora, E., and Díez, E. (1996) Enantioselective nucleophilic formylation and cyanation of conjugated enones via Michael addition of formaldehyde SAMP-hydrazone. *J. Am. Chem. Soc.*, **118**, 7002–7003; (b) Vazquez, J., Prieto, A., Fernández, R., Enders, D., and Lassaletta, J.M. (2002) Asymmetric Michael addition of formaldehyde *N,N*-dialkylhydrazones to alkylidene malonates. *Chem. Commun.*, 498–499; (c) Vázquez, J., Cristea, E., Díez, E., Lassaletta, J.M., Prieto, A., and Fernández, R. (2005) Michael addition of chiral formaldehyde *N,N*-dialkylhydrazones to activated cyclic alkenes. *Tetrahedron*, **61**, 4115–4128.

49. Herrera, R.P., Monge, D., Martín-Zamora, E., Fernández, R., and Lassaletta, J.M. (2007) Organocatalytic conjugate addition of formaldehyde *N,N*-dialkylhydrazones to β,γ-unsaturated α-keto esters. *Org. Lett.*, **9**, 3303–3306.

50. Ordóñez, M., Rojas-Cabrera, H., and Cativiela, C. (2009) An overview of stereoselective synthesis of α-aminophosphonic acids and derivatives. *Tetrahedron*, **65**, 17–49.

51. Palacios, F., Alonso, C., and de los Santos, J.M. (2005) Synthesis of β-aminophosphonates and phosphinates. *Chem. Rev.*, **105**, 899–931.

52. Merino, P., Marqués-López, E., and Herrera, R.P. (2008) Catalytic enantioselective hydrophosphonylation of aldehydes and imines. *Adv. Synth. Catal.*, **350**, 1195–1208.

53. (a) Wang, J., Heikkinen, L.D., Li, H., Zu, L., Jiang, W., Xie, H., and Wang, W. (2007) Quinine-catalyzed enantioselective Michael addition of diphenyl phosphite to nitroolefins: synthesis of chiral precursors of α-substituted β-aminophosphonates. *Adv. Synth. Catal.*, **349**, 1052–1056; (b) Terada, M., Ikehara, T., and Ube, H. (2007) Enantioselective 1,4-addition reactions of diphenyl phosphite to nitroalkenes catalyzed by an axially chiral guanidine. *J. Am. Chem. Soc.*, **129**, 14112–14113; (c) Zhu, Y., Malerich, J.P., and Rawal, V.H. (2010) Squaramide-catalyzed enantioselective Michael addition of diphenyl phosphite to nitroalkenes. *Angew. Chem. Int. Ed.*, **49**, 153–156.

54. Enders, D., Saint-Dizier, A., Lannou, M.-I., and Lenze, A. (2006) The Phospha-Michael addition in organic synthesis. *Eur. J. Org. Chem.*, 29–49.

55. (a) Bartoli, G., Bosco, M., Carlone, A., Locatelli, M., Mazzanti, A., Sambri, L., and Melchiorre, P. (2007) Organocatalytic asymmetric hydrophosphination of niroalkenes. *Chem. Commun.*, 722–724; (b) Carlone, A., Bartoli, G., Bosco, M., Sambri, L., and Melchiorre, P. (2007) Organocatalytic asymmetric hydrophosphination of α,β-unsaturated aldehydes. *Angew. Chem. Int. Ed.*, **46**, 4504–4506; (c) Ibrahem, I., Rios, R., Vesely, J., Hammar, P., Eriksson, L., Himo, F., and Córdova, A. (2007) Enantioselective organocatalytic hydrophosphination of α,β-unsaturated aldehydes. *Angew. Chem. Int. Ed.*, **46**, 4507–4510; (d) Maerten, E., Cabrera, S., Kjaersgaard, A., and Jørgensen, K.A. (2007) Organocatalytic asymmetric direct phosphonylation of α,β-unsaturated aldehydes: mechanism, scope, and application in synthesis. *J. Org. Chem.*, **72**, 8893–8903.

56. Alcaine, A., Marqués-López, E., Merino, P., Tejero, T., and Herrera, R.P. (2011)

Thiourea catalyzed organocatalytic enantioselective Michael addition of diphenyl phosphite to nitroalkenes. *Org. Biomol. Chem.*, **9**, 2777–2783.

57. (a) Okino, T., Hoashi, Y., and Takemoto, Y. (2003) Enantioselective Michael reaction of malonates to nitroolefins catalyzed by bifunctional organocatalysts. *J. Am. Chem. Soc.*, **125**, 12672–12673; (b) Okino, T., Nakamura, S., Furukawa, T., and Takemoto, Y. (2004) Enantioselective aza-Henry reaction catalyzed by a bifunctional organocatalyst. *Org. Lett.*, **6**, 625–627; (c) Okino, T., Hoashi, Y., Furukawa, T., Xu, X., and Takemoto, Y. (2005) Enantio- and diastereoselective Michael reaction of 1,3-dicarbonyl compounds to nitroolefins catalyzed by a bifunctional thiourea. *J. Am.Chem. Soc.*, **127**, 119–125.

58. Frisch, M.J., Trucks, G.W., Schlegel, H.B., Scuseria, G.E., Robb, M.A., Cheeseman, J.R., Scalmani, G., Barone, V., Mennucci, B., Petersson, G.A., Nakatsuji, H., Caricato, M., Li, X., Hratchian, H.P., Izmaylov, A.F., Bloino, J., Zheng, G., Sonnenberg, J.L., Hada, M., Ehara, M., Toyota, K., Fukuda, R., Hasegawa, J., Ishida, M., Nakajima, T., Honda, Y., Kitao, O., Nakai, H., Vreven, T., Montgomery, Jr., J.A., Peralta, J.E., Ogliaro, F., Bearpark, M., Heyd, J.J., Brothers, E., Kudin, K.N., Staroverov, V.N., Kobayashi, R., Normand, J., Raghavachari, K., Rendell, A., Burant, J.C., Iyengar, S.S., Tomasi, J., Cossi, M., Rega, N., Millam, N.J., Klene, M., Knox, J.E., Cross, J.B., Bakken, V., Adamo, C., Jaramillo, J., Gomperts, R., Stratmann, R.E., Yazyev, O., Austin, A.J., Cammi, R., Pomelli, C., Ochterski, J.W., Martin, R.L., Morokuma, K., Zakrzewski, V.G., Voth, G.A., Salvador, P., Dannenberg, J.J., Dapprich, S., Daniels, A.D., Farkas, Ö., Foresman, J.B., Ortiz, J.V., Cioslowski, J., and Fox, D.J. (2009) *Gaussian 09, Revision A.1*, Gaussian, Inc., Wallingford CT.

59. Marqués-López, E., Merino, P., Tejero, T., and Herrera, R.P. (2009) Catalytic enantioselective aza-Henry reactions. *Eur. J. Org. Chem.*, 2401–2420.

60. Bernadi, L., Bonini, B.F., Capitò, E., Dessole, G., Comes-Franchini, M., Fochi, M., and Ricci, A. (2004) Organocatalyzed solvent-free aza-Henry reaction: a breakthrough in the one-pot synthesis of 1,2-diamines. *J. Org. Chem.*, **69**, 8168–8171.

61. Ballini, R. and Petrini, M. (2004) Recent synthetic development in the nitro to carbonyl conversion (Nef reaction). *Tetrahedron*, **60**, 1017–1047.

62. (a) Westermann, B. (2003) Asymmetric catalytic aza-Henry reactions leading to 1,2-diamines and 1,2-diaminocarboxylic acids. *Angew. Chem. Int. Ed.*, **42**, 151–153; (b) Ting, A. and Schaus, S.E. (2007) Organocatalytic asymmetric Mannich reactions: new methodology, catalyst design and synthetic applications. *Eur. J. Org. Chem.*, 5797–5815.

63. (a) Song, C.E. (ed.) (2009) *Cinchona Alkaloids in Synthesis and Catalysis: Ligands, Immobilization and Organocatalysis*, Wiley-VCH Verlag GmbH, Weinheim; (b) Marcelli, T. and Hiemstra, H. (2010) Cinchona alkaloids in asymmetric organocatalysis. *Synthesis*, 1229–1279; (c) Ting, A., Goss, J.M., McDougal, N.T., and Schaus, S.E. (2010) Brønsted base catalysts. *Top. Curr. Chem.*, **291**, 145–200.

64. Bernardi, L., Fini, F., Herrera, R.P., Ricci, A., and Sgarzani, V. (2006) Enantioselective aza-Henry reaction using Cinchone organocatalysts. *Tetrahedron*, **62**, 375–380.

65. Pettersen, D., Piana, F., Bernardi, L., Fini, F., Fochi, M., Sgarzani, V., and Ricci, A. (2007) Organocatalytic asymmetric aza-Michael reaction: enantioselective addition of *O*-benzylhydroxylamine to chalcones. *Tetrahedron Lett.*, **48**, 7805–7808.

9
Electrosynthesized Structured Catalysts for H_2 Production

Patricia Benito, Francesco Basile, Giuseppe Fornasari, Marco Monti, Erika Scavetta, Domenica Tonelli, and Angelo Vaccari

9.1
Introduction

Structured catalysts [1], which were initially developed for automobile exhaust treatment, are attracting interest in other gas/solid or liquid/solid processes such as the treatment of NOx, diesel oxidation, catalytic combustion, volatile organic compound (VOC) elimination, reforming and partial oxidation of hydrocarbons, preferential oxidation of CO, Fischer–Tropsch, hydrogenation, water splitting, and oxidative dehydrogenation of propane [2–14]. The main advantages of structured packings over conventional packed-bed reactors are the high geometric surface area of the catalyst, the low pressure drop, the high mechanical resistance, and stability. Considering the distribution of the active phase, they have been classified as "incorporated" or "coated" [15], depending on whether the active phase is distributed over the whole support or just on its surface. Nowadays, most of the structured catalysts consist of a catalyst-coated ceramic (Al_2O_3, cordierite, and SiC) or metallic (Al, Ni, FeCrAlloy, Inconel, and AISI 304 stainless steel) supports, adopting several shapes such as honeycomb, foil, corrugated sheet, gauze, foam, fiber, or knitted wire packing. The material selection is based on its thermal and mechanical stability under reaction conditions [16]. For instance, metals are preferred when enhanced heat transfer and mechanical robustness are required; moreover, because of their good ductility, thin walls and complex shapes can be obtained. The shape of the support, particularly the geometry of the pores, determines the flow path and therefore the mass and heat transfer [17]. Honeycombs with straight square, triangular, or round parallel channels lead to a laminar flow and some mass transfer limitations; furthermore, no axial mixing of the gas is allowed. In corrugated sheets, several configurations have been adopted to modulate the flow path and to obtain an open or closed cross-flow structure. Complex supports such as foams, felts, and fibers with an open-pore structure exhibit a significant degree of interconnectivity and void space, which results in lower pressure drop, enhanced mass and heat transfer, and a considerable degree of radial mixing [8, 18].

New Strategies in Chemical Synthesis and Catalysis, First Edition. Edited by Bruno Pignataro.

In particular, structured catalysts based on metal supports are promising in heat-transfer-limited processes, such as the endothermic steam reforming (SR) of methane (Eq. (9.1)), and in those where the heat removal is important, such as the exothermic catalytic partial oxidation (CPO) of methane (Eq. (9.2)). In the SR process, the heat supplied to the catalytic bed can be decreased, and the lifetime of the reactor tubes increased. Moreover, since the amount of catalyst and the contact time are reduced, compact reactors have been developed for on-site H_2 production in fuel cells and in microplants [19]. On the other hand, in the CPO process the formation of hot spots could be reduced, decreasing safety risks. Since SR and CPO are H_2 production processes, the improvement of the catalyst would lead to an enhancement in the H_2 economy [9].

$$CH_4 + H_2O \rightleftharpoons CO + 3H_2 \qquad \Delta H^0{}_{298K} = 206\ \text{kJ mol}^{-1} \tag{9.1}$$

$$CH_4 + O_2 \longrightarrow CO + 2H_2 \qquad \Delta H^0{}_{298K} = -36\ \text{kJ mol}^{-1} \tag{9.2}$$

9.2
Preparation of Structured Catalysts

The replacement of conventional catalysts by structured ones depends on the development of active and stable catalytic films. Because of the low amount of catalyst, a coating with highly active catalysts is required. The amount and dispersion of the active phase can be controlled by changing either the thickness of the coating or the metal loading. Furthermore, the performance of the catalyst is determined by the morphology and stability of the film. All these features strongly depend on the coating method.

Many methods have been reported in the literature to prepare structured catalysts [15, 20, 21]. They involve the coating of a readymade support or catalyst, as well as the synthesis of the catalyst directly on the surface of the structured support. These methods are used to coat both ceramic and metal supports. When dealing with metal supports, special attention must be paid since the low interaction between the support and the catalyst (usually a ceramic material) and the mismatch between thermal expansion coefficients may lead to the formation of creeps and peel-off of the catalytic film. In order to increase the adhesion, the metal support is usually subjected to thermal treatment, anodic oxidation, or chemical treatment. Thermal treatment generates an oxide scale of Al_2O_3 for the FeCrAlloy [22] and manganese and chromium oxides for AISI 304 [23]. This scale not only increases the adhesion of the coating but also protects the metal from corrosion. Anodic oxidation is mainly performed on Al supports in order to obtain an alumina layer [24, 25], but it has been also applied to a zirconium plate to produce zirconia [26]. In some cases, the deposition of a binder such as γ-Al_2O_3 is carried out to increase the specific surface area of the support [27, 28].

Washcoating [15], plasma spraying [29], remote plasma-enhanced chemical vapor deposition [30], and sol–gel [31] are used to deposit a readymade catalyst or support. Among all of them, washcoating is by far the most widely used. A support is immersed for a short period in a stabilized slurry of the catalyst (a colloid or a sol can be also used), then it is withdrawn, the excess slurry is blown out, and finally the liquid is removed by drying. A calcination step is usually performed to better anchor the catalyst to the support. The quality of the coating (homogeneity, specific load, and adherence) depends on the slurry's properties (stability and viscosity) and the procedure used to remove the excess slurry (air blowing or centrifugation). The properties of the slurry depend on the solid concentration, particle size distribution, pH, and the presence of binders or surfactants [15, 32–39]. Therefore a large number of parameters must be controlled to achieve a reproducible coating. Moreover, when the active phase must be incorporated in a subsequent step, such as impregnation, special precautions must be taken to avoid problems of heterogeneity [40–42].

As an alternative to washcoating, the catalyst can be directly synthesized on the support. Synthetic methods used for the synthesis of powder catalysts have been extended to the in situ synthesis of the catalytic coating: hydrothermal treatment to coat zeolites on ceramic and metallic supports [43]; deposition–precipitation to obtain $LaMnO_3$ on cordierite monoliths washcoated with a lanthanum-stabilized γ-Al_2O_3 [44] or ZrO_2 [45]; spray pyrolysis to produce perovskites [46]; solution combustion synthesis of perovskites and mixed oxides on metallic fibers [47]; citrate and urea combustion method to load CuO–CeO_2 on meso/macroporous alumina monoliths [48]; and the reflux synthesis of manganese oxides on austenitic stainless steel [49].

9.3 Electrosynthesis

Electrochemical deposition is a unique technique for coating electrically conducting supports. It is achieved by passing an electric current between two or more electrodes immersed in a plating solution, which contains the salts of the cations to be deposited. The electrosynthesis is usually performed in a potentiostatic or galvanostatic mode, although a potentiodynamic operation mode has also been reported [50–52]. The working electrode is the conductive material to be coated and it can act as the anode (anodic deposition) or the cathode (cathodic deposition). The plating solution is usually an aqueous solution, but nonaqueous media such as organic solvents or molten NaOH [53] can be also used. The composition, morphology, and texture of the deposited films are tuned by means of the potential, current density, deposition time, and composition of the plating solution.

Metal or alloy coatings have been largely produced [53–55]. The metallic precursors are electrochemically redox-active species that are reduced, and the deposition of metal particles of different shape takes place on the surface of the electrode. On the other hand, oxides, hydroxides, or peroxides are synthesized on conductive materials by anodic or cathodic electrochemical synthesis [56, 57]. In this case, two distinctions can be made [58]: (i) the metal precursors suffer from an electrochemical change in order to convert them from soluble to insoluble species, mainly taking place at the anode [59]; (ii) or the electrochemically active species are not the metals but other species exchanging charge with the electrode. H_2O [60], NO_3^- [56], H_2O_2 [61], or dissolved O_2 [62] is used to precipitate hydroxides, peroxides, and, sometimes directly, the oxide. Low-crystallized compounds are usually obtained by the base electrogeneration method, but in some cases the deposition leads to the formation of crystalline thin films under near-room temperatures.

Metal hydroxide thin films and powders produced by this technique have been explored extensively as materials in energy storage (often using cobalt or nickel precursors). Currently, the method is being applied to the synthesis of a wide range of hydroxides/oxides on several supports (platinum, stainless steel, carbon nanotubes, and indium–tin oxide) [63]. $Mg(OH)_2$ [60, 64], Al_2O_3 [65], CeO_2 [66, 67], ZrO_2 [68], FeOOH, iron oxides (Fe_2O_3 [69] and Fe_3O_4 [70]), hydrotalcite-type (HT) compounds [71], TiO_2 [72, 73], RuO_2 [72–74], PbO_2 [75], ZnO [76, 77], perovskites [78–80], $La(OH)_3$ [81, 82], high-temperature oxide superconductors, and composite materials such as $PbO_2–TiO_2$ [83] have been electrosynthesized. By means of electrosynthesis, not only the thickness of the film but also the shape of the single deposited particles has been controlled. The addition of templates [84], polyelectrolyte polymers [85], capping agents such as ammonium fluoride or pyridine [86], and Brij56 surfactant as structure-directing [87] or the selective etching of the nanorods [81] give rise to nanorods, nanotubes, or nanodot arrays. Moreover, binders such as chitosan [69] or poly(diallyldimethylammonium chloride) (PDDA) [88] improve the adhesion and cracking resistance in the as-deposited films.

Electrosynthesis has been extended to the preparation of HT compounds on FeCrAlloy foams [89–92] to obtain structured catalysts for H_2 production by the SR and CPO of methane. HTs are lamellar compounds with the general chemical formula $[M^{2+}{}_{1-x}M^{3+}{}_x(OH)_2](A^{n-})_{x/n}nH_2O$ [93–95], which have been largely used for the preparation of H_2 production catalysts [96]. In particular, Mg/Al compounds have been synthesized where part of divalent cations are replaced by Ni^{2+} or part of the trivalent ones by noble metals, such as Rh^{3+} [96]. After calcination at a high temperature, basic oxides with relatively high surface area are obtained wherein the active species are well dispersed. The most largely applied synthesis method to obtain this kind of compounds is the coprecipitation of the salts of the metals by adding a base, and the electrosynthesis has been used to prepare electrodes modified with HT compounds [71, 97–99]. The aim of this chapter is to give an overview of the results obtained during the preparation of Ni/Al and Rh/Mg/Al compounds on FeCrAlloy foams.

9.4 Electrosynthesis of Hydrotalcite-Type Compounds

9.4.1 Experimental

Ni/Al and Rh/Mg/Al HT compounds are electrochemically synthesized on metal foams in a single-compartment, three-electrode cell containing a 0.03 M solution

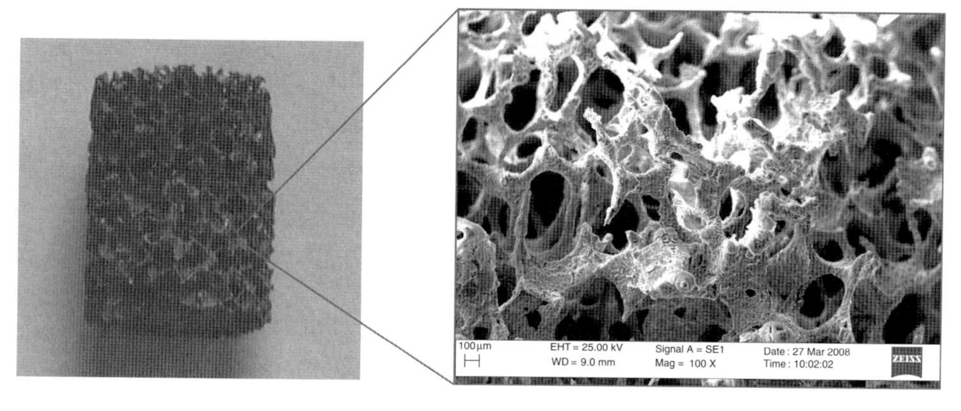

Figure 9.1 Details of a bare FeCrAlloy foam.

of the nitrates of the cations and a 0.3 M solution of KNO_3 as supporting electrolyte. The atomic ratio of the HT cations in the solutions is Ni/Al = 75/25 and Rh/Mg/Al = 11/70/19. The synthesis is performed at room temperature by applying a cathodic potential to the working electrode: 80 ppi FeCrAlloy foam pellets (Figure 9.1) or a plate. The dimensions of the foam pellets are 1.2×1 cm for the Ni/Al catalysts and 0.8×1 cm for the Rh/Mg/Al catalysts. Before the electrochemical synthesis, the foam pellets are rinsed with ethanol and water. The electric contact is made by placing a platinum wire through the foam. The electrode potentials are measured versus an aqueous saturated calomel electrode (SCE) (i.e., reference electrode or RE). The counter electrode, consisting in a Pt gauze, is placed around the foam (at a distance of about 0.5 cm), in order to ensure an optimal cell geometry and minimize potential gradients. The applied cathodic potential varies from −0.9 to −1.3 V versus SCE and the deposition time ranges from 600 to 1800 s (Table 9.1). After washing and drying at 40 °C, the coated foam pellets are weighed. Catalysts are obtained by calcination at 900 °C for 12 h (heating rate 10 °C min^{-1}).

Table 9.1 Summary of the HT samples prepared by electrosynthesis.

Composition	Potential vs SCE (V)	Time (s)	HT weight (%)
Ni/Al	−0.9	600	nd
Ni/Al	−0.9	1800	1.2
Ni/Al	−1.2	600	1.5
Ni/Al	−1.2	1000	2.2
Rh/Mg/Al	−1.2	1000	<1
Rh/Mg/Al	−1.3	1000	5

nd, not determined.

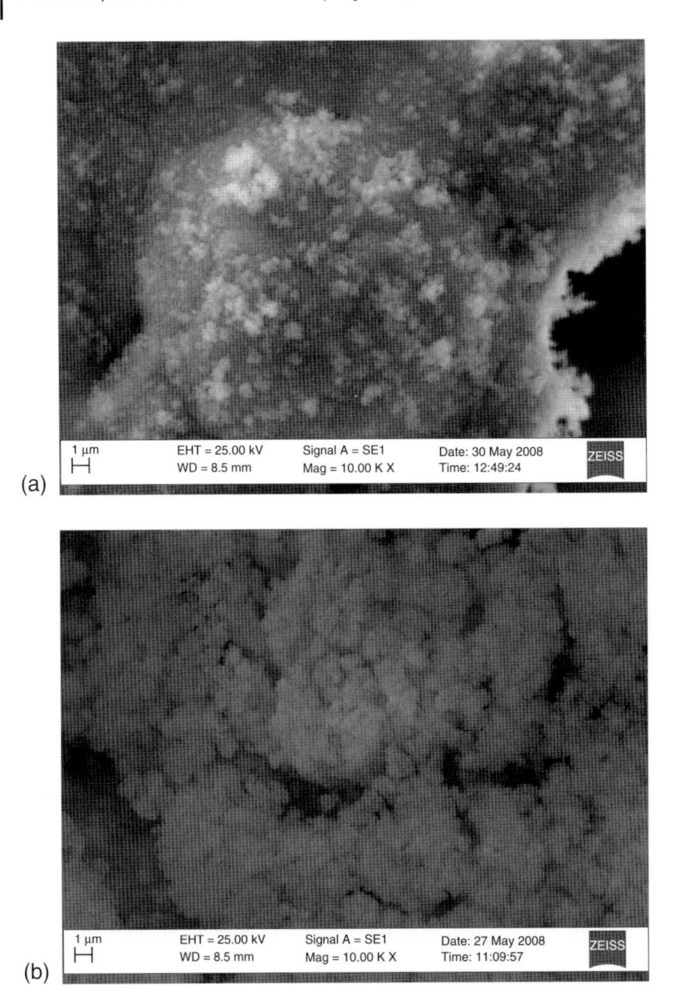

Figure 9.2 SEM images of the Ni/Al samples prepared at (a) −0.9 V for 1800 s and (b) −1.2 V for 1000 s.

SR tests are carried out under industrial-like conditions in an INCOLOY 800 HT reactor (i.d. 1.2 cm), $P = 20$ bar, steam/carbon = 1.7, $T_{oven} = 900\,^\circ C$, and contact time (τ) = 4 s (calculated at the operative conditions) [90]. Six foam pellets (1.2 × 1 cm) are loaded in the isothermal zone of the reactor, while the top and bottom sections are filled with an inert material. CPO tests are performed in a quartz reactor (i.d. 0.8 cm) operating at atmospheric pressure. Two cylinders of the foams (8 × 10 mm) are loaded in the isothermal zone of the reactor. The oven temperature is kept at 750 °C. Two gas hourly space velocity (GHSV) values are used: 28 000 and 120 000 h^{-1}; and two gas mixtures are fed: $CH_4/O_2/He = 2/1/20$ and 2/1/4 v/v [91]. Before testing, catalysts are activated by in situ reduction at 900 °C (Ni/Al) or 750 °C (Rh/Mg/Al) for 2 h with a H_2/N_2 equimolar flow (7 l h^{-1}). GHSV values are calculated by using the total gross volume of the foam pellets.

9.4.2
Ni/Al and Rh/Mg/Al HT Compounds on FeCrAlloy Foams

The film of solid deposited on the surface of the supports is obtained by applying a cathodic potential to foam pellets immersed in a nitrate solution containing the cations (Me^{n+}) required to form the Ni/Al or Rh/Mg/Al compounds. After the cathodic pulse, a series of reactions may take place at the electrode/electrolyte interface [56]:

$$H^+ + e^- \rightarrow H_{ads} \text{ or } H_{abs} \tag{9.3}$$

$$2H^+ + 2e^- \rightarrow H_2 \qquad E^\circ = 0.0V \tag{9.4}$$

$$NO_3^- + 2H^+ + 2e^- \rightarrow NO_2^- + H_2O \qquad E^\circ = 0.934V \tag{9.5}$$

$$NO_3^- + 10H^+ + 8e^- \rightarrow NH_4^+ + 3H_2O \qquad E^\circ = 0.36V \tag{9.6}$$

$$2H_2O + 2e^- \rightarrow H_2 + 2OH^- \qquad E^\circ = -0.828V \tag{9.7}$$

$$NO_3^- + H_2O + 2e^- \rightarrow NO_2^- + 2OH^- \qquad E^\circ = 0.01V \tag{9.8}$$

$$NO_3^- + 7H_2O + 8e^- \rightarrow NH_4^+ + 10OH^- \qquad E^\circ = -0.12V \tag{9.9}$$

$$Me^{n+} + ne^- \rightarrow Me^0 \tag{9.10}$$

$$Me^{n+} + nOH^- \rightarrow Me(OH)n \tag{9.11}$$

The nitrate reduction is considered the most significant reaction contributing to the formation of OH^- species near the working electrode [56]. The cations react with the OH^- species forming the hydroxides, which subsequently precipitate on the surface of the foam (heterogeneous nucleation) (Eq. (9.11)). Together with the electrochemical reactions involving the solvent and the nitrates, the reduction of the cations to the metal (Eq. (9.10)) may become competitive for easily reducible metals.

The OH^- formation rate and the length of the potentiostatic electrosynthesis are dependent on the working electrode (material and shape) as well as on the potential applied. Furthermore, it should be also taken into account that the precipitation of HT compounds takes place in a selected pH range, which is dependent on the composition of the layers. Outside this range, the precipitation of side phases such as the single hydroxides or the redissolution of the precipitated solid may take place [93]. Therefore, the optimization of the potential is of paramount importance to control the coverage and composition of the film.

The pH value in the vicinity of the foam pellets is estimated by using acid–base indicators during the application of a cathodic potential to the metal foam immersed in a KNO_3 solution [90]. In this way, the pH at the beginning of the synthesis can be measured. A stable pH value is reached in the vicinity of the foam at about 50 s after the application of the cathodic potential; by prolonging the time, the pH does not increase, but the generation of the basic media is extended throughout the bulk of the solution. The more negative the applied potential, the more basic the medium close to the electrode, since the nitrate reduction is enhanced. The pH reached at −0.9 V versus SCE is in the 6.5 and 7.5 range; at −1.2 V versus SCE a value of about 8.7–9.6 is obtained, whereas by applying a −1.3 V versus SCE potential the

Table 9.2 M^{2+}/M^{3+} atomic ratio in the coating of the precursors obtained by EDS.

Composition	Potential vs SCE (V)	Time (s)	M^{2+}/M^{3+} atomic ratio	
			Zone 1	Zone 2
Ni/Al	−0.9	600	<0.1	0.6
Ni/Al	−0.9	1800	1.0	2.0
Ni/Al	−1.2	600	2.1	2.2
Ni/Al	−1.2	1000	3.2	3.4
Rh/Mg/Al	−1.2	1000	<1 (2.0)	2.9 (0.1)
Rh/Mg/Al	−1.3	1000	1.7 (0.21)	3.5 (0.06)

Zone 1: flat strut, Zone 2: tip of the strut. The values in brackets correspond to the Rh/Mg ratio.

pH reaches values close to 9.0–11.0. However, during the potentiostatic synthesis, the diffusion rates of reactant molecules and the conductivity of the metallic foam can change as the solid film grows, and thus it could imply a decrease in the cell current and a modification of the pH. Moreover, gradients of potential are generated in the foam, locally delaying the nitrate reduction and modifying the pH values. Lastly, some differences are also observed between the inner and outer parts of the foams.

The properties of the film, namely, the homogeneity, thickness, and composition, depend on the potential applied and synthesis time, but the same trend is observed for all the samples. During the synthesis of Ni/Al compounds at −0.9 V versus SCE, the potential gradients require that the precipitation first takes place on the surface flaws of the foam pellets (or tips) where a higher potential is generated. The homogeneity of the film is improved by performing the synthesis for longer periods (1800 instead of 600 s). However, the low pH reached at this potential leads to the precipitation of small particles of the hydroxides with lower solubility constants, that is, $Al(OH)_3$ or AlOOH (Figure 9.2a). The coated solids contain only a small amount of nickel in comparison to that in the initial solution. The Ni/Al ratio (Table 9.2) increases with synthesis time (from 600 to 1800 s) and it is larger in the better covered struts; but the values are far from that in the original solution. The application of a more cathodic potential (−1.2 V vs SCE) increases the rate of generation of the basic media; however, as during the synthesis at −0.9 V, a short deposition time (600 s) induces the formation of the film mainly on the tips of the foam struts; after prolonging the time up to 1000 s, a better coverage is obtained as shown in Figure 9.2b. The Ni/Al ratio is around 2.1 at 600 s and reaches a value of ~3.4, which is closer to the theoretical one, at 1000 s (Table 9.2), indicating that the conditions required to coprecipitate Ni^{2+} and Al^{3+} ions as an HT phase are achieved. Some small cracks are developed in foam coated for 1000 s, probably related to the volume shrinkage during the drying step.

The replacement of Ni^{2+} by Mg^{2+} cations in the structure increases the precipitation pH of the HT phase [93]. Therefore the electrosynthesis is performed

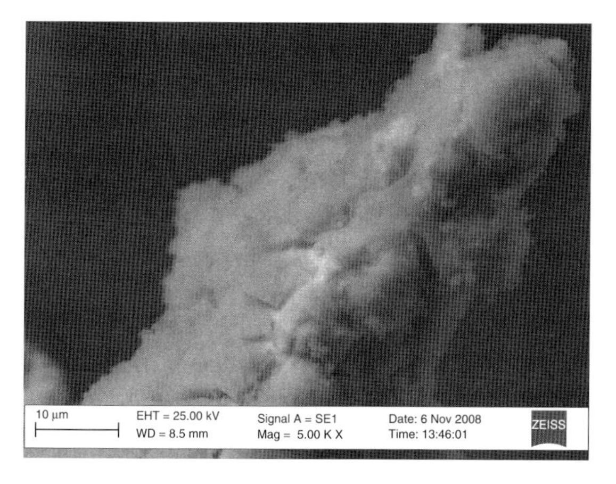

Figure 9.3 SEM image of the Rh/Mg/Al-coated foam obtained after electrosynthesis at −1.2 V for 1000 s.

at −1.2 and −1.3 V versus SCE in order to reach a higher pH in the vicinity of the foam. As commented previously, a drawback of the synthesis involving easily reducible cations, such as Rh^{3+}, is the deposition of metal particles on the working electrode surface. In fact, the potential gradients do not ensure that in some zones the precipitation of the rhodium as hydroxide occurs. In order to overcome this problem, the pH of the plating solution must be carefully adjusted. According to the Pourbaix diagram, the formal potential of reduction decreases by increasing the pH; however, as the pH increases, the precipitation of the aluminum hydroxides in the bulk may also take place. A value of 3.8 fulfills the above requirements under the working conditions used in this study. By applying a −1.2 V versus SCE potential, a deposition time of 1000 s does not provide a homogeneous coverage of the surface, and the deposition takes place preferentially in the most exposed areas (Figure 9.3). Furthermore, as during the synthesis of Ni/Al compounds at low potential, the solid is richer in Al^{3+} and Rh^{3+} cations than in Mg^{2+} ions (Table 9.2). As deposition potential becomes more negative (−1.3 V vs SCE), the current increases and leads to a faster electrosynthesis rate, as well as to a better coverage and a higher incorporation of Mg^{2+} ions in the solid (Table 9.2). However, the increase of the coating thickness brings about the formation of deep cracks. The formation of the cracks can be mainly related to the shrinkage due to capillary forces of the solid during the drying step. However, it should be also kept in mind that, at high potential, the cathodic reduction of the water (water electrolysis) can take place with the formation of H_2 bubbles, which may disturb the coating and lead to the formation of creeps.

The formation of a poorly crystallized HT phase containing nitrate anions in the interlayer is confirmed by X-ray diffraction (XRD) patterns of the Ni/Al and Rh/Mg/Al powders synthesized on FeCrAlloy plates at −1.2 and −1.3 V versus SCE for 1000 s [90, 91]. Nevertheless, the brucite, $Mg(OH)_2$, phase is segregated in the Rh/Mg/Al compounds.

The formation of HT compounds on the surface of the foam may take place through a one- or two-step mechanism. At low potential (−0.9 V vs SCE) or just at the beginning of the synthesis at higher potentials (−1.2 and −1.3 V vs SCE), Al^{3+} and/or Rh^{3+} cations precipitate as hydroxide or oxyhydroxide phases. The increase of the pH allows the coprecipitation of the HT compound. The process may be completed by the redissolution and reprecipitation of the hydroxides previously precipitated. It may be assumed that, during the electrosynthesis, the foam surface is firstly covered with a thin solid layer, after which micro and macro aggregates begin to grow. In a precipitation process, the size of the particles is dependent on the nucleation and growth rate. In the work reported here, during the electrosynthesis at low potentials, the growth rate should be higher than the nucleation rate and the particle size should decrease with increasing applied potential. Contrarily, the morphology and size of the particles are similar regardless of the potential applied; it points to the fact that, under the electrosynthesis conditions, the generation reaction rate may be high and the rate of nucleation may exceed that of particle growth [100]. Lastly, it should be remarked that coated solids are not electrical conductors, water molecules and NO_3^- species should go through the solid phase toward the electrode, and the NO_2^- and OH^- ions that are formed at the inner interface go in the opposite direction toward the electrolyte. The lamellar structure of the HTs with large basal spacing would favor this process; but the possible delay or inhibition of the deposition due to the coating resistivity should be considered.

9.4.3 Catalysts

Coated metal foams are calcined at high temperature (900 °C) in order to obtain the final catalyst. During this thermal treatment, both the coating and the foam undergo some changes. The thermal decomposition of HT compounds [93–95] involves the removal of physisorbed and intercalated water molecules in a first step. The dehydroxylation and decomposition of the interlayer anions as CO_2 and NO_x occur at moderate temperatures (ca. 400–600 °C), resulting in the formation of a mixed oxide. At higher temperatures (above 750–800 °C), the mixed oxide further transforms into a more compact structure composed of oxide (MgO and NiO) and spinel-type ($NiAl_2O_4$ and $MgAl_2O_4$) phases. On the other hand, concerning the metal foam changes, the inward diffusion of O_2 to the FeCrAlloy foam takes place, with the movement of aluminum from the inner part to the outer surface of the metal. A reactive γ-Al_2O_3 is formed in the first stages, which then transforms into an α-Al_2O_3 scale.

These thermal transformations lead to modifications in the morphology, composition, and structure of the coating. The shrinkage in volume of the catalyst and the differences in the thermal expansion coefficients of the FeCrAlloy and the ceramic solid may lead to the formation of cracks. The possible reaction between the alumina coming from the foam and the cations of the HT phase could form additional amounts of spinel-type phase and increase the compatibility and

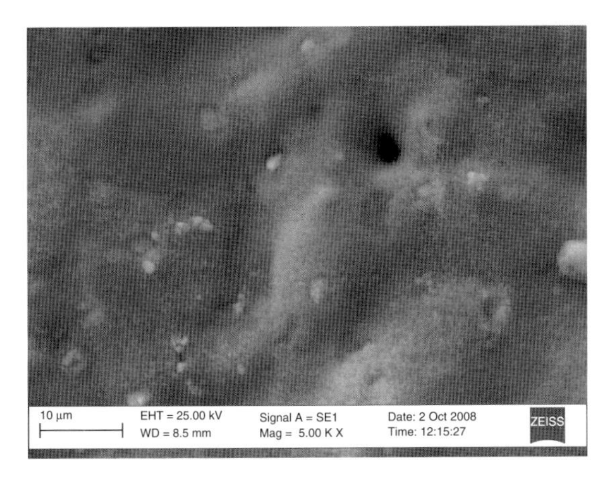

Figure 9.4 SEM image of the Ni/Al sample prepared at −0.9 V for 1800 s and calcined at 900 °C for 12 h.

interaction between the catalyst and the support. These changes depend on the coverage of the surface and the composition of the solids before calcination.

The coating of the foams covered by an Al-rich solid before calcination consists of a homogeneous film of needle-like particles, due to the deposited solid and the alumina coming from the foam (Figure 9.4). On the other hand, when HT compounds are deposited, either for Ni/Al and Rh/Mg/Al catalysts, small particles, more compact than in the precursor, coat the foam surface. The degree of coverage depends on the homogeneity and thickness of the HT film in the precursors. During the calcination of the Ni/Al sample prepared at −1.2 V versus SCE for 600 s, the low coating degree and thermal stresses result in cracking and spalling, leaving some parts of the metallic foam pellets uncovered (Figure 9.5). The surface

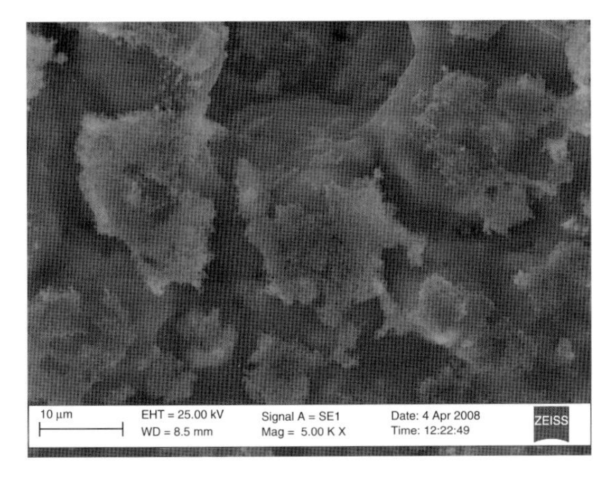

Figure 9.5 SEM image of the Ni/Al catalyst obtained by calcination at 900 °C for 12 h of the sample synthesized at −1.2 V for 600.

of the Ni/Al and Rh/Mg/Al samples prepared for 1000 s shows some cracks, where alumina coming from the oxidation of the FeCrAlloy foam grows. Lastly, the film of the Rh/Mg/Al catalyst prepared at −1.3 V versus SCE is formed by surface flakes and interconnected cracks. Alumina particles are also observed in the cracks where Rh^{3+} ions are present; the thick layer may enhance the thermal stresses and therefore the creep formation. Despite the cracks, the coatings of the Ni/Al and Rh/Mg/Al catalysts obtained at −1.2 and −1.3 V versus SCE are quite stable even after ultrasound treatment in petroleum ether for 40 min at room temperature, as evidenced by scanning electron microscopy (SEM) images [90]; this feature is also confirmed by the characterization of the catalysts after catalytic tests (see below). In this case, the weight loss during the ultrasound treatment cannot be considered because of the low amount of solid deposited and a small breakage of the foam itself during the ultrasonication.

Temperature-programmed reduction (TPR) profiles of the coated foams suggest the presence of NiO and $NiAl_2O_4$ phases in the Ni/Al catalysts prepared at −1.2 V versus SCE for 1000 s. This was further confirmed by μ-XRD measurements using synchrotron radiation [92]. Concerning Rh-based catalysts, the H_2 consumption peak at circa 500 °C indicates the formation of highly stabilized Rh^{3+} species. The specific surface area values of the coated samples (foam + coating) are very low, between 0.5 and 1.0 $m^2\ g^{-1}$, due to the small amount of deposited catalyst. In fact, layer thicknesses of around 1–4 μm are obtained for the catalysts prepared at −1.2 and −1.3 V versus SCE for 1000 s. The thickest coatings correspond to the best covered zones, usually the tips and the outer part of the foam cylinders.

9.4.4 Steam Reforming and Catalytic Partial Oxidation of Methane

Electrosynthesized catalysts are tested in the endothermic SR (Ni/Al) and in the exothermic CPO (Rh/Mg/Al) of methane in order to study the effect of the synthesis parameters on the activity and stability of the catalysts. Methane conversion, syngas selectivity, and deactivation with time-on-stream are correlated to the number of available Ni^0 and Rh^0 sites as well as to the homogeneity and thickness of the catalytic film.

Ni/Al catalysts are reduced before the reforming tests; however, an activation is observed regardless of the synthetic parameters during the first hours of time-on-stream. The Ni/Al catalyst prepared at −0.9 V versus SCE for a short time, that is, 600 s, shows a low CH_4 conversion (34%) and deactivates with time-on-stream under the industrial-like conditions (Figure 9.6). The poor performances are related to the low amount of nickel active sites, which is a consequence of the small amount of solid deposited and the low Ni/Al ratio in the solid obtained by electrosynthesis. The formation of carbon is also observed in this catalyst. By increasing the synthesis time, the larger amount of electrosynthesized solid, the higher Ni/Al ratio, as well as the more homogeneous coating result in enhanced activity (CH_4 conversion reaches 50%) and stability. The activity in

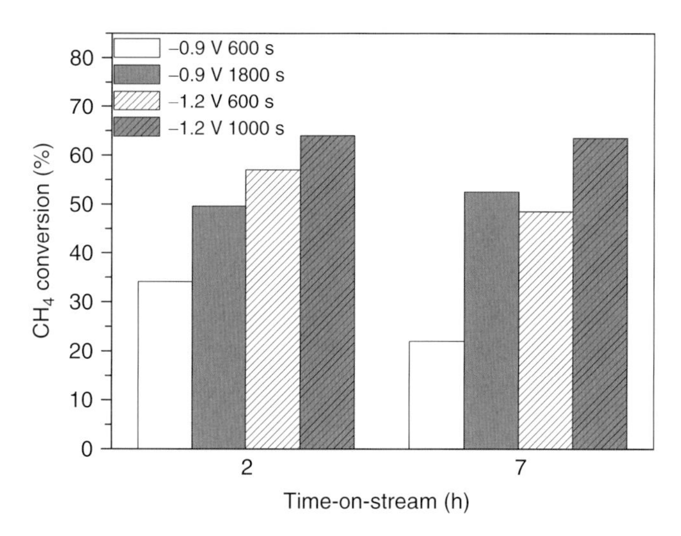

Figure 9.6 Methane conversion during steam reforming with Ni/Al catalysts prepared at several cathodic potentials and time. $P = 20$ bar, $T_{oven} = 920\,^{\circ}C$, $\tau = 4$ s, $S/C = 1.7$.

the reforming of methane is further increased by performing the electrosynthesis at a more cathodic potential, −1.2 V versus SCE. Even at short times, 600 s, the conversion achieved is 57%; however, the large number of cracks and the presence of uncovered zones in the catalyst lead to the catalyst deactivation. In fact, the surface of the used catalyst is not homogeneous, and three main regions are observed: some uncoated struts; some "flakes" of catalyst with Ni^0 particles supported on them; and uniformly sized Ni^0 particles on an alumina phase, corresponding to the zones where a larger amount of catalyst has been deposited. Finally, the catalyst prepared at −1.2 V versus SCE for 1000 s, which shows the best coverage and the Ni/Al ratio closer to that expected, is the one showing higher activity and stability, comparable to a Ni-based commercial-type catalyst [89]. The surface of this catalyst is constituted by Ni^0 particles supported on an alumina layer.

For Rh/Mg/Al compounds (Figure 9.7), when the synthesis of the precursor is performed at −1.2 V versus SCE, low conversion and selectivity values are obtained probably due to the small number of available Rh^0 active sites. Furthermore, the activity depends on the reaction conditions – the catalyst is activated by feeding the $CH_4/O_2/He = 2/1/20$ v/v gas mixture at 120 000 h^{-1}, whereas it is deactivated by performing the tests with the concentrated gas mixture ($CH_4/O_2/He = 2/1/4$ v/v) at the same GHSV value. By increasing the potential (−1.3 V vs SCE), despite the cracks observed in the coating, high conversion and selectivity values are reached regardless of the GHSV value and the dilution factor of the gas mixture. The catalytic layer seems to be stable, and no detachment is observed.

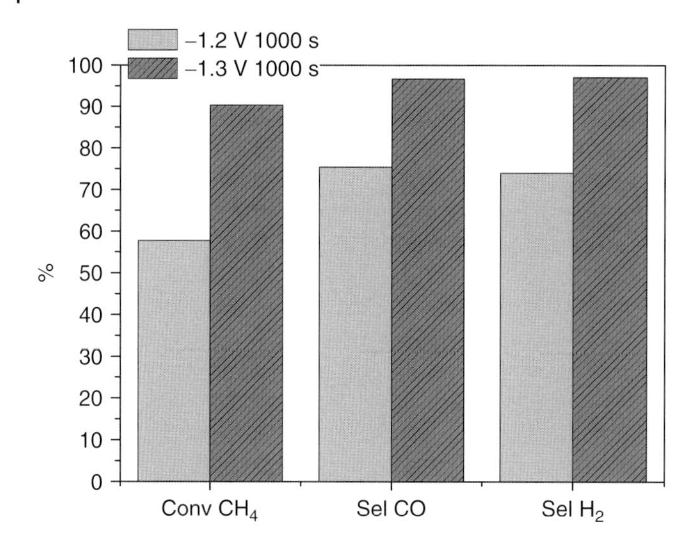

Figure 9.7 Catalytic performances of the Rh/Mg/Al compounds in the catalytic partial oxidation of methane. T_{oven} = 750 °C, GHSV = 28 000 h^{-1}, $CH_4/O_2/He$ = 2/1/20 v/v.

9.5 Summary and Outlook

Summarizing, the electrosynthesis shows several advantages for the coating of metallic supports.

1) It is a soft technique [101]: the atom source is an aqueous solution, the deposition temperature is close to the ambient, and the energy source is an electrical current.
2) The instrumentation is simple, widely available, and inexpensive.
3) Metal supports of different shapes (even complex supports) or selected areas can be coated in short times.
4) The tuning of the electrochemical parameters results in a tight and facile control of phase composition, size, mass, thickness, crystallographic morphology, and deposition rate of the films.
5) The precipitation in the bulk of the solution is avoided.
6) Last but not least, it is easy for large-scale adaptation and industrial implementation.

This study demonstrated the feasibility of the electrosynthesis (or electrodeposition) of HT compounds to coat metallic foams in order to prepare structured catalysts. The coating growth and composition are controlled by varying the deposition time, applied potential, and the bath composition. The technique could be extended to the electrosynthesis of other compounds (hydroxides or oxides) on metal supports with different shapes, from plates to more complex geometries such as gauzes, fibers, and membranes.

References

1. Cybulski, A. and Moulijn, J.A. (eds) (2005) *Structured Catalysts and Reactors*, CRC Taylor & Francis, Boca Raton, FL.
2. Geus, J.W. and van Giezen, J.C. (1999) *Catal. Today*, **47**, 169–180.
3. Heck, R.M. and Farrauto R.J. (eds) (1995) *Catalytic Air Pollution Control: Commercial Technology*, John Wiley & Sons, Inc., New York.
4. Farrauto, R.J. (1997) *React. Kinet. Catal. Lett.*, **60**, 233–241.
5. Kapteijn, F., Nijhuis, T.A., Heiszwolf, J.J., and Moulijn, J.A. (2001) *Catal. Today*, **66**, 133–144.
6. Tomasic, V. and Jovic, F. (2006) *Appl. Catal. A: Gen.*, **311**, 112–121.
7. Boger, T., Heibel, A.K., and Sorensen, C.M. (2004) *Ind. Eng. Chem. Res.*, **43**, 4602–4611.
8. Twigg, M.V. and Richardson, J.T. (2002) *Trans. IChemE*, **80** (Part A), 183–189.
9. Farrauto, R.J., Liu, Y., Ruettinger, W., Ilinich, O., Shore, L., and Giroux, T. (2007) *Catal. Rev.*, **49**, 141–196.
10. Hilmen, A.M., Bergene, E., Lindvag, O.A., Schanke, D., Eri, S., and Holmen, A. (2001) *Catal. Today*, **69**, 227–232.
11. Kapteijn, F., de Deugd, R.M., and Moulijn, J.A. (2005) *Catal. Today*, **105**, 350–356.
12. Pérez-Cadenas, A.F., Zieverink, M.P., Kapteijn, F., and Moulijn, J.A. (2005) *Catal. Today*, **105**, 623–628.
13. Agrafiotis, C., Roeb, M., Konstandopoulos, A.G., Nalbandian, L., Zaspalis, V.T., Sattler, C., Stobbe, P., and Steele, A.M. (2005) *Sol. Energy*, **79**, 409–421.
14. Löfberg, A., Giornelli, T., Paul, S., and Bordes-Richard, E. (2011) *Appl. Catal. A: Gen.*, **391**, 43–51.
15. Ávila, P., Montes, M., and Miró, E.E. (2005) *Chem. Eng. J.*, **109**, 11–36.
16. Trimm, D.L. (1995) *Catal. Today*, **26**, 231–238.
17. Pangarkar, K., Schildhauer, T.J., van Ommen, J.R., Nijenhuis, J., Kapteijn, F., and Moulijn, J.A. (2008) *Ind. Eng. Chem. Res.*, **47**, 3720–3751.
18. Fornasiero, P., Montini, T., Graziani, M., Zilio, S., and Succi, M. (2008) *Catal. Today*, **137**, 475–482.
19. Ferreira-Aparicio, P., Benito, M.J., and Sanz, J.L. (2005) *Catal. Rev.*, **47**, 491–588.
20. Meille, V. (2006) *Appl. Catal. A: Gen.*, **315**, 1–17.
21. Nijhuis, T.A., Beers, A.E.W., Vergunst, T., Hoek, I., Kapteijn, F., and Moulijn, J.A. (2001) *Catal. Rev.*, **43**, 345–380.
22. Wu, X., Weng, D., Zhao, S., and Chen, W. (2005) *Surf. Coat. Technol.*, **190**, 434–439.
23. Martínez, L.M., Sanz, O., Domínguez, M.I., Centeno, M.A., and Odriozola, J.A. (2009) *Chem. Eng. J.*, **148**, 191–200.
24. Burgos, N., Paulis, M., and Montes, M. (2003) *J. Mater. Chem.*, **13**, 1458–1467.
25. Sanz, O., Martínez, L.M., Echave, F.J., Domínguez, M.I., Centeno, M.A., Odriozola, J.A., and Montes, M. (2009) *Chem. Eng. J.*, **151**, 324–332.
26. Fukuhara, C. and Kawamorita, K. (2009) *Appl. Catal. A: Gen.*, **370**, 42–49.
27. Valentini, M., Groppi, G., Cristiani, C., Levi, M., Tronconi, E., and Forzatti, P. (2001) *Catal. Today*, **69**, 307–314.
28. Meille, V., Pallier, S., Santa Cruz Bustamante, G.V., Roumanie, M., and Reymond, J.-P. (2005) *Appl. Catal. A: Gen.*, **286**, 232–238.
29. Rönkkönen, H., Klemkaite, K., Khinsky, A., Baltušnikas, A., Simell, P., Reinikainen, M., Krause, O., and Niemelä, M. (2010) *Fuel*, **90**, 1076–1089.
30. Essakhia, A., Löfberg, A., Supiot, P., Mutel, B., Paula, S., Le Courtois, V., and Bordes-Richard, E. (2010) *Stud. Surf. Sci. Catal.*, **175**, 17–24.
31. Giornelli, T., Löfberg, A., and Bordes-Richard, E. (2005) *Thin Solid Films*, **479**, 64–72.
32. Agrafiotis, C., Tsetsekou, A., and Ekonomakou, A. (1999) *J. Mater. Sci. Lett.*, **18**, 1421–1424.
33. Agrafiotis, C. and Tsetsekou, A. (2000) *J. Eur. Ceram. Soc.*, **20**, 815–824.

34. Pérez-Cadenas, A.F., Kapteijn, F., and Moulijn, J.A. (2007) *Appl. Catal. A: Gen.*, **319**, 267–271.
35. Stefanescu, A., van Veen, A.C., Mirodatos, C., Beziat, J.C., and Duval-Brunel, E. (2007) *Catal. Today*, **125**, 16–23.
36. Cristiani, C., Visconti, C.G., Finocchio, E., Stampino, P.G., and Forzatti, P. (2009) *Catal. Today*, **147S**, S24–S29.
37. Meille, V., Pallier, S., and Rodriguez, P. (2009) *Colloid Surf. A*, **336**, 104–109.
38. Almeida, L.C., Echave, F.J., Sanz, O., Centeno, M.A., Odriozola, J.A., and Montes, M. (2010) *Stud. Surf. Sci. Catal.*, **175**, 25–33.
39. Aguero, F.N., Barbero, B.P., Almeida, L.C., Montes, M., and Cadús, L.E. (2011) *Chem. Eng. J.*, **166**, 218–223.
40. Vergunst, T., Kapteijn, F., and Moulijn, J.A. (2001) *Appl. Catal. A: Gen.*, **213**, 179–187.
41. Pereda-Ayo, B., López-Fonseca, R., and González-Velasco, J.R. (2009) *Appl. Catal. A: Gen.*, **363**, 73–80.
42. Villegas, L., Massets, F., and Guilhaume, N. (2007) *Appl. Catal. A: Gen.*, **320**, 43–55.
43. Ulla, M.A., Miró, E., Mallada, R., Coronas, J., and Santamaría, J. (2004) *Chem. Commun.*, 528–529.
44. Cimino, S., Pirone, R., Lisi, L., Turco, M., and Russo, G. (2000) *Catal. Today*, **59**, 19–31.
45. Cimino, S., Pirone, R., and Lisi, L. (2002) *Appl. Catal. B: Environ.*, **35**, 243–254.
46. Ugues, D., Specchia, S., and Saracco, G. (2004) *Ind. Eng. Chem. Res.*, **43**, 1990–1998.
47. Specchia, S. and Toniato, G. (2009) *Catal. Today*, **147S**, S99–S106.
48. Gu, C., Lu, S., Miao, J., Liu, Y., and Wang, Y. (2010) *Int. J. Hydrogen Energy*, **35**, 6113–6122.
49. Frías, D.M., Nousir, S., Barrio, I., Montes, M., Martínez, L.M., Centeno, M.A., and Odriozola, J.A. (2007) *Appl. Catal. A: Gen.*, **325**, 205–212.
50. Kleiman-Shwarsctein, A., Huda, M.N., Walsh, A., Yan, Y., Stucky, G.D., Hu, Y.-S., Al-Jassim, M.M., and McFarland, E.W. (2010) *Chem. Mater.*, **22**, 510–517.
51. Pathan, H.M., Min, S.-K., Jung, K.-D., and Joo, O.-S. (2006) *Electrochem. Commun.*, **8**, 273–278.
52. El-Deaba, M.S. and Ohsaka, T.J. (2008) *Electrochem. Soc.*, **155**, D14–D21.
53. Simka, W., Puszczyk, D., and Nawrat, G. (2009) *Electrochim. Acta*, **54**, 5307–5319.
54. Jason Riley, D. (2002) *Curr. Opin. Colloid Interface Sci.*, **7**, 186–192.
55. Bicelli, L.P., Bozzini, B., Mele, C., and D'Urzo, L. (2008) *Int. J. Electrochem. Sci.*, **3**, 356–408.
56. Therese, G.H.A. and Kamath, P.V. (2000) *Chem. Mater.*, **12**, 1195–1204.
57. Prasad, B.E. and Kamath, P.V. (2008) *J. Am. Ceram. Soc.*, **91**, 3870–3874.
58. Kovtyukhova, N.I. and Mallouk, T.E. (2010) *Chem. Mater.*, **22**, 4939–4949.
59. Tench, D. and Warren, L.F. (1983) *J. Electrochem. Soc.*, **130**, 869–872.
60. Dinamani, M. and Kamath, P.V. (2004) *J. Appl. Electrochem.*, **34**, 899–902.
61. Pauporte, T. and Lincot, D. (2001) *J. Electrochem. Soc.*, **148**, C310–C314.
62. Peulon, S. and Lincot, D. (1996) *Adv. Mater.*, **8**, 166–169.
63. Yan, J., Zhou, H., Yu, P., Su, L., and Mao, L. (2008) *Electrochem. Commun.*, **10**, 761–765.
64. Zou, G., Chen, W., Liu, R., and Xu, Z. (2008) *Mater. Chem. Phys.*, **107**, 85–90.
65. Sankara Narayanan, T.S.N. and Seshadri, S.K. (2000) *J. Mater. Sci. Lett.*, **19**, 1715–1718.
66. Switzer, J.A. (1987) *Am. Ceram. Soc. Bull.*, **66**, 1521–1524.
67. Fu, Y., Wie, Z.D., Ji, M.B., Li, L., Shen, P.K., and Zhang, J. (2008) *Nanoscale Res. Lett.*, **3**, 431–434.
68. Stefanov, P., Stoychev, D., Valov, I., Kakanakova-Georgieva, A., and Marinov, Ts. (2000) *Mater. Chem. Phys.*, **65**, 222–225.
69. Nagarajan, N. and Zhitomirsky, I. (2006) *J. Appl. Electrochem.*, **36**, 1399–1405.
70. Franger, S., Berthet, P., and Berthon, J. (2004) *J. Solid State Eletrochem.*, **8**, 218–223.
71. Scavetta, E., Ballarin, B., Giorgetti, M., Carpani, I., Cogo, F., and Tonelli, D. (2004) *J. New Mater. Electrochem. Syst.*, **7**, 43–50.

72. Zhitomirsky, I. (1998) *Mater. Lett.*, **33**, 305–310.
73. Roy, B.K., Zhang, G., Magnuson, R., Poliks, M., and Chow, J. (2010) *J. Am. Ceram. Soc.*, **93**, 774–781.
74. Patake, V.D., Lokhande, C.D., and Joo, O.S. (2009) *Appl. Surface Sci.*, **255**, 4192–4196.
75. Inguanta, R., Vergottini, F., Ferrara, G., Piazza, S., and Sunseri, C. (2010) *Electrochim. Acta*, **55**, 8556–8562.
76. Izaki, M. and Omi, T. (1996) *Appl. Phys. Lett.*, **68**, 2439–2440.
77. Otani, S., Katayama, J., Umemoto, H., and Matsuoka, M. (2006) *J. Electrochem. Soc.*, **153**, C551–C556.
78. Matsumoto, Y., Morikawa, T., Adachi, H., and Hombo, J. (1992) *J. Mater. Res. Bull.*, **27**, 1319–1327.
79. Quarez, E., Roussel, P., Pérez, O., Leligny, H., Bendraoua, A., and Mentré, O. (2004) *Solid State Sci.*, **6**, 931–938.
80. Therese, G.H.A., Dinamani, M., and Kamath, P.V. (2005) *J. Appl. Electrochem.*, **35**, 459–465.
81. Therese, G.H.A. and Kamath, P.V. (1999) *Chem. Mater.*, **11**, 3561–3564.
82. Zheng, D., Shi, J., Lu, X., Wang, C., Liu, Z., Liang, C., Liu, P., and Tong, Y. (2010) *CrystEngComm*, **12**, 4066–4070.
83. Amadellia, R., Samiolo, L., Velichenko, A.B., Knysh, V.A., Luk'yanenko, T.V., and Danilov, F.I. (2009) *Electrochim. Acta*, **54**, 5239–5245.
84. Lai, M. and Jason Riley, D. (2008) *J. Colloid Interface Sci.*, **323**, 203–212.
85. Zhitomirsky, I. (2004) *J. Appl. Electrochem.*, **34**, 235–240.
86. Jiao, S., Xu, L., Hu, K., Li, J., Gao, S., and Xu, D. (2010) *J. Phys. Chem. C*, **114**, 269–273.
87. Zhou, W.-J., Zhang, J., Xue, T., Zhao, D.-D., and Li, H.-l. (2008) *J. Mater. Chem.*, **18**, 905–910.
88. Zhitomirsky, I. and Petric, A. (2000) *J. Mater. Chem.*, **10**, 1215–1218.
89. Basile, F., Benito, P., Del Gallo, P., Fornasari, G., Gary, D., Rosetti, V., Scavetta, E., Tonelli, D., and Vaccari, A. (2008) *Chem. Commun.*, 2917–2919.
90. Basile, F., Benito, P., Fornasari, G., Rosetti, V., Scavetta, E., Tonelli, D., and Vaccari, A. (2009) *Appl. Catal. B: Environ.*, **91**, 563–572.
91. Basile, F., Benito, P., Fornasari, G., Monti, M., Scavetta, E., Tonelli, D., and Vaccari, A. (2010) *Catal. Today*, **157**, 183–190.
92. Basile, F., Benito, P., Bugani, S., De Nolf, W., Fornasari, G., Janssens, K., Morselli, L., Scavetta, E., Tonelli, D., and Vaccari, A. (2010) *Adv. Funct. Mater.*, **20**, 4117–4126.
93. Cavani, F., Trifirò, F., and Vaccari, A. (1991) *Catal. Today*, **11**, 173–301.
94. Trifirò, F. and Vaccari, A. (1996) in *Comprehensive Supramolecular Chemistry* (eds J.L. Atwood, J.E.D. Davies, D.D. MacNicol, and F. Vögtle), Pergamon Press, Oxford, pp. 251–291.
95. Rives, V. (ed.) (2001) *Layered Double Hydroxides: Present and Future*, Nova Science Publishers, New York.
96. Basile, F., Benito, P., Fornasari, G., and Vaccari, A. (2010) *Appl. Clay Sci.*, **48**, 250–259.
97. Ballarin, B., Berrettoni, M., Carpani, I., Scavetta, E., and Tonelli, D. (2005) *Anal. Chim. Acta*, **538**, 219–224.
98. Scavetta, E., Ballarin, B., Berrettoni, M., Carpani, I., Giorgetti, M., and Tonelli, D. (2006) *Electrochim. Acta.*, **51**, 3305–3311.
99. Scavetta, E., Mignani, A., Prandstraller, D., and Tonelli, D. (2007) *Chem. Mater.*, **19**, 4523–4529.
100. Hamlaoui, Y., Pedraza, F., and Tifouti, L. (2008) *Corros. Sci.*, **50**, 2182–2188.
101. Rouxel, J., Tournoux, M., and Brec, J. (1993) *Soft Chemistry Routes to New Materials*, Trans Tech Publications, Switzerland.

10
Microkinetic Analysis of Complex Chemical Processes at Surfaces

Matteo Maestri

Notation

D_i	diffusivity of species i in the mixture
L	reactor length
MW_i	molecular weight
R_I	inner radius
R_O	outer radius
R_i^{hom}	production rate of species i due to gas-phase chemistry
R_i^{het}	production rate of species i due to surface chemistry
T	temperature
u	axial velocity
z	reactor axial coordinate

Greek letters

ρ	density
ω	mass fraction

10.1
Introduction

During most of the twentieth century, the main purpose of kinetic modeling was basically the quantification of experimental data for the purpose of chemical reactor analysis and design. In this respect, the approach to heterogeneous catalysis was based on the concept of proposing a rate equation and fitting it to the data by means of effective rate constants. Typically, the rate equation was the result of a hypothesized mechanism and the rate constant was chosen by best fit [1]. This approach has allowed the development of valuable tools for process analysis, design, and optimization, and most of the catalytic processes running today have been (successfully) developed on the basis of these approaches. Nevertheless, their

New Strategies in Chemical Synthesis and Catalysis, First Edition. Edited by Bruno Pignataro.

use under different conditions than those used in the parameterization is known to potentially lead to wrong predictions (e.g., rate-determining step (RDS) may change, effective constants may become inaccurate). Moreover, with the focus on an effective description of existing experimental data, such models do not provide any insights into the underlying catalytic mechanism [2].

In recent years, much attention has been paid to the physical meaning of the rate constant and to the fundamentals of catalysis, by exploiting the great advances in surface science and physical chemistry [2–5]. As pointed out by Michel Boudart in a very seminal and inspiring paper [6], this new concept has opened the way toward a revolution in catalytic kinetics by leading to a transition from a *rate-equation* to a *rate-constant* based approach. In essence, rather than fitting a hypothesized rate equation to experimental data, attention has focused on kinetic models that attempt to incorporate the basic surface chemistry involved in the catalytic reaction. Essential to this approach is the application of fundamental theoretical concepts for explaining the catalytic phenomenon in chemical physical terms, avoiding the qualitative ideas typical of the classical catalytic kinetics, that cannot be directly translated into theoretically accessible phenomena [3].

The possibility of modeling in detail and at the fundamental level the surface chemistry has reinterpreted the kinetic models as tool not only for the analysis and design of chemical reactors but also for the consolidation of fundamental knowledge about a catalytic process under different operating conditions [7]. Catalysis at the very heart is a kinetic phenomenon, and kinetic models, if properly incorporate the basic surface chemistry, can help to elucidate the key features of surface reactivity. This new perspective opens up the way toward the replacement of the existing trial-and-error type approaches for the development of new catalysts with rational design based on atomic-scale understanding of the catalytic phenomenon. Following this view, the great challenge ahead is to invert the classical kinetic problem by designing materials that can provide the "best" electronic structure to catalyze a particular reaction under a specific set of operating conditions [6, 8]. In this respect, to allow nano-engineering of the catalysts, it is of pivotal importance to develop fundamental knowledge of the factors determining the catalytic activity and selectivity [2, 6, 8, 9].

Central to the quest toward this atomic-scale understanding of a catalytic process is the identification of the *dominant reaction mechanism* that establishes under a particular set of operating conditions. In essence, the dominant reaction mechanism consists of the dominant chemical pathways and intermediate species through which the reactants convert into the products under specific conditions. These dominant chemical pathways are the result of the interplay among a much larger number of chemical events that can *potentially* occur at the catalyst surface at specific conditions of temperature, pressure, and composition. These operating conditions are dictated at the reactor scale by the transport phenomena of mass, energy, and momentum. Therefore, if the reaction mechanism (i.e., the whole set of potential chemical events at the catalytic surface) represents *what can happen* and is specific to the catalyst, the dominant reaction mechanism represents *what actually happens* and turns out to be the result of different phenomena occurring at different time and length scales (from the catalyst to the reactor).

Given the length and time scales involved, along with the potentially huge number of chemical events that can occur at the catalyst surface, the identification of the dominant reaction mechanism is a very challenging problem, especially for processes of real technological interest. Therefore, it is of particular relevance to develop efficient methodologies to link insights across all the relevant time and length scales [10–13].

In this chapter, an overview of the state-of-the-art approach to the microkinetic analysis of complex chemical processes is provided. First, a general overview of the hierarchical multiscale approach for the microkinetic modeling and analysis of complex chemical processes at surfaces is given. Then, as show-case, results on the microkinetic modeling and analysis of the partial oxidation of methane on Rh catalysts is reviewed.

10.2 Time and Length Scales in Heterogeneous Catalysis

As reported in Figure 10.1, in a catalytic process different time and length scales are involved and the coupling between them is crucial in determining the actual behavior of the system [14].

The nature of the physics is different in the three different regimes: at the microscale, one has to deal with kinetic parameters of the elementary steps. This is related to the making and breaking of chemical bonds and ultimately to the behavior of the electrons and the interactions between atoms and molecules. At the meso-scale the interplay among all these potential elementary steps determines

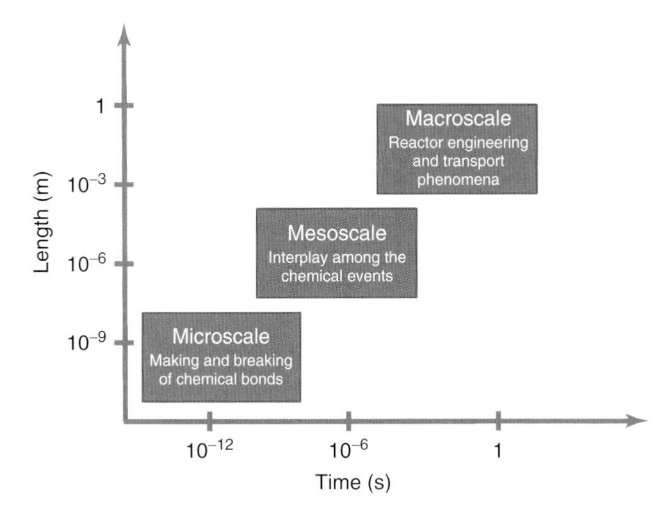

Figure 10.1 Schematic of the time and length scales involved in a catalytic process.

which of them will become dominant. These phenomena are governed by the electronically determined microscopic parameters at the microscale at the specific conditions of pressure, temperature, and composition dictated by the macroscale [15]. In essence, the microscale deals with what *can* happen. However, whether a particular chemical event is actually relevant for the overall process will be determined by the conditions dictated by the physics in the other two regimes.

Therefore, if one aims at understanding the *actual* behavior of the system by the identification of the dominant reaction mechanism, one has to combine and bridge among the three relevant regimes. This basically means that the dominant reaction mechanism is a *multiscale property* of the system. In this respect, the main difficulty is related to the very large difference in the involved time and length scales. As an example [14], molecular processes at surfaces proceed on a length scale of 0.1 nm, electrons move and adjust to perturbations in the femtosecond time scale, and atoms vibrate on a time scale of picoseconds. However, for catalysis, the relevant time scale is of the order of microseconds or even seconds. For example, a single dissociation event takes some femtoseconds ($\sim 10^{-15}$ s), and diffusion over a distance of 10 nm may take a time of the order of several picoseconds ($\sim 10^{-12}$ s). In order to evaluate turn-over rates, one needs to perform a statistical average over many chemical events and consequently has to span a time period of several microseconds ($\sim 10^{-6}$ s). In addition, the catalytic surface is an open system, in the sense that reactants and products continuously adsorb and desorb at the gas–surface interface. Thus, transport phenomena of mass, energy, and momentum between the surface and the flowing gas have to be accounted for at the macroscopic regime, which in turn have characteristic times up to the order of seconds [16]. For instance, in a turbulent reacting flow, depending on the Reynolds number, the flow time scales range from the Kolmogorov time scale ($\sim 10^{-4} - 10^{-3}$ s) through the turbulent time scale ($\sim 10^{-4} - 1$ s) up to the mean residence time in the reactor ($\sim$1–10 s). As a whole, this implies that a multiscale simulation has to bridge 16 orders of magnitude in the time scale. Therefore, if on one side it is essential to accurately describe the insights at each scale, on the other side it is crucial to embed and link all of them. This ultimately calls for methodological approaches that efficiently integrate the various levels of theory into one multiscale simulation.

There are two important approaches to multiscale modeling [17]. The first one is based on a unidirectional "bottom-up" approach. The information gathered at the small scale is passed into the next. This is usually the case with a well-defined and controlled system [18]: such an approach concentrates therefore on achieving an accurate, first-principles description of all individual elementary processes, and on combining them appropriately within a thermodynamic, statistical, or continuum mechanic framework. The second class is specific to problems that are complex both in terms of model description and dimension, with a strong coupling between scales. This is usually typical of processes of real technological interest. In these cases, the information flow has to be bidirectional, and hierarchical "top-down" approaches are employed, according to a typical *engineering science* concept [19]. This is basically the ability to move from a complex problem that cannot be solved (but needs to be) to a meta-problem which can be solved, without losing the essential important features of the full problem. Typically, a hierarchical combination of methodologies is employed for the identification of the appropriate meta-problem, by erasing from the real problem the steps that are – for the specific conditions – not relevant. The next section focuses on this hierarchical "top-down" approach for the development of microkinetic models.

10.3 Hierarchical Multiscale Approach for Microkinetic Model Development

The main idea behind the hierarchical approach is to tackle the problem with methods of increasing accuracy. Lower (but controlled, see Section 10.3.1.2) accuracy methods are used to identify the key issues of the full problem (otherwise nontractable). These key issues are then refined with higher and computationally demanding methods, which necessarily need to focus on narrowly defined aspects and specific steps of the overall mechanism.

A schematic of this approach is shown in Figure 10.2 for the identification of the dominant reaction mechanism and the development of microkinetic models. First, an estimation of the kinetic parameters (step 1, Figure 10.2) of all the potential elementary reactions at the microscale is performed, using efficient but computationally nondemanding methodologies. These typically comprise semiempirical theories, with selected parameters calculated with first-principles methods. Then, such microkinetic model is used in conjunction with surface and reactor modeling (step 2) in order to identify the dominant reaction mechanism and compare it with selected experiments. The comparison between model predictions and experimental evidence allows the identification of the key features (step 3) that need to be refined or analyzed with higher accuracy (but computationally demanding) methods (step 4). As such, they are most effectively utilized, since there is detailed guidance as to what the most relevant issues are. Following this schematic, in

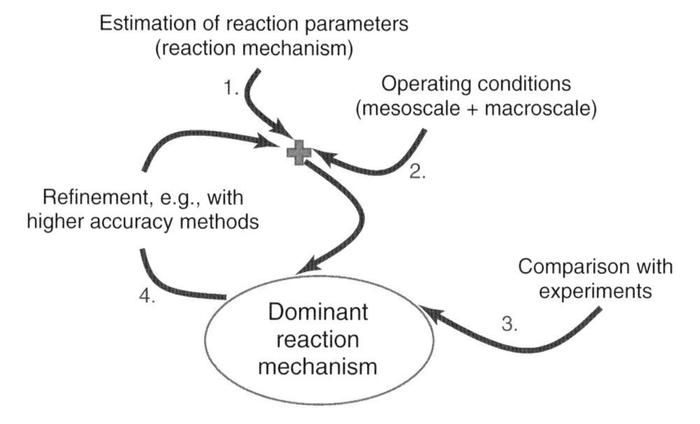

Figure 10.2 Hierarchical multiscale methodology for the identification of the dominant reaction mechanism.

the rest of this section, the main features and framework of the state-of-the-art approach to microkinetic modeling and analysis of complex catalytic processes are presented.

10.3.1 Microkinetic Model Development

The first very important step is the appropriate description of all the potential elementary steps at the microscale (step 1 in Figure 10.2). In principle, this comprises adsorption, desorption, and reaction of reactants, products, and intermediates. Therefore, the construction of a microkinetic model relies on a comprehensive sequence of elementary reactions and the estimation of the associated rate parameters. This approach does not make any a priori assumption about the elementary steps or the surface coverages and all the potential paths should be included at this stage [6]. First-principles electronic structure theories offer a unique possibility for the accurate calculation of the kinetic parameters of elementary reactions. Nevertheless, complex processes of real technological interest may involve hundreds of reaction steps, each of which needs to be evaluated at different conditions of surface coverage and structure. The severe limitations imposed by the huge computational costs of even the most efficient present-day density functional theory (DFT) calculations make an exhaustive first-principles computation of all reaction steps prohibitive but for selected model studies [20–25]. Therefore, the critical step is the (accurate) estimation of the rate parameters for all the elementary reaction pathways. The major challenge is to achieve a compromise between accuracy and computational time. As a result, simpler and less demanding methodologies to calculate the rate parameters are necessary. As a consequence, rate parameters in a microkinetic model are usually fitted, extracted from experiments, or estimated using semiempirical methods. However, each approach has limitations, which makes the development of predictive microkinetic modeling a difficult task.

Among a variety of approaches reported in the literature, the most common one is to take reaction parameters from wherever available, for example, first-principles calculations, surface science experiments, or fits of selected experimental data (e.g., [26, 27]). While this type of models can lead to a good prediction of experiments under specific conditions, there are a number of clear drawbacks, the most prominent being that the model in principle could be thermodynamically inconsistent and that the underlying chemistry may not be correctly accounted for.

To overcome these limitations and to make possible the development of physically sound microkinetic models, Vlachos and coworkers have proposed a hybrid theoretical approach for the development of a predictive surface reaction mechanism [17, 28–31]. This approach consists of a combination of semiempirical techniques with DFT-based heats of chemisorption or reaction energetics in the zero-coverage limit and unity bond index-quadratic exponential potential (UBI-QEP) [32] to account for adsorbate–adsorbate interactions for all steps. In addition, transition state theory (TST) is used to obtain an order-of-magnitude estimate for the pre-exponential factors. Moreover, because of the lack of thermodynamic information concerning the adsorbed species, thermodynamic consistency of microkinetic models of catalytic processes is not a trivial task. Parameters must be subjected to specific constraints in order to meet the thermodynamic consistency [30]. For instance, consider a general reversible elementary reaction in the mechanism (hereafter * implies an adsorbed species)

$$\mathrm{A}^* + \mathrm{B}^* \longleftrightarrow \mathrm{C}^* + \mathrm{D}^* \tag{10.1}$$

The reaction rates for the forward and backward reactions are given by

$$r_\mathrm{f} = k_\mathrm{f}[\mathrm{A}^*][\mathrm{B}^*] \tag{10.2}$$

$$r_\mathrm{b} = k_\mathrm{b} = \frac{k_\mathrm{f}}{K_\mathrm{c}}[\mathrm{C}^*][\mathrm{D}^*] \tag{10.3}$$

The equilibrium constant is related to the thermodynamic properties as

$$K_\mathrm{c} = \exp\left(\frac{-\Delta G^0}{RT}\right) = \exp\left(\frac{\Delta S^0}{R}\right)\cdot\exp\left(\frac{-\Delta H^0}{RT}\right) \tag{10.4}$$

where G, S and H are free Gibbs energy, entropy and enthalpy in the reference state, R and T are the universal gas constant and the temperature, respectively. Often, thermodynamic parameters for the surface-adsorbed species are not available and the equilibrium constant cannot be computed from the above expression. In such a case, the reversible reaction (10.1) is written as a pair of irreversible reactions:

$$\mathrm{A}^* + \mathrm{B}^* \longrightarrow \mathrm{C}^* + \mathrm{D}^* \tag{10.5}$$

$$\mathrm{C}^* + \mathrm{D}^* \longrightarrow \mathrm{A}^* + \mathrm{B}^* \tag{10.6}$$

The forward and backward reaction rate constants have the standard Arrhenius form

$$k_{\mathrm{if/b}} = A_{\mathrm{o,if/b}}\left(\frac{T}{T_0}\right)^{\beta_{\mathrm{if/b}}}\exp\left(\frac{-E_{\mathrm{if/b}}}{RT}\right) \tag{10.7}$$

where the subscripts f and b indicate forward and backward reaction, respectively. These forward and backward rate constants do not vary independently, but are subject to thermodynamic constraints of Eq. (10.4). The entropic and enthalpic components of these constraints are expressed as

$$\exp\left(\frac{\Delta S^0}{R}\right)\cdot\exp\left(\frac{-\Delta H^0}{RT}\right) = \underbrace{\frac{A_{0,\mathrm{if}}}{A_{0,\mathrm{ib}}}\left(\frac{T}{T_0}\right)^{\beta_{\mathrm{if}}-\beta_{\mathrm{ib}}}}_{\text{Entropic}}\cdot\underbrace{\exp\left(-\frac{E_{\mathrm{if}}-E_{\mathrm{bf}}}{RT}\right)}_{\text{Enthalpic}} \tag{10.8}$$

The difference in activation energies equals the heat of reaction, while the ratio of the temperature-adjusted pre-exponential factors equals the change in entropy.

$$\Delta H^0 = E_{\mathrm{if}} - E_{\mathrm{bf}} \tag{10.9}$$

$$\frac{\Delta S^0}{R} = \ln\left(\frac{A_{0,\mathrm{if}}}{A_{0,\mathrm{ib}}}\right) + (\beta_{\mathrm{if}} - \beta_{\mathrm{ib}})\ln\left(\frac{T}{T_0}\right) \tag{10.10}$$

Enthalpy and entropy are both state functions. They depend only on the initial and the final state and are independent of the path taken to reach the final state. This results in additional constraints on the enthalpy and entropy of reactions, which have to be met for the entire mechanism to be thermodynamically consistent. For example, the reaction (10.5) may be written as

$$\begin{array}{ccccccc} A^* & + & B^* & \longrightarrow & C^* & + & D^* \\ \downarrow & & \downarrow & & \uparrow & & \uparrow \\ A & + & B & \longrightarrow & C & + & D \end{array} \tag{10.11}$$

Since enthalpy is a state function, the heat of surface reaction (10.5) depends on the heats of chemisorption of the adsorbed species and the heat of the equivalent gas phase reaction, that is,

$$\Delta H = E_{b,A} + E_{b,B} - E_{b,C} - E_{b,D} + Q_r^{gas} \tag{10.12}$$

where Q_r^{gas} is the heat of reaction for $A + B \rightarrow C + D$ and $E_{b,X}$ is the heat of chemisorption of surface species X^*. In this particular example, the adsorption and desorption reactions form the "basis set" [30]. These additional constraints on heat and entropy of reactions have to be satisfied for all reactions within a microkinetic mechanism for it to be thermodynamically consistent.

An important outcome of the Vlachos' methodology is that the final parameters of the entire microkinetic model are thermodynamically consistent both enthalpically and entropically [30]. In particular, criteria are derived when adsorption/desorption steps constitute the basis of a surface reaction mechanism. Following this approach, semiempirical techniques, such as UBI-QEP, can be combined with optimization to tackle the thermodynamic consistency of surface reaction mechanisms.

10.3.1.1 **Prediction of Activation Energies Using the UBI-QEP Semiempirical Method**

The Vlachos' methodology heavily resorts to the UBI-QEP semiempirical method. The major advantage of the UBI-QEP method is the use of analytical expressions to estimate activation energies. The only information required is the heats of chemisorption and heat of dissociations of the species involved in the reaction [32]. The main assumptions can be summarized as follows:

1) The forces are spherical (i.e., depend on distance only, r).
2) In a many-body system, the two-body interactions are described by a quadratic potential of an exponential function of the distance, called the *bond index* $x(r) = \exp(-(r - r_0)/b)$.
3) The total energy of the many-body system is constructed from additive two-body contributions under the heuristic assumption that the total bond index of the system is conserved at unity.

Under these assumptions, one arrives at simple analytical expressions for the total energy of the system as a function of the different bond indices. As an example, for a diatomic molecule AB interacting with the solid surface

$$E_{AB}(x_A, x_B, x_{AB}) = E_{b,A}(x_A^2 - 2x_A) + E_{b,B}(x_B^2 - 2x_B) + D_{AB}(x_{AB}^2 - 2x_{AB}) \tag{10.13}$$

with the constraint $x_A + x_B + x_{AB} = 1$. Here, $E_{b,A}$, $E_{b,B}$, and D_{AB} are the heats of chemisorption of the atomic species A and B at the surface, as well as the dissociation energy of AB in the gas phase, respectively. Constrained minimization of this expression leads to the minimum energy path (MEP) for the interaction and dissociation of AB at the surface:

$$E_{AB}^{MEP}(x_{AB}) = [P + D_{AB}]\, x_{AB}^2 + 2\,[P - D_{AB}]\, x_{AB} + [P - E_{b,A} - E_{b,B}] \tag{10.14}$$

with $P = E_{b,A}E_{b,B}/\left(E_{b,A} + E_{b,B}\right)$. The minimum of this path is related to the heat of chemisorption of AB at the surface as

$$\min_{x_{AB}}\left(E_{AB}^{MEP}\right) = -E_{b,AB} - D_{AB} \tag{10.15}$$

A first estimate for the transition state (TS) of the dissociation reaction at the surface is given by equating it with the position on the MEP that corresponds to the fully dissociated limit ($x_{AB} = 0$):

$$E_{AB\rightarrow A+B}^{TS} = E_{AB}^{MEP}\big|_{x_{AB}=0} = [P - E_{b,A} - E_{b,B}] \tag{10.16}$$

The difference between this TS estimate and the MEP minimum provides a simple algebraic expression for the activation barrier purely in terms of the thermodynamic parameters of the species involved in the reaction. Equating the TS with the fully dissociated state will typically overestimate the true barrier. Therefore, to account for the nature of the TS, $\Delta E_{AB\rightarrow A+B}^{UBI\text{-}QEP}$ is scaled with an empirical parameter ϕ, which would be close to unity for a "late" and close to zero for an "early" TS [33, 34].

$$\Delta E_{AB\rightarrow A+B}^{UBI\text{-}QEP} = \phi\left[\left(P - E_{b,A} - E_{b,B}\right) + D_{AB} + E_{b,AB}\right] \tag{10.17}$$

From this, the activation energy for the reverse process follows straightforwardly from the conservation of energy:

$$\Delta E_{A+B\rightarrow AB}^{UBI\text{-}QEP} = \Delta E_{AB\rightarrow A+B}^{UBI\text{-}QEP} + \left(D_{AB} + E_{b,AB} - E_{b,A} - E_{b,B}\right) \tag{10.18}$$

Since the semiempirical method lacks any insight into the TS, ϕ is empirically set to 0.5 in the standard formulation, as the arithmetic mean between early and late TS.

In the common approach to UBI-QEP in microkinetic modeling, all thermodynamic parameters entering Eq. (10.17) would be determined, for example, by DFT, which is still computationally much less intense than the explicit DFT calculation of the activation energy [35]. This allows for a tremendous reduction of the computational costs and makes the estimation of activation energies tractable even for a very large reaction network.

10.3.1.2 First-Principles Assessment of the UBI-QEP Semiempirical Method

If, on one side, this method makes tractable the microkinetic modeling of large reaction systems, on the other side one has to be aware of the accuracy of such semiempirical methods. To test the accuracy of this approach, Maestri and Reuter [36] have performed a first-principles assessment of the semiempirical method. In particular, when benchmarked against an extensive DFT dataset for a range of surface catalytic reactions in the context of water–gas shift (WGS)

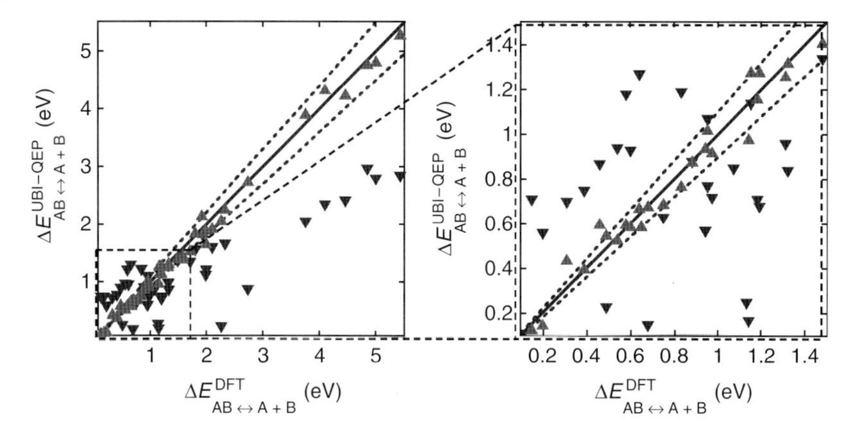

Figure 10.3 Benchmark of UBI-QEP derived against DFT activation energies for various dissociation reactions at Rh(111) and Pt(111) surfaces and at different oxygen coverages. Inverted triangles: standard UBI-QEP, triangles: modified UBI-QEP proposed in [36]. The dashed lines mark a ±10% corridor around the DFT references. (Figure adapted from [36]. © 2011 Wiley-VCH.)

conversion and steam reforming (SR) at Rh(111) and Pt(111), the dominant formulation of the UBI-QEP method is found to exhibit large errors, with individual barriers deviating by more than 100% from their reference value (Figure 10.3). This analysis traces this back to the blindness of UBI-QEP with respect to the nature of the involved transition states and to an intrinsic inconsistency in the established parameterization procedure. A simple modification of this procedure that maintains the dependence on bond distances underlying the UBI-QEP idea but simultaneously recovers the consistency with the reference thermochemistry leads to a significant improvement in the accuracy of the predicted barriers [36].

This modification has been discussed in details in [36] for the example of dissociation reactions, but is easily generalized to other classes of reactions. Insight into the nature of the transition state needs to enter the scheme through the TS parameter ϕ. Such insight is established for many classes of reactions, but can alternatively come from selected first-principles calculations or fit of selected experiments. As a result, once the parameter ϕ has been determined, for example, by a reference DFT-calculation, then the lateral interactions in the activation energies can be computed only as a function of the lateral interactions in the heats of chemisorption. For the considered dataset, the UBI-QEP barriers according to the new parameterization provided by the modified scheme fall consistently within a window of ±10% around the DFT reference data [36], as summarized in Figure 10.3.

With no intention to be fully quantitative, such semiempirical estimates for the barriers can provide at controlled uncertainty most useful insights into complex reaction networks, where an exhaustive first-principles treatment is prohibitive. As shown in Section 10.3.3, the computationally undemanding barrier estimates may serve as initial approximate inputs for the identification of the dominant reaction pathways, which require refined kinetic parameters. The identification of such key features is performed via comparison and analysis of selected experiments. Therefore, the microkinetic model has to be integrated with a proper model of the meso- and macroscale.

10.3.2 Meso-Scale and Macroscale: Reaction and Reactor Engineering

In the previous section, an effective methodology for the derivation of fundamental microkinetic models has been presented. Such models account for the intrinsic reaction rate. The observed reaction rates and the dominant reaction mechanism are determined by the interplay between the intrinsic rate (microkinetic) and the phenomena that are taking place both at the surface and in the reactor.

At the catalytic surface, each of the potential chemical events occurs at a specific rate. To model this, the conventional approach is to employ a mean-field description of the catalytic surface. This approach relies on the assumption of a perfect and rapid mixing of the reactants, products, and intermediates on the surface. Therefore, one can express the probability that reactants are in the right position to react (e.g., nearest neighbors) as the product of the corresponding surface coverage. It has been shown [37] that, in some catalytic reactions, a perfect mixing of the species on the surface is not achieved and interaction between molecules can lead to their segregation. In such situation, the mean-field assumptions are no longer valid, and one needs to account for interactions and diffusion of the molecules on the surface for the calculation of the reaction rate. This can be done by employing modern nonequilibrium statistical mechanics techniques, such as kinetic Monte Carlo simulations [15, 17].

As a whole, the rate of each chemical event at the surface is strongly influenced by the conditions of pressure, temperature, and composition at the gas–surface interface. These conditions are dictated by the physics that takes place at the reactor scale. In principle, the number of types and configurations of chemical reactors is very large. Nevertheless, the modeling of reactors is not based on the specific apparatus and configuration, rather on the characteristic phenomena that occur in the reactor. Such phenomena can be basically broken down in transport of mass, energy, and momentum [38]. At the reactor scale, transport phenomena determine the macroscopic distribution of the velocity of the fluid (hydrodynamics), temperature, and composition. At the length of the characteristic dispersion of the reactor (e.g., solid particles, liquid droplets, and gas bubbles), they are responsible of the mixing and of the conditions of temperature and concentrations at the interface between the phases. The modeling of reactors is therefore based on balance equations describing the conservation of mass, energy, and momentum. This reduces the apparent diversity of reactors into a small number of model reactors. For details and a comprehensive survey on chemical reactor modeling, the reader is referred to [39]. The form and the complexity of these equations depend on the employed model. A brief summary of these models for fixed bed reactors is reported in Table 10.1.

Table 10.1 Classification of models for fixed-bed reactors.

		Pseudo-homogenous models		Heterogeneous models
One-dimensional	✓	Ideal PFR reactor (basic)	✓	+ Interfacial gradients
	✓	+ Axial mixing	✓	+ Intraparticle gradients
Two-dimensional	✓	+ Radial mixing	✓	+ Radial mixing

The behavior of the macroscale is mainly determined by hydrodynamics and in this respect the basic case (Table 10.1) is to use a plug-flow-reactor (PFR) model. This basically turns out to be an idealized picture of the motion of the fluid: all the fluid elements proceed in the reactor with same velocity and along parallel streamlines. Whether the presence of the catalyst is accounted for explicitly or not leads to the classification of the models in pseudo-homogeneous and heterogeneous (i.e., separate conservation equations for the fluid and catalyst) models. By increasing the complexity and (in principle) the accuracy of the model, one can move to two-dimensional (2D) or 3D models [40] and consider deviations from the PFR by accounting for axial and radial mixing.

The integration of complex microkinetic models with reactor modeling – even under a mean-field approach for the surface – is mainly limited and applied to 1D reacting models. Two-dimensional reacting flow simulation with detailed chemistry is still not trivial in computational resources and numerical challenges (see e.g., [40]). The direct integration of kinetic Monte Carlo simulations with reactor modeling is still very challenging and demanding due to the very large difference in time scales [15, 18].

10.3.3 Hierarchical Multiscale Refinement of the Microkinetic Model

The combination of the microkinetic model and the reactor model (steps 1 and 2 in Figure 10.2) allows the simulation of selected experimental data and the identification of the key issues that need to be refined, for example, with higher hierarchy (and more demanding) methods. In doing so, one can selectively fine-tune the parameters of the microkinetic model. In fact, these parameters are known within an uncertainty range that arises both from the intrinsic uncertainty of the calculation methods and from the catalyst model: for example, presence of defects on the surface can considerably affect activation energies and make the prediction of reaction parameters for surface reactions very challenging [41]. It was found that this uncertainty can markedly affect the model predictions, related not only to unsatisfactory quantitative agreement between model predictions and experimental data but also to incorrect prediction of the macroscopic kinetic behavior [42].

In order to overcome these problems, the parameters must be refined within the uncertainty range using kinetically relevant experimental data. To date, this is the most viable method to make the microkinetic model quantitatively predictive and reliable. However, because of the very large number of parameters of microkinetic models, a "blind" optimization based on best fit of selected experimental data could be very inefficient, leading to models that are either valid over a limited range of operating conditions or incapable of being quantitatively predictive.

In order to refine the parameters through selected experimental data and make the scheme reliable and predictive, one has to follow the hierarchical multiscale refinement of the original mechanism (Section 10.3.1), grounded on the information provided by the microkinetic analysis of the experimental data (step 3, Figure 10.2).

In particular, for each experiment included in the refinement process, first a model-driven analysis is performed. This entails the identification of the RDS and the main reaction paths in the flow from the reactants to the products. On the basis of the information from the microkinetic analysis, it is possible to infer the dominant reaction mechanism and check the qualitative consistency of the model response with respect to key experimental observations. As an example, by the identification of the RDS and the main reaction paths, one could rationalize the effect of reactants and products on the overall reaction rate predicted by the model and compare them with the experimental trend. As a result, the identified reaction parameters are modified within the uncertainty range in order to get the qualitative agreement with the experiments. Then, sensitivity analysis (SA) with respect to pre-exponentials and activation energies of each reaction and heats of chemisorption of the species is performed aiming at the identification of the important parameters to be modified to get a quantitative match with the experimental data. During this process, the identified kinetic parameters are modified within their range of uncertainty. Since the number of the dominant chemical pathways is considerably lower than the whole number of elementary steps included in the microkinetic model, such pathways can be also refined with explicit DFT calculation of the activation energies and transition states [43–45]. Finally, the modified parameters are tested with respect to additional experiments as validation. In doing so, this proposed refinement methodology is not just a systematic optimization of model parameters with respect to a set of experimental data, but rather an interplay between chemistry analysis and quantitative comparison with the experimental data. This methodology turns out to be the most suitable one for the incorporation of the experimental information in the parameters of the microkinetic model, without affecting its fundamental nature. However, such an approach is feasible only if a comprehensive set of kinetically relevant experimental data is available.

10.4 Show Case: Microkinetic Analysis of CH_4 Partial Oxidation on Rh

Complementing the preceding general introduction to hierarchical multiscale microkinetic modeling of chemical catalytic processes, as a show case, the results on the microkinetic analysis of the catalytic partial oxidation of methane (CPOX) on Rh are reviewed in the next sections.

Among short-contact-time reforming technologies that convert fossil fuels and biofuels to syngas and hydrogen, CH_4 CPOX is particularly appealing for decentralized energy production due to the compactness of reactors and the fast response to transients. Though extremely flexible, the process is highly complex. CH_4 CPOX short-contact-time reactors work under severe conditions, with very high gas hourly space velocity and temperatures, complex fluid pattern, and strong coupling of heat and mass transfer with surface kinetics. These severe operating conditions pose a challenge for operation and it is therefore critical to develop fundamental mathematical models to assist with the CPOX reactor analysis and design.

First, the development of the microkinetic model is presented. Then, it is shown how one can integrate such a model with reactor modeling in order to perform a microkinetic analysis of the process and interpret the complex experimental evidence. This allows for the comprehension of the molecular-level mechanisms underlying the observed macroscopic phenomena.

10.4.1
Microkinetic Model for the Conversion of CH_4 to Syngas

A thermodynamically consistent microkinetic model for the conversion of methane to syngas on Rh-based catalysts was developed [42]. The detailed mechanism consists of 82 elementary reactions and 13 surface species. Reaction steps and corresponding rate parameters are reported in [42]. This set of reactions includes the elementary steps that are characteristic of several relevant reacting subsystems: for example, H_2 and CO oxidation, chemistry of the coupling between CO and H_2, CH_x pyrolysis, and oxidation.

The parameters of the reaction mechanism have been derived according to the hierarchical multiscale methodology presented in Section 10.3: activation energies are predicted using the UBI-QEP theory; coverage effects are accounted for using DFT; and pre-exponentials are calculated using TST. Overall, it was verified that the model is able to predict correctly the behavior of several reacting mixtures, including CH_4/H_2O (SR) and CH_4/CO_2 (dry reforming, DR) reforming, CH_4/O_2 (CPOX), H_2/O_2, and CO/O_2, CO/H_2O (WGS), under a wide range of operating conditions.

10.4.2
Microkinetic Analysis of Isothermal CPOX Data in Annular Reactor

The possibility of modeling the surface chemistry in detail allows us to propose an explanation of the main molecular pathways involved in the process, leading to a deeper understanding of the underlying catalytic mechanisms. Within this scope, Maestri *et al.* have used the microkinetic model in combination with detailed

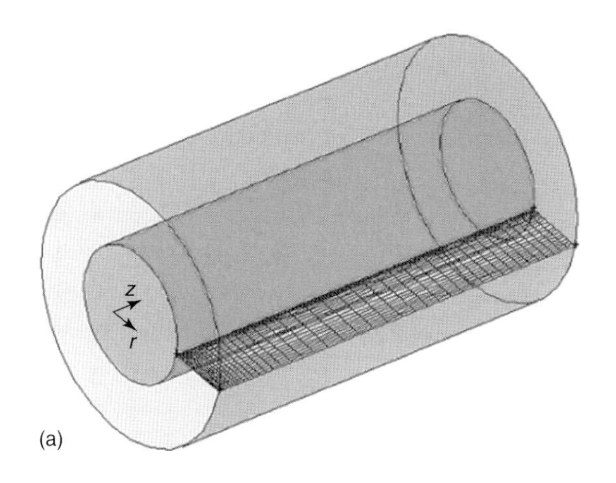

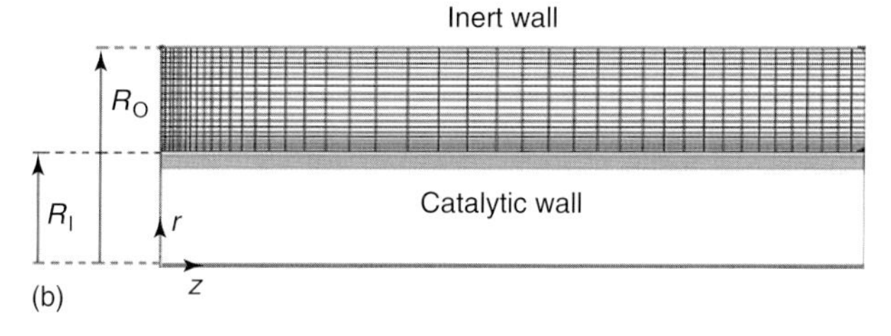

Figure 10.4 Sketch of the annular reactor. The catalyst is deposited on the external wall of the inner tube. The reacting fluid flows through the duct between the inner and the outer tubes. (Figure adapted from [48]. © 2008 Elsevier.)

reactor modeling for the microkinetic analysis of CH_4 CPOX in annular reactor under quasi-isothermal conditions [46, 47].

For the description of the macro-scale regime a 2D model of the annular reactor has been considered [48]. In particular, they have developed a dynamic, isothermal, isobaric, 2D model of a catalytic annular reactor. A schematic of the annular reactor is shown in Figure 10.4.

The model includes the species mass balance equation for each gaseous species:

$$\rho\frac{\partial\omega_i}{\partial t} = -\frac{u(\vartheta)}{L}\rho\frac{\partial\omega_i}{\partial z^*} + \frac{D_i}{R_O^2\,(1-R^*)^2}\rho\left(\frac{\partial^2\omega_i}{\partial\vartheta^2} + \frac{R^*-1}{1-\vartheta\,(1-R^*)}\frac{\partial\omega_i}{\partial\vartheta}\right) + \frac{D_i}{L^2}\rho\frac{\partial^2\omega_i}{\partial z^{*2}} \tag{10.19}$$

where

$$\vartheta = \frac{R_O - r}{R_O - R_I}$$

$$z^* = \frac{z}{L}$$

$$R^* = \frac{R_I}{R_O}$$

with ω_i is the species mass fraction. Given the very dilute conditions of the system [47], the molecular weight has been considered constant [48]. For gas-phase species, the following boundary conditions were set:

Inert wall ($\vartheta = 0$):

$$\left.\frac{\partial \omega_i}{\partial \vartheta}\right|_{\vartheta=0} = 0, \forall z^* \tag{10.20}$$

Catalytic wall ($\vartheta = 1$), for gas species:

$$a_{\text{reactor}} \frac{D_i \rho}{R_O (1 - R^*)} \left.\frac{\partial \omega_i}{\partial \vartheta}\right|_{\vartheta=1} = a_{\text{catalyst}} R_i^{\text{het}} \text{MW}_i, \quad \forall z^*, \tag{10.21}$$

Danckwert's boundary conditions were set at the reactor inlet ($z^* = 0$):

$$\left.\frac{D_i}{L \cdot u} \frac{\partial \omega_i}{\partial z^*}\right|_{z^*=0+} = \left(\omega_{i,0+} - \omega_{i,0}\right), \quad \forall \vartheta \tag{10.22}$$

and zero fluxes at the reactor outlet ($z^* = 1$):

$$\left.\frac{\partial \omega_i}{\partial z^*}\right|_{z^*=1} = 0, \quad \forall \vartheta \tag{10.23}$$

For adsorbed (surface) species:

$$\frac{\partial \theta_i}{\partial t} = \frac{R_i^{\text{het}}}{\Gamma_{\text{Rh}}}, \quad \forall z^* \tag{10.24}$$

along with the site conservation balance. Here θ_i is the site fraction of the ith species and Γ_{Rh} is the Rh site density. A fully developed laminar flow velocity profile in an annular duct has been assumed [38, 49]:

$$u(\vartheta) = -\frac{c_1 R_O^2}{4} \left[1 - \left(1 - \vartheta\left(1 - R^*\right)\right)^2 + 2r_m^{*2} \ln\left(1 - \vartheta\left(1 - R^*\right)\right)\right] \tag{10.25}$$

where

$$c_1 = -\frac{8\overline{u}}{R_O^2 \left(1 + R^{*2} - 2r_m^*\right)} \tag{10.26}$$

$$r_m^* = \left[\frac{1 - R^{*2}}{2 \ln\left(1/R^*\right)}\right]^{1/2} \tag{10.27}$$

where $\overline{u} = f(T)$ is the mean velocity. Ideal gas law has been considered for the evaluation of ρ. Physicochemical and transport properties were estimated according to the CHEMKIN correlations [50]. In particular, for the evaluation of the transport properties, the mixture-average model of the CHEMKIN package has been used.

Figure 10.5 compares model predictions and data, and gives an example of how well the model is capable of describing CPOX (integral) data. Their analysis [46] can be summarized as follows. At low temperatures, O_2 is only partially converted and deep oxidation occurs (CO_2 and H_2O are the main products); at higher temperatures, O_2 is completely converted, syngas is produced, and steam and CO_2 are being consumed. The integral data indicate methane steam and DR at moderate and high temperatures, that is, the indirect route to syngas. However, the

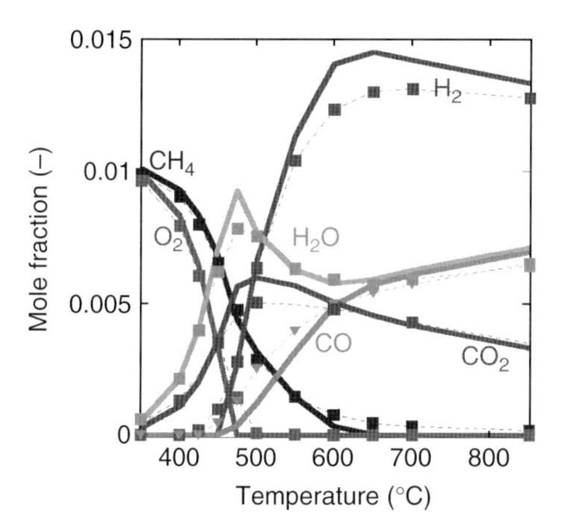

Figure 10.5 CH_4 CPOX on Rh. Mole fractions: CH_4: 0.01, O_2: 0.01, balance N_2: gas hourly space velocity 2.9 × 10^6 Nl kg_{cat}^{-1} h^{-1}. Annular reactor: I.D. 4 mm, O.D. 5 mm, and length 22 mm. Symbols: experimental data; solid lines: model predictions. (Figure adapted from [46]. © 2009 Springer.)

simulations reveal that at intermediate temperatures, a stratification of the reactor (Figure 10.6) may happen, with up to three reaction zones.

Upstream (zone 1), the most abundant reactive intermediate (MARI) on the catalyst surface is O*. Under these conditions, methane is completely oxidized to combustion products. Reaction path analysis (RPA) (Figure 10.7) indicates that, following CH_4 dissociative adsorption, CH_x^* is consumed via oxidative dehydrogenation ($CH_x^* + O^* \rightarrow CH_{x-1}^* + OH^*$ or $CH_x^* + OH^* \rightarrow CH_{x-1}^* + H_2O^*$), whereas the parallel pyrolytic path ($CH_x^* + ^* \rightarrow CH_{x-1}^* + H^*$) is negligible.

This behavior arises because the activation energies of the oxidative paths (via OH* and O*) are energetically favored at high O* coverages. C*, which is formed via oxidative dehydrogenation, is oxidized (with O*) to CO*. CO* is oxidized faster to CO_2^* than desorbing, according mainly to

$$CO^* + OH^* \longrightarrow COOH^* + ^* \longrightarrow CO_2{}^* + H^* \tag{10.28}$$

Similarly, H^*, which is formed via oxidative dehydrogenation of CH_4, is consumed via oxidation:

$$O^* + H^* \longrightarrow OH^* + ^* \tag{10.29}$$

$$H^* + OH^* \longrightarrow H_2O^* + ^* \tag{10.30}$$

giving eventually gaseous water. Equations (10.29) and (10.30) dominate over the H^* desorption. H^* and CO^* are key surface intermediates, but their oxidation (Eqs. (10.28–10.30)) is faster at high O^* coverages than their desorption, leading to deep CH_4 oxidation. Under these conditions, the RDS (identified via SA [51, 52]) is the CH_4 dissociative adsorption. As O^* is being depleted, more Rh sites

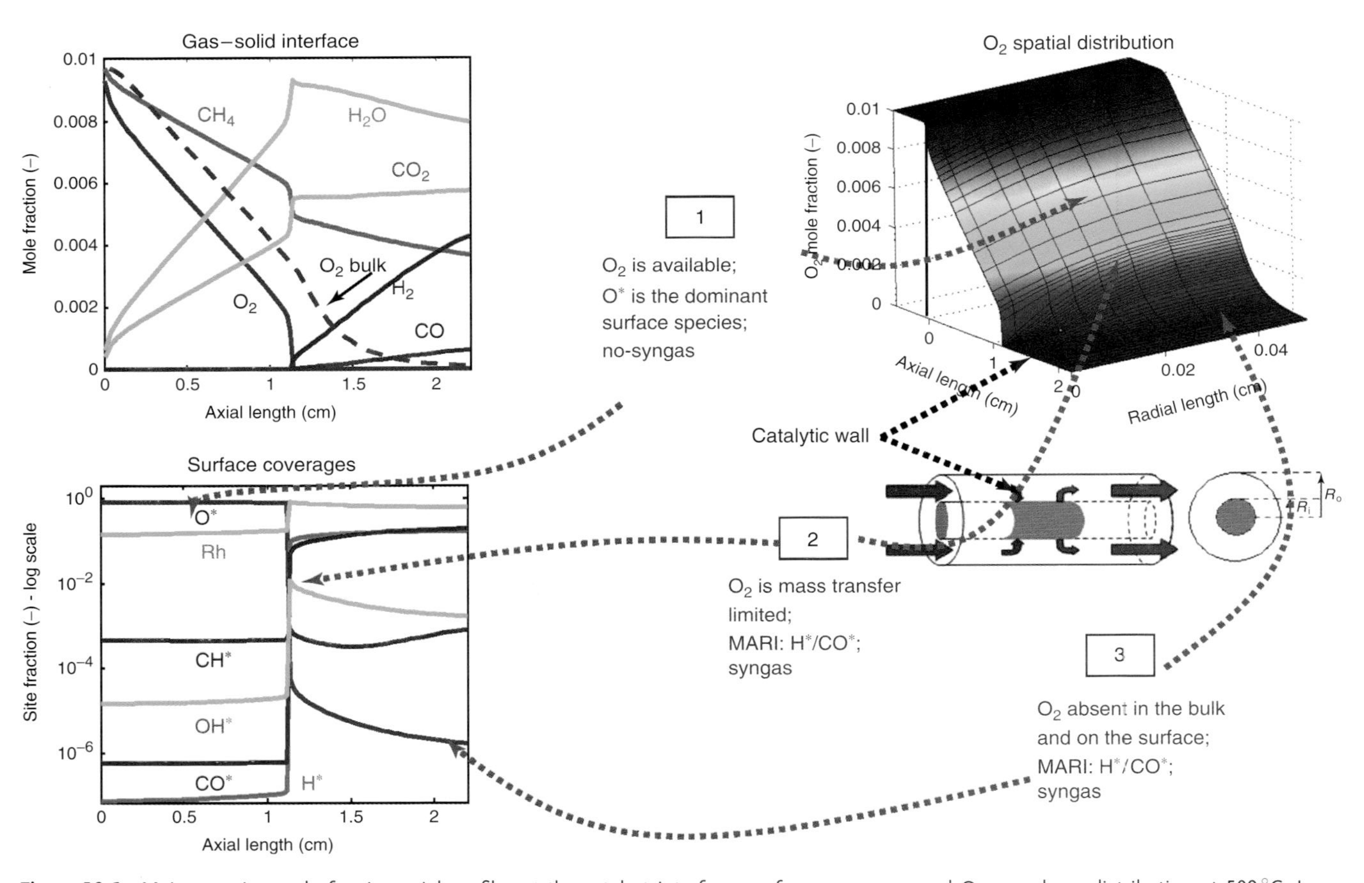

Figure 10.6 Major species mole fraction axial profiles at the catalyst interface, surface coverages, and O_2 gas-phase distribution at 500 °C. In the Gas–solid interface panel the bulk gas-phase O_2 mole fraction is also reported (dashed line). The identified characteristic zones are highlighted. Parameters are as in Figure 10.5.

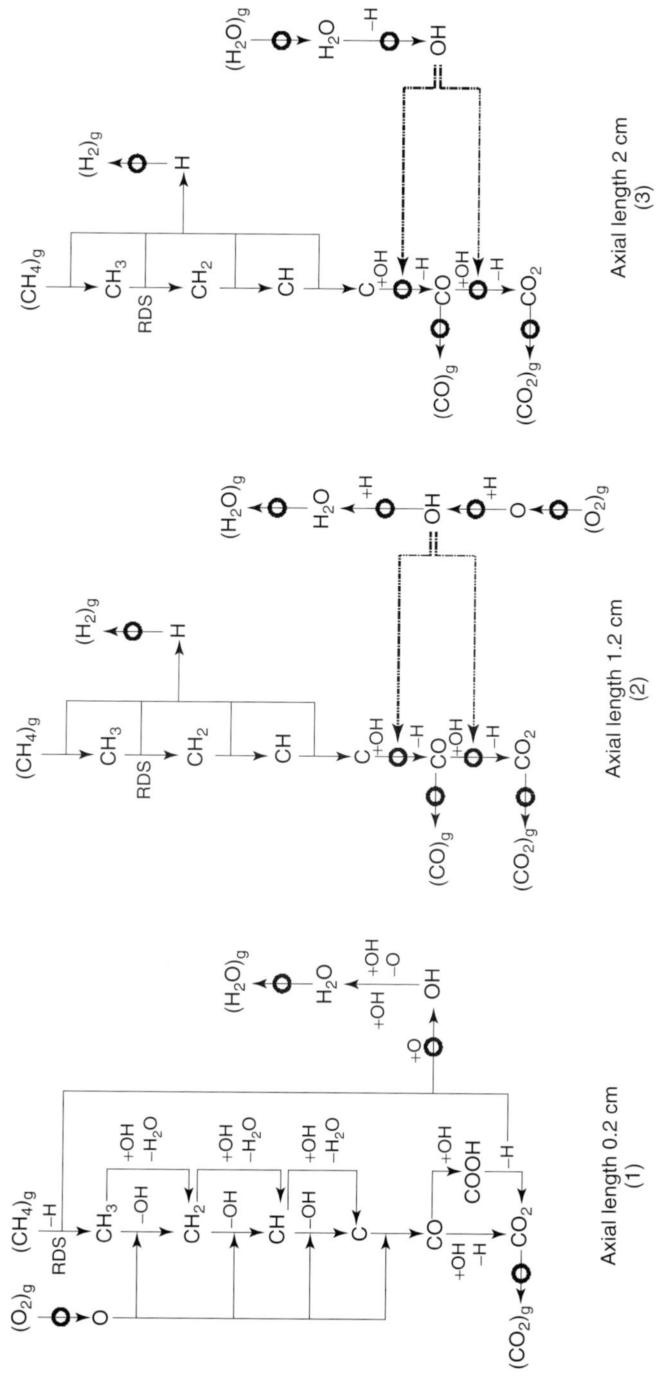

Figure 10.7 Reaction path analysis at different axial coordinates. Conditions are as in Figure 10.5. Circled-arrows indicate quasi-equilibrated reactions. (Figure adapted from [46]. © 2009 Springer.)

become available and the rate of CH_4 consumption increases both because of mass action kinetics and the O^* coverage dependence of the energetics of the oxidative paths.

Once O^* has been consumed, H^* and CO^* become most abundant and the activation energies of the oxidative paths become comparable with the pyrolytic ones (zones 2 and 3 in Figures 10.6 and 10.7). In addition, the surface turns out to be reasonably clean (Rh vacancies > 70%, Figure 10.6). As a result, the dominant pathways on the catalyst in zones 2 and 3 change completely. In zone 3, there is no gaseous O_2 left and OH^* forms from adsorption of H_2O and its subsequent decomposition to OH^* and H^*. C^*, which is formed via pyrolysis ($CH_4 \rightarrow C^* + 4\,H^*$), is oxidized to CO^* and then to CO_2^* via OH^*.

Overall, syngas forms via the indirect route and SR and WGS (close to equilibrium) are the overarching reactions [53], in good agreement with independent kinetic studies [54]. In zone 2, the gaseous oxygen concentration is nearly zero at the catalytic surface and, as in zone 3, H^* and CO^* dominate on the surface. However, owing to interphase mass transfer limitations, gaseous oxygen is still present in the bulk of the channel. O_2 from the bulk of the gas phase leaks to the catalytic surface and activates different reaction paths: OH^* forms from the reaction between H^* (from CH_4 pyrolysis) and O^* (from dissociation of molecular oxygen that slowly diffuses from the gas phase). A fraction of OH^* reacts with H^* giving H_2O^*, leading to gaseous H_2O, and with CO^* giving eventually rise to CO_2. The slowdown of CO^* and H^* oxidation reactions makes desorption of syngas competitive to its combustion. As a result, syngas is produced via a "direct route," since H_2O adsorption is not needed. In this second zone, a mixed mode with competition between the direct path of methane to syngas and syngas combustion is found. The lack of excess adsorbed oxygen activates the pyrolytic chemistry over the oxidative chemistry. Consequently, zone 2 has many features (in terms of elementary steps) of the SR and WGS zone 3 with noticeable differences, for example, OH^* forms from $H^* + O^*$ rather than from H_2O [53].

An additional key observation of the model analysis is that the main oxidizer is OH^* rather than O^*. This is different from most of the mechanisms reported in the literature, where it is usually *assumed* that the main oxidizer is O^* (coming from $OH^* + ^* \rightarrow H^* + O^*$) and oxidation via OH^* is not considered. According to these calculations, the OH^* dissociation to O^* and H^* is slow, and the parallel consumption paths (e.g., oxidation of C^* via OH^*) turn out to be favored at the investigated conditions.

As a whole, the microkinetic analysis revealed that the partial oxidation of methane does not occur at the molecular level at the surface. Rather, it turns out to be the result of a combination of different reacting systems, giving rise to up to three reaction zones: a deep combustion of methane, followed by a zone where direct formation of syngas in parallel with catalytic combustion occurs, and finally an SR and WGS zone, when oxygen is no longer available. Simulations in the absence of mass transfer have indicated that zone 2 exists only because of mass transfer limitations, and the extent of various zones depends mainly on temperature. Therefore, the analysis revealed that the interplay between mass transfer effects and surface chemistry is crucial in dictating the dominant reaction pathways leading to syngas.

These findings underscore the importance of the long-postulated interplay of transport phenomena and surface chemistry, pointing out that an accurate description of transport effects is required for the identification of the dominant reaction mechanism and the description of the related experimental data. In particular, these results have reconciled, for the first time in the literature, the apparent long-debated mechanistic differences regarding direct versus indirect pathways.

The extent of the three zones in the reactor strictly depends on temperature. At low temperatures, O^* dominates and complete combustion prevails (zone 1). At higher temperatures, zones 2 and 3 prevail and O_2 consumption can be mass-transfer-controlled. This is usually the case for the steady-state operation of adiabatic CPOX reformers [23–27]. In order to elucidate this, the microkinetic analysis has been extended to autothermal CPOX reformers.

10.4.3
Microkinetic Analysis of Autothermal CPOX Data on Foams

The analysis has been extended to steady-state autothermal CH_4 CPOX data on foams [46, 55]. In particular, a fully predictive microkinetic analysis of CH_4 CPOX spatially resolved experiments on Rh-based foams has been performed by addressing in detail both transport and chemical phenomena. See [55] for details. Figure 10.8 compares the spatially resolved experimental data and model predictions. The numerical results show that the model is able to quantitatively account for the axial evolution of the species and the temperature profiles of the solid and the gas phases within the catalyst volume. In line with the isothermal data analysis at higher temperatures, a mixed mode (direct production to syngas plus combustion; zone 2) followed by the indirect path (SR/WGS; zone 3) is seen (zone 1 does not exist). Therefore, it was confirmed that the consumption of O_2 is strictly governed by mass transfer, and that the co-presence of O_2 and syngas in the bulk of the gas-phase is exclusively due to mass transfer limitations. Since O_2 consumption – which governs the heat release on the catalytic surface – is always mass-transfer-controlled, thus one could design the reactor to fine-tune the optimal temperature profile [56–58].

As an example, Figure 10.9 shows the simulated effect of varying the foam pore density while keeping constant the Rh loading: the increase in interphase mass transfer resistance results in a more gradual dosing of O_2 along the axial coordinate and consequently in the reduction of the hot-spot temperature, which is anticipated to have a positive effect on catalyst stability [58]. The more gradual dosing of O_2 results in a decrease of H_2 and CO selectivity at short distances: nevertheless, given the high activity of Rh, the outlet composition is usually close to equilibrium and, consequently, no significant effect on the integral H_2 and CO selectivity is observed [56].

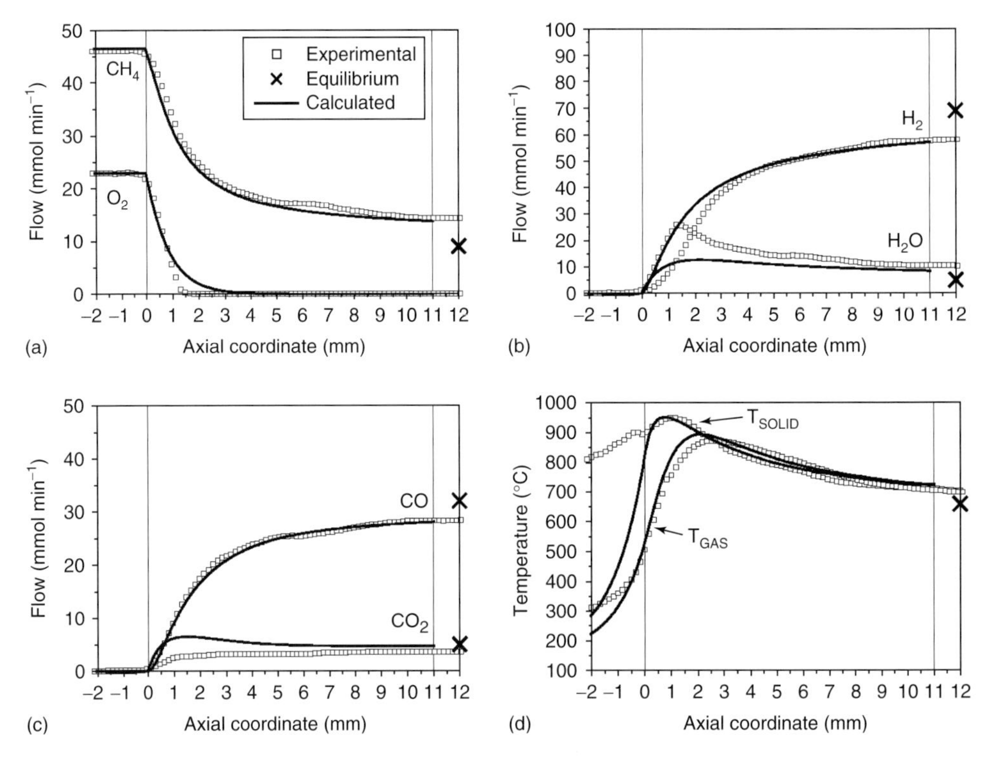

Figure 10.8 Comparison between model prediction (solid lines) and data (symbols) for a CH_4 CPOX experiment over the 5% Rh catalyst. Operating conditions: $CH_4 = 20\%$v/v, O/C = 1, Flow = 5 slpm, $T_{IN} = 150\,°C$, $P = 1$ atm. (a) CH_4 and O_2 flow rates. (b) H_2 and H_2O flow rates. (c) CO and CO_2 flow rates. (d) Gas and surface temperature profiles. (Adapted from [55]. © 2010 Elsevier.)

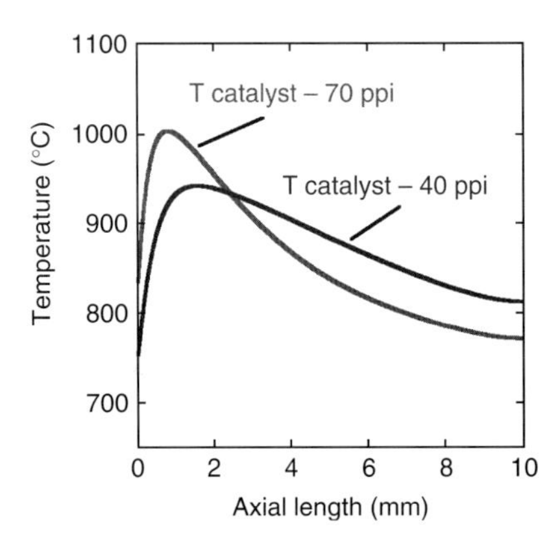

Figure 10.9 Catalyst temperature profiles for foams at different pore densities. (Figure adapted from [46]. © 2009 Springer.)

10.5 Conclusions

The possibility of modeling in detail and at the fundamental level the surface chemistry has reinterpreted the kinetic models as tools not only for the analysis and design of chemical reactors but also for the consolidation of fundamental knowledge about a catalytic process under different operating conditions. In this respect, the atomic-scale understanding of the factors determining the catalytic activity and selectivity is a requirement to allow the development of new and improved processes, especially in energy-related fields. In particular, of pivotal importance is the identification of the dominant reaction mechanism that is established in the reactor at specific operating conditions of temperature, pressure, and concentration. Different from the reaction mechanism (i.e., the whole set of elementary steps that can occur at the catalyst surface) that is a property of the microscale, the dominant reaction mechanism is the result of the interplay among the specific phenomena occurring at the different scales. Consequently, it is a multiscale property of the system. This implies that, if, on one side, the accurate description of each scale is a requirement, on the other side, it is not sufficient for the understanding of the actual behavior of the system. Therefore, the development of efficient methodologies to properly link between the scales is essential, where several orders of magnitude in the time and length scales have to be bridged. This is particularly challenging for complex processes of real technological interests, where a potentially huge number of chemical events can occur at the microscale with a strong coupling between the scales. Given the complexity both in the model description and in the problem dimension, hierarchical multiscale approaches turn out to be the most effective way to deal with this challenge. The main idea behind this approach is to tackle the problem with methods of increasing accuracy and computational demand. Lower accuracy and computationally nondemanding methods are used to identify the key issues of the full problem which is otherwise nontractable. Then, these key features are refined, for example, with higher accuracy and computationally demanding methods, which necessarily need to focus on narrowly defined aspects and specific steps of the full problem. In this respect, it is very important that the control of the error of the lower accuracy methods used to cope with the full problem for a correct identification of the key features be refined with higher accuracy methods. Within this scope, it has been shown that the modified parameterization of the UBI-QEP semiempirical method with DFT [36] can be effectively employed at the lower hierarchical level.

In particular, the microkinetic analysis of CH_4 CPOX reviewed here has provided an effective show case of how the detailed understanding of the molecular level mechanisms underlying the macroscopic phenomena can be achieved by the proposed hierarchical multiscale approach. Specifically, this multiscale study has substantially contributed to the understanding of the molecular level mechanisms underlying the macroscopic phenomena in the CH_4 activation to syngas as well as hydrogen on Rh which is relevant to decentralized hydrogen production for energy applications. The microkinetic analysis revealed that the partial oxidation of methane on Rh does not occur at the molecular level on the surface. Rather, it turns out to be the result of a combination of different reacting systems. The analysis revealed that the interplay between mass transfer effects and surface chemistry is crucial in dictating the dominant reaction pathways leading to syngas. On one hand, these findings have been of primary interest for the understanding and interpretation of the complex experimental evidence, thus enabling conflicting reports in the literature to be reconciled. On the other hand, they are of direct use for the rational design and scaling-up of short-contact-time reformers for small-scale production of hydrogen in the sustainable energy field.

As to the future perspectives of hierarchical multiscale modeling, the main and crucial challenge is the realistic description of the microscale, both in terms of chemical steps and relevant intermediates and in terms of model description (i.e., active sites, structure sensitivity). This is crucial because a nonrigorous or erroneous description of the microscale can lead to a dramatically wrong interpretation of the observed macroscopic phenomena. In this respect, the trend must be toward the effective inclusion of theoretically sound methods (quantum and molecular dynamics calculations), avoiding ad hoc and blind fitting of the energetics of the single steps. In addition, support effects, multiple types of sites, model beyond a mean-field approximation, and the dynamics of catalyst nanoparticles with changing reaction environment are important aspects to be included to further improve the understanding of reaction mechanisms at surfaces.

Acknowledgments

This chapter is based on the work that I have done for the last five years at Politecnico di Milano, Italy (2005 and 2008), University of Delaware, USA (2006–2007), and the Fritz–Haber Institut, Berlin, Germany (2009–2010). Much of the information reviewed here would have been impossible without the stimulating environment offered by these world-leading institutions. In particular, I acknowledge the stimulating collaboration and enlightening discussions with my mentors: Enrico Tronconi, Alessandra Beretta, Gianpiero Groppi, and Pio Forzatti (Politecnico di Milano), Dion Vlachos (University of Delaware), and Karsten Reuter (Fritz–Haber Institut).

I gratefully acknowledge MIUR (Italy) and the Alexander von Humboldt Foundation (Germany) for funding.

References

1. Hougen, O. and Watson, K. (1947) *Chemical Processes Principles: Kinetics and Catalysis*, vol. **3**, John Wiley & Sons, Inc., London.
2. Dumesic, J.A., Huber, G.W., and Boudart, M. (2008) Rates of catalytic reactions, in *Handbook of Heterogeneous Catalysis* (ed. G. Ertl *et al.*), Wiley-VCH Verlag GmbH, New York, 1445–1462.
3. Schlogl, R. (2001) Theory in heterogeneous catalysis – an experimentalist's view. *Cattech*, **5** (3), 146–170.
4. Stoltze, P. and Nørskov, J.K. (2008) Theoretical modeling of catalytic reactions, in *Handbook of Heterogeneous*

Catalysis (ed. G. Ertl *et al.*), Wiley-VCH Verlag GmbH, New York, 1479–1492.
5. Ertl, G. (2008) Reactions at surfaces: from atoms to complexity (Nobel lecture). *Angew. Chem. Int. Ed.*, **47** (19), 3524–3535.
6. Boudart, M. (2000) From the century of the rate equation to the century of the rate constants: a revolution in catalytic kinetics and assisted catalyst design. *Catal. Lett.*, **65** (1–3), 1–3.
7. Gokhale, A.A., Kandoi, S., Greeley, J.P., Mavrikakis, M., and Dumesic, J.A. (2004) Molecular-level descriptions of surface chemistry in kinetic models using density functional theory. *Chem. Eng. Sci.*, **59** (22–23), 4679–4691.
8. Nørskov, J.K., Bligaard, T., Rossmeis, J., and Christensen, C.H. (2009) Towards the computational design of solid catalysts. *Nat. Chem.*, **1**, 37–46.
9. Campbell, C.T. (1994) Future directions and industrial perspectives micro- and macro- kinetics: their relationship in heterogeneous catalysis. *Top. Catal.*, **1**, 353–366.
10. Deutschmann O. (ed.) (2011) *Modeling of Heterogeneous Catalytic Reactions: From the Molecular Process to the Technical System*, Wiley-VCH Verlag GmbH, Weinheim.
11. Hansen, N., Krishna, R., van Baten, J.M., Bell, A.T., and Keil, F.J. (2009) Analysis of diffusion limitation in the alkylation of benzene over H-ZSM-5 by combining quantum chemical calculations, molecular simulations, and a continuum approach. *J. Phys. Chem. C*, **113** (1), 235–246.
12. Hansen, N., Krishna, R., van Baten, J.M., Bell, A.T., and Keil, F.J. (2010) Reactor simulation of benzene ethylation and ethane dehydrogenation catalyzed by ZSM-5: A multiscale approach. *Chem. Eng. Sci.*, **65** (8), 2472–2480.
13. Moulijn, J., Perez-Ramirez, J., van Diepen, A., Kreutzer, M., and Kapteijn, F. (2003) Catalysis engineering on three levels. *Int. J. Chem. React. Eng.*, **1**, 1–17.
14. Stampfl, C., Ganduglia-Pirovano, M.V., Reuter, K., and Scheffler, M. (2002) Catalysis and corrosion: the theoretical surface-science context. *Surf. Sci.*, **500** (1–3), 368–394.
15. Reuter, K. (2009) First-principles kinetic Monte Carlo simulations for heterogeneous catalysis: concepts, status and frontiers, in *Modeling Heterogeneous Catalytic Reactions: From the Molecular Process to the Technical System* (ed. O. Deutschmann), Wiley-VCH Verlag GmbH, Weinheim, 71–112.
16. Fox, R.O. (2003) *Computational Models for Turbulent Reacting Flows*, Cambridge University Press, Cambridge.
17. Raimondeau, S. and Vlachos, D.G. (2002) Recent developments on multiscale, hierarchical modeling of chemical reactors. *Chem. Eng. J.*, **90** (1–2), 3–23.
18. Meskine, H., Matera, S., Scheffler, M., Reuter, K., and Metiu, H. (2009) Examination of the concept of degree of rate control by first-principles kinetic Monte Carlo simulations. *Surf. Sci.*, **603** (10–12), 1724–1730.
19. Astarita, G. and Ottino, J.M. (1995) 35 years of BSL. *Ind. Eng. Chem. Res.*, **34** (10), 3177–3184.
20. Gokhale, A.A., Dumesic, J.A., and Mavrikakis, M. (2008) On the mechanism of low-temperature water gas shift reaction on copper. *J. Am. Chem. Soc.*, **130** (4), 1402–1414.
21. Honkala, K., Hellman, A., Remediakis, I.N., Logadottir, A., Carlsson, A., Dahl, S., Christensen, C.H., and Norskov, J.K. (2005) Ammonia synthesis from first-principles calculations. *Science*, **307** (5709), 555–558.
22. Inderwildi, O.R., Jenkins, S.J., and King, D.A. (2006) An unexpected pathway for the catalytic oxidation of methylidyne on Rh (111) as a route to syngas. *J. Am. Chem. Soc.*, **129**, 1751–1759.
23. Reuter, K., Frenkel, D., and Scheffler, M. (2004) The steady-state of heterogeneous catalysis, studied by first-principles statistical mechanics. *Phys. Rev. Lett.*, **93**, 116105.
24. Hansen, N., Kerber, T., Sauer, J., Bell, A.T., and Keil, F.J. (2010) Quantum chemical modeling of benzene ethylation over h-ZSM-5 approaching chemical accuracy: a hybrid MP2:DFT study. *J. Am. Chem. Soc.*, **132** (33), 11525–11538.

25. Heyden, A., Peters, B., Bell, A.T., and Keil, F.J. (2005) Comprehensive DFT study of nitrous oxide decomposition over Fe-ZSM-5. *J. Phys. Chem. B*, **109** (5), 1857–1873.
26. Schwiedernoch, R., Tischer, S., Correa, C., and Deutschmann, O. (2003) Experimental and numerical study on the transient behavior of partial oxidation of methane in a catalytic, monolith. *Chem. Eng. Sci.*, **58** (3–6), 633–642.
27. Xu, J., Clayton, R., Balakotaiah, V., and Harold, M.P. (2008) Experimental and microkinetic modeling of steady-state NO reduction by H_2 on $Pt/BaO/Al_2O_3$ monolith catalysts. *Appl. Catal. B: Environ.*, **77** (3–4), 395–408.
28. Aghalayam, P., Park, Y.K., and Vlachos, D.G. (2000) Construction and optimization of complex surface-reaction mechanisms. *AIChE J.*, **46** (10), 2017–2029.
29. Mhadeshwar, A.B. and Vlachos, D.G. (2005) Hierarchical multiscale mechanism development for methane partial oxidation and reforming and for thermal decomposition of oxygenates on Rh. *J. Phys. Chem. B*, **109** (35), 16819–16835.
30. Mhadeshwar, A.B., Wang, H., and Vlachos, D.G. (2003) Thermodynamic consistency in microkinetic development of surface reaction mechanisms. *J. Phys. Chem. B*, **107** (46), 12721–12733.
31. Vlachos, D.G., Mhadeshwar, A.B., and Kaisare, N.S. (2006) Hierarchical multiscale model-based design of experiments, catalysts, and reactors for fuel processing. *Comput. Chem. Eng.*, **30** (10–12), 1712–1724.
32. Shustorovich, E. and Sellers, H. (1998) The UBI-QEP method: a practical theoretical approach to understanding chemistry on transition metal surfaces. *Surf. Sci. Rep.*, **31** (1–3), 5–119.
33. Hammond, G.S. (1955) A correlation of reaction rates. *J. Am. Chem. Soc.*, **77** (2), 334–338.
34. Leffler, J.E. (1953) Parameters for the description of transition states. *Science*, **117** (3039), 340–341.
35. Shustorovich, E. (2007) Chemisorption energetics and surface reactivity: UBI-QEP versus DFT projections. *Russ. J. Phys. Chem. B*, **1**, 307–329.
36. Maestri, M. and Reuter, K. (2011) Semi-empirical rate constants for complex chemical kinetics: first-principles assessment and rational refinement. *Angew. Chem. Int. Ed.*, **50**, 1194–1197.
37. Temel, B., Meskine, H., Reuter, K., Scheffler, M., and Metiu, H. (2007) Does phenomenological kinetics provide an adequate description of heterogeneous catalytic reactions? *J. Chem. Phys.*, **126**, 204711.
38. Bird, R.B., Stewart, W.E., and Lightfoot, E.N. (2002) *Transport Phenomena*, 2nd edn, John Wiley & Sons, Inc., New York.
39. Froment, G.B. and Bischoff, K.B. (1990) *Chemical Reactor Analysis and Design*, John Wiley & Sons, Inc., New York.
40. Mladenov, N., Koop, J., Tischer, S., and Deutschmann, O. (2010) Modeling of transport and chemistry in channel flows of automotive catalytic converters. *Chem. Eng. Sci.*, **65** (2), 812–826.
41. Liu, Z.P. and Hu, P. (2003) General rules for predicting where a catalytic reaction should occur on metal surfaces: a density functional theory study of C-H and C-O bond breaking/making on flat, stepped, and kinked metal surfaces. *J. Am. Chem. Soc.*, **125** (7), 1958–1967.
42. Maestri, M., Vlachos, D.G., Beretta, A., Groppi, G., and Tronconi, E. (2009) A C_1 microkinetic model for the conversion of methane to syngas on Rh/Al2O3. *AIChE J.*, **55**, 993–1008.
43. Dumesic, J.A., Rudd, D.F., Aparicio, L.M., Rekoske, J.E., and Trevino, A.A. (1993) *The Microkinetics of Heterogeneous Catalysis*, American Chemical Society, Washington, DC.
44. Bolhuis, P., Chandler, D., Dellago, C., and Geissler, P. (2002) Transition path sampling: throwing ropes over rough mountain passes, in the dark. *Ann. Rev. Phys. Chem.*, **53**, 291–317.
45. Weinan, E. and Vanden-Eijnden, E. (2010) Transition-path theory and path-finding algorithms for the study of rare events. *Ann. Rev. Phys. Chem.*, **61**, 391–420.
46. Maestri, M., Vlachos, D.G., Beretta, A., Forzatti, P., Groppi, G., and Tronconi, E. (2009) Dominant reaction pathways in

the catalytic partial oxidation of methane on Rh. *Top. Catal.*, **52**, 1983–1988.

47. Donazzi, A., Beretta, A., Groppi, G., and Forzatti, P. (2008) Catalytic partial oxidation of methane over a 4% Rh/alpha-Al_2O_3 catalyst part I: kinetic study in annular reactor. *J. Catal.*, **255** (2), 241–258.
48. Maestri, M., Beretta, A., Faravelli, T., Groppi, G., Tronconi, E., and Vlachos, D.G. (2008) Two-dimensional detailed modeling of fuel rich H_2 combustion over Rh/Al_2O_3 catalyst. *Chem. Eng. Sci.*, **63**, 2657–2669.
49. Shah, R.K. and London, A.L. (1978) *Laminar Flow Forced Convection in Ducts*, Academic Press, New York.
50. Kee, R.J., Rupley, F.M., and Miller, J.A. (1987) *The Chemkin Thermodynamic Database*, Sandia National Laboratories.
51. Campbell, C.T. (2001) Finding the rate-determining step in a mechanism – comparing DeDonder relations with the "degree of rate control". *J. Catal.*, **204** (2), 520–524.
52. Stegelmann, C., Andreasen, A., and Campbell, C.T. (2009) Degree of rate control: how much the energies of intermediates and transition states control rates. *J. Am. Chem. Soc.*, **131** (37), 13563–13563; **131**, 8077, (2009).
53. Maestri, M., Vlachos, D.G., Beretta, A., Groppi, G., and Tronconi, E. (2008) Steam and dry reforming of methane on Rh: microkinetic analysis and hierarchy of kinetic models. *J. Catal.*, **259**, 211–222.
54. Wei, J.M. and Iglesia, E. (2004) Structural requirements and reaction pathways in methane activation and chemical conversion catalyzed by rhodium. *J. Catal.*, **225** (1), 116–127.
55. Donazzi, A., Maestri, M., Micheal, B., Beretta, A., Forzatti, P., Groppi, G., Tronconi, E., Schmidt, L.D., and Vlachos, D.G. (2010) Microkinetic modeling of spatially resolved autothermal CH4 catalytic partial oxidation experiments over Rh-coated foams. *J. Catal.*, **275**, 270–279.
56. Beretta, A., Groppi, G., Lualdi, M., Tavazzi, I., and Forzatti, P. (2009) Experimental and modeling analysis of methane partial oxidation: transient and steady-state behavior of Rh-coated honeycomb monoliths. *Ind. Eng. Chem. Res.*, **48** (8), 3825–3836.
57. Maestri, M., Beretta, A., Groppi, G., Tronconi, E., and Forzatti, P. (2005) Comparison among structured and packed-bed reactors for the catalytic partial oxidation of CH_4 at short contact times. *Catal. Today*, **105** (3–4), 709–717.
58. Tavazzi, I., Beretta, A., Groppi, G., Maestri, M., and Tronconi, E. (2007) Experimental and modeling analysis of the effect of catalyst aging on the performance of a short contact time adiabatic CH4-CPO reactor. *Catal. Today*, **129** (3–4), 372–379.

11
Synthetic Potential behind Gold-Catalyzed Redox Processes

Cristina Nevado and Teresa de Haro

11.1
Introduction

Gold has drifted through history as a noble metal: an attractive ornamentation material and a currency of high value, but an inert element in terms of chemical reactivity. Surprisingly though, the last 20 years have witnessed an exponential growth of gold catalysis, both heterogeneous and homogeneous. Chemically speaking, we are thus immersed in the "gold-rush" era.

The focus of this chapter is to review gold-catalyzed redox processes and their synthetic impact. Because the field is rather broad, we will just focus on homogeneous catalysis [1], skipping the heterogeneous methodologies [2]. We believe that it is also important to define here what we understand by "gold-catalyzed redox processes." On one hand, we will refer to processes in which the *redox pair Au(I)/Au(III) is proposed to be involved in the catalytic cycle*, but we will also consider processes in which *a formal redox process has occurred in the substrate with either Au(I), Au(III), or even radical species* involved.

Because of the fast development of gold catalysis, a selection of the most representative examples has been made. We hope that, even if key contributions have been left out, those that are included give the reader an overview of the targets already achieved as well as the new trends and challenges still ahead in the fascinating area of gold-catalyzed redox processes.

11.2
Gold-Catalyzed Reactions Involving Oxygen Functionalities

11.2.1
Oxidation of Alkanes

The oxygenation of inert C–H bonds has been, and still is, one of the Holy Grails for chemists. The major challenges associated with the development of efficient processes stem from both the unreactive nature of C–H bonds and the lack of

New Strategies in Chemical Synthesis and Catalysis, First Edition. Edited by Bruno Pignataro.

$$[Au^{I}] + H_2O_2 \longrightarrow [Au^{I}\text{–}OOH] + H^+ \longrightarrow \underset{\mathbf{1}}{[Au^{III}{=}O]} + H_2O$$

$$\underset{\mathbf{1}}{[Au^{III}{=}O]} + R\text{–}H \longrightarrow \underset{\mathbf{2}}{[Au^{II}\text{–}OH]} + R\cdot$$

$$R\cdot + O_2 \longrightarrow R\text{–}OO\cdot \xrightarrow{\mathbf{2}} \underset{\mathbf{3}}{R\text{–}OOH} + [Au^{III}{=}O]$$

Scheme 11.1

selectivity due to the harsh reaction conditions that are usually employed. In this context, the oxidation of methane (CH_4) has become the benchmark process for the development of new late-transition-metal-catalyzed methodologies to activate C–H bonds. The first successful examples employed fuming sulfuric acid at 263 °C with $HgSO_4$ to produce oxygenated and sulfonated derivatives in moderate yields [3]. Periana demonstrated that the reaction relied on a Hg(II)/Hg(I) redox couple [4]. The Pt(II)/Pt(IV) redox system can also oxidize Csp^3–H bonds to form alcohols, as reported by Shilov and Shul'pin [5]. Because of the isoelectronic nature of Au(I) and Au(III) with Hg(II) and Pt(II), respectively, gold was also sought as a potential catalyst for these transformations. The major breakthrough came from the work of Shilov and Shul'pin, who managed to oxidize alkanes to alkyl hydroperoxides with $NaAuCl_4$ and PPh_3AuCl as catalysts in the presence of H_2O_2 and acetonitrile as solvent [6]. The proposed mechanism involves gold-oxo and peroxo complexes as key oxidizing species (together with HO• and possibly also RO•) so that, overall, a Au(I)/Au(III) manifold operates in these transformations. As shown in Scheme 11.1, Au(I) reacts with H_2O_2 to form a high-valent gold-oxo derivative (**1**), which attacks the C–H bond of an alkane (R–H), leading to the formation of an alkyl radical (R•) and Au(II)–OH species **2**. Further reaction of the alkyl radical with oxygen delivers the observed alkyl hydroperoxide (**3**) (Scheme 11.1).

Further support for a two-electron Au(I)/Au(III) catalytic cycle was obtained by Periana and coworkers in the selective oxidation of methane to methanol using a solution of H_2SeO_4 in 96% sulfuric acid [7]. On the basis of both experimental and computational studies, a catalytic cycle is proposed involving the formation of Au–CH_3 species, oxidation of the metal, and product elimination, although the order of these events might depend on whether methane is activated on Au(I) by oxidative addition (Scheme 11.2, pathway a) or on Au(III) by electrophilic substitution (pathway b). Although the majority of gold is present as Au(III), a reaction starting with oxidative addition of CH_4 on Au(I) seems to be energetically favored and cannot be ruled out.

11.2.2
Oxidation of Alcohols to Carbonyl Compounds

Gupta and coworkers reported in 2001 the oxidation of alcohols with gold(III) salts as stoichiometric oxidants [8]. However, the development of new catalytic oxidation methods using environmentally benign oxidants fostered further efforts

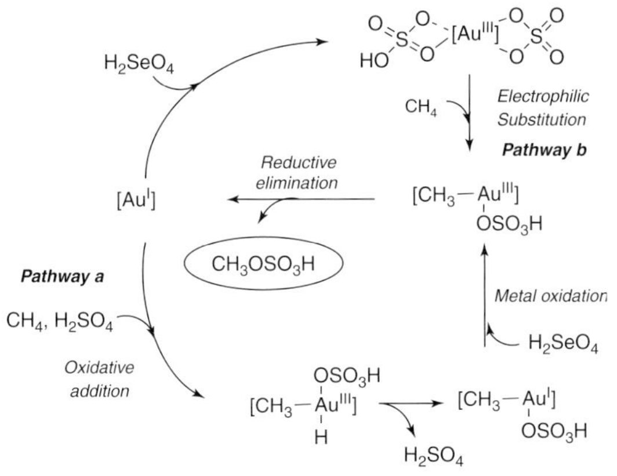

Scheme 11.2

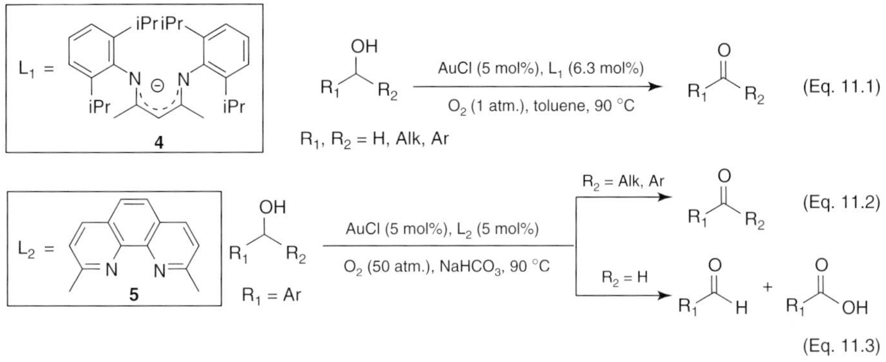

Scheme 11.3

in this field. Shi and coworkers showed that primary and secondary alcohols could be selectively transformed into aldehydes and ketones, respectively, using AuCl and β-ketimidate anionic ligands (**4**), and dioxygen (O_2) as oxidant to effect the catalyst turnover (Scheme 11.3, Eq. (11.1)) [9]. The negative charge on the ligand enhances its donating ability, whereas the flanking groups on the nitrogen atoms provide steric hindrance to leave open coordination sites on the metal center to bind an oxygen transfer reagent and also the substrate, thus facilitating the oxidation process. Alternatively, when neocuproine (**5**) was used as ligand, AuCl was able to oxidize secondary benzylic alcohols into the corresponding ketones at high oxygen pressure (Scheme 11.3, Eq. (11.2)). In the case of primary benzylic alcohols, aldehydes and/or carboxylic acids are obtained depending on the substrate (Scheme 11.3, Eq. (11.3)) [10]. Although these processes are still not fully understood, in both cases the gold–ligand complex could be detected by electrospray ionization mass spectrometry (ESI-MS) at different stages of the reaction, supporting the idea of mononuclear gold complexes as the efficient catalytic species in solution.

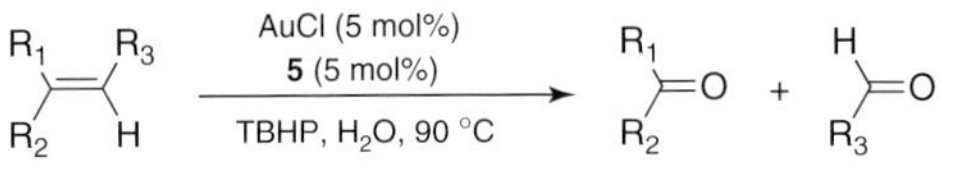

Scheme 11.4

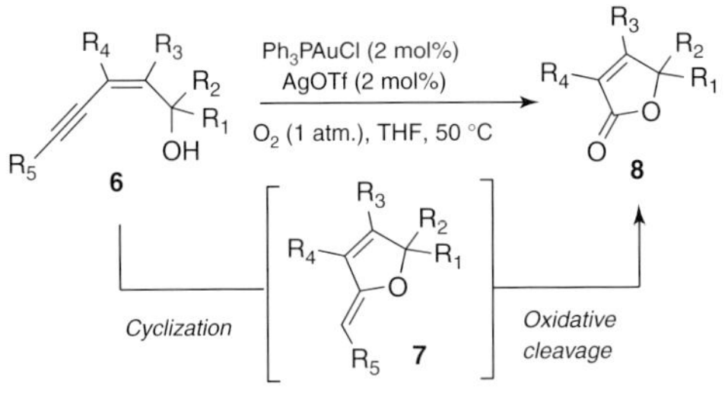

Scheme 11.5

11.2.3 Oxidation of Alkenes

The gold-catalyzed oxidative cleavage of alkenes to give carbonyl compounds has also been successfully developed in the presence of cuproine ligands with *tert*-butyl hydroperoxide (TBHP) as stoichiometric oxidant in aqueous media (Scheme 11.4) [11].

As an example of gold's ability to increase molecular complexity, a cationic gold complex generated in situ ($Ph_3PAuOTf$) has been used by Liu and coworkers to develop a catalytic cyclization/oxidative cleavage cascade. (*Z*)-Enynols (**6**) were transformed into butenolides (**8**) using dioxygen as oxidant. The reaction intermediates were dienes (**7**), which could be independently oxidized under the reaction conditions (Scheme 11.5) [12].

In the transformations shown in Schemes 11.4 and 11.5, the presence of a radical scavenger such as 2,6-di-*tert*-butyl-*p*-cresol or (4-hydroxy-(2,2,6,6)-tetramethylpiperidin-1-yl)oxyl (TEMPO) completely inhibits the reaction, suggesting that radical species might be involved in these processes, at least during the first stages of the catalytic cycle. However, in the oxidative cleavage of alkenes, depicted in Scheme 11.4, the homogenous (Neo)AuCl and $[(Neo)Au(styrene)]^+$ complexes could be detected by ESI-MS throughout the reaction, highlighting the inherent mechanistic complexity associated to gold-catalyzed oxidative processes. Interestingly, a mechanism involving first epoxidation of the olefin followed by ring opening and oxidation to the carbonyl compound could be ruled out in the above-mentioned cases.

Olefin epoxidation in the presence of gold is, nevertheless, not unprecedented. Cinellu and coworkers reported that dimeric Au(III)–oxo complexes (proposed intermediates in a large number of Au-mediated oxidation processes) with bipyridine as ligands (**9**) could transfer the oxo group to highly reactive olefins such as norbornene, as shown in Scheme 11.6 [13]. In fact, the

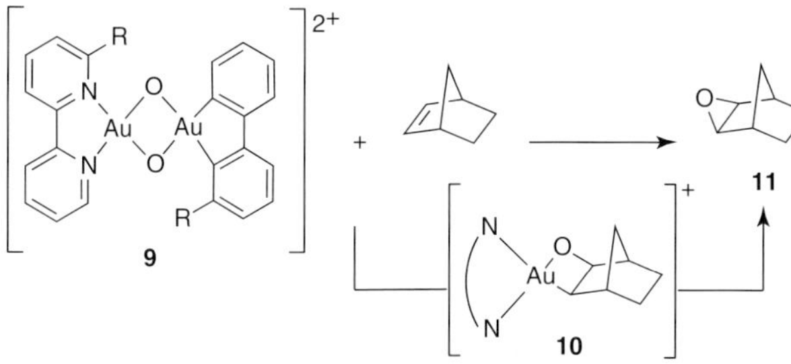

Scheme 11.6

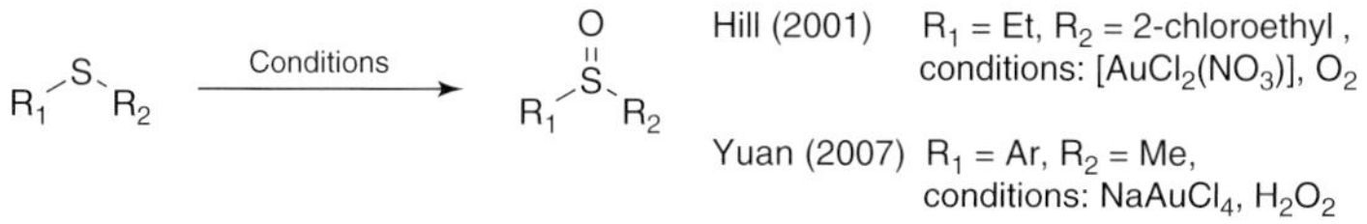

Hill (2001) R_1 = Et, R_2 = 2-chloroethyl, conditions: $[AuCl_2(NO_3)]$, O_2

Yuan (2007) R_1 = Ar, R_2 = Me, conditions: $NaAuCl_4$, H_2O_2

Scheme 11.7

aura-oxa-cyclobutane intermediates **10** operating in this transformation could be isolated, thus opening avenues for a better mechanistic understanding of Au-catalyzed oxidation processes.

11.2.4 Oxidation of Sulfides to Sulfoxides

In 2001, Hill reported that $[AuCl_2(NO_3)]$ catalyzed the oxidation of 2-chloroethyl ethyl sulfide (CEES) to the corresponding sulfoxide in a dioxygen atmosphere at room temperature [14]. This catalytic system is several orders of magnitude faster than the most reactive Ru(II) and Ce(IV) catalysts. In the case of gold, the thioether, in the rate-determining step, is proposed to reduce Au(III) to Au(I), which is quickly reoxidized by O_2 to Au(III) without any metallic gold formation (Scheme 11.7). A few years later, Yuan and coworkers extended the reaction to aromatic aliphatic tioethers using $NaAuCl_4$ as catalyst and hydrogen peroxide as oxidant [15]. The reaction proceeded in good yields and chemoselectivities and since the catalyst was stable, it could be reused at in least six cycles, increasing the overall efficiency of the process (Scheme 11.7).

11.2.5 Oxidation of Gold-Carbene Intermediates

The oxidation of metal-carbenoid species generated from alkynes precursors to form carbonyl compounds has also attracted considerable attention [16]. In most reports metal-carbene species are used stoichiometrically because of the inevitable oxidation of the metal in the presence of the oxidizing agent. However, catalytic oxidations

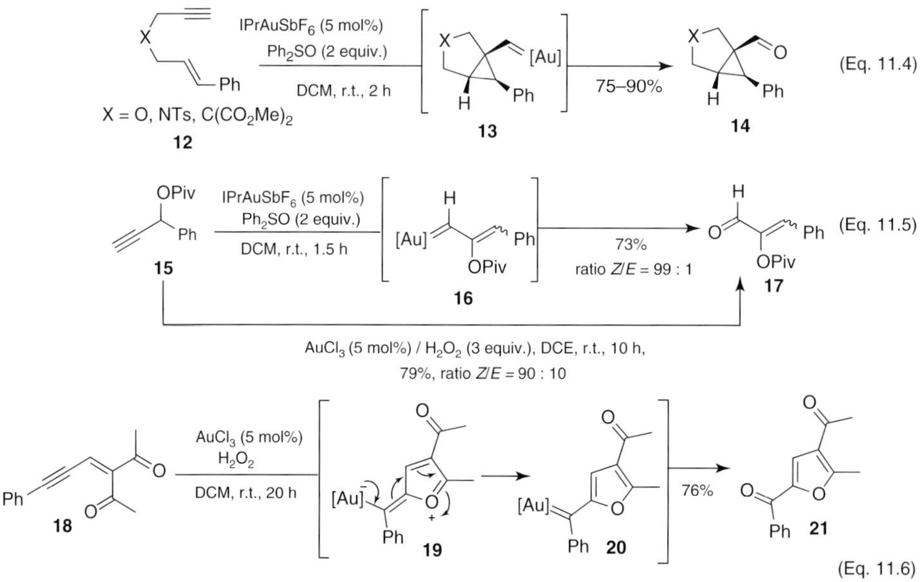

Scheme 11.8

of metal-carbenoid intermediates have also been disclosed [17]. In 2007, Toste and coworkers developed a series of gold(I)-catalyzed oxidative rearrangement reactions involving enynes (**12**) and propargylic acetates (**15**) using diphenyl sulfoxide as stoichiometric oxidant to obtain cyclopropyl- (**14**) and α,β-unsaturated aldehydes (**17**) respectively (Scheme 11.8, Eqs. (11.4) and (11.5)) [18]. In the latter case, $AuCl_3$ as catalyst and H_2O_2 as oxidant can also be used [19] and this protocol has been successfully applied to the oxidative cyclization of 3-(1-alkynyl)-2-alkene-1-ones (**18**) to afford trisubstituted 2-acylfurans (**21**) (Scheme 11.8, Eq. (11.6)). These reactions highlight the possibility of oxygen atom transfer from a nucleophilic oxidant to electrophilic gold-carbene intermediates such as **13**, **16**, and **20**.

Further instances of combining gold catalysts and H_2O_2 as oxidant have been reported. For example, naphthylaldehyde (**24**) can be obtained through a gold-catalyzed oxidative cyclization of 2-ethenyl-1-(prop-2′-yn-1′-ol)-benzene (**22**), which proceeds via the gold-carbene intermediate **23** (Scheme 11.9) [20].

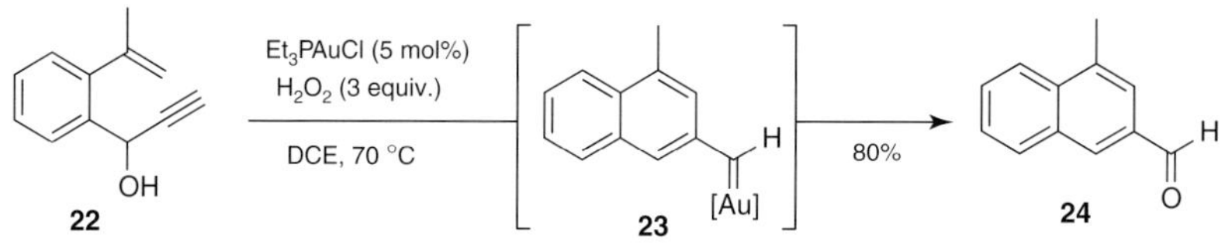

Scheme 11.9

11.2.6
Substrates as Internal Oxidants

The addition of a nucleophile onto a gold-activated alkyne (**25**) affords a gold intermediate, which can, a priori, be viewed either as a vinyl–metal complex (**26**) or as a metal carbene **27** (Scheme 11.10, Eq. (11.7)). The use of an oxidant, which can play both the role of a nucleophile and latent leaving group, can render synthetically versatile α-carbonyl gold carbenes (**28**), analogous to those traditionally formed in situ from α-diazocarbonyl compounds (Scheme 11.10, Eq. (11.8)) [21].

Toste and Zhang pioneered the development of efficient synthetic methods based on this concept. In 2007, Toste reported the rearrangement of homopropargyl and propargyl arylsulfoxides to benzothiepinones (**31**), benzothiopines (**34**), and α-thioenones (**36**) catalyzed by gold (Scheme 11.11) [22]. The intermediate α-carbonyl gold-carbenes were formed through an oxygen atom transfer from the sulfoxide, which underwent a formal intramolecular C–H insertion. When the alkyne is terminal or substituted with an electron withdrawing group, *5-exo-dig*

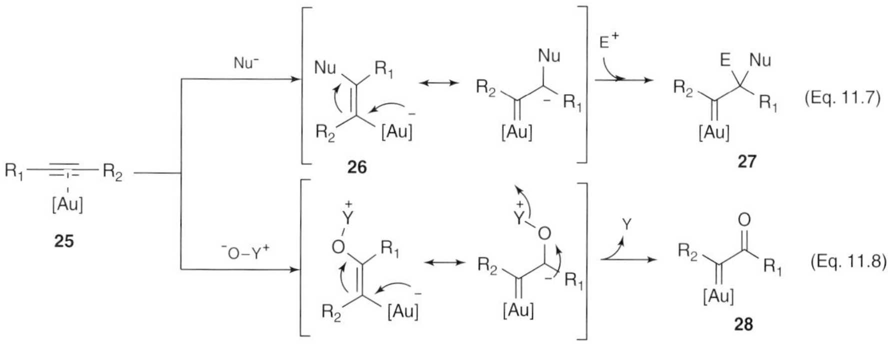

Scheme 11.10

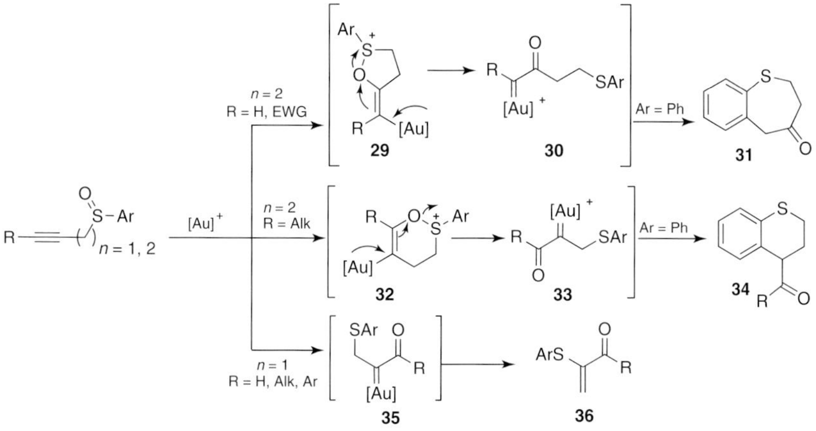

Scheme 11.11

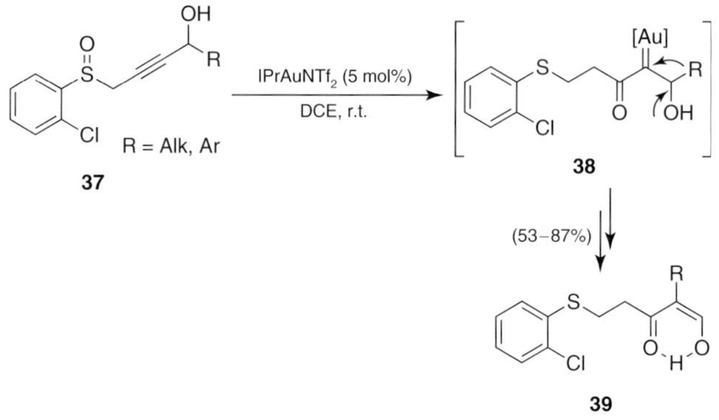

Scheme 11.12

cyclization is favored forming **29** and then **30**. However, if the alkyne is substituted with an alkyl group, 6-*endo-dig* cyclization occurred via **32** and **33**. When propargyl sulfoxides are used, a 5-*endo-dig* cyclization/S–O cleavage sequence is proposed to yield carbene (**35**), which upon 1,2-hydrogen migration forms the corresponding thienones (**36**).

At the same time, Zhang and coworkers reported a related intramolecular redox system using sulfinyl propargyl alcohols (**37**), which upon activation of the alkyne and rearrangement delivered the intermediate **38**. Subsequent pinacol rearrangement afforded α, γ-diketones (**39**) in good yields (Scheme 11.12) [23].

In situ generated tertiary aniline *N*-oxide and tertiary aliphatic amine *N*-oxide can also be used as internal oxidants to form tetrahydrobenz[*b*]azepin-4-ones [24] and piperidin-4-one, respectively [25]. Other successful internal oxidants are nitrones and nitro groups. The former was used to generate azomethine ylides for subsequent cycloaddition [26], whereas the latter was used for the synthesis of azacyclic compounds [27]. In all these methods, the oxidants are tethered to the C–C triple bond at optimal distances. The intramolecular requirement imposes significant structural constraints on both substrates and products, which limits the synthetic potential of this methodology. Intermolecular processes were sought as an atom-economical alternative to methods based on diazo compounds. Zhang and coworkers used terminal propargyl (**40**) and homopropargyl alcohols (**41**) in the presence of pyridine *N*-oxide as external oxidant for the synthesis of oxetan-3-ones (**42**) and dihydrofuranones (**43**) (Scheme 11.13) [28, 29]. Pyridine *N*-oxide has also

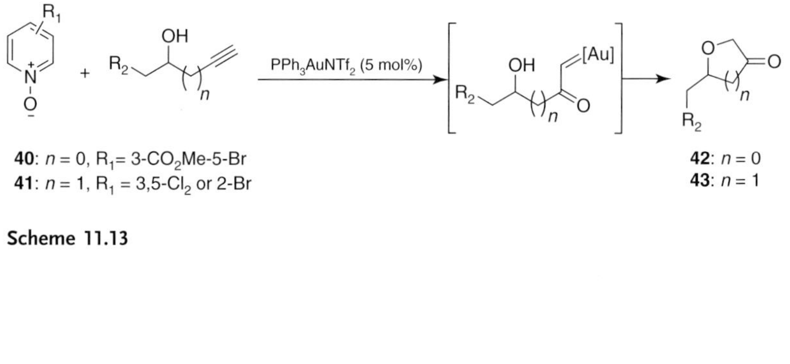

40: $n = 0$, R_1= 3-CO_2Me-5-Br
41: $n = 1$, R_1 = 3,5-Cl_2 or 2-Br

42: $n = 0$
43: $n = 1$

Scheme 11.13

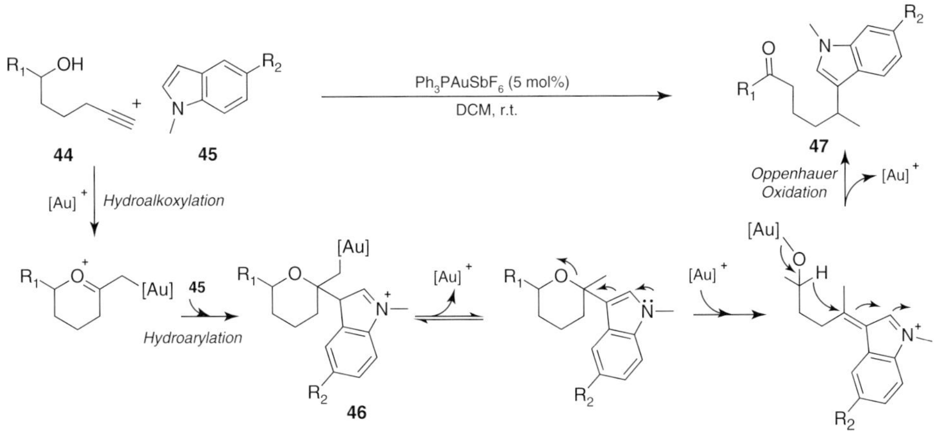

Scheme 11.14

been used for gold catalyzed intermolecular oxidation of ynamines and ynol ether. The oxidation of internal alkynes with high regioselectivities has been also reported using 8-alkylquinoline *N*-oxide as external oxidant in the absence of acidic additives [30].

In most of the above-mentioned processes, the Y–O moiety of the oxidant is reduced and the carbon-carbon triple bond is oxidized. However, Barluenga and coworkers reported in 2009 a gold-catalyzed cascade reaction between alkynols (**44**) and indoles (**45**) involving a conceptually different redox process (Scheme 11.14) [31]. The reaction starts with an intramolecular hydroalkoxylation followed by an intermolecular hydroarylation to give intermediate **46**. In the final step, an unusual intramolecular Oppenauer-type oxidation occurs to give 5-heteroaryl-substituted ketones **47**.

11.3
Gold-Catalyzed Reactions Involving Nitrogen Functionalities

Electron-rich aromatics can undergo auration in the presence of stoichiometric amounts of gold(III) complexes to form Ar–Au(III) species [32]. In 2007, He reported a method to incorporate C–N bonds not only into aromatic rings but also at benzylic positions in the presence of catalytic amounts of $AuCl_3$ at room temperature (Scheme 11.15) [33]. The in situ generated aryl-gold(III) species (**48**) is proposed to react with a source of electrophilic nitrogen such as PhI=NNs (Ns = *p*-nitrosulfonyl) to give **49**. Neither metallic salts nor HCl could perform the reaction, thus ruling out the hypothesis of a Lewis acid-based chemistry with gold(III) activating PhI=NNs. The nitrogen atom on the carbon of the aryl-gold(III) species results in the formation of the Csp^2-N bond product **50**, regenerating $AuCl_3$. This mechanism can also explain the activation of weak C–H benzylic bonds: gold(III) can displace a weak benzylic proton to form a Csp^3-gold(III) bond (**51**), which

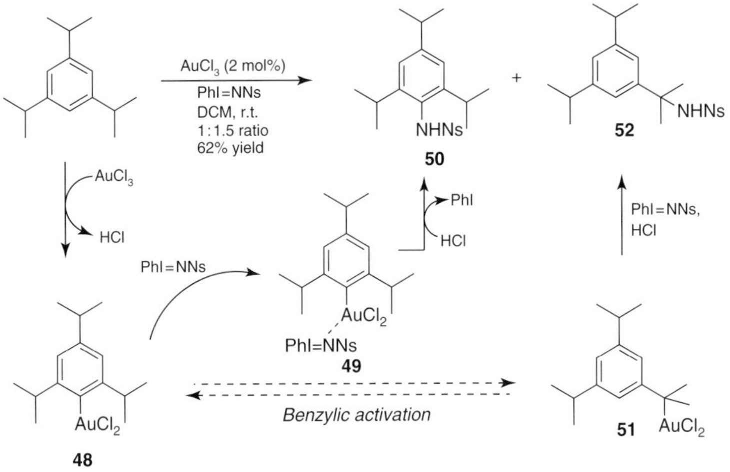

Scheme 11.15

subsequently undergoes the nitrene insertion to give product **52**. The aromatic metalation does not seems to be the rate-determining step and the aryl-gold(III) species (**48**) may be formed prior to the gold(III) migration to the benzylic position. In a conceptually related process, He reported a gold-catalyzed olefin aziridination that uses the commercially available oxidant $PhI(OAc)_2$ and different sulfonamides [34]. This process avoids the use of the sulfonyliminoiodinane-type nitrenoids such as PhI=NTs as a nitrogen source, in contrast to most of the previously reported olefin aziridination reactions.

11.4 Gold-Catalyzed Reactions Involving C–C Bond Formation

11.4.1 Ethynylation Reactions

As described in the previous section, the direct functionalization of electron-rich aromatic rings in the presence of gold(III) salts is a synthetically useful process. In the presence of stabilizing ligands such as 2,6-lutidine, Fuchita and coworkers were able to isolate and characterize this type of complexes. Thus, *p*-xylene was aurated to give **53**, which reacted in the presence of phenyl acetylene to give 1,4-dimethyl-2-(phenylethynyl)benzene (**54**) (Scheme 11.16, Eq. (11.9)) [35]. In view of the fact that this method requires a stoichiometric amount of gold, Nevado and coworkers developed a catalytic version using PPh_3AuCl and $PhI(OAc)_2$ as stoichiometric oxidant to obtain the desired turnover [36]. The synthetic utility of the latter approach stems from the possibility to react electron-rich aromatics (**55**) and electron-deficient alkynes (**56**), both substrates that traditionally react sluggishly under the conditions of the classical Pd-catalyzed Sonogashira reaction

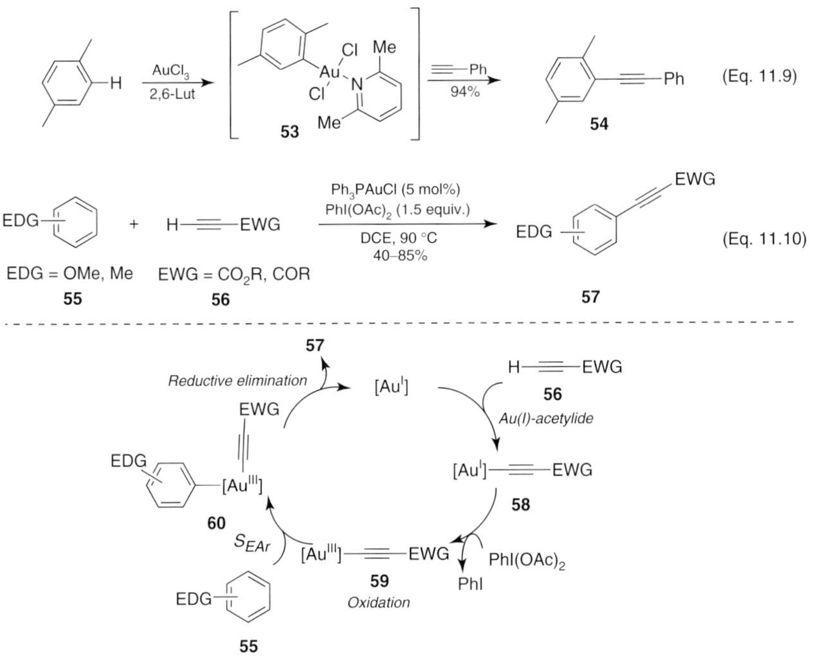

Scheme 11.16

[37] to give cross-coupling products **57** (Scheme 11.16, Eq. (11.10)). A possible mechanistic rationale for this transformation invokes a Au(I)/Au(III) catalytic cycle. After formation of gold(I)-acetylide (**58**), oxidation by $PhI(OAc)_2$ affords the gold(III)–acetylide complex (**59**). Deuterium labeling studies suggested that a Friedel-crafts type reaction occurs on the electrophilic Au(III) center to form intermediate **60**, which, after reductive elimination, generates the ethynylated adduct and regenerates the active catalytic species (Scheme 11.16, bottom).

Waser and coworkers have also reported a gold-catalyzed alkynylation of indoles and thiophenes (Scheme 11.17) [38]. In this case, preformed alkynyl-iodonium derivatives were used [39]. The presence of a halogen atom in the aromatic substrates is tolerated, thus highlighting the orthogonality of the method compared to classical Pd-catalyzed cross-coupling reactions. For example, 5-bromo-indole reacted with 1-[(triisopropylsilyl)ethynyl]-1,2-benziodoxol-3(1*H*)-one (TIPS-EBX, **61**) in the presence of 1 mol% of AuCl to give alkynylated product **62** in 93% yield.

A catalytic cycle is proposed involving first the addition of the indole to the activated triple bond of the hypervalent iodine derivative to form a vinyl-gold intermediate (**63** or regioisomer **64** in Scheme 11.18). The end of the catalytic cycle comprises a β-elimination or an α-elimination/1,2-shift sequence to give 2-iodobenzoic acid and the alkynylated product **65**. In this case, no Au(I)/Au(III) redox couple is invoked, although an alternative pathway analogous to that shown in Scheme 11.16, in which alkynyl-Au(III) species are formed followed by indole metalation and reductive elimination, might be also plausible. At this

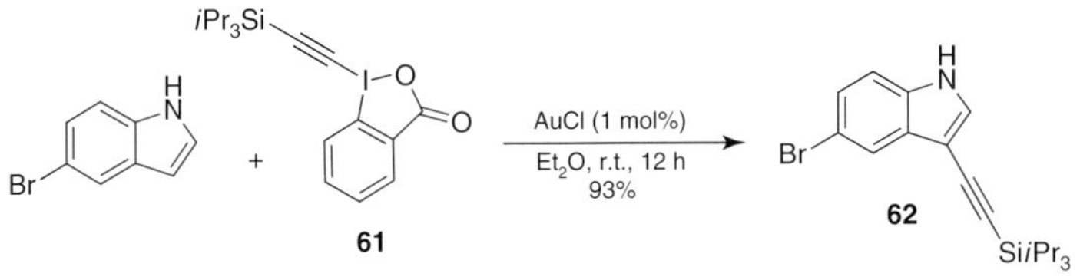

Scheme 11.17

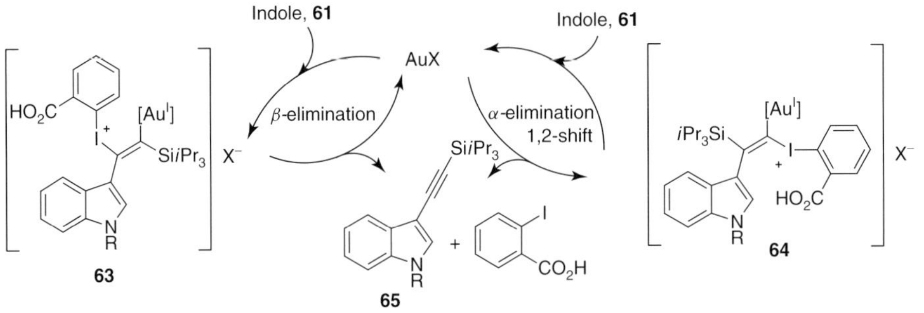

Scheme 11.18

stage, experimental evidence does not suffice to distinguish between these two hypotheses.

Governeur has combined allenoates (**66**) and terminal alkynes (**67**) in a gold-catalyzed cyclization/oxidative alkynylation cascade to give β-alkynyl-δ-butenolides (**68**) (Scheme 11.19) [40]. Two alternative mechanisms can be proposed. Either gold(I) acetylides (**69**) are oxidized to the corresponding Au(III) species **70**, triggering the cyclization of the allenoate (**66**) to give **73**, or, alternatively, Au(I) mediates the cyclization of the allenoate to give **71**, which is oxidized to **72**, which then reacts with the alkyne (**67**) (or acetylide **69**) to give intermediate **73**. In both cases, Au(I) species, upon oxidation in the presence of

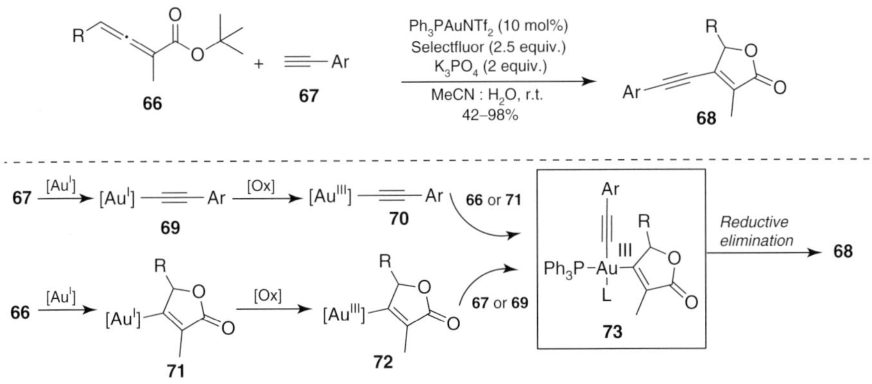

Scheme 11.19

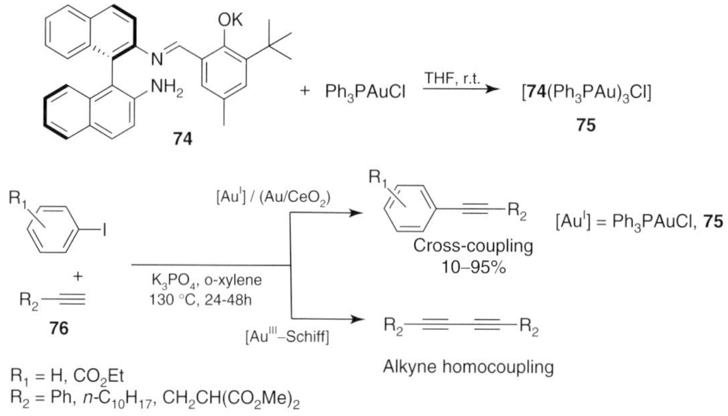

Scheme 11.20

a strong oxidant such as Selectfluor, are proposed to deliver Au(III) intermediates, triggering the formation of new Csp^2-Csp bond via reductive elimination on **73**.

A more conservative approach to trigger alkynylation processes involves the replacement of palladium for gold in the classical Sonogashira cross-coupling between aromatic halides and terminal alkynes [41]. Corma and coworkers reported a gold, Pd-free, Sonogashira reaction using gold metal nanoparticles supported on nanocrystalline ceria (Au/CeO_2) as well as mononuclear Schiff-base gold(I) complex (**75**) and commercially available Ph_3PAuCl [42]. Corma reports that Au(III) complexes catalyzed the undesired alkyne homocoupling whereas Au(I) complexes seem to be responsible for the formation of the desired Sonogashira cross-coupling products (Scheme 11.20). Wang and coworkers also reported a tandem Sonogashira/cyclization reaction of 2-amino-1-haloarenes with terminal alkynes to give indoles using a gold(I) iodide catalyst with the bidentate 1,10-bis(diphenylphosphino)ferrocene (dppf) ligand in 1 : 1 molar ratio [43]. The authors mention that any ratio of dppf to Au(I) less that 1 : 1 resulted in the reaction not going to completion.

The mechanism in these transformations is proposed to follow the same steps commonly accepted for Pd salts in which oxidative addition on the aryl halide occurs on gold(I) species thus generating a Au(III) intermediate **77**, which undergoes transmetalation with a gold acetylide (**78**) [44] to afford, upon reductive elimination on **79**, the cross-coupling products (**80**) (Scheme 11.21).

However, in a recent work, Echavarren and Espinet [45] questioned this hypothesis, showing the inability of gold(I) mononuclear complexes to undergo an oxidative addition. Careful experimental work showed that, indeed, traces of Pd present even in the commercially available gold would be responsible of the observed cross-coupling products, as addition of small amounts of Pd to the gold-catalyzed reaction increased the yield in the Sonogashira product. At the same time, Lambert *et al.* [46] reported that Ph_3PAuCl was catalytically inert in this cross-coupling. Only after more than 100 h, slow conversion is observed, likely due to the formation of Au nanoparticles in solution upon decomposition of the

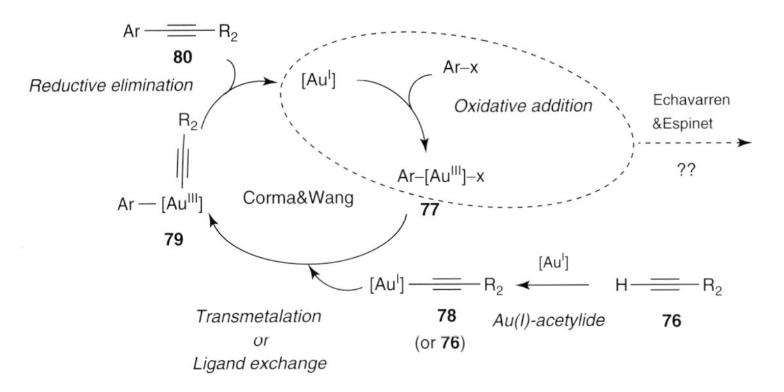

Scheme 11.21

initial catalyst. The activation energy for the oxidative addition on these elusive species is much lower than that expected for the mononuclear complexes [47]. However, further experimental work will be needed to ensure the reproducibility of the reaction when mononuclear gold complexes are used.

11.4.2 Homocoupling Reactions

In the late 1990s, several dimerization processes were reported using stoichiometric amount of gold complexes [48]. A representative example is the dimerization of nucleobases mediated by stoichiometric amounts of $NaAuCl_4$, which was reported by Lippert in 1999 [49].

Almost 10 years later, Beller and Tse identified a general gold-catalyzed oxidative coupling of arenes using tetrachloroauric acid as catalyst and $PhI(OAc)_2$ as stoichiometric oxidant [50]. As an example, 1,4-dimethylbenzene reacts to give 2,2′,5,5′-tetramethylbiphenyl in 74% isolated yield (Scheme 11.22). Interestingly, the reaction requires an acidic solvent and does not require any Ag salt to enhance the reactivity. The authors proposed that the reaction proceeds via a double C–H functionalization, which is in principle the most efficient process for the synthesis of biaryls, Au(III) being the active catalytic redox state. A typical electrophilic aromatic substitution pattern was observed so that participation of a free cationic radical can be ruled out. However, the possibility of an auration reaction, a Friedel–Crafts type substitution, or even the participation of coordinated radical cation cannot be ruled out.

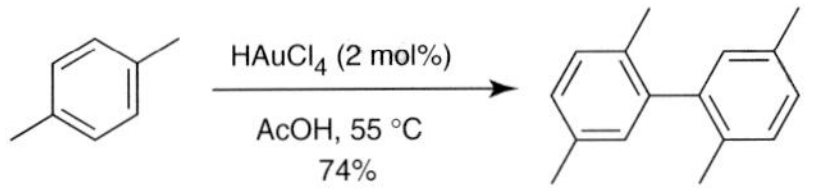

$HAuCl_4$ (2 mol%)
AcOH, 55 °C
74%

Scheme 11.22

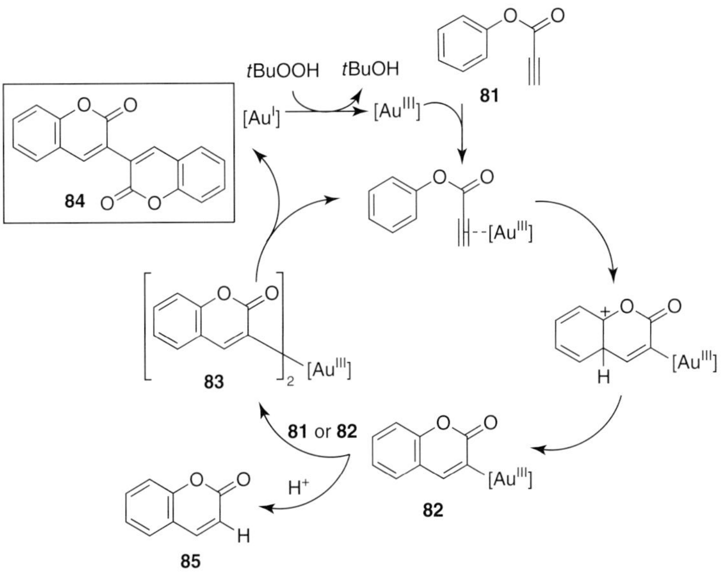

Scheme 11.23

More elaborated homocoupling products can be obtained by combining cyclizations with Au(I)/Au(III) catalytic cycles. Stoichiometric examples were reported by Hashmi *et al.* [51], for example, during the study of the gold(III)-catalyzed cyclization of tertiary allenyl carbinols. Dimeric dihydrofurans could be isolated in small yields as a result of a stoichiometric process on the gold salt. A catalytic manifold was devised by Wegner and coworkers in the cyclization of arylpropionic esters (**81**) (Scheme 11.23). Bis-coumarins (**84**) can be formed in moderate to good yields in the presence of $HAuCl_4$ and *t*BuOOH as oxidant [52]. The reaction, which has also been extended to benzofurans, is proposed to occur through a first cyclization step catalyzed by Au(III) to give **82** after a re-aromatization step. Compound **82** can work as a catalyst for another cyclization of **81** or react with itself through a ligand exchange process to give intermediate **83**, which upon reductive elimination delivers the observed homocoupling product **84** and Au(I), which is re-oxidized in the presence of *tert*-butylhydroperoxide. Alternatively, protonolysis of **82** can explain the formation of coumarin (**85**) as a major byproduct in this process.

Zhang has developed a highly efficient gold-catalyzed oxidative dimerization of propargylic acetates (**86**) pioneering the use Selectfluor as oxidant (Scheme 11.24) [51, 53]. Au(I) catalyzes the rearrangement to the corresponding vinyl-gold(I)-keto species (**87**), which are oxidized with Selectfluor to Au(III) complexes (**88**). Upon transmetalation with **87** and reductive elimination on the Au(III) center, dimerization products (**90**) are obtained (Scheme 11.24).

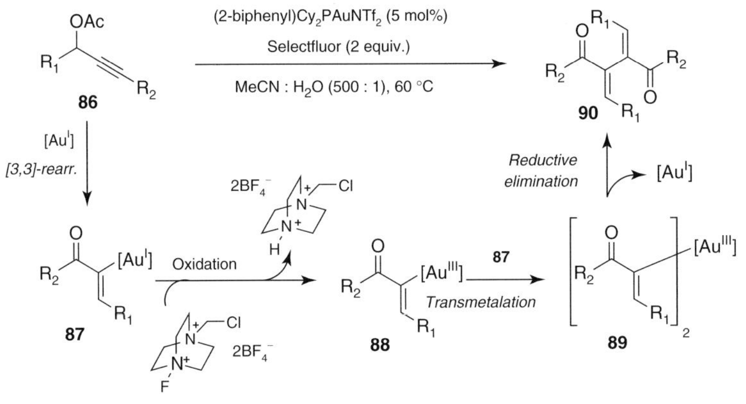

Scheme 11.24

11.4.3 Cross-Coupling Involving B and Si Reagents

Further inspiration has been drawn from Pd(0)/Pd(II) catalytic cycles, to develop gold-catalyzed cross-coupling processes. After developing a gold(III)-catalyzed homocoupling of boronic acids and esters [54], Corma and coworkers studied the gold(I)-catalyzed cross-coupling of aromatic halides with aromatic boronates to give nonsymmetric biaryl compounds in a Suzuki-type process [55]. Schiff-gold(I) complex **75** and also complex **91**, bearing an unsymmetrical *N*-heterocyclic carbene (NHC) ligand, proved to be a very efficient catalyst for this transformation (Scheme 11.25). As already described in the Sonogashira reaction, the proposed catalytic cycle would involve the classical oxidative addition, transmetalation, and reductive elimination sequence. However, after the studies of Echavarren and Espinet [45] and Lambert *et al.* [46] on the first step of the catalytic cycle, a mechanistic explanation for these transformations might be kept on hold.

Zhang and coworkers devised an efficient synthesis of α-arylenones through an oxidative cross-coupling of propargyl acetates and boronic acids [56]. This transformation takes advantage of the recognized ability of gold to, upon coordination to the triple bond, trigger the migration of the acetoxy group (see Scheme 11.24). The vinyl-gold(I) intermediate **87** generated in situ is oxidized in the presence of

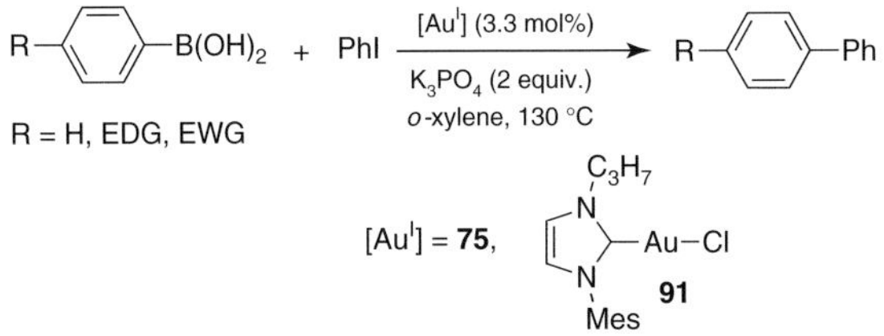

Scheme 11.25

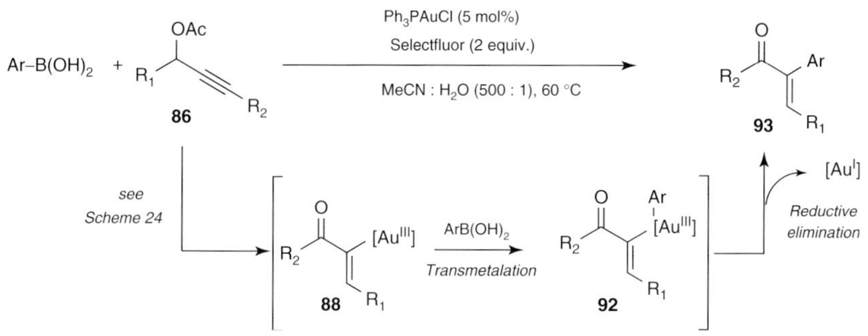

Scheme 11.26

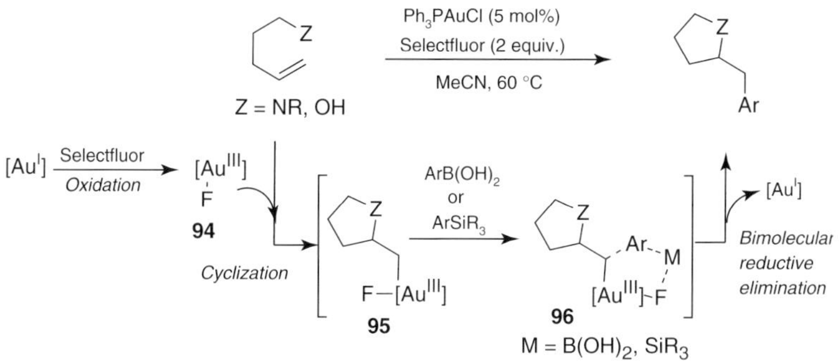

Scheme 11.27

Selectfluor to a vinyl-gold(III) intermediate **88**, which undergoes, similar to Pd(II) intermediates, a transmetalation process in the presence of boronic acids to give **92**, so that, upon reductive elimination, a new Csp^2-Csp^2 bond in **93** is formed (Scheme 11.26).

Different synthetic applications of the [Au(I)]/Selectfluor system in the oxidative carbohetero-functionalization of alkenes have been reported by Zhang *et al.* [57], Toste *et al.* [58], and Lloyd-Jones *et al.* [59] (Scheme 11.27). In these transformations, activation of the double bond seems to occur via Au(III) species (**94**), which are formed in the first step of the catalytic cycle from the reaction of gold(I) complexes with Selectfluor. Next, the intramolecular attack of a N/O nucleophile affords an alkyl-gold(III) intermediate **95**, which in the presence of aryl boronic acids or aryl silanes reacts through a bimolecular reductive elimination to give the new Csp^3-Csp^2 bonds. An alternative mechanism would involve a transmetalation event on **94** or **95** to give Au(III)–Ar species prior to the reductive elimination, although experimental evidence seem to disfavor this hypothesis. Intermolecular versions of these reactions have also been simultaneously reported [59, 60].

11.5
Gold-Catalyzed Reactions Involving Alkene Difunctionalization

The ability of gold to activate double bonds has also been further exploited to trigger new alkene difunctionalization avoiding the use of metallic (B, Si, etc.) counterparts. Muñiz and coworkers reported a gold-catalyzed intramolecular diamination of alkenes related to those based on Pd(II)/Pd(IV) catalytic cycles [61]. The experimental study of the individual steps along the catalytic cycle indicates that, after a *trans*-5-*exo*-hydroamination of the olefin triggered by Au(I), intermediate **97** is formed. Oxidation takes place to form Au(III)-alkyl species (**98**) followed by an intramolecular alkyl–nitrogen bond formation via S_N2 reaction, in which Au(III) behaves as a leaving group (Scheme 11.28) [62]. Nevado and coworkers expanded this concept by developing a flexible method that allows the incorporation of external nucleophiles such as water, alcohols, or even acetonitrile to give amino alcohols (**101**), amino ethers (**102**), or amino amides (**103**) in excellent yields. In contrast to the previous case, the reaction is highly 6-*endo*-regioselective. Although a reaction mechanism can be proposed invoking a 6-*endo* cyclization, there is an alternative reaction path in which, upon 5-*exo*-cyclization mode and oxidation of the metal, the N atom displaces the Au(III) center to form an aziridinium intermediate **100**, which reacts with the external nucleophile to give the observed products. This would also explain the high stereoselectivity observed in the process.

11.6
Gold-Catalyzed Reactions Involving Halogen Functionalities

In the last years, *N*-chlorosuccinimide, *N*-bromosuccinimide, and *N*-iodosuccinimide (NCS, NBS, and NIS, respectively) have been used in

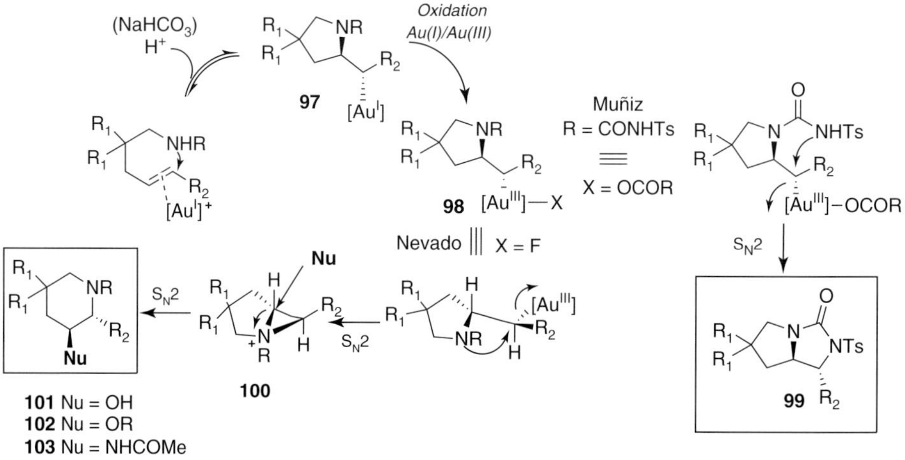

(NaHCO$_3$) H$^+$; **97**; Oxidation Au(I)/Au(III); **98** [Au^{III}]—X; Muñiz R = CONHTs, X = OCOR; Nevado X = F; S_N2; **99**; **100**; **101** Nu = OH; **102** Nu = OR; **103** Nu = NHCOMe

Scheme 11.28

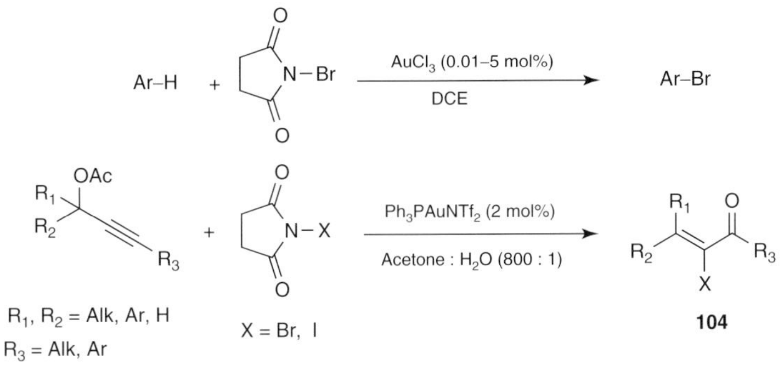

Ar–H + NBS → Ar–Br [$AuCl_3$ (0.01–5 mol%), DCE]

R_1, R_2 = Alk, Ar, H; R_3 = Alk, Ar; X = Br, I; $Ph_3PAuNTf_2$ (2 mol%), Acetone : H_2O (800 : 1); **104**

Scheme 11.29

gold-catalyzed halogenation of arenes, via activation of both the aromatic ring and the NXS reagents [63]. They have been also employed in the gold-catalyzed synthesis of *Z*- and *E*-α-haloenones (**104**) from readily accessible propargylic acetates and alcohols (Scheme 11.29) [64].

Gold-catalyzed reactions for the construction of Csp^2-F bonds have been also investigated because of the rather unique properties of fluorinated compounds. Sadighi *et al.* [65] and Miller's [66] groups focused on the hydrofluorination of alkynes using $Et_3N{\bullet}3HF$ as a fluorinating agent to generate vinyl fluorides (Scheme 11.30, Eq. (11.11)). In contrast, the groups of Gouverneur *et al.* [67] and de Haro and Nevado [68] have used the fluoro-deauration of vinyl gold species with electrophilic sources of fluorine to give Csp^2-F bonds. In a representative example, the 6-*endo-dig* cyclization of difluorinated ynones (**105**) conducted in the presence of gold and Selectfluor affords trifluorinated dihydropyranone (**107**), although protonation of **106** becomes a competitive process in this case (Scheme 11.30, Eq. (11.12)) [69].

Xu and Hammond reported a one-pot tandem addition/oxidative coupling/fluorination reaction using readily available alkynes as starting materials.

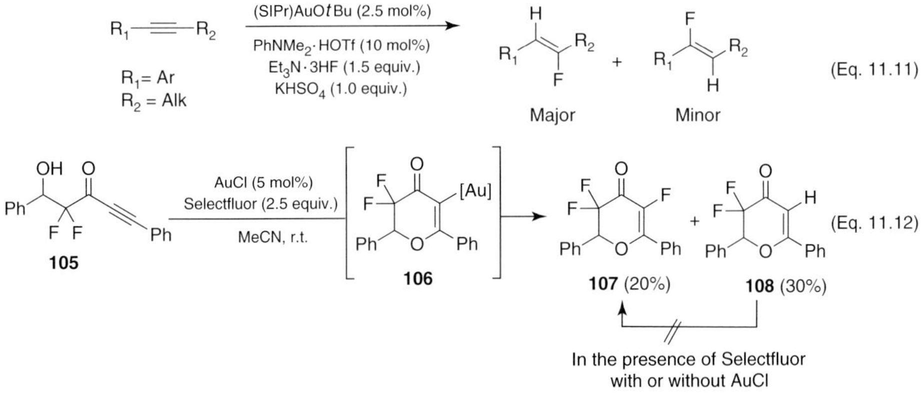

(SIPr)AuO*t*Bu (2.5 mol%), $PhNMe_2 \cdot HOTf$ (10 mol%), $Et_3N \cdot 3HF$ (1.5 equiv.), $KHSO_4$ (1.0 equiv.); R_1 = Ar, R_2 = Alk; Major; Minor (Eq. 11.11)

AuCl (5 mol%), Selectfluor (2.5 equiv.), MeCN, r.t.; **105**; **106**; **107** (20%); **108** (30%); In the presence of Selectfluor with or without AuCl (Eq. 11.12)

Scheme 11.30

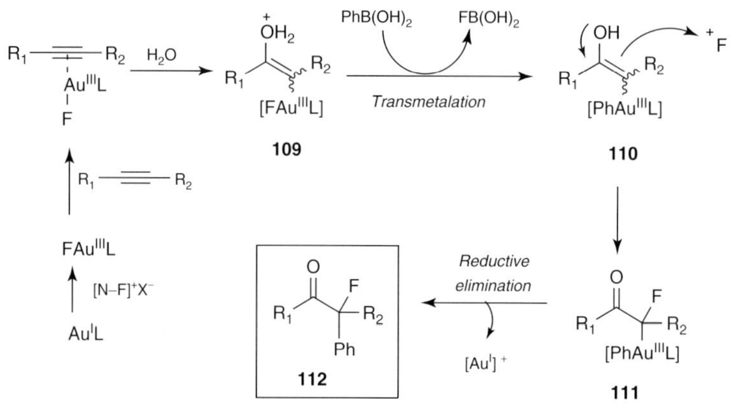

Scheme 11.31

They proposed the initial oxidation of Au(I) to Au(III) in the presence of Selectfluor. Then, water attacks the gold-activated alkyne to generate vinyl-gold species **109**, which undergo transmetalation with the phenylboronic acid to form the vinyl–gold complex **110**. This intermediate reacts with selectfluor to give the gold species **111**, which upon reductive elimination afford the observed α-fluoroketones (**112**) (Scheme 11.31) [70].

11.7
Summary and Outlook

Over the past decades, gold has come to be recognized as a powerful Lewis acid able to activate unsaturated moieties triggering a wide palette of transformations. In contrast to other transition metals, the most significant limitation of homogeneous gold catalysts seemed to be the poor ability of gold to change oxidation states during catalytic cycles. The development of gold catalysts triggering redox processes on the substrates and, more recently, the utilization of external stoichiometric oxidants to trigger Au(I)/Au(III) catalytic cycles, have broadened the scope of gold catalysis. Discoveries of new reactivity are rapidly growing, opening up new pathways to tackle unprecedented transformations that increase molecular complexity.

References

1. For previous accounts on similar topics, see: (a) Li, Z., Brouwer, C., and He, C. (2008) Gold-catalyzed organic transformations. *Chem. Rev.*, **108** (8), 3239–3265; (b) Skouta, R. and Li, C.-J. (2008) Gold-catalyzed reactions of C–H bonds. *Tetrahedron*, **64** (22), 4917–4938; (c) Boorman, T.C. and Larrosa, I. (2010) Gold-mediated C–H bond functionalisation. *Chem. Soc. Rev.*, doi: 10.1039/c0cs00098a
2. Bond, G.C., Louis, C., and Thompson, D.T. (2006) *Catalysis by Gold*, Catalytic Science Series, Vol. **6**, Imperial College Press; (b) Hashmi, A.S.K. and Hutchings, G.J. (2006) Gold catalysis.

Angew. Chem. Int. Ed., **45** (47), 7896–7936.

3. Snyder, J.C. and Grosse, A.V. (1950) Reaction of methane with sulfur trioxide. US Patent 2, 493, 038.
4. Periana, R.A., Taube, D.J., Evitt, E.R., Löffler, D.G., Wentrcek, P.R., Voss, G., and Masuda, T. (1993) A mercury-catalyzed, high-yield system for the oxidation of methane to methanol. *Science*, **259**, 340–343.
5. Shilov, A.E. and Shul'pin, G.B. (1997) Activation of C–H bonds by metal complexes. *Chem. Rev.*, **97** (8), 2879–2932.
6. Shul'pin, G.B., Shilov, A.E., and Süss-Fink, G. (2001) Alkane oxygenation catalyzed by gold complexes. *Tetrahedron Lett.*, **42** (41), 7253–7256.
7. (a) Jones, C., Taube, D., Ziatdinov, V.R., Periana, R.A., Nielsen, R.J., Oxgaard, J., and Goddard, W.A. (2004) Selective oxidation of methane to methanol catalyzed, with CH activation, by homogeneous, cationic gold. *Angew. Chem. Int. Ed.*, **43** (35), 4626–4629; (b) De Vos, D.E. and Sels, B.F. (2005) Gold redox catalysis for selective oxidation of methane to methanol. *Angew. Chem. Int. Ed.*, **44** (1), 30–32.
8. Pal, B., Sen, P.K., and Sen Gupta, K.K. (2001) Reactivity of alkanols and aryl alcohols towards tetrachloroaurate(III) in sodium acetate–acetic acid buffer medium. *J. Phys. Org. Chem.*, **14** (5), 284–294.
9. Guan, B., Xing, D., Cai, G., Wan, X., Yu, N., Fhang, Z., Yang, L., and Shi, Z. (2005) Highly selective aerobic oxidation of alcohol catalyzed by a gold(I) complex with an anionic ligand. *J. Am. Chem. Soc.*, **127** (51), 18004–18005.
10. Li, H., Guan, B., Wang, W., Xing, D., Fang, Z., Wan, X., Yang, L., and Shi, Z. (2007) Aerobic oxidation of alcohol in aqueous solution catalyzed by gold. *Tetrahedron*, **63** (35), 8430–8434.
11. Xing, D., Guan, B., Cai, G., Fang, Z., Yang, L., and Shi, Z. (2006) Gold(I)-catalyzed oxidative cleavage of a C–C double bond in water. *Org. Lett.*, **8** (4), 693–696.
12. Liu, Y., Song, F., and Guo, S. (2006) Cleavage of a carbon-carbon triple bond via gold-catalyzed cascade cyclization/oxidative cleavage reactions of (Z)-enynols with molecular oxygen. *J. Am. Chem. Soc.*, **128** (35), 11332–11333.
13. Cinellu, M.A., Minghetti, G., Cocco, F., Stoccoro, S., Zucca, A., and Manassero, M. (2005) Reactions of gold(III) oxo complexes with cyclic alkenes. *Angew. Chem. Int. Ed.*, **44** (42), 6892–6895.
14. Boring, E., Geletti, Y.V., and Hill, C.L. (2001) Homogeneous catalyst for selective O_2 oxidation at ambient temperature. Diversity-based discovery and mechanistic investigation of thioether oxidation by the Au(III)Cl_2NO_3(thioether)/O_2 System. *J. Am. Chem. Soc.*, **123** (8), 1625–1635.
15. Yuan, Y. and Bian, Y. (2007) Gold(III) catalyzed oxidation of sulfides to sulfoxides with hydrogen peroxide. *Tetrahedron Lett.*, **48** (48), 8518–8520.
16. For oxidation of metal–carbenes with decomposition of metal complexes, see selected examples: (a) Barluenga, J., Bernad, P.L., Concellon, J.M., Pinera-Nicolas, A., and Carcia-Granda, S. (1997) High-diastereoselective and enantioselective cyclopropanation of α,β-unsaturated fischer carbene complexes: synthesis of chiral 1,2-disubstituted and 1,2,3-trisubstituted cyclopropanes. *J. Org. Chem.*, **62** (20), 6870–6875; (b) Barrett, A.G., Mortier, J., Sabat, M., and Sturgess, M.A. (1988) Iron(II) vinylidenes and chromium carbene complexes in beta.-lactam synthesis. *Organometallics*, **7** (12), 2553–2561; (c) Liang, W.-K., Li, W.-T., Peng, S.-M., Wang, S.-L., and Liu, R.-S. (1997) Tungsten(II)-carbene complex functions as a dicationic synthon: efficient constructions of furan and pyran frameworks from readily available α,δ - and α,ε -alkynols. *J. Am. Chem. Soc.*, **119** (19), 4404–4412; (d) Erker, G. and Sosna, F. (1990) Conversion of metallacyclic zirconoxycarbene complexes yielding conventional Fischer-type carbene complexes or metal-free organic products. *Organometallics*, **9** (6), 1949–1953; (e) Quayle, P., Rahman, S., and Ward, E.L.M. (1994) Transition metal promoted acetylene isomerisation reactions in

organic synthesis. *Tetrahedron Lett.*, **35** (22), 3801–3804; (f) Miki, K., Yokoi, T., Nishino, F., Ohe, K., and Uemura, S. (2002) Synthesis of 2-pyranylidene or (2-furyl)carbene–chromium complexes from conjugated enyne carbonyl compounds with $Cr(CO)_5(THF)$. *J. Organomet. Chem.*, **645** (1–2), 228–234.

17. (a) Trost, B.M. and Rhee, Y.H. (1999) Ruthenium-catalyzed cycloisomerization-oxidation of homopropargyl alcohols. A new access to γ-butyrolactones. *J. Am. Chem. Soc.*, **121** (50), 11680–11683; (b) Shin, S., Gupta, A.K., Rhim, C.Y., and Oh, C.H. (2005) Rhodium-catalyzed tandem cyclization–cycloaddition reactions of enynebenzaldehydes: construction of polycyclic ring systems. *Chem. Commun.*, 4429–4431.
18. Witham, C.A., Mauleon, P., Shapiro, N.D., Sherry, B.D., and Toste, F.D. (2007) Gold(I)-catalyzed oxidative rearrangements. *J. Am. Chem. Soc.*, **129** (18), 5838–5839.
19. Wang, T. and Zhang, J. (2010) Synthesis of 2-acylfurans from 3-(1-alkynyl)-2-alken-1-ones via the oxidation of gold–carbene intermediates by H_2O_2. *Dalton Trans.*, **39**, 4270–4273.
20. Taduri, B.P., Abu Sohel, S.M., Cheng, H.-M., Lin, G.-Y., and Liu, R.-S. (2007) Pt- and Au-catalyzed oxidative cyclization of 2-ethenyl-1-(prop-2′-yn-1′-ol)benzenes to naphthyl aldehydes and ketones: catalytic oxidation of metal-alkylidene intermediates using H_2O and H_2O_2. *Chem. Commun.*, 2530–2532.
21. For gold-catalyzed reaction of α-diazoesters, see: (a) Fructos, M.R., Belderrain, T.R., de Frémont, P., Scott, N.M., Nolan, S.P., Díaz-Requejo, M.M., and Pérez, P.J. (2005) A gold catalyst for carbene-transfer reactions from ethyl diazoacetate. *Angew. Chem. Int. Ed.*, **44** (33), 5284–5288; (b) Fructos, M.R., de Frémont, P., Nolan, S.P., Díaz-Requejo, M.M., and Pérez, P.J. (2006) Alkane carbon-hydrogen bond functionalization with (NHC)MCl precatalysts (M = Cu, Au; NHC = N-Heterocyclic Carbene). *Organometallics*, **25** (9), 2237–2241.
22. Shapiro, N.D. and Toste, F.D. (2007) Rearrangement of alkynyl sulfoxides catalyzed by gold(I) complexes. *J. Am. Chem. Soc.*, **129** (14), 4160–4161.
23. Li, G. and Zhang, L. (2007) Gold-catalyzed intramolecular redox reaction of sulfinyl alkynes: efficient generation of α-oxo gold carbenoids and application in insertion into RCO bonds. *Angew. Chem. Int. Ed.*, **46** (27), 5156–5159.
24. Cui, L., Zhang, G., Peng, Y., and Zhang, L. (2009) Gold or no gold: one-pot synthesis of tetrahydrobenz[b]azepin-4-ones from tertiary N-(But-3-ynyl)anilines. *Org. Lett.*, **11** (6), 1225–1228.
25. Cui, L., Peng, Y., and Zhang, L. (2009) A two-step, formal [4 + 2] approach toward piperidin-4-ones via Au catalysis. *J. Am. Chem. Soc.*, **131** (24), 8394–8395.
26. Yeom, H.-S., Lee, J.-E., and Shin, S. (2008) Gold-catalyzed waste-free generation and reaction of azomethine ylides: internal redox/dipolar cycloaddition cascade. *Angew. Chem. Int. Ed.*, **47** (37), 7040–7043.
27. Jadhav, A.M., Bhunia, S., Liao, H.-Y., and Liu, R.-S. (2011) Gold-catalyzed stereoselective synthesis of azacyclic compounds through a redox/[2 + 2 + 1] cycloaddition cascade of nitroalkyne substrates. *J. Am. Chem. Soc.*, doi: 10.1021/ja110514s
28. (a) Ye, L., Cui, L., Zhang, G., and Zhang, L. (2010) Alkynes as equivalents of α-diazo ketones in generating α-oxo metal carbenes: A gold-catalyzed expedient synthesis of dihydrofuran-3-ones. *J. Am. Chem. Soc.*, **132** (10), 3258–3259; (b) Ye, L., He, W., and Zhang, L. (2010) Gold-catalyzed one-step practical synthesis of oxetan-3-ones from readily available propargylic alcohols. *J. Am. Chem. Soc.*, **132** (25), 8550–8551.
29. Davis, P.D., Cremonesi, A., and Martin, N. (2011) Site-specific introduction of gold-carbenoids by intermolecular oxidation of ynamides or ynol ethers. *Chem. Commun*, **47**, 379–381.
30. Lu, B., Li, C., and Zhang, L. (2010) Gold-catalyzed highly regioselective oxidation of C–C triple bonds without acid additives: propargyl moieties as masked

α,β-unsaturated carbonyls. *J. Am. Chem. Soc.*, **132** (40), 14070–14072.

31. Barluenga, J., Fernández, A., Rodríguez, F., and Fañanás, F. (2009) A gold-catalyzed cascade reaction involving an unusual intramolecular redox process. *Chem. Eur. J.*, **15** (33), 8121–8123.
32. (a) Shi, Z. and He, C. (2004) Efficient functionalization of aromatic C–H bonds catalyzed by gold(III) under mild and solvent-free conditions. *J. Org. Chem.*, **69** (11), 3669–3671; (b) Shi, Z. and He, C. (2004) Direct functionalization of arenes by primary alcohol sulfonate esters catalyzed by gold(III). *J. Am. Chem. Soc.*, **126** (42), 13596–13597.
33. Li, Z., Capretto, D.A., Rahaman, R.O., and He, C. (2007) Gold(III)-catalyzed nitrene insertion into aromatic and benzylic C–H groups. *J. Am. Chem. Soc.*, **129** (40), 12058–12059.
34. Li, Z., Ding, X., and He, C. (2006) Nitrene transfer reactions catalyzed by gold complexes. *J. Org. Chem.*, **71** (16), 5876–5880.
35. Fuchita, Y., Utsonomiya, Y., and Yasutake, M. (2001) Synthesis and reactivity of arylgold(III) complexes from aromatic hydrocarbons via C–H bond activation. *J. Chem. Soc., Dalton Trans.*, 2330–2333.
36. de Haro, T. and Nevado, C. (2010) Gold-catalyzed ethynylation of arenes. *J. Am. Chem. Soc.*, **132** (5), 1512–1513.
37. (a) Sonogashira, K., Tohda, Y., and Hagihara, N. (1975) A convenient synthesis of acetylenes: catalytic substitutions of acetylenic hydrogen with bromoalkenes, iodoarenes and bromopyridines. *Tetrahedron Lett.*, **16** (50), 4467–4470; (b) Sonogashira, K. (1998) in *Metal-Catalyzed Cross-Coupling Reactions* (eds F. Diederich and P.J. Stang), Wiley-VCH Verlag GmbH, Weinheim, pp. 203–229.
38. (a) Brand, J., Charpentier, J., and Waser, J. (2009) Direct alkynylation of indole and pyrrole heterocycles. *Angew. Chem. Int. Ed.*, **48** (49), 9346–9349; (b) Brand, J.P. and Waser, J. (2010) Direct alkynylation of thiophenes: cooperative activation of TIPS–EBX with gold and brønsted acids. *Angew. Chem. Int. Ed.*, **49** (40), 7304–7307.
39. Brand, J.P., Fernández González, D., Nicolai, S., and Waser, J. (2011) Benziodoxole-based hypervalent iodine reagents for atom-transfer reactions. *Chem. Commun.*, **47**, 102–115.
40. Hopkinson, M.N., Ross, J.E., Giuffredi, G.T., Gee, A.D., and Gouverneur, V. (2010) Gold-catalyzed cascade cyclization-oxidative alkynylation of allenoates. *Org. Lett.*, **12** (21), 4904–4907.
41. Plenio, H. (2008) Katalysatoren für die Sonogashira-kupplung – unedle metalle auf dem vormarsch. *Angew. Chem.*, **120** (37), 7060–7063.
42. González-Arellano, C., Abad, A., Corma, A., García, H., Iglesias, M., and Sánchez, F. (2007) Catalysis by gold(I) and gold(III): a parallelism between homo- and heterogeneous catalysts for copper-free Sonogashira cross-coupling reactions. *Angew. Chem. Int. Ed.*, **46** (9), 1536–1538.
43. Li, P., Wang, L., Wang, M., and You, F. (2008) Gold(I) iodide catalyzed sonogashira reactions. *Eur. J. Org. Chem.*, (35), 5946–5951.
44. Copper was replaced by gold in Sonogashira coupling: (a) Jones, L.A., Sanz, S., and Laguna, M. (2007) Gold compounds as efficient co-catalysts in palladium-catalysed alkynylation. *Catal. Today*, **122**, 403–406; However several contradictory reports have been recently published: (b) Panda, B. and Sarkar, T.K. (2010) On the catalytic duo $PdCl_2(PPh_3)_2/AuCl(PPh_3)$ that cannot effect a Sonogashira-type reaction: a correction. *Tetrahedron Lett.*, **51** (2), 301–305.
45. Lauterbach, T., Livendahl, M., Roselln, A., Espinet, P., and Echavarren, A.M. (2010) Unlikeliness of Pd-free gold(I)-catalyzed Sonogashira coupling reactions. *Org. Lett.*, **12** (13), 3006–3009.
46. Kyriakou, G., Beaumont, S.K., Humphrey, S.M., Antonetti, C., and Lambert, R.M. (2010) Sonogashira coupling catalyzed by gold nanoparticles: does homogeneous or heterogeneous catalysis dominate? *ChemCatChem*, **2** (11), 1444–1449.

47. Corma, A., Juárez, R., Boronat, M., Sánchez, F., Iglesias, M., and García, H. (2011) Gold catalyzes the Sonogashira coupling reaction without the requirement of palladium impurities. *Chem. Commun.*, **47**, 1446–1448.
48. (a) Constable, E.C. and Sousa, L.R. (1992) Metal-ion dependent reactivity of 2-(2′-thienyl)pyridine (Hthpy). *J. Organomet. Chem.*, **427** (1), 125–139; (b) Bennett, M.A., Hockless, D.C.R., Rae, A.D., Welling, L.L., and Willis, A.C. (2001) Carbon-Carbon coupling in dinuclear cycloaurated complexes containing bridging 2-(diphenylphosphino)phenyl or 2-(diethylphosphino)phenyl. Role of the axial ligand and the fine balance between gold(II)-gold(II) and gold(I)-gold(III). *Organometallics*, **20** (1), 79–87; (c) Sahoo, A.K., Nakamura, Y., Aratani, N., Kim, K.S., Noh, S.B., Shinokubo, H., Kim, D., and Osuka, A. (2006) Synthesis of brominated directly fused diporphyrins through gold(III)-mediated oxidation. *Org. Lett.*, **8** (18), 4141–4144.
49. Zamora, F., Amo-Ochoa, P., Fischer, B., Schimanski, A., and Lippert, B. (1999) 5,5′-Diuracilyl species from uracil and $[AuCl_4]^-$: nucleobase dimerization brought about by a metal. *Angew. Chem. Int. Ed.*, **38** (15), 2274–2275.
50. (a) Kar, A., Mangu, N., Kaiser, H.M., Beller, M., and Tse, M.K. (2008) A general gold-catalyzed direct oxidative coupling of non-activated arenes. *Chem. Commun.*, 386–388; (b) Kar, A., Mangu, N., Kaiser, H.M., and Tse, M.K. (2009) Gold-catalyzed direct oxidative coupling reactions of non-activated arenes. *J. Organomet. Chem.*, **694** (4), 524–537.
51. (a) Hashmi, A.S.K., Blanco, M.C., Fischer, D., and Bats, J.W. (2006) Gold catalysis: evidence for the in-situ reduction of gold(III) during the cyclization of allenyl carbinols. *Eur. J. Org. Chem.*, (6), 1387–1389; (b) Hashmi, A.S.K., Ramamurthi, T.D., and Rominger, F. (2009) Synthesis, structure and reactivity of organogold compounds of relevance to homogeneous gold catalysis. *J. Organomet. Chem.*, **694**, 592–597.
52. (a) Wegner, H., Ahles, S., and Neuburger, M. (2008) A new gold-catalyzed domino cyclization and oxidative coupling reaction. *Chem. Eur. J.*, **14** (36), 11310–11313; (b) Auzias, M.G., Neuburger, M., and Wegner, H.A. (2010) 3,3′-Bis(arylbenzofurans) via a gold-catalyzed domino process. *Synlett*, (16), 2443–2448; (c) Wegner, H.A. (2009) Oxidative coupling reactions with gold. *CHIMIA*, **63**, 45–49.
53. Cui, L., Zhang, G., and Zhang, L. (2009) Homogeneous gold-catalyzed efficient oxidative dimerization of propargylic acetates. *Bioorg. Med. Chem. Lett.*, **19** (14), 3884–3887.
54. González-Arellanoa, C., Corma, A., Iglesias, M., and Sánchez, F. (2005) Homogeneous and heterogenized Au(III) Schiff base–complexes as selective and general catalysts for self-coupling of aryl boronic acids. *Chem. Commun.*, 1990–1992.
55. (a) González-Arellanoa, C., Corma, A., Iglesias, M., and Sánchez, F. (2006) Gold (I) and (III) catalyze Suzuki cross-coupling and homocoupling, respectively. *J. Catal.*, **238** (2), 497–501; (b) González-Arellano, C., Corma, A., Iglesias, M., and Sánchez, F. (2008) Soluble gold and palladium complexes heterogenized on MCM–41 are effective and versatile catalysts. *Eur. J. Inorg. Chem.*, (7), 1107–1115; (c) Corma, A., Gutiérrez-Puebla, E., Iglesias, M., Monge, A., Pérez-Ferreras, S., and Sánchez, F. (2006) New heterogenized gold(I)-heterocyclic carbene complexes as reusable catalysts in hydrogenation and cross-coupling reactions. *Adv. Synth. Catal.*, **348** (14), 1899–1907; (d) Corma, A., González-Arellanoa, C., Iglesias, M., Pérez-Ferreras, S., and Sánchez, F. (2007) Heterogenized gold(I), gold(III), and palladium(II) complexes for C–C bond reactions. *Synlett*, (11), 1771–1774.
56. Zhang, G., Peng, Y., Cui, L., and Zhang, L. (2009) Gold-catalyzed homogeneous oxidative cross-coupling reactions. *Angew. Chem.*, **121** (17), 3158–3161; *Angew. Chem. Int. Ed.*, **48** (17), 3112–3115.
57. Zhang, G., Cui, L., Wang, Y., and Zhang, L. (2010) Homogeneous

gold-catalyzed oxidative carbohetero-functionalization of alkenes. *J. Am. Chem. Soc.*, **132** (5), 1474–1475.

58. (a) Brenzovich, W.E., Benitez, D., Lackner, A.D., Shunatona, H.P., Tkatchouk, E., Goddard, W.A., and Toste, F.D. (2010) Gold-catalyzed intramolecular aminoarylation of alkenes: C–C bond formation through bimolecular reductive elimination. *Angew. Chem. Int. Ed.*, **49** (32), 5519–5522; (b) Brenzovich, W.E. Jr., Brazeau, J.-F., and Toste, F.D. (2010) Gold-catalyzed oxidative coupling reactions with aryltrimethylsilanes. *Org. Lett.*, **12** (21), 4728–4731.
59. Ball, L.T., Green, M., Lloyd-Jones, G.C., and Russell, C.A. (2010) Arylsilanes: application to gold-catalyzed oxyarylation of alkenes. *Org. Lett.*, **12** (21), 4724–4727.
60. Melhado, A.D., Brenzovich, W.E. Jr., Lackner, A.D., and Toste, F.D. (2010) Gold-catalyzed three-component coupling: oxidative oxyarylation of alkenes. *J. Am. Chem. Soc.*, **132** (26), 8885–8887.
61. Iglesias, A. and Muñiz, K. (2009) Oxidative interception of the hydroamination pathway: a gold-catalyzed diamination of alkenes. *Chem. Eur. J.*, **15** (40), 10563–10569.
62. (a) de Haro, T. and Nevado, C. (2011) Flexible gold-catalyzed regioselective oxidative difunctionalization of unactivated alkenes. *Angew. Chem. Int. Ed.*, **50** (4), 906–910; (b) Hopkinson, M.N., Tessier, A., Salisbury, A., Giuffredi, G.T., Combettes, L.E., Gee, A.D., and Gouverneur, V. (2010) Gold-catalyzed intramolecular oxidative cross-coupling of nonactivated arenes. *Chem. Eur. J.*, **16** (16), 4739–4743.
63. Mo, F., Yan, J., Qiu, D., Li, F., Zhang, Y., and Wang, J. (2010) Gold-catalyzed halogenation of aromatics by N-halosuccinimides. *Angew. Chem. Int. Ed.*, **49** (11), 2028–2032.
64. (a) Yu, M., Zhang, G., and Zhang, L. (2007) Gold-catalyzed efficient preparation of linear α-iodoenones from propargylic acetates. *Org. Lett.*, **9** (11), 2147–2150; (b) Yu, M., Zhang, G., and Zhang, L. (2009) Gold-catalyzed efficient preparation of linear α-haloenones from propargylic acetates. *Tetrahedron*, **65** (9), 1846–1855; (c) Ye, L. and Zhang, L. (2009) Practical synthesis of linear α-iodo/bromo-α,β-unsaturated aldehydes/ketones from propargylic alcohols via Au/Mo bimetallic catalysis. *Org. Lett.*, **11** (16), 3646–3649; (d) Kirsch, S., Binder, J., Crone, B., Duschek, A., Haug, T., Liébert, C., and Menz, H. (2007) Catalyzed tandem reaction of 3-silyloxy-1,5-enynes consisting of cyclization and pinacol rearrangement. *Angew. Chem. Int. Ed.*, **46** (13), 2310–2313; (e) Wang, D., Ye, X., and Shi, X. (2010) Efficient synthesis of *E*-α-haloenones through chemoselective alkyne activation over allene with triazole-Au catalysts. *Org. Lett.*, **12** (9), 2088–2091; (f) Wang, Y., Lu, B., and Zhang, L. (2010) The use of Br/Cl to promote regioselective gold-catalyzed rearrangement of propargylic carboxylates: an efficient synthesis of *(1Z, 3E)*-1-bromo/chloro-2-carboxy-1,3-dienes. *Chem. Commun.*, **46**, 9179–9181.
65. Akana, J.A., Bhattacharyya, K.X., Müller, P., and Sadighi, J.P. (2007) Reversible C-F bond formation and the Au-catalyzed hydrofluorination of alkynes. *J. Am. Chem. Soc.*, **129** (25), 7736–7737.
66. Gorske, B.C., Mbofana, C.T., and Miller, S.J. (2009) Regio- and stereoselective synthesis of fluoroalkenes by directed Au(I) catalysis. *Org. Lett.*, **11** (19), 4318–4321.
67. (a) Hopkinson, M.N., Giuffredi, G.T., Gee, A.D., and Gouverneur, V. (2010) Gold-catalyzed diastereoselective synthesis of α-fluoroenones from propargyl acetates. *Synlett*, 2737–2742; (b) Brown, J. and Gouverneur, V. (2009) Transition-metal-mediated reactions for C-F bond construction: the state of play. *Angew. Chem. Int. Ed.*, **48** (46), 8610–8614.
68. de Haro, T. and Nevado, C. (2011) Domino gold-catalyzed rearrangement and fluorination of propargyl acetates. *Chem. Commun.*, **47**, 248–249.
69. Schuler, M., Silva, F., Bobbio, C., Tessier, A., and Gouverneur, V. (2008) Gold(I)-catalyzed alkoxyhalogenation of

β-hydroxy-α,α-difluoroynones. *Angew. Chem. Int. Ed.*, **47** (41), 7927–7930.

70. (a) Wang, W., Jasinski, J., Hammond, G.B., and Xu, B. (2010) Fluorine-enabled cationic gold catalysis: functionalized hydration of alkynes. *Angew. Chem. Int. Ed.*, **49** (40), 7247–7252; (b) de Haro, T. and Nevado, C. (2010) Gold-catalyzed synthesis of α-fluoro acetals and α-fluoro ketones from alkynes. *Adv. Synth. Catal.*, **352** (16), 2767–2772.

12
Transition-Metal Complexes in Supported Liquid Phase and Supercritical Fluids – A Beneficial Combination for Selective Continuous-Flow Catalysis with Integrated Product Separation

Ulrich Hintermair, Tamilselvi Chinnusamy, and Walter Leitner

12.1
Strategies for Catalyst Immobilization Using Permanent Separation Barriers

There is sustained interest in developing methods to separate homogeneous catalysts from a reaction mixture [1]. Motivation to recover and reuse organometallic catalysts stems not only from economic arguments on catalyst efficiency but also from regulations regarding product purity [2]. The integration of a permanent separation barrier for the discrimination of catalyst and products inside a reactor allows for straightforward continuous operation, which is a highly desirable mode of operation for reasons of process control and efficiency [3]. To this end, multiphasic systems consisting of a molecular catalyst in a product-separable fluid phase or on the surface of a solid support have proven to be viable approaches for organometallic catalysis [4]. Phase boundaries are convenient separation strategies that may be introduced by, for example, inorganic oxide materials, organic polymers, water, and specific solvents such as fluorous phases or ionic liquids (ILs) [1]. However, most multiphasic systems seeking to *bridge the gap between homogeneous and heterogeneous catalysis* [5] also combine some of their respective disadvantages at the same time. When using molecular catalysts in multiphasic systems, accessibility, characterizability, and tunability of the once homogeneous catalysts are reduced by various degrees to enable effective catalyst retention in continuous operation [3]. One promising approach to compromise accessibility and variability of immobilized molecular catalysts with effective retention is to use them in supported liquid phases (SLPs); dispersion of a concentrated catalyst solution on the surface of a porous support combines the respective advantages of liquid- and solid-phase immobilization (Figure 12.1 and Table 12.1) [6].

On the molecular level, SLPs represent a "gentle" immobilization technique because it immobilizes the solvent, and not the catalyst. It is also beneficial to the macroscale process scheme because the bulk properties of the catalyst material are dominated by the solid support, and thus the solvent is kept entirely on the meso-scale where it permits homogeneous catalytic turnover [3].

New Strategies in Chemical Synthesis and Catalysis, First Edition. Edited by Bruno Pignataro.

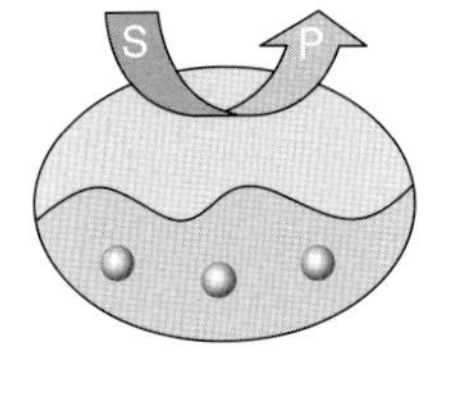

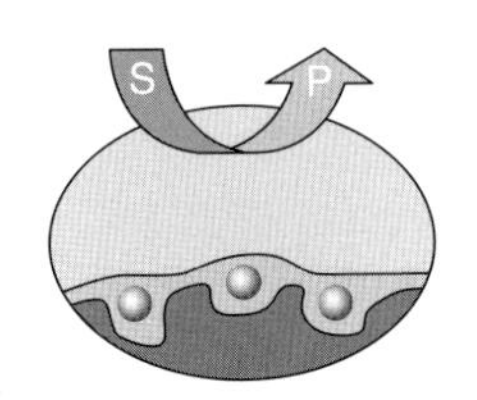

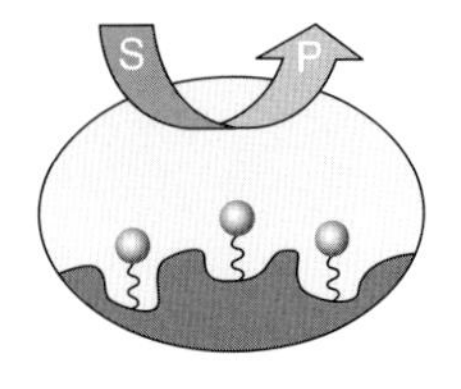

Figure 12.1 Permanent separation strategies based on phase boundaries to discriminate reaction products from the organometallic catalyst.

Table 12.1 General characteristics of the immobilization approaches shown in Figure 12.1.

Approach	Molecular entity of catalyst	Transport limitation	Engineering of continuous process	Reactor volume (process efficiency)
Liquid-phase immobilization	Preserved	Liquid diffusion (Hatta [7])	Extensive	Large (low STYs)
Solid-phase immobilization	Restricted	Pore diffusion (Thiele [7])	Straightforward	Small (high STYs)
Supported liquid-phase immobilization	Preserved	Adjustable (*vide infra*)	Straightforward	Small (high STYs)

12.2 Supported Liquid-Phase Catalysts Based on Organic Solvents (SLP)

Besides a few from the late 1930s [8] on the use of supported Brønsted acids for olefin polymerizations, the idea to use a supported liquid for the immobilization of catalytically active solutions of transition-metal compounds originated from independent reports of industrial research laboratories at Johnson Matthey [9] and Monsanto [10] in 1966. It was the very same year Osborn and Wilkinson [11] reported on the preparation and catalytic properties of $[RhCl(PPh_3)_3]$, which makes the technique of SLPs the oldest by-design approach to continuous-flow multiphasic homogenous catalysis in the open literature (Figure 12.2).

Ethylene glycol solutions of $RhCl_3$ hydrate were supported on porous silicates by wet impregnation with CH_3OH, and drying yielded free-flowing powders containing the catalyst solution in the pores of the support material [9]. The gas-phase isomerization of pentenes was studied as model reaction. Despite good initial activity, a progressive deactivation was detected in pulsed continuous-flow

The Use of Supported Solutions of Rhodium Trichloride for Homogeneous Catalysis

A practical limitation to performing homogeneously catalyzed reactions in the liquid phase is the difficulty of removing the product continuously. No equivalent problem arises with heterogeneously catalyzed reactions which are conveniently performed by causing the gaseous or liquid reagents to flow through a bed of the solid catalyst. With homogeneous catalysis we have, however, a number of potential advantages over the heterogeneous catalysis of the same reaction, in particular the possibility of greater selectivity and of more efficient use of the metal atoms. We have attempted to combine the desirable features of both systems by using catalytic solutions supported within a porous solid, after the manner of gas-liquid chromatography.

Figure 12.2 Introduction of the 1966 paper on SLP catalysts from Johnson Matthey [9].

mode. In their conclusion, the authors stated that "*. . . the application of this method to other possibly more amenable and commercially important systems will readily be conceived,*" and a patent was filed the same year [9].

Researchers at Monsanto [10] modeled the kinetic behavior of SLP catalysts in detail, and derived mathematical functions describing the diffusion resistance of gaseous substrates to the homogeneous catalysts in the SLP. They found relations very similar to the classical pore diffusion limitation characteristically encountered in heterogeneous catalysis (Figure 12.3).

For classical heterogeneous catalysts, the extent of the pore-diffusion-limited regime of a given reaction is determined by the pore structure of the solid (Thiele modulus [13]). Importantly, for SLP catalysts this regime was found to be a function

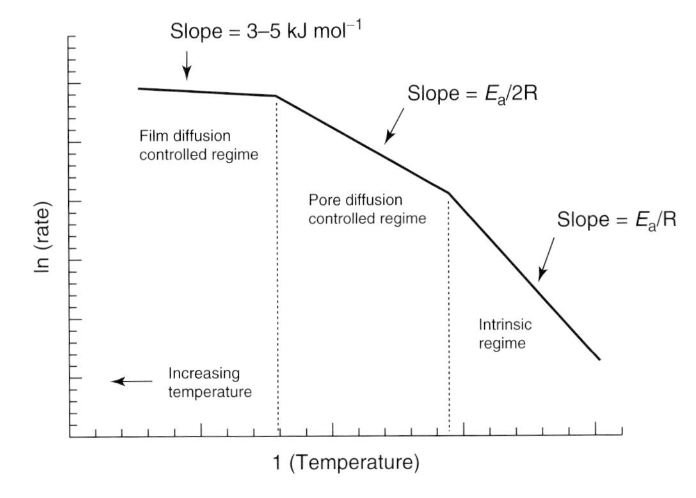

Figure 12.3 General kinetic behavior of heterogeneous catalysts over inverse temperature (E_A = activation energy). At higher temperatures the reaction becomes limited by mass transport [12].

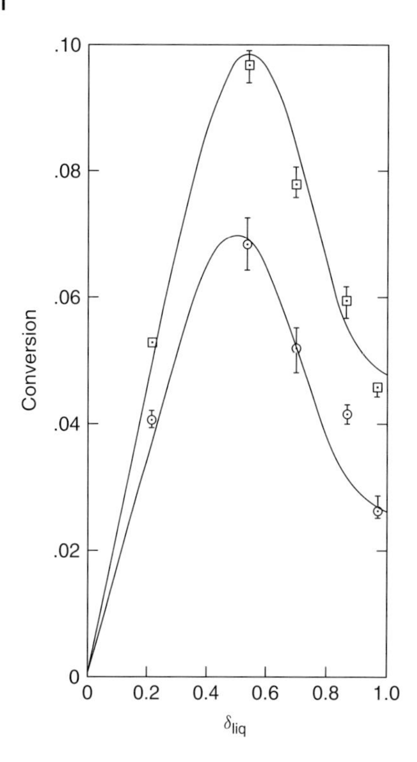

Figure 12.4 Steady-state conversion of continuous propylene hydroformylation in SLP (theoretically modeled curves with experimental points for two different flow rates including experimental error margins) as function of pore filling δ_{liq} [14].

of the liquid loading (defined either as weight percentage liquid or as pore filling, i.e., liquid volume/total pore volume). This prediction was verified experimentally for the continuous gas-phase hydroformylation of propylene with $[RhCl(PPh_3)_2CO]$ immobilized in butylbenzylphthalate on silica gel (Figure 12.4) [14].

The decrease in catalyst performance above intermediate levels of loadings was interpreted as the onset of diffusion limitation through the SLP. In these experiments, a presaturator bed (butylbenzylphthalate on silica gel) and an adsorber bed (dry silica gel) were placed before and after the SLP catalyst, respectively. From post-reaction gravimetric analysis, it was concluded that little to no exchange of SLP between the beds occurred under reaction conditions, confirming effective retention of catalyst and SLP [14]. In 1973, Rinker [15] proposed an experimental method for locating the optimum pore filling of SLP catalysts for minimum mass transfer limitation; the time-dependent measurement of gas uptake of nonreactive SLP with various liquid loadings under isobaric reaction conditions yielded diffusion rates as function of pore filling.

After these reports, many industrially employed solid catalysts were reexamined, and some were found to be SLP-type systems under reaction conditions. An extensive review summarizing the state of the art of 1978 [16] also includes early examples of *molten salt SLP catalysts* [17]. Under consideration of different attractive interactions between solids and liquids (capillary and adhesive forces), the microscopic distribution of the SLP on the surface of the porous support materials

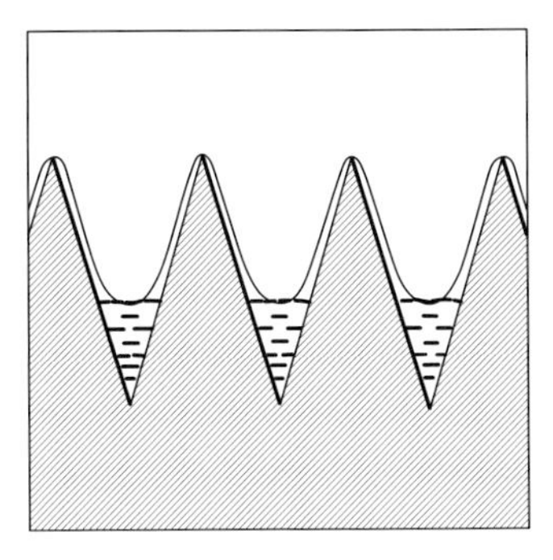

Figure 12.5 Idealized wetting of a porous surface (capillary forces cause pore filing up to a critical diameter above which surface wetting, caused by adhesive forces, may exist) [18].

was discussed in the same paper. In analogy with chromatography, a wetting model was suggested (Figure 12.5), and an analytical function for the maximum radius of liquid-filled pores was derived on the basis of equal chemical potential of the surface film and pore filling [18].

As suggested earlier, the diffusion limitation of SLP was confirmed to be a function of liquid loading also in these models. However, at the time it was concluded that "... *the problem of obtaining an adequate liquid distribution model is, however, so complicated that no theory has yet been proposed by which experimental observations can be predicted from first principles with any degree of confidence*" [16].

In a series of seminal papers, Scholten [19–24] studied the gas-phase hydroformylation of propylene with $[HRh(PPh_3)_3CO]$ in SLP in detail. The supported catalyst materials were analyzed by multiple techniques including differential scanning calorimetry (DSC), microscopy, porosimetry, gas adsorption, and IR spectroscopy [19]. Intrinsic reaction rates, activation energies, and diffusion effects were experimentally measured and theoretically modeled [23]. Different substrates, additives, and ligands (including phosphines, arsines, and amines) were screened [22]; the respective liquid loading and reaction conditions optimized [20]; and support surfaces modified [21]. Under optimized conditions, short activation periods and suppressed formation of aldol-condensation side products could be achieved: over 800 h of stable continuous operation at time of flights (TOFs) exceeding 2000 h^{-1} with 99.5% selectivity to aldehydes at l/b ratios of up to 8.8 was achieved with neat molten PPh_3 as SLP on mesoporous silica.

Shortly after, C=O hydrogenations of ketones and aldehydes with homogeneous ruthenium complexes in SLP were reported [25]. High-boiling substrates such as cycloheptanone could be passed over the SLP bed below their boiling point using H_2 as strip gas. The transport phenomena occurring in porous SLP catalysts were investigated in more detail by Rinker [26], who derived more elaborate models than initially proposed and also verified the predictions experimentally.

Engineering aspects of SLP catalysts in continuous-flow mode and the influence of various reactor configurations on the stability of such systems were studied by Hesse [27]. It was shown that presaturation of the mobile gas phase with the supported solvent at reaction temperature may compensate for progressive solvent loss of the SLP [28], and thus prevent deactivation through catalyst precipitation. Over 700 h of stable propene hydroformylation was demonstrated in a fluidized bed reactor [29]. The optimum loading phenomenon was reinvestigated [30], and the influence of the pore structure [31] and pore size distribution on the optimum value was modeled by three-dimensional simulations [32].

Using again more sophisticated models, recent simulations distinguished uniform from nonuniform film distributions [33]. The theoretical findings were successfully validated against earlier literature results. In 2003, it was demonstrated that the progressive solvent loss encountered in long-term continuous gas-phase application may also be compensated by the use of reversed flow techniques [34].

12.3
Supported Aqueous-Phase Catalysts (SAP)

In 1989, the concept of SLP was extended to supported aqueous-phase (SAP) catalysis [35]. While classical SLP catalysts relied on a difference in volatility as separation barrier, the retention of SAP catalysts was based on polarity differences. Thereby, conversion of liquid substrates became accessible, provided they were of sufficient hydrophobicity. In this respect, the large surface area and short diffusion pathways appeared particularly advantageous over bulk aqueous–organic systems.

The hydroformylation of oleyl alcohol with TPPTS-modified rhodium complexes in SAP on porous glasses with hydrophilic surfaces was demonstrated batchwise with nondetectable rhodium leaching [36]. It was proposed that the catalysis proceeded just at the aqueous–organic interface. A strong dependence of the activity, selectivity, and stability of the SAP catalysts in the hydroformylation of various substrates on the water loading used (added via vapor condensation after impregnation of the neat catalyst complex on the support) was noted, with an optimum water content at 4–12 wt% [36]. On the basis of detailed NMR analysis, this particular system was shortly after revealed to be not a genuine SAP catalyst, but rather a surface-adsorbed catalyst operating in the organic substrate phase (Figure 12.6) [37]. Hydrogen bonding of the surface silanols to the sulfonate groups in a minimal aqueous film was derived as a more realistic description. This proposition was later verified independently for similar complexes [38], and developed further into an immobilization strategy of its own [39]. The altered selectivity of the aqueous system was ascribed to water-mediated hydrogen bonding between the sulfonato groups of neighboring TPPTS on Rh, forming a weakly associated multidentate ligand scaffold [40]. The rhodium-catalyzed hydroformylation of propene in SAP was recently reinvestigated

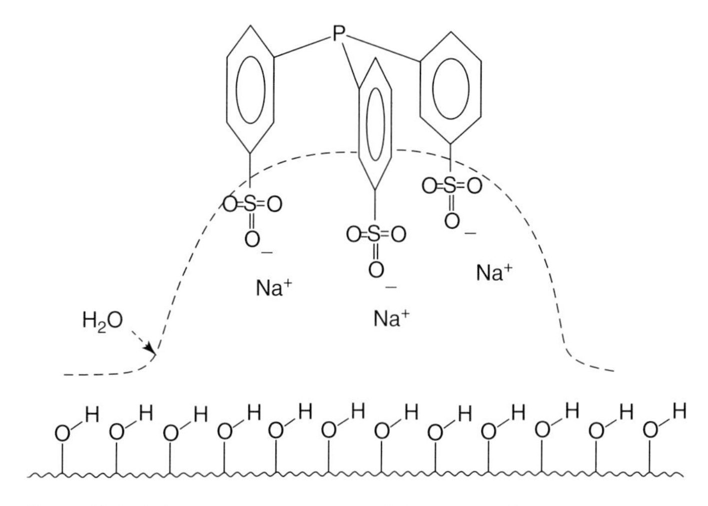

Figure 12.6 Schematic representation of the proposed immobilization mode of homogeneous catalysts bearing sulfonated ligands on hydrated surfaces [37].

with regard to different ligands, ligand-to-metal ratios, support materials, and reaction conditions [41].

SAP catalysis has also been used for the selective hydrogenation of α,β-unsaturated aldehydes with Ru catalysts on silica [42]. Asymmetric C=C hydrogenation using a sulfonated Ru–BINAP complex was shown to proceed with up to 70% ee in SAP [43]. The same system was subsequently transferred "back" to SLP using anhydrous ethylene glycol instead of water, which increased enantioselectivity to 96% ee, which is the same value achieved in a homogeneous solution in absence of a support [44]. For the hydroformylation of acrylic acid esters with Rh SAP catalysts, higher activities than in bulk biphasic systems were found [45]. Using an optimum loading of 37 wt%, the TOFs of the SAP catalyst reached values as high as 2370 h^{-1}, roughly one order of magnitude higher than under bulk biphasic conditions.

Wacker oxidation of liquid olefins with O_2 and Pd/Cu catalysts in SAP has also been performed successfully [46]. The lower activity as compared to the bulk aqueous system was explained on basis of restricted mobility of the two metal catalysts, which need to interact during the catalytic cycle. Pd-catalyzed allylic alkylations have also been studied in SAP [47]. Catalyst leaching and Pd black formation were claimed to be suppressed in SAP, but decreased activities were still observed during batchwise recycling.

Arai [48] reported an interesting example of cascade catalysis by multiple organometallic complexes in SLP/SAP, which proved incompatible in homogeneous solution. The simultaneous selective hydrogenation of two different substrates in a mixture with two SAP catalysts inside one reactor was demonstrated, as well as a catalytic sequence of Heck coupling and hydroformylation with two different SAP catalysts. However, only one recycling experiment was conducted for the hydrogenation system, and none for the SLP cascade.

12.4
Supported Ionic Liquid-Phase Catalysts (SILP)

With the rising interest in ILs as solvents for organometallic catalysis in the late 1990s [49], the extension of the SLP concept to supported ionic liquid-phase (SILP) systems became manifest. The advantage over organic SLP or SAP systems was thought to be that SILP would benefit from both polarity and volatility barriers, and would thus permit liquid- as well as gas-phase applications.

12.4.1
Synthetic Methods

Different synthetic strategies to SILP-type materials have been developed, which can be categorized into three different methods:

1) Deposition of IL/catalyst solutions on the surface of a porous support via wet impregnation (physisorbed SILP);
2) Chemical functionalization of a surface with a component of the IL (chemically anchored SILP); and
3) Sol–gel synthesis of a porous solid in presence of the IL/catalyst solution (ionogels).

Method 1 is the classical approach developed for SLP materials, and represents the most versatile strategy. The preparation procedure is straightforward and well reproducible, and a wide range of different components can efficiently be combined and screened. The more sophisticated method 2 may enhance the affinity (retention) of SILP and catalyst on the support, and extends the range of support materials to nonporous structures such as organic polymers, metal fibers, or carbon nanotubes (CNTs). It is typically used in conjunction with method 1 to introduce more IL and the catalyst. Capillary forces are of little relevance to method 2, and high surface coverage is typically achieved. Method 3, a material science approach, had been developed independently from SLP methodology in the early 1990s [50]. It requires a robust catalyst that survives the material synthesis procedure (sol–gel chemistry [51] of inorganic oxides, metal-organic frameworks, etc.), and poses more difficulties on post-synthetic characterization. The SILP containing the catalyst is encapsulated in the inner pores of the solid by various degrees, depending on the structure of the material. Although examples of successful application in catalysis have been reported [52], these materials will not be discussed further here as the microenvironment around the catalyst is distinct from the film-like situation in SLP systems.

12.4.2
Characteristics

As a result of multiple intermolecular forces (electrostatic attraction, H-bonding, π and hydrophobic interactions), ILs are highly ordered solvents [53] which may display specific interactions with solutes [54]. As to which degree the bulk solvent properties of ILs are altered upon their deposition on the surface of a support is a challenging question which, despite some efforts, has not yet been fully resolved [55]. So far, mainly model systems on idealized surfaces have been investigated with, for example, X-ray reflectivity [56], sum-frequency vibrational spectroscopy [57, 58], electrical impedance spectroscopy [59], and X-ray photoelectron spectroscopy (XPS) [60], all under analytic conditions. The difference to reactive systems is highlighted by recent findings on surface rearrangements of ILs containing polar transition-metal complexes [61]. The use of XPS for studying IL interfaces, including also some reactive systems, has recently been reviewed [62]. From most spectroscopic studies, a different orientation of the IL layer in direct contact with the surface and the upper layers was generally deduced. However, implications for catalytic systems under reactive conditions (that is, in particular, under gas pressure) are still difficult to rationalize. Interestingly, strong, directed hydrogen bonding of water in wet ILs to both surface silanols and NTf_2 anions [63] was observed by sum-frequency vibrational spectroscopy [57].

Owing to very high surface to volume ratios, fluids confined in nanospaces experience changes in some of their physical properties (Gibbs–Thompson effect) [64]. Extraordinarily high melting point depressions of some ILs deposited on porous supports may be taken as indication for specific IL–surface interactions in addition to physical confinement effects. Upon deposition on dehydroxylated porous hydrophilic silicates, carefully dried hydrophobic ILs experienced even higher melting point depressions than water, with an inverse relationship between pore size and melting point depression (Figure 12.7) [65].

Recently, the thermal stability of BMIM PF_6 was also reported to be influenced by the presence of silica surfaces [66]. Thermogravimetry analysis (TGA) scans

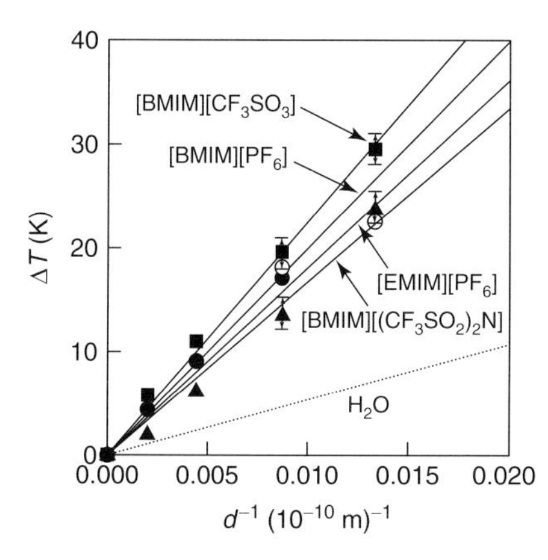

Figure 12.7 Melting point depression of ILs as measured by DSC over inverse pore diameter of the porous glass support [65].

and high vacuum distillation experiments showed that thermal decomposition of the IL was lowered by more than 100 °C when deposited on dehydroxylated silica or alumina as compared to the bulk IL. X-ray diffractograms below the melting point of the IL revealed that the crystalline order of BMIM PF_6 was disturbed by the silica surface to yield an entirely amorphous composite material. Furthermore, solid-state magic-angle spinning nuclear magnetic resonance (SS-MAS NMR) analysis indicated SiOH–anion interactions.

Catalytically active SILP materials containing $[Pd(DPPF)(CF_3CO_2)_2]$ and CF_3SO_3H in imidazolium ILs on fumed amorphous silica were analyzed in detail by multiple techniques [67]. From N_2 adsorption isotherms and transmission electron microscopy (TEM) images, it was concluded that pores up to 9 nm radius were flooded by the IL, whereas larger pores were only surface-covered. IR spectroscopy suggested that all surface silanols were involved in H-bonding, indicating complete surface coverage. From linewidth analysis of SS-MAS NMR spectra, it was concluded that the mobility of the imidazolium cations and the Pd complex were reduced in the SILP material. The formation of ordered solvent cages around the organometallic complex was suggested as possible explanation for this observation at the high molar ratio of Pd to IL (1 : 25–33) used in these experiments.

The transport properties of hydrophilic and hydrophobic solutes through SILP membranes with hydrophobic ILs have also been studied [68]. Furthermore, an attempt was made to quantify the degree of polarity change that imidazolium ILs undergo when covalently immobilized on different polymers [69]. By comparing the π^* values [70] of the polymer SILPs with different organic solvents and the blank polymers, increased micropolarity of the SILP materials was inferred. However, direct comparison of IL in bulk form and as SILP is currently not available.

Para-hydrogen induced polarization (PHIP) NMR has been used as *in situ* method for the continuous-flow hydrogenation of propyne with Pd nanoparticles in an IL on activated carbon cloth as support [71]. In comparison with catalysts without IL, it was found that the presence of the latter lowered the activity, but at increased selectivity and greatly enhanced polarization transfer. Interpretation of the observed effects was not attempted.

Despite their relative young history, SILP-type materials have already found applications in different areas [72]. Engineering applications such as gas separation processes [73] or selective absorption techniques [74] will not be considered in this chapter. Selected examples of catalytic applications will be discussed briefly, wherein the focus will be on molecularly defined organometallic catalysts. Therefore, biocatalysis [75], organocatalysts [76], metal nanoparticles [77], and heterogeneous catalysts with IL coatings [78] will not be considered. Polymer-supported ILs are included only when of relevance to molecular transition-metal catalysis.

12.4.3
Gas-Phase Applications

The first intentional organometallic SILP catalysts were prepared by Carlin [79]. $[Rh(nbd)(PPh_3)_2]^+$ in different imidazolium ILs was deposited on porous polyvinylfluoride by wet impregnation. The materials were used for the continuous hydrogenation of propylene and ethylene, with olefins and hydrogen fed from different sides of the polymer membrane. Various IL anions were screened, and activity trends could be related to gas solubilities. Dimerization of ethylene with a $[NiCl_2(Pcy_3)]_2/EtAlCl_2$ catalyst in chloroaluminate ILs on polyethersulfone membranes was also reported in the same year [80].

Continuous-flow rhodium-catalyzed hydroformylation of propene with SILP catalysts based on rhodium–phosphine was reported by Wasserscheid, Riisager, and Fehrmann [81]. Different ligands, ILs, and IL loadings on silica were screened for optimum activity and stability. At 100 °C, the best system at low IL loadings of 11 wt% (8% pore filling) based on the sulfo–xantphos ligand achieved a TOF of 37 h^{-1} with selectivities of l/b up to 23. After 4 h of stable operation, deactivation set in regardless of which IL or loading was used. The system was developed further and improved by optimizing the pretreatment of the support surface (dehydroxylation at 500 °C) [82]. With SS-MAS NMR, it could be shown that some of the excess of free phosphine got protonated by the silanols groups of the support. In detailed investigations, the Wasserscheid group [83] improved the materials systematically to reach stabilities over 200 h at TOFs around 100 h^{-1}, and transferred the process to gradient-free loop reactors to facilitate kinetic analysis for the determination of activation energies and reaction orders.

Continuous methanol carbonylation with CH_3I co-feed using a Monsanto-type catalyst in SILP was also reported by the same group [84]. Moderate rates and selectivities were obtained, and stable operation was possible for the first 90 min only.

Kiwi-Minsker [85] reported on rhodium-based hydrogenation catalysts in SILP on microstructured support materials consisting of plates of sintered metal fibers coated with carbon nanofibers. TOFs of 250 h^{-1} with stability over 6 h were achieved in the continuous hydrogenation of 1,3-cyclohexadiene. The stability of catalyst performance was tentatively ascribed to the high thermal conductivity of the support, suppressing hot-spot formation during the exothermic hydrogenation reaction.

Low-temperature water gas shift (WGS) catalysis with SILP materials was reported by Wasserscheid [86]. The initial catalyst systems showed limited activity and stability, but could be greatly improved through tuning of the support basicity and use of preformed Ru catalysts. Stable operation over 100 h with unprecedented high activity for molecular WGS catalysts could eventually be achieved at temperatures as low as 120 °C [87].

12.4.4
Liquid-Phase Applications

Hölderich described Lewis acidic ILs with chloroaluminate anions deposited on various porous oxide supports, which were used as catalysts for Friedel–Crafts alkylations of aromatic compounds [88]. Covalent surface attachment via condensation of a siloxane functionality in the alkyl chain of the cation was used (SILP method 2). The range of materials was broadened to different Lewis-acidic anions, and more catalytic reactions were screened [89]. However, liberation of HCl from reaction of the silanol groups with the chloroaluminates led to partial destruction of the oxide structure during the preparation of these materials. Continuous liquid-phase application showed limited stability, with obvious deactivation within a few hours [90]. Similar materials bearing chlorostannate ILs were used by the Landau group [91] as catalyst for the Prins condensation of isobutene with formaldehyde, showing moderate stability in repetitive batch experiments.

Friedel–Crafts isopropylation of cumene catalyzed by Lewis-acidic SILP materials, similar to those used by Hölderich, was compared under liquid-phase and gas-phase conditions [92]. Besides different regioselectivities of alkylation, the SILP materials were more active than the catalyst in bulk IL biphasic application. The loading was varied and the support surface pretreated with the chloroaluminate IL in dichloromethane to activate the surface for enhanced acidity. Although $AlCl_3$ leaching in the percentage range was observed in the liquid phase, the materials could be recycled four times batchwise without alteration of activity but with changing selectivity. In continuous gas-phase application, the materials could be used for more than 200 h on stream with stable performance [93]. The quality of the reagents with respect to high-boiling impurities and residual water content were found to be of crucial importance.

Hagiwara [94] described [3 + 2] Huisgen cycloadditions of benzylazide with acetylenes catalyzed by copper bromide catalysts in BMIM PF_6 on reverse-phase silica with mercaptopropyl surface functionalization. The catalysts were reused six times using a reaction medium of aqueous EtOH with addition of NEt_3. This solvent mixture would be expected to have a good solubility for the IL and the catalyst, albeit no quantitative data on cross contamination or leaching are available.

Mehnert reported on solution-phase applications of SILP catalysts based on typical organometallic complexes for rhodium-catalyzed olefin hydroformylation [95] and hydrogenation [96]. Both chemically anchored imidazolium fragments (method 2) and purely physisorbed SILP (method 1) were prepared using BMIM PF_6 and BMIM BF_4. For the hydroformylation of 1-hexene, a 10 : 1 mixture of TPPTS and $[Rh(acac)(CO)_2]$ was used as catalyst with 25 wt% IL loading and applied in neat liquid substrate. At 100 °C, the SILP system achieved nearly three times the rate of the unsupported bulk biphasic system (TOF = 3900 h^{-1} vs 1380 h^{-1}), which was attributed to enhanced catalyst accessibility in SILP. However, at conversions >50%, leaching of IL into the liquid product phase entrained up to 2% of the catalyst per run. The hydrogenation SILP catalyst containing $[Rh(nbd)(PPh_3)_2]$ PF_6 in BMIM PF_6 on silica did not suffer from such strong depletion effects because of the cationic catalytic active species and because the liquid substrate/product phase remained unpolar throughout the reaction. No metal could be detected in the product phase (<0.03 ppm), and the SILP catalyst could be recycled for at least 18 times without apparent deactivation in batch mode. Electron microscopy confirmed the absence of larger rhodium clusters (>6 Å), and variation of hydrogen pressure revealed first-order kinetics, which was taken as strong evidence for genuine homogeneous catalysis in SILP.

Serp [97] reported on functionalized CNTs as tailored support materials for hydrogenation catalysts in SILP. Post-synthetic grafting of imidazolium fragments via amide linkers on carboxylate CNTs yielded structured supports, which were coated with BMIM PF_6 containing $[Rh(nbd)(PPh_3)_2]$ PF_6 at various loadings (method 2). The materials were analyzed by different techniques and shown to afford much increased reaction rates (TOF 2880 h^{-1}) in the hydrogenation of 1-hexene as compared to the same catalyst in SILP on oxidic support materials such as silica, titania, zirconia, or alumina, and also activated carbon. The high thermal conductivity as well as the open channel structures of the CNT were used as explanation for this effect. The materials were recycled five times without loss of activity and undetectable rhodium leaching into the organic phase.

Ring-closing metathesis and cross-metathesis catalyzed by Grubbs' catalyst in HMIM PF_6 on silica was reported by Hagiwara *et al.* [98]. Six recycling runs with decreasing activity were demonstrated. The materials were later used for the synthesis of macrocyclic lactones [99]. Ring-closing metathesis has been performed also with an imidazolium-tagged Grubbs' catalyst in BMIM PF_6 on a polyimide nanomembrane [100]. A decreased filtration resistance was observed after coating the membrane with IL, which was interpreted as due to the changed surface polarity. Stepwise catalysis in toluene and filtration showed pronounced deactivation already in the third cycle. Catalyst leaching or IL cross contamination were not quantified.

Catalytic hydroamination of phenylacetylene with rhodium, palladium, copper, and zinc complexes in SILP physisorbed according to method 1 on silylated diatomaceous earth (amorphous silica) was reported by Müller [101]. Selectivities and activities of the complexes in SILP were found to be higher than under homogeneous conditions.

Hagiwara [102] reported on palladium complexes immobilized in BMIM PF_6 physisorbed on silica. At 150 °C, the materials were highly active in Heck coupling reactions in dodecane as solvent and with tertiary amines as base. TOFs reached 8000 h^{-1} and batchwise recycling was demonstrated for five runs. Decreasing activity could be restored by washing the SILP catalyst with aqueous NaOH. Nevertheless, the formation of small amounts of homogeneous palladium species leaching into the reaction medium under the reaction conditions could not be fully excluded.

Luis [103] used imidazolium-functionalized Merrifield resins as support for palladium catalysts. In this case, the support acted as a *catch and release* reservoir for the active catalyst: Pd(0) species partitioned into the substrate phase at $T > 100$ °C, catalyzing Heck coupling reactions. At lower temperatures, the Pd was stabilized through *N*-heterocyclic carbene (NHC) complexation by the imidazolium functions of the support after the reaction. Different precursors, reaction conditions, IL anions, and bases were screened, and the kinetics of the reaction investigated. The optimum system could be recycled five times in dimethylformamide (DMF). As the catalyst was active in the solution phase rather than in the SILP material, this strategy does not impart a permanent separation barrier and is therefore not suitable for continuous application.

Allylic substitution with Pd catalysts in SILP on preconditioned chitosan as support has been reported [104]. Drying of the support by supercritical carbon dioxide ($scCO_2$) extraction yielded superior catalyst performance than freeze-drying from aqueous solutions. By performing the catalysis in a neat substrate, leaching issues could not be resolved (up to 9% per run), but stability was rather poor during batchwise recycling experiments (complete deactivation from fifth cycle). Asymmetric versions were also demonstrated, but without recycling experiments.

Vankelecom [105] described the first chiral organometallic complexes in a SILP catalyst on a polar polymer as support. Enantiomerically pure Ru–BINAP complexes in BMIM PF_6 were used on poly(diallyldimethylammonium chloride) (method 2) for the asymmetric C=O hydrogenation of methylacetoacetate. At 60 °C, activities of the SILP catalysts were up to three times higher than the bulk IL organic biphasic media with identical levels of enantioselectivity (97% ee). The catalysts were reused once in batchwise recycling, but catalyst retention or leaching was not quantified.

Müller [106] reported on the enantioselective C=O hydrogenation of acetophenone with ruthenium and rhodium–BINAP catalyst in SILP. While no enantioselectivity was observed for the reaction in CH_3OH, up to 74% ee were achieved with Rh–BINAP and K_2CO_3 in a tetraalkylphosphonium carboxylate IL physisorbed on silica. However, no comparison of the same system in neat IL without support was available, and it was speculated that intensified substrate–catalyst interactions induced by the thin films of the SILP might be responsible for this effect.

The same reaction was investigated with chiral diamines as ligands for Ru catalysts in SILP [107]. Different support materials were examinated, and all SILP catalysts preserved the bulk-solution selectivity of the reaction (78% ee). Catalysis and recycling were conducted with isopropyl alcohol as solvent, making conclusive decisions about the actual reaction phase difficult.

For the asymmetric transfer hydrogenation of acetophenone with ruthenium catalysts on SBA-15, reaction rates were found to increase by a factor of 10 when the surface was functionalized with tetraalkylammonium species as compared to hydrophobic alkyl capping [108]. Enhanced availability of the polar reducing agent (aqueous sodium formate) at the catalytic site was used as explanation for this accelerating effect.

Highly enantioselective epoxidation of methylstyrene with Mn–salen complexes in BMIM PF_6 on IL-modified MCM-41 prepared by method 2 has been described [109]. At undisturbed activities, enantiomeric excess values were reported to reach 99% in SILP, whereas only 50% was achieved in a homogeneous solution. The reactions were carried out in CH_2Cl_2, however, which is fully miscible with the IL used and also a good solvent for the catalyst. No data on catalyst recycling were provided.

Halligudi [110] performed oxidative kinetic resolution of a range of secondary alcohols with chiral Mn–salen complexes in SILP. Imidazolium-functionalized

MCM-41, SBA-15, and amorphous silica were tested on a support and different additives (halide salts) screened. Up to 99% ee could be achieved over five recyclings in a multiphasic reaction system comprising water, hexane, and the SILP catalyst.

Jacobs [111] reported on continuous liquid-phase application of chiral Cr–salen complexes in SILP physisorbed on silica. Asymmetric ring opening of epoxides with $TMSN_3$ was performed, yielding good enantioselectivities in the range of 65–96% ee and reaching a turnover number (TON) of 314. Catalyst leaching up to 1% was observed with hexane as the mobile phase.

Mayoral [112] observed different stereoselectivities in SILP as compared to bulk solution catalysis using laminar clays such as laponite as support materials (method 1). Both the cis/trans ratio as well as the enantioselectivities of the cyclopropanation of styrene with diazoacetate using bis(oxazoline)–copper complexes showed a dependence on the nature of the support as well as the loading with BMIM PF_6. Small variations of selectivities were observed at different stages of conversion, and recycling of the materials led to varying selectivities over time.

Hardacre [113] performed asymmetric Mukaiyama aldol reactions catalyzed by similar bis(oxazoline) –copper complexes in SILP and compared it to the nonsupported IL. In bulk EMIM NTf_2, the Lewis-acid-promoted condensation of methylpyruvate and phenyl–TMS–ethene was at least 60 times faster than in CH_2Cl_2, with complete retention of enantioselectivity (82% ee). However, hydrolysis of the TMS–ether resulted in about 10% lower chemoselectivities. On adsorbing the IL on either imidazolium-functionalized or plain silica, this side reaction was effectively suppressed, and the SILP catalyst combined high activity with good enantioselectivity when applied in a biphasic system with Et_2O. While neutral copper catalysts leached out to 19%, an imidazolium-modified ligand afforded significantly enhanced retention. However, deactivation was still noticeable after the fifth recycling.

12.5
Supported Liquid-Phase Catalysts and Supercritical Fluids

From the examples given above, it becomes obvious that SLP and in particular SILP catalysts are versatile systems applicable to many different reactions. However, like all "soft" immobilization techniques, their stability is critically dependent on the choice of the mobile phase. They may well be applied with either gas or liquid phases, but each approach suffers from distinctive limitations: gas-phase processes allow excellent retention of the catalysts, but are restricted to volatile substrates, thermally robust catalysts, and reaction systems that do not form products or side products that are of distinctively lower volatility than the substrates. With the exception of asymmetric C=O hydrogenation, which has been described with SILP catalysts in the gas phase at $T > 100\,^{\circ}C$[114], the thermal process window excludes stereoselective transformations of complex nonvolatile molecules, one of the key skills of molecular transition-metal catalysts. Liquid-phase applications, on the other hand, allow broader reaction and substrate spectra, but leaching (chemical solubility or physical abrasion) of the SILP and catalyst severely limits the lifetime and overall efficiency of the system. This practical complication on the macroscale process level is rarely addressed in the literature describing SILP catalysts.

In order to overcome these limitations, it seems attractive to consider supercritical fluids (SCFs) as the mobile phase. SCFs combine the transport properties of gases with the solvent properties of liquids for many organic molecules [115]. They are, however, unable to dissolve ILs [116] or other low-volatile liquids [117] and are very poor solvents for many organometallic catalysts [118]. Thus, they have the potential to capitalize on the respective advantages of gas- and liquid-phase processing in combination with SILP systems. While a number of examples are known that exploit these properties in liquid/SCF biphasic systems [119], SLP systems have been combined with $scCO_2$ only in very few cases up to now.

Arai [120] reported on Ru–TPPTS complexes in SAP on silica for cinnamaldehyde hydrogenation with $scCO_2$. Higher activity than with toluene as the second phase was observed, but no recycling was attempted. Up to know, this is the only report on SAP catalysts in combination with $scCO_2$. Given the promising applications documented for $H_2O/scCO_2$ systems including the formation of carbonic and percarbonic acid in such media [119], there seems to be considerable room for future developments in this area.

Leitner [121] investigated Pd nanoparticles supported on poly(ethylene glycol) (PEG) modified silica for the selective aerobic oxidation of alcohols in $scCO_2$ (Figure 12.8). Materials based on physisorbed PEG-750 were compared to silica modified with covalently attached PEG chains. The latter provided significantly enhanced stability in catalysis. Continuous application with $scCO_2$ as the mobile phase showed stable performance over at least 30 h, confirming effective integrated separation of products and catalyst in the fixed-bed reactor setup.

For the same reaction, Pagliaro [122] described sol–gel entrapped imidazolium ILs containing perruthenate in silica as highly active and selective catalysts. Application with $scCO_2$ allowed efficient oxidation of nonvolatile alcohols. Batchwise

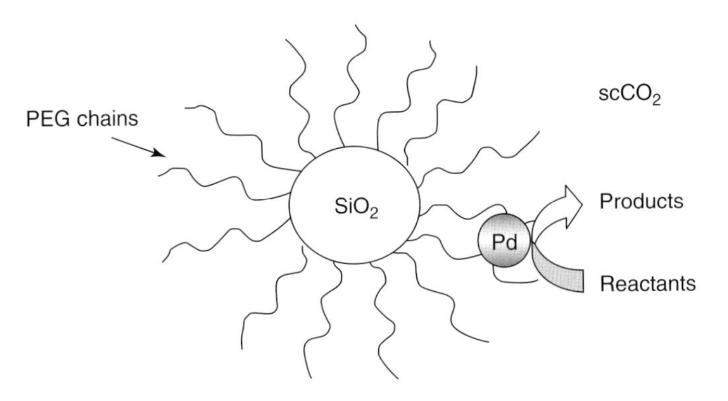

Figure 12.8 Pd nanoparticles stabilized on PEG-modified silica for alcohol oxidation using $scCO_2$ [121].

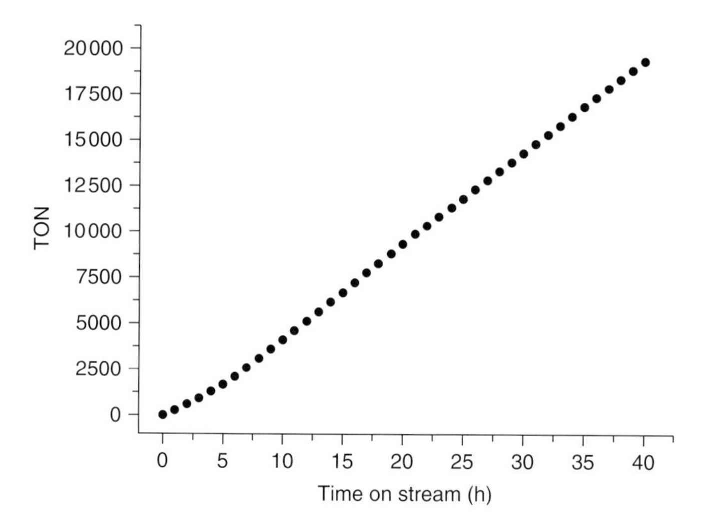

Figure 12.9 1-Octene conversion to aldehydes using SILP catalysts with continuous CO_2 flow [123].

recycling was attempted, but practical difficulties of material loss were encountered preventing more than three repetitive runs.

Cole-Hamilton [123] applied organometallic complexes in SILP for continuous catalysis with $scCO_2$. The hydroformylation of 1-octene with Rh–TPPMS complexes in OMIM NTf_2 on dehydroxylated silica (method 1) proceeded with even higher activity than in bulk IL/$scCO_2$ media, and the system was perfectly stable for at least 40 h (Figure 12.9). Rh leaching levels were as low as 0.5 ppm in the organic product fraction recovered from $scCO_2$ by decompression. Stability and response times of the system were sufficient to use onstream parameter variations for more detailed investigations with statistical methods. By varying the IL loading, substrate flow, and syngas pressure, it was shown that at low loadings (29 wt% IL) higher syngas pressures decreased the rate (in accordance with the intrinsic reaction kinetics), while at high loadings (44 wt% IL) higher syngas pressure increased the hydroformylation rate (Figure 12.9). In conjunction with phase behavior observations, it was concluded that the reaction proceeded best in an expanded liquid substrate phase rather than in a single supercritical phase, and the IL film thickness became limiting for the gas availability at the catalytic centers at high loadings [124].

The group of Leitner developed a highly efficient example of enantioselective continuous-flow catalysis using chiral transition-metal complexes in SILP [125]. Rhodium catalysts comprising chiral ligands of the QUINAPHOS family were immobilized in SILP for the asymmetric C=C hydrogenation of dimethyl itaconate as typical example for a prochiral solid substrate. The reaction was conducted in continuous-flow mode with $scCO_2$ using a small and flexible setup (Figure 12.10) [126]. At quantitative conversion, up to 99 ee was achieved and 65 h of continuous operation demonstrated, although at slightly reduced selectivity (70–75% ee) after

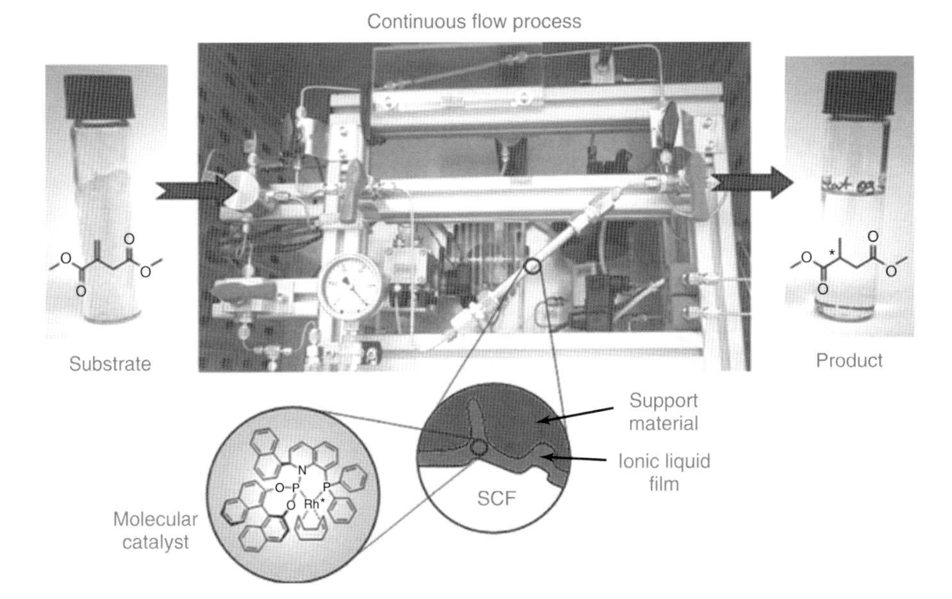

Figure 12.10 Continuous flow organometallic catalysis using SILP catalysts with $scCO_2$ [3].

10 h on stream. Remarkably, more than 100 000 TON of the chiral transition-metal complex was reached. Productivities surpassed values of 150 kg product per gram rhodium and the process operated at a space time yield (STY) of 0.3 $kg\,l^{-1}\,h^{-1}$. Rhodium leaching levels were below the detection limit of 1 ppm as judged by inductively coupled plasma-optical emission spectroscopy (ICP-OES) of both the product fractions and the spent SILP catalyst. This SILP $scCO_2$ concept provided about one order of magnitude higher STYs than other immobilization techniques such as self-supported catalysis in continuous liquid-phase application [127].

Very recently, Cole-Hamilton and coworkers [128] extended the approach to Ru-catalyzed alkene metathesis using SILP catalysts in the continuous-flow mode with compressed CO_2 (Figure 12.11). Self-metathesis of methyl oleate with ion-tagged ruthenium catalyst in BMIM NTf_2 on dehydroxylated silica (method 1) proceeded with high reactivity (TON $>$10 000) for at least 9 h, although slight loss of activity over time was observed. Ruthenium-leaching levels were in the range of 10 ppm in the self-metathesis product. In the best case, 6 g of substrate per hour could be converted into products using a 9 ml reactor.

12.6 Conclusion

From the examples summarized in this chapter, it becomes evident that SLPs represent a versatile and successful approach to organometallic catalyst recovery and recycling with the particularly attractive option of using molecular or nanoscale

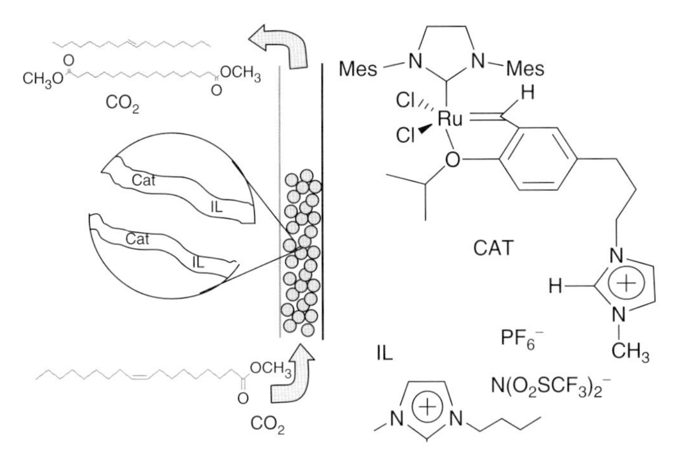

Figure 12.11 Conceptual visualization of the SILP process for the self-metathesis of methyl oleate catalyzed by boomerang ruthenium catalyst [128].

catalysts in continuous-flow mode. The concept of SLPs appears very attractive for ILs because they provide a good medium for homogeneous organometallic catalysis while being of extremely low volatility, a perfect match for the requirements of the liquid component in supported systems under continuous application. Other liquids of low volatility such as PEG can also be envisaged. Transfer of the knowledge from SLP and SAP catalyst studies to these systems is desirable for an advanced understanding of these materials.

The combination of SLP catalysts with compressible gases in form of SCFs or expanded liquid phases offers particularly promising possibilities of broadening the window of applicable reactions and increasing overall process efficiency. For example, the effective use of both solvent and catalyst in SILP catalysts is enhanced as compared to bulk IL systems, because the catalyst is very close to a large interface in the ultrathin film on the surface of the solid material. Recovery of products from the mobile phase, which is free of any organic solvent, may be conveniently achieved by depressurization or temperature swings [129].

Importantly, the SLP strategies preserve the full spectrum of organometallic coordination chemistry, an important feature that is often severely restricted when catalysts are directly immobilized on the surface of a solid support. Thus, the concepts of molecular catalyst design can be directly coupled to attractive engineering perspectives with continuous-flow SLP systems. Online catalyst analysis and reaction monitoring will greatly improve the understanding of these catalyst materials under reaction conditions and can enable self-regulating and even self-optimizing continuous-flow systems [130]. Therefore, SLP/SCF systems appear as ideal candidates for fully integrated processes that selectively produce a single product in essentially pure form. Exciting combinations of selective product extraction [131] and continuous-flow cascade catalysis await to be realized on the basis of these developments.

References

1. Cornils, B., Herrmann, W.A., Horváth, I.T., Leitner, W., Mecking, S., Olivier-Bourbigou, H., and Vogt, D. (2005) *Multiphase Homogeneous Catalysis*, 1st edn, Wiley-VCH Verlag GmbH, Weinheim.
2. Cole-Hamilton, D.J. and Tooze, R.P. (2006) *Catalyst Separation, Recovery and Recycling – Chemistry and Process Design*, 1st edn, Springer, Dordrecht.
3. Hintermair, U., Franciò, G., and Leitner, W. (2011) *Chem. Commun.*, **47**, 3691–3701.
4. Cole-Hamilton, D.J. (2003) *Science*, **299**, 1702–1706.
5. Copéret, C., Chabanas, M., Petroff Saint-Arroman, R., and Basset, J.-M. (2003) *Angew. Chem.*, **115**, 164–191.
6. Lindner, E., Schneller, T., Auer, F., and Mayer, H.A. (1999) *Angew. Chem. Int. Ed.*, **38**, 2155–2174.
7. Bird, R.B., Stewart, W.E., and Lightfoot, E.N. (2005) *Transport Phenomena*, 2nd edn, John Wiley & Sons, Inc., New York.
8. Moravec, R.Z., Schelling, W.T., and Oldershaw, C.F. (1939) (N. V. de Bataafsche Petroleum Maatschappij). GB 511566.
9. Acres, G.J.K., Bond, G.C., Cooper, B.J., and Dawson, J.A. (1966) *J. Catal.*, **6**, 139–141.
10. Rony, P.R. (1968) *Chem. Eng. Sci.*, **23**, 1021–1034.
11. Osborn, J.A., Jardine, F.H., Young, J.F., and Wilkinson, G. (1966) *J. Chem. Soc. A.*, 1711–1732.
12. Dumesic, J.A., Huber, G.W., and Boudart, M. (2008) in *Handbook of Heterogeneous Catalysis*, 2nd edn, vol. 1 (eds G. Ertl, H., Knözinger, F., Schüth, and J., Weitkamp), Wiley-VCH Verlag GmbH, Weinheim, pp. 1–15.
13. Thiele, E.W. (1939) *Ind. Eng. Chem.*, **31**, 916–920.
14. Rony, P.R. (1969) *J. Catal.*, **14**, 142–147.
15. Abed, R. and Rinker, R.G. (1973) *J. Catal.*, **31**, 119–126.
16. Villadsen, J. and Livbjerg, H. (1978) *Catal. Rev. Sci. Eng.*, **17**, 203–272.
17. Kenney, C.N. (1975) *Catal. Rev. Sci. Eng.*, **11**, 197–224.
18. Giddings, J.C. (1962) *Anal. Chem.*, **34**, 458–465.
19. Gerritsen, L.A., Van Meerkerk, A., Vreugdenhil, M.H., and Scholten, J.J.F. (1980) *J. Mol. Catal.*, **9**, 139–155.
20. Gerritsen, L.A., Herman, J.M., Klut, W., and Scholten, J.J.F. (1980) *J. Mol. Catal.*, **9**, 157–168.
21. Gerritsen, L.A., Herman, J.M., and Scholten, J.J.F. (1980) *J. Mol. Catal.*, **9**, 241–256.
22. Gerritsen, L.A., Klut, W., Vreugdenhil, M.H., and Scholten, J.J.F. (1980) *J. Mol. Catal.*, **9**, 257–264.
23. Gerritsen, L.A., Klut, W., Vreugdenhil, M.H., and Scholten, J.J.F. (1980) *J. Mol. Catal.*, **9**, 265–274.
24. Pelt, H.L., Verburg, R.P.J., and Scholten, J.J.F. (1985) *J. Mol. Catal.*, **32**, 77–90.
25. Strohmeier, W., Graser, B., Marcec, R., and Holke, K. (1981) *J. Mol. Catal.*, **11**, 257–262.
26. (a) Datta, R. and Rinker, R.G. (1985) *J. Catal.*, **95**, 181–192; (b) Datta, R., Savage, W., and Rinker, R.G. (1985) *J. Catal.*, **95**, 193–201; (c) Datta, R., Rydant, J., and Rinker, R.G. (1985) *J. Catal.*, **95**, 202–208.
27. Stegmueller, R. and Hesse, D. (1988) *Chem.-Ing.-Tech.*, **60**, 1074–1075.
28. Richers, U. and Hesse, D. (1992) *Chem.-Ing.-Tech.*, **64**, 633–634.
29. Brüsewitz, R. and Hesse, D. (1991) *DECHEMA Monogr.*, **122**, 283–299.
30. Hesse, D. and Hoffmeister, M. (1987) *Chem. Ing. Tech.*, **59**, 520–521.
31. Hoffmeister, M. and Hesse, D. (1990) *Chem. Eng. Sci.*, **45**, 2575–2580.
32. Gottsleben, F., Hoffmeister, M., and Hesse, D. (1991) *DECHEMA Monogr.*, **122**, 269–282.
33. (a) Abramova, L.A., Baranov, S.P., and Dulov, A.A. (2000) *Appl. Catal., A*, **193**, 243–250; (b) Abramova, L.A., Baranov, S.P., and Dulov, A.A. (2000) *Appl. Catal. A*, **193**, 251–256.
34. Beckmann, A. and Keil, F.J. (2003) *Chem. Eng. Sci.*, **58**, 841–847.

35. Arhancet, J.P., Davis, M.E., Merola, J.S., and Hanson, B.E. (1989) *Nature*, **339**, 454–455.
36. Arhancet, J.P., Davis, M.E., Merola, J.S., and Hanson, B.E. (1990) *J. Catal.*, **121**, 327–339.
37. Horváth, I.T. (1990) *Catal. Lett.*, **6**, 43–48.
38. Bianchini, C., Burnaby, D.G., Evans, J., Frediani, P., Meli, A., Oberhauser, W., Psaro, R., Sordelli, L., and Vizza, F. (1999) *J. Am. Chem. Soc.*, **121**, 5961–5971.
39. Barbaro, P. and Liguori, F. (2009) *Chem. Rev.*, **109**, 515–529.
40. Horváth, I.T., Kastrup, R.V., Oswald, A.A., and Mozeleski, E.J. (1989) *Catal. Lett.*, **2**, 85–90.
41. Riisager, A., Eriksen, K.M., Hjortkjær, J., and Fehrmann, R. (2003) *J. Mol. Catal. A.*, **193**, 259–272.
42. Fache, E., Mercier, C., Pagnier, N., Despeyroux, B., and Panster, P. (1993) *J. Mol. Catal.*, **79**, 117–131.
43. Wan, K.T. and Davis, M.E. (1994) *J. Catal.*, **148**, 1–8.
44. Wan, K.T. and Davis, M.E. (1994) *Nature*, **370**, 449–450.
45. Fremy, G., Carpentier, J.-F., Castanet, Y., Monflier, E., and Mortreux, A. (1995) *Angew. Chem.*, **107**, 1608–1610.
46. Arhancet, J.P., Davis, M.E., and Hanson, B.E. (1991) *Catal. Lett.*, **11**, 129–136.
47. dos Santos, S., Tong, Y., Quignard, F., Choplin, A., Sinou, D., and Dutasta, J.P. (1998) *Organometallics*, **17**, 78–89.
48. Bhanage, B.M., Fujita, S.-I., Yoshida, T., Sano, Y., and Arai, M. (2003) *Tetrahedron Lett.*, **44**, 3505–3507.
49. (a) Chauvin, Y., Mussmann, L., and Olivier, H. (1996) *Angew. Chem. Int. Ed. Engl.*, **34**, 2698–2700; (b) Welton, T. (1999) *Chem. Rev.*, **99**, 2071–2083; (c) Wasserscheid, P. and Keim, W. (2000) *Angew. Chem. Int. Ed.*, **39**, 3772–3789.
50. Rosenfeld, A., Avnir, D., and Blum, J. (1993) *J. Chem. Soc., Chem. Commun.*, 583–584.
51. (a) Karout, A. and Pierre, A.C. (2009) *Catal. Commun.*, **10**, 359–361; (b) Vioux, A., Viau, L., Volland, S., and Le, B.J. (2010) *C. R. Chim.*, **13**, 242–255.
52. (a) Gelman, F., Blum, J., and Avnir, D. (2000) *J. Am. Chem. Soc.*, **122**, 11999–12000; (b) Murphy, E.F., Schmid, L., Bürgi, T., Maciejewski, M., Baiker, A., Günther, D., and Schneider, M. (2001) *Chem. Mater.*, **13**, 1296–1304; (c) Gelman, F., Blum, J., and Avnir, D. (2002) *J. Am. Chem. Soc.*, **124**, 14460–14463; (d) Shi, F., Zhang, Q., Li, D., and Deng, Y. (2005) *Chem.–Eur. J.*, **11**, 5279–5288; (e) Craythorne, S.J., Anderson, K., Lorenzini, F., McCausland, C., Smith, E.F., Licence, P., Marr, A.C., and Marr, P.C. (2009) *Chem.–Eur. J.*, **15**, 7094–7100; (f) Karimi, B., Elhamifar, D., Clark, J.H., and Hunt, A.J. (2010) *Chem. Eur. J.*, **16**, 8047–8053; (g) Marr, A.C. and Marr, P.C. (2010) *Dalton Trans.*, **40**, 20–26; (h) Liu, L., Ma, J., Xia, J., Li, L., Li, C., Zhang, X., Gong, J., and Tong, Z. (2011) *Catal. Commun.*, **12**, 323–326.
53. Dupont, J. (2004) *J. Braz. Chem. Soc.*, **15**, 341–350.
54. (a) Crowhurst, L., Mawdsley, P.R., Perez-Arlandis, J.M., Salter, P.A., and Welton, T. (2003) *Phys. Chem. Chem. Phys.*, **5**, 2790–2794; (b) Hardacre, C., Holbrey, J.D., Nieuwenhuyzen, M., and Youngs, T.G.A. (2007) *Acc. Chem. Res.*, **40**, 1146–1155; (c) Castner, E.W., Wishart, J.F., and Shirota, H. (2007) *Acc. Chem. Res.*, **40**, 1217–1227.
55. Riisager, A. and Fehrmann, R. (2008) in *Ionic Liquids in Synthesis (Green Chemistry)*, 2nd edn, vol. **2** (eds P. Wasserscheid and T. Welton), Wiley-VCH Verlag GmbH, Weinheim, pp. 527–558.
56. Carmichael, A.J., Hardacre, C., Holbrey, J.D., Nieuwenhuyzen, M., and Seddon, K.R. (2001) *Mol. Phys.*, **99**, 795–800.
57. Fitchett, B.D. and Conboy, J.C. (2004) *J. Chem. Phys. B*, **108**, 20255–20262.
58. (a) Rollins, J.B., Fitchett, B.D., and Conboy, J.C. (2007) *J. Chem. Phys. B*, **111**, 4990–4999; (b) Aliaga, C. and

Baldelli, S. (2008) *J. Chem. Phys. C*, **112**, 3064–3072.
59. Fortunato, R., Branco, L.C., Afonso, C.A.M., Benavente, J., and Crespo, J.G. (2006) *J. Membr. Sci.*, **270**, 42–49.
60. (a) Fortunato, R., Afonso, C.A.M., Benavente, J., Rodriguez-Castellón, E., and Crespo, J.G. (2005) *J. Membr. Sci.*, **256**, 216–223; (b) Gottfried, J.M., Maier, F., Rossa, J., Gerhard, D., Schulz, P.S., Wasserscheid, P., and Steinrück, H.P. (2006) *Z. Phys. Chem.*, **220**, 1439–1453.
61. Kolbeck, C., Paape, N., Cremer, T., Schulz, P.S., Maier, F., Steinrück, H.-P., and Wasserscheid, P. (2010) *Chem.–Eur. J.*, **16**, 12083–12087.
62. Lovelock, K.R.J., Villar-Garcia, I.J., Maier, F., Steinrück, H.-P., and Licence, P. (2010) *Chem. Rev.*, **110**, 5158–5190.
63. Cammarata, L., Kazarian, S.G., Salter, P.A., and Welton, T. (2001) *Phys. Chem. Chem. Phys.*, **3**, 5192–5200.
64. Christenson, H.K. (2001) *J. Phys.: Condens. Matter*, **13**, R95–R133.
65. Kanakubo, M., Hiejima, Y., Minami, K., Aizawa, T., and Nanjo, H. (2006) *Chem. Commun.*, 1828–1830.
66. Rodriguez-Perez, L., Coppel, Y., Favier, I., Teuma, E., Serp, P., and Gomez, M. (2010) *Dalton Trans.*, **39**, 7565–7568.
67. Sievers, C., Jimenez, O., Müller, T.E., Steuernagel, S., and Lercher, J.A. (2006) *J. Am. Chem. Soc.*, **128**, 13990–13991.
68. Fortunato, R., Afonso, C.A.M., Reis, M.A.M., and Crespo, J.G. (2004) *J. Membr. Sci.*, **242**, 197–209.
69. Burguete, M.I., Galindo, F., Garcia-Verdugo, E., Karbass, N., and Luis, S.V. (2007) *Chem. Commun.*, 3086–3088.
70. Kamlet, M.J., Abboud, J.L., and Taft, R.W. (1977) *J. Am. Chem. Soc.*, **99**, 6027–6038.
71. Kovtunov, K.V., Zhivonitko, V.V., Kiwi-Minsker, L., and Koptyug, I.V. (2010) *Chem. Commun.*, **46**, 5764–5766.
72. (a) Mehnert, C.P. (2005) *Chem.–Eur. J.*, **11**, 50–56; (b) Riisager, A., Fehrmann, R., Haumann, M., and Wasserscheid, P. (2006) *Eur. J. Inorg. Chem.*, **4**, 695–706; (c) Zhao, F., Fujita, S.-I., and Arai, M. (2006) *Curr. Org. Chem.*, **10**, 1681–1695; (d) Krull, F.F., Medved, M., and Melin, T. (2007) *Chem. Eng. Sci.*, **62**, 5579–5585; (e) Gu, Y. and Li, G. (2009) *Adv. Synth. Catal.*, **351**, 817–847; (f) Van Doorslaer, C., Wahlen, J., Mertens, P., Binnemans, K., and De Vos, D. (2010) *Dalton Trans.*, **39**, 8377–8390.
73. Neves, L.A., Crespo, J.G., and Coelhoso, I.M. (2010) *J. Membr. Sci.*, **357**, 160–170.
74. Kohler, F., Roth, D., Kuhlmann, E., Wasserscheid, P., and Haumann, M. (2010) *Green Chem.*, **12**, 979–984.
75. (a) Lozano, P., Garcia-Verdugo, E., Piamtongkam, R., Karbass, N., De Diego, T., Burguete, M.I., Luis, S.V., and Iborra, J.L. (2007) *Adv. Synth. Catal.*, **349**, 1077–1084; (b) Lozano, P., García-Verdugo, E., Karbass, N., Montague, K., De Diego, T., Burguete, M.I., and Luis, S.V. (2010) *Green Chem.*, **12**, 1803–1810.
76. Hagiwara, H., Kuroda, T., Hoshi, T., and Suzuki, T. (2010) *Adv. Synth. Catal.*, **352**, 909–916.
77. Huang, J., Jiang, T., Gao, H., Han, B., Liu, Z., Wu, W., Chang, Y., and Zhao, G. (2004) *Angew. Chem.*, **116**, 1421–1423.
78. Kernchen, U., Etzold, B., Korth, W., and Jess, A. (2007) *Chem. Eng. Technol.*, **30**, 985–994.
79. Cho, T.H., Fuller, J., and Carlin, R.T. (1998) *High Temp. Mater. Proc.*, **2**, 543–558.
80. Carlin, R.T., Cho, T.H., and Fuller, J. (1998) *Proc. - Electrochem. Soc.*, **98-11**, 180–186.
81. (a) Riisager, A., Wasserscheid, P., Van Hal, R., and Fehrmann, R. (2003) *J. Catal.*, **219**, 452–455; (b) Riisager, A., Eriksen, K.M., Wasserscheid, P., and Fehrmann, R. (2003) *Catal. Lett.*, **90**, 149–153.
82. (a) Riisager, A., Fehrmann, R., Haumann, M., Gorle, B.S.K., and Wasserscheid, P. (2005) *Ind. Eng. Chem. Res.*, **44**, 9853–9859; (b) Riisager, A., Fehrmann, R.,

Flicker, S., van Hal, R., Haumann, M., and Wasserscheid, P. (2005) *Angew. Chem. Int. Ed.*, **44**, 815–819.

83. Haumann, M., Jakuttis, M., Werner, S., and Wasserscheid, P. (2009) *J. Catal.*, **263**, 321–327.
84. Riisager, A., Jorgensen, B., Wasserscheid, P., and Fehrmann, R. (2006) *Chem. Commun.*, 994–996.
85. Ruta, M., Yuranov, I., Dyson, P.J., Laurenczy, G., and Kiwi-Minsker, L. (2007) *J. Catal.*, **247**, 269–276.
86. Werner, S., Szesni, N., Fischer, R.W., Haumann, M., and Wasserscheid, P. (2009) *Phys. Chem. Chem. Phys.*, **11**, 10817–10819.
87. Werner, S., Szesni, N., Kaiser, M., Fischer, R.W., Haumann, M., and Wasserscheid, P. (2010) *ChemCatChem*, **2**, 1399–1402.
88. DeCastro, C., Sauvage, E., Valkenberg, M.H., and Hölderich, W.F. (2000) *J. Catal.*, **196**, 86–94.
89. Valkenberg, M.H., deCastro, C., and Hölderich, W.F. (2000) *Topics in Catalysis*, **14**, 139–144.
90. Valkenberg, M.H., deCastro, C., and Hölderich, W.F. (2002) *Green Chem.*, **4**, 88–93.
91. Jyothi, T.M., Kaliya, M.L., and Landau, M.V. (2001) *Angew. Chem.*, **113**, 2965–2968.
92. Joni, J., Haumann, M., and Wasserscheid, P. (2009) *Adv. Synth. Catal.*, **351**, 423–431.
93. Joni, J., Haumann, M., and Wasserscheid, P. (2009) *Appl. Catal., A*, **372**, 8–15.
94. Hagiwara, H., Sasaki, H., Hoshi, T., and Suzuki, T. (2009) *Synlett*, **4**, 643–647.
95. Mehnert, C.P., Cook, R.A., Dispenziere, N.C., and Afeworki, M. (2002) *J. Am. Chem. Soc.*, **124**, 12932–12933.
96. Mehnert, C.P., Mozeleski, E.J., and Cook, R.A. (2002) *Chem. Commun.*, 3010–3011.
97. Rodriguez-Perez, L., Teuma, E., Falqui, A., Gomez, M., and Serp, P. (2008) *Chem. Commun.*, 4201–4203.
98. Hagiwara, H., Okunaka, N., Hoshi, T., and Suzuki, T. (2008) *Synlett*, **2008**, 1813–1816.
99. Hagiwara, H., Nakamura, T., Okunaka, N., Hoshi, T., and Suzuki, T. (2010) *Helv. Chim. Acta*, **93**, 175–182.
100. Keraani, A., Rabiller-Baudry, M., Fischmeister, C., and Bruneau, C. (2010) *Catal. Today*, **156**, 268–275.
101. Breitenlechner, S., Fleck, M., Müller, T.E., and Suppan, A. (2004) *J. Mol. Catal. A.*, **214**, 175–179.
102. Hagiwara, H., Sugawara, Y., Isobe, K., Hoshi, T., and Suzuki, T. (2004) *Org. Lett.*, **6**, 2325–2328.
103. Burguete, M.I., García-Verdugo, E., Garcia-Villar, I., Gelat, F., Licence, P., Luis, S.V., and Sans, V. (2010) *J. Catal.*, **269**, 150–160.
104. Moucel, R., Perrigaud, K., Goupil, J.-M., Madec, P.-J., Marinel, S., Guibal, E., Gaumont, A.-C., and Dez, I. (2010) *Adv. Synth. Catal.*, **352**, 433–439.
105. Wolfson, A., Vankelecom, I.F.J., and Jacobs, P.A. (2003) *Tetrahedron Lett.*, **44**, 1195–1198.
106. Fow, K.L., Jaenicke, S., Müller, T.E., and Sievers, C. (2008) *J. Mol. Catal. A: Chem.*, **279**, 239–247.
107. Lou, L.-L., Peng, X., Yu, K., and Liu, S. (2008) *Catal. Commun.*, **9**, 1891–1893.
108. Bai, S., Yang, H., Wang, P., Gao, J., Li, B., Yang, Q., and Li, C. (2010) *Chem. Commun.*, **46**, 8145–8147.
109. Lou, L.-L., Yu, K., Ding, F., Zhou, W., Peng, X., and Liu, S. (2006) *Tetrahedron Lett.*, **47**, 6513–6516.
110. Sahoo, S., Kumar, P., Lefebvre, F., and Halligudi, S.B. (2009) *Appl. Catal., A*, **354**, 17–25.
111. Dioos, B.M.L. and Jacobs, P.A. (2006) *J. Catal.*, **243**, 217–219.
112. Castillo, M.R., Fousse, L., Fraile, J.M., Garcia, J.I., and Mayoral, J.A. (2007) *Chem.–Eur. J.*, **13**, 287–291.
113. Doherty, S., Goodrich, P., Hardacre, C., Parvulescu, V., and Paun, C. (2008) *Adv. Synth. Catal.*, **350**, 295–302.
114. (a) Haumann, M., Öchsner, E., and Wasserscheid, P. (2009) DE 102009011815, Germany, p. 13; (b) Öchsner, E. (2010) Entwicklung neuartiger Katalysatorsysteme zur effizienten Herstellung chiraler Verbindungen in der Gasphasenhydrierung. PhD thesis, Friedrich-Alexander-Universität Erlangen/Nürnberg (Erlangen).

115. Leitner, W. and Jessop, P.G. (2010) *Handbook of Green Chemistry*, Supercritical Solvents, Vol. **4**, Wiley-VCH Verlag GmbH, Weinheim.
116. Blanchard, L.A., Hancu, D., Beckman, E.J., and Brennecke, J.F. (1999) *Nature*, **399**, 28–29.
117. Leitner, W. (2000) *Nature*, **405**, 129–130.
118. Leitner, W. (2002) *Acc. Chem. Res.*, **35**, 746–756.
119. Hintermair, U., Leitner, W., and Jessop, P.G. (2010) in *Handbook of Green Chemistry*, Supercritical Solvents, Vol. **4** (eds W. Leitner and P. Jessop), Wiley-VCH Verlag GmbH, Weinheim, pp. 103–188.
120. Bhanage, B.M., Shirai, M., Arai, M., and Ikushima, Y. (1999) *Chem. Commun.*, 1277–1278.
121. Hou, Z., Theyssen, N., and Leitner, W. (2007) *Green Chem.*, **9**, 127–132.
122. Ciriminna, R., Hesemann, P., Moreau, J.J.E., Carraro, M., Campestrini, S., and Pagliaro, M. (2006) *Chem.–Eur. J.*, **12**, 5220–5224.
123. Hintermair, U., Zhao, G., Santini, C.C., Muldoon, M.J., and Cole-Hamilton, D.J. (2007) *Chem. Commun.*, 1462–1464.
124. Hintermair, U., Gong, Z., Serbanovic, A., Muldoon, M.J., Santini, C.C., and Cole-Hamilton, D.J. (2010) *Dalton Trans.*, **39**, 8501–8510.
125. Hintermair, U., Höfener, T., Pullmann, T., Franciò, G., and Leitner, W. (2010) *ChemCatChem*, **2**, 150–154.
126. Hintermair, U., Roosen, C., Kaever, M., Kronenberg, H., Thelen, R., Aey, S., Leitner, W., and Greiner, L. (2011) *Org. Process Res. Dev.*, **15** (6) 1275–1280.
127. Shi, L., Wang, X., Sandoval, C.A., Wang, Z., Li, H., Wu, J., Yu, L., and Ding, K. (2009) *Chem.–Eur. J.*, **15**, 9855–9867.
128. Duque, R., Öchsner, E., Clavier, H., Caijo, F., Nolan, S.P., Mauduit, M., and Cole-Hamilton, D.J. (2011) *Green Chem.*, **13**, 1187–1195.
129. Harwardt, T., Franciò, G., and Leitner, W. (2010) *Chem. Commun.*, **46**, 6669–6671.
130. (a) McMullen, J.P. and Jensen, K.F. (2010) *Org. Proc. Res. Dev.*, **14**, 1169–1176; (b) Parrott, A.J., Bourne, R.A., Akien, G.R., Irvine, D.J., and Poliakoff, M. (2011) *Angew. Chem. Int. Ed.*, **50** (16), 3788–3792.
131. Koch, T.J., Desset, S.L., and Leitner, W. (2010) *Green Chem.*, **12**, 1719–1721.

Part III
Combinatorial and Chemical Biology

New Strategies in Chemical Synthesis and Catalysis, First Edition. Edited by Bruno Pignataro.

13
Inhibiting Pathogenic Protein Aggregation: Combinatorial Chemistry in Combating Alpha-1 Antitrypsin Deficiency

Yi-Pin Chang

13.1
Introduction

This chapter mainly presents one of the projects in the course of my PhD study under the supervision of Prof. Yen-Ho Chu (Department of Chemistry and Biochemistry, National Chung Cheng University, Taiwan) and in collaboration with Dr. Wun-Shaing Wayne Chang (National Institute of Cancer Research, National Health Research Institutes, Taiwan) and Dr. Ravi Mahadeva (Department of Medicine, University of Cambridge, UK). A combinatorial approach was developed and identified for the most potent small-peptide inhibitor of pathogenic α_1-antitrypsin (AT) polymerization, to the best of our knowledge. The other two projects involved the employment of surface plasmon resonance (SPR) to study protein–ligand interactions of streptavidin and VanX between combinatorially selected cyclopeptides and dipeptide phosphonates, respectively, and will be only described briefly. Currently, I am exploring glycobiology and working as a Postdoctoral Research Assistant with Prof. Benjamin G. Davis (Department of Chemistry, University of Oxford, UK).

Proteins are the major components of the molecular machinery in nature, and many of them self-assemble into sophisticated quaternary structures. These complexes form the basis of numerous biological processes and perform a vast array of biological functions that make life possible. In contrast, protein misfolding and aggregation can be lethal, which are responsible for many human disorders that have been recognized as conformational diseases [1]. The gradual accumulation of extracellular fibrils or intracellular inclusions is the cause of a variety of neurodegenerative and non-neuropathic disorders such as Alzheimer's disease (AD), Huntington's disease, Parkinson's disease, amyloidosis, and serpinopathies [2]. α_1-Antitrypsin deficiency (AATD) is the best characterized serpinopathies and the only known genetic disorder that leads to chronic obstructive pulmonary disease (COPD). Dysfunctional mutations of AT result in abnormal protein aggregation in cells and then cause lung and liver diseases, eventually. According to a World Health Organization (WHO) meeting report (available from the WHO web site), the frequency of AATD in Europe and North America is comparable to that of cystic

New Strategies in Chemical Synthesis and Catalysis, First Edition. Edited by Bruno Pignataro.

fibrosis, at 1 in 7000 to 1 in 20 000. Recent epidemiologic estimates suggest that the frequency is high and most populations are affected, but no cure is available for AATD to date. A memorandum of the aforementioned WHO meeting for AATD emphasized the healthcare problem and underscored the need to develop effective therapy. For the past few decades, the crystal structure, major substrate, inhibitory mechanism, pathogenic mechanism, and clinical manifestation of AT have been revealed by several medical research groups. Several therapeutic models have also been suggested including inhibition of the pathogenic polymerization of AT. The strategy was promising, but the lack of molecular diversity and effective screening platform precluded the discovery of potent inhibitors. The gap between medical doctors and chemists came closer when Dr. Wun-Shaing Wayne Chang bridged Dr. Ravi Mahadeva to my PhD supervisor's group. To cope with this clinical problem, we set out to introduce combinatorial chemistry, which we believe is the established technique best suited for the identification of anti-AT polymerization ligands.

The renaissance of combinatorial chemistry has intimate relations with the development of solid-phase synthesis. In 1963, the Nobel Laureate Bruce Merrifield [3] pioneered solid-phase synthesis and described the preparation of a tetrapeptide. The growing peptide chain was covalently linked to an insoluble resin bead and elongated stepwise. Subsequently, combinatorial synthesis mushroomed in parallel in several labs and these approaches emerged almost simultaneously. Leznoff [4, 5] synthesized small molecules by solid-phase synthesis. Frank [6] came up with the idea to synthesize nucleotides and later peptides on circles of cellulose paper. Geysen [7, 8] performed solid-phase synthesis on polymeric rods (pins) arranged in a 96-well microtiter plate format, whereas Houghten [9] carried out peptide synthesis on "tea bags" (resin beads in meshed polypropylene packets). Furka [10] devised the concept of split-and-mix method, and it was applied to the solid-phase synthesis of peptide libraries. Lam and coworkers [11] conceived the method to prepare peptide libraries and screening on beads, whereas Houghten [12] used the split-and-mix approach and tea-bag chemistry to make soluble-peptide libraries. The Ugi [13] multicomponent reaction was a forerunner of the current solution-phase combinatorial chemistry and has also been employed in solid-phase synthesis. Chemists endeavored to produce more molecules in a short period and the advances in synthetic organic chemistry have made this task feasible. The major components and challenges in combinatorial chemistry are the design of a library, development of screening method, characterization of lead compounds, and further rational design of the leads.

On the basis of the structural information and mechanism of disease, several small-peptide libraries were prepared and screened initially to explore the binding interfaces of the wild-type and mutant AT. This preliminary screening was critical to define the structural requirements for the construction of larger libraries. In addition, several attempts have been made and a conformation-sensitive assay was eventually developed, which played a crucial role in the combinatorial approach. Through the evolution of libraries from libraries, potent anti-AT polymerization inhibitors were identified systematically. The combinatorially selected inhibitor was further characterized by biophysical, cellular, and computational methods. These biochemical data indicate a potential therapeutic strategy to ameliorate AT-related emphysema and cirrhosis. Additionally, because AT is the archetypical member of the serpin superfamily, the developed combinatorial approach is not limited to AT, but is also applicable to other serpinopathies and even other conformational diseases.

13.2 α_1-Antitrypsin Deficiency

Protein misfolding and the associated uncontrolled/unwanted self-aggregation is the starting point of many human diseases. The resulting ordered and stable protein aggregates can accumulate in a variety of cells, tissues, and organs and are known to cause Alzheimer's disease, amyloidosis associated with hemodialysis, Creutzfeld–Jacob disease, dementia with Lewy bodies, diffuse Lewy bodies disease, fatal familial insomnia, Gerstmann–Straussler–Schneiker syndrome, Hallervorden–Spatz disease, Huntington disease, Kuru, Lewy bodies variant of AD, light chain-associated amyloidosis, light chain deposition disease, multiple system atrophy, neuronal intranuclear inclusion disease, Parkinson's disease, spinal and bulbar muscular atrophy, and spinocerebellar ataxia [14]. Apart from these clinical conditions, there is a unique category of disorders derived from the serpin (*ser*ine *p*rotease *in*hibitor) superfamily and termed *serpinopathies* accordingly [15]. Among the members of serpin, AT represents the archetype and is associated with several lung and liver manifestations [16–18]. These pathological conditions are known collectively as *AATD*, which was first described in 1963 and is suspected to be responsible for the premature death of Frédéric François Chopin [19–21]. Epidemiologic estimates suggest that 1 in 1600 to 1 in 5000 individuals are affected in most populations, and approximately 200 000 Americans and Europeans suffer from the severe type of AATD [22]. Nevertheless, the variable clinical manifestations of AATD are commonly underdiagnosed or misdiagnosed [23]. The slow accumulation of these aggregates and the acceleration of their formation by stress explain the characteristic late onset of the conformational disease [24]. The understanding of the mechanism of diseases at the molecular level can shed light on these biological problems and open prospects of rational approaches to therapy [25].

13.2.1 α_1-Antitrypsin and Serpin

AT is the most abundant protease inhibitor in the circulation, and is primarily synthesized and secreted by the hepatocytes in the liver and lesser amounts by enterocytes, mononuclear phagocytes, neutrophils, and respiratory epithelium [26–28]. The structure of native AT is illustrated in Figure 13.1. This 52 kDa single-polypeptide-chain glycoprotein consists of 394 amino acids and serves as a suicide substrate to inhibit the cognate enzyme neutrophil elastase (NE) [29]. Insufficient circulating level of AT is unable to neutralize NE and predispose

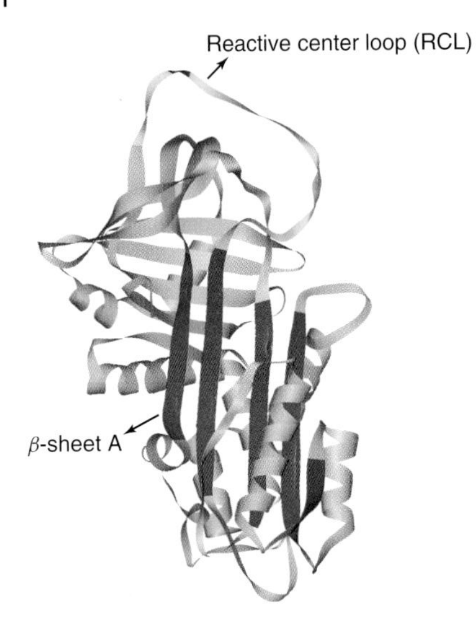

Figure 13.1 Structure of α_1-antitrypsin (PDB code: 1QLP). The reactive central loop and the β-sheet A are marked in yellow and red, respectively. The five strands in the β-sheet are termed *s1A*, *s2A*, *s3A*, *s5A*, and *s6A* (from left to right). The s4A site is blank in the native form of α_1-antitrypsin.

toward bronchiectasis, emphysema, and vasculitis [30–32]. It exists in a number of genetic forms, and the MM variant is the most common one. Point mutations in the AT gene such as the I (Arg39Cys), Mmalton (ΔPhe52), S (Gly264Val), Siiyama (Ser53Phe), and Z (Glu342Lys) alleles can result in mild to severe plasma deficiency of AT and manifest as lung and/or liver diseases [33]. Among these alleles, the Z variant of AT causes the most significant deficiency (about 90%) compared with the wild-type M allele. It has been estimated that at least 116 million carriers are with the mild combination alleles of MS and MZ, while 3.4 million individuals are with the SS, SZ, and ZZ severe alleles worldwide [34]. Moreover, AATD is not limited to Caucasian and may be one of the most common but underdiagnosed hereditary disorders in the world. The molecular basis of AATD lies in the structural rearrangement of the mutant protein and is the key to developing therapeutic approaches.

Approximately 1500 serpins (36 serpins in human) have been identified in different biological kingdoms, and their function, mechanism, and pathogenesis are directly related to the metastable protein structures [35]. Regardless of the relatively low similarity of the primary structures, the secondary and tertiary structures of serpins exhibit great homology and share a conserved core domain, which include the three dominant β-sheets (A, B, and C) and nine α-helices (A through I) [36]. Apart from these defined secondary structures, a distinct loop can be found that is known to adopt a variety of conformations in different serpins. In the native state of inhibitory serpins such as AT, a unique 17-residue reactive center loop (RCL) is exposed whose function is to entrap its targeted protease NE.

Following the cleavage of the scissile bond P1–P1′ (Met358–Ser359) in RCL by the protease, the cleaved RCL inserts between strands 3 and 5 of the β-sheet A as an additional strand 4 (s4A). This event concomitantly translocates the NE more than 70 Å to the opposite end of AT with the formation of a 1 : 1 enzyme:inhibitor complex, and renders the β-sheet A fully antiparallel [37–41]. This is in contrast to most of the other classes of protein protease inhibitors, which employ a lock-and-key arrangement to form tight but reversible noncovalent complexes between proteases and their corresponding inhibitors. The inhibitory mechanism of serpins basically consists of two stages: (i) rapid assembly of a noncovalent Michaelis complex, attack of the scissile bond to form a tetrahedral intermediate, and the subsequent cleavage to give the covalent acyl-enzyme intermediate; and (ii) insertion of the cleaved RCL into β-sheet A and stabilization of the second tetrahedral intermediate through the attack of water [42–46]. The most striking characteristics of the serpin inhibitory mechanism are the dramatic conformational change and irreversible and suicidal nature of inhibition, in which the RCLs of serpins play an essential role. The understanding of the molecular events of serpins is important for the subsequent design of a combinatorial library.

13.2.2
The Polymerization Pathways of Serpins

It has been a mystery how serpins aggregate, and a few polymerization pathways have been suggested. Most likely, the mobile RCLs are at the junctions of the aggregates. The pathogenic polymers of serpins derived from the intermolecular linkages between RCLs and β-sheets constitute *loop-sheet polymerization* [47]. In spite of the pivotal role of RCLs in the inhibitory function, a single-nucleotide mutation may perturb the protein structure and render the β-sheet susceptible to be inserted by the RCL of another molecule of serpin. For example, it has been suggested that the salt bridge Glu342–Lys290 of AT is disrupted by the Z mutation, in which the glutamic acid at position 342 has been substituted by lysine [48, 49]. The substitution of the hinge residue could not only extend the RCL but also expand the β-sheet as a donor and receptor for polymerization, respectively [50]. The assembled dimer is available for linkages by other molecules through its protruded RCL and unoccupied β-sheet and serves as the root for the formation of oligomers and eventually polymers. This spontaneous loop–sheet polymerization has been widely accepted as the molecular basis of Z-AT-associated liver and lung diseases. Intracellular polymers and inclusion bodies of Mmalton, Siiyama, and Z alleles have been identified within the endoplasmic reticulum (ER) of hepatocytes, where AT is synthesized and cause liver damages [50–52]. In addition, the ER retention of polymeric AT results in the secretory defection and causes plasma deficiency, which renders the lungs vulnerable to elastolytic damage [35, 50]. The imbalance between AT and NE leads to emphysema and other lung diseases [53–55]. Given the conserved structure and the inhibitory function of serpins, the similar mechanism of disease is not an exclusive of AT. Other pathogenic polymerization of serpins has also been reported, including α_1-antichymotrypsin (ACT;

Leu55Pro and Pro228Ala), antithrombin (ATIII; Pro54Thr and Asn158Asp), C1 inhibitor (Phe52Ser, Pro54Leu, Ala349Thr, Val366Met, Phe370Ser, and Pro391Ser), neuroserpin (Ser49Pro) in association with emphysema, thrombosis, angioedema, and dementia, respectively [56–59]. The process of neuroserpin polymerization and the resulting inclusion body was also recognized, and is termed familial encephalopathy with neuroserpin inclusion bodies (FENIB) [59].

Crystallographic structures of serpins provided perceptions of the serpin polymerization pathways. It appears that the nature of serpin RCLs accounts not only for their function but also for the pathogenesis of serpinopathies. The structure of RCLs varies among serpins, such as the helical loop in ovalbumin [60], the β-strand in the latent form of plasminogen activator inhibitor-1 (PAI-1) and antithrombin [61, 62], and the distorted helix in α_1-ACT [63]. Apart from the canonical stable structure, the RCL of AT can adopt a β-strand conformation and is the cornerstone of the pathogenic polymerization [64]. A few crystal structures and molecular models have been reported that provided insights into the propagation of polymers. The suggested model of extended AT polymers was in agreement with the flexibility of inclusions observed by electron microscopy [51, 64]. In addition, the structure of cleaved AT polymer has given implications for the mechanism of conformational diseases [65]. While the classical loop–sheet polymerization pathway remained an attractive explanation for the pathogenesis of serpinopathies, the most recent crystal structure of a stable antithrombin dimer suggested a new concept for the molecular basis of serpin polymerization [66]. Rather than the single RCL insertion, a β-hairpin of RCL and the adjacent strand 5 participate the domain swapping to form the dimerized structure. The new domain-swapping structure can be easily adopted and gives clues for how other serpins aggregate [67]. Taken together, the answers to the long-sought-after serpin puzzles lie in those unique RCLs among serpins [68–70].

13.2.3
Emerging Therapeutic Strategies

Several promising strategies for AATD have been suggested in the past few years, which include gene therapy [71, 72], promotion of hepatic AT secretion [73, 74], inhibition of NE [75–78], prolongation of AT half-life [79], and inhibition of AT polymerization. The last strategy has demonstrated, at least *in vitro*, that serpin polymerization can be attenuated by cavity-filling stabilization, chemical chaperones, small-molecule binding, and peptide annealing. Cavity-filling mutations of AT (T114F or G117F) on strand 2 of β-sheet A could stabilize the protein and retard polymer formation [80]. In addition, these identified cavities are potential targets for rational drug design [81–84]. It has also been shown that high concentrations of carbohydrates, glycerol, and trimethylamine *N*-oxide reduce the polymerization rate of serpins [85–88]. By acting as chaperone, 4-phenyllbutyrate (4-PBA) mediated a significant increase in Z-AT secretion from cells in tissue-culture and transgenic mice [74]. However, 4-PBA did not increase AT levels in human plasma and caused symptomatic and metabolic side effects in patients with AATD [89]. A few small molecules were identified by in silico screening with 1.2 million commercial drug-like compounds and found to inhibit the polymerization of Z-AT [90]. Finally, the peptide annealing method, conceived from the idea of binding serpins with synthetic peptides containing RCL sequences, has led to the development of combinatorial library screening. A number of potent ligands were found to bind to Z-AT tightly and block the pathogenic polymerization. To date, no cure or effective treatment for patients with AATD is available yet; only supplemental treatments temporarily alleviate the symptoms. In broad terms, conventional therapy includes standard interventions for generic COPD and augmentation therapy [23]. The efficacy for treatment of the lung disease is under detailed evaluation [91, 92]. For patients with severe conditions, lung or liver transplantation is the last resort [18, 93]. The predicament of AATD underscores the need to develop effective treatment.

13.3 Targeting the s4A Site with the Peptide Annealing Method

The peptide annealing method is basically "an eye for an eye" strategy. In view of the fact that the inhibitory mechanism of serpins is conducted by the incorporation of the RCL into the corresponding β-sheet, it appears that an RCL fragment alone is capable of binding to a certain site and thus intervening in the propagation of polymerization. The concept of using RCL peptide analogs to block polymerization by direct competition for binding in the s4A position is therefore logical. Initially, a few functional and structural studies were carried out by the insertion of synthetic RCL peptides, which can be taken as the proof of concept of the method. By mimicking the inhibitory mechanism of serpin, it has been shown by several groups that serpin polymerization can be attenuated by competing RCL peptides and derivatives. On the basis of this concept, we have developed a combinatorial approach targeting the s4A site in the β-sheet A of AT. The combinatorially selected peptides not only annealed to the susceptible target but also arrested the pathogenic polymerization. The results revealed that most likely the s4A position of AT is the "Achilles" heel' of AATD and is a promising target for high-throughput screening and structure-based drug design [94].

13.3.1 Functional and Structural Studies of RCLs

Ever since Laurell and Eriksson first described AATD in 1963, researchers have been fascinated by the molecular events related to serpins until today. The major clinical features of AATD was established by a survey of 1500 serum protein electrophoresis gels, in which three of the five gels with the absence of the α_1 band were from patients with early onset emphysema [19]. The advances in biophysics, cell biology, and structural biology have revealed the crystal structure, major substrate, inhibitory mechanism, pathogenic mechanism, and clinical manifestation of AT during the past few decades. All in all, serpin RCLs play crucial roles in many aspects, and segments of RCLs have been synthesized to explore the structural and functional nature of serpins. In retrospect, quite a few RCL peptides (mostly 8–16-mer) of AT, ATIII, and PAI-1 were synthesized to probe the conformational stability, inhibitory activity, physical properties, polymerization mechanism, and structural transition of serpins in the 1990s [95–108]. The detection of the binary complex (BC) formation was basically done by monitoring the change of molecular weight, conformational change, and enzymatic activity and was extensively achieved by spectroscopic techniques (circular dichroism and tryptophan fluorescence), polyacrylamide gel electrophoresis, PAGE (SDS and native PAGE), and AT activity assessment (residual tryptic activity), respectively. An adequate screening approach remains to be devised at this stage.

These RCL peptides were useful for functional and structural studies, but were not suitable for the development of mimetics from a drug discovery perspective. Firstly, the binding affinity of these peptides is poor, typically required high concentrations (100–200-fold molar excess) and long periods (usually days) of incubation to form BCs. In addition, these RCL-derived peptides are promiscuous and could not only bind to their corresponding protein but also be cross-recognized by other serpins, such as an ATIII RCL peptide binds to ATIII, ACT, and AT [106, 108–112]. The unpredictable specificity, poor affinity, and large size preclude these peptides from being therapeutically significant. On the other hand, it established the prototype of the screening approach and inspired the peptide annealing method. A benchmark peptide remained to be identified at this stage.

13.3.2
Smaller RCL-Derived and Non-RCL Serpin-Binding Peptides

No significant progress was made until a 6-mer RCL peptide was used to explore the subtle structural differences between the pathogenic Z-AT and normal M-AT [110]. The assessment of peptide binding and BC formation was achieved by intrinsic tryptophan fluorescence and native PAGE as in previous related works. In addition, doping urea as the additive in native PAGE was able to improve the readout and distinguish the bound and unbound protein unambiguously. The effect of the chaotropic reagent was imperative for this conformation-sensitive gel electrophoresis and generated distinct band shifts. Clear BC of the hexapeptide and Z-AT was observed on 8 M urea PAGE, and no polymers of Z-AT were found on native PAGE. Even though the sequence was derived from the AT RCL, the short peptide preferentially annealed to Z-AT and did not significantly bind to other serpins (ATIII, ACT, and PAI-1) that bear the conserved tertiary structure. The binding specificity of peptide is critical because cross-binding could not only hamper the biological function but also accelerate the pathogenic polymerization of other serpins [109]. Taken together, these findings highlighted the implication for the prevention of Z-AT-related cirrhosis yet without intervention of other serpins.

From the point of view of drug discovery, serpin-binding peptides may serve as templates for mimetic design. Defining the structural requirements for binding is beneficial in fine-tuning both affinity and specificity. It was reported that the conserved P8 threonine and a key hydrogen-bond network centered on His334 that bridges strands 3 and 5 of the A-sheet play crucial roles in maintaining the metastable conformation of serpins [111]. The recombinant variants of AT (H334A and H334S) have a large decrease in melting point (T_m) and more readily form polymers or annealed by RCL peptides. To further investigate how small peptides interact with serpins, the binding of around 40 RCL and exogenous (randomly selected from commercial sources) peptides were assessed for the structural study [112]. Among these peptides, the threonine equivalent to the P8 site was found to facilitate the entry and anchoring of peptide annealing into the s4A position. The crystal structures of the ATIII ternary complexes also demonstrated that the insertion of hydrophobic side chains into the P4 and P6 sites of s4A is important. The acquired structural information represents a step forward toward the rational design of small molecules.

13.4 Expanding the Molecular Diversity

The search of serpin-binding peptides has been restricted to limited sources of RCL derivatives, basically randomly selection and to some extent rational design. The scarcity of compounds for screening has precluded the identification of potent ligands. In addition, the lack of appropriate screening platform with higher throughput and efficacy was one of the main challenges in this area. To this end, combinatorial chemistry was first introduced not only to expand the molecular diversity but also to increase the throughput of screening. The significance of combinatorial chemistry lies in the unprecedented molecular diversity that originated from the solid-phase synthesis of peptides and thrived on the construction of peptide libraries [3, 113]. The screening and characterization of the immense number of compounds in combinatorial library require adequate analytical methods and is often done on a case-by-case basis. For example, throughput, nondestructive analysis, in situ monitoring, and non-hindrance of protein–ligand interaction are some important concerns of library screening. Initially, several small peptide libraries were synthesized, which facilitated the development of the screening method. The conformation-sensitive 8 M urea PAGE was used throughout the course of the combinatorial approach. The alanine scanning experiment was first started to systematically probe the structural requirements for effective binding based on the previous benchmark peptide Ac-FLEAIG-OH. The subsequent truncation library not only led to the identification of a strong Z-AT-binding peptide Ac-FLAA-NH_2 but also defined the scaffold (acetylated 4-mer peptide amide) for subsequent synthesis of the combinatorial library. A combinatorial screening strategy, namely, iterative deconvolution, was employed to determine the optimal residues in peptide libraries. The deconvolution procedure is a strategy

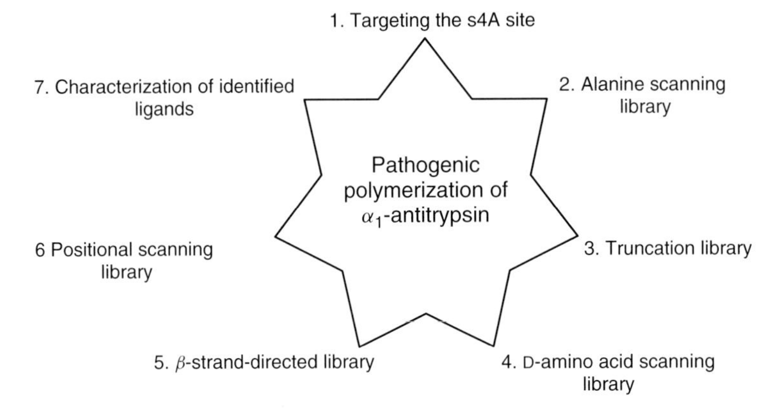

Figure 13.2 Combinatorial approach (evolution of libraries) targeting at the s4A site to arrest the pathogenic polymerization of α_1-antitrypsin.

that, by screening the activity of each predefined sublibrary, defines the synthesis of the next-generation libraries for subsequent screening. The stepwise cycles of resynthesis and screening were repeated until the optimal peptide sequences for binding to AT were identified. Finally, one tetrapeptide (Ac-TTAI-NH_2) emerged from the β-strand-directed library of 10 000 peptides, which preferentially annealed to and blocked the pathogenic polymerization of AT. The overall strategy of the combinatorial approach is illustrated in Figure 13.2.

13.4.1
Alanine Scanning, Truncation, and D-Amino Acid Scanning Libraries

The RCL peptide (Ac-FLEAIG-OH; P7–P2) of AT was selected as the benchmark and starting point of the combinatorial approach. First, this hexapeptide binds to Z-AT and inhibits its polymerization. Second, no obvious conservation is found in the corresponding P7–P2 region in the RCL sequence of human serpins [114]. The diverse region in human serpins is expected to play a pivotal role in both binding affinity and specificity for each protease inhibitor system. Using the sequence corresponding to this region as a template to develop derivatives and probe the potential recognition site is a promising approach to obtain structural information for peptide annealing. Additionally, prior characterization of lead compounds is generally useful for the design of combinatorial libraries. As a result, we took advantage of the established technique termed *alanine scanning*, as it has been used extensively to identify critical residues for protein conformation, function, and stability [115]. The alanine scanning library of the benchmark peptide was first prepared to investigate the binding interface of AT. The smallest chiral amino acid alanine was used to substitute each non-alanine residue based on the hexapeptide one at a time to construct the small library for screening. The only alanine in the hexapeptide was replaced by glycine. This library allowed us to rapidly determine each individual residue's contribution to the binding and identified two extra hexapeptides (E3A: Ac-FLAAIG-OH and I5A: Ac-FLEAAG-OH) which bound to M and Z-AT [116]. The hydrophobic residues appeared to be critical at the corresponding P6 (leucine and alanine) and P4 (alanine) sites, and the achiral glycine was not favored in the middle of the peptide sequence. The result indicates that fully extended peptides are favored in the binding interface.

The two identified hexapeptides were used to develop their corresponding truncation libraries. Truncation libraries allow the determination of the minimum length required for specific bioactivity [117, 118]. The incremental truncation of the two identified hexapeptides led to the identification of the two daughter peptides Ac-FLEAA-NH_2 and Ac-FLAA-NH_2, which bound avidly to Z-AT and blocked the polymerization [116]. Consequently, a 4-mer peptide was found to be sufficient to interact with AT. The effect of D-amino acids on the identified peptide Ac-FLAA-NH_2 was also evaluated. Under physiologic conditions peptides are susceptible to proteolysis and are usually considered as poor drug candidates. Owing to the resistance to proteolysis, D-amino acid substitution is regularly used to resist proteolytic degradation [119]. In addition, substitution of D-amino acids introduces a conformational change in the peptide backbone, which is useful for structural studies. The D-amino acids were not only substituted one at a time but also fully substituted to give the peptide Ac-$_D$F$_D$L$_D$A$_D$A-NH_2 for screening. None of the D-substituted peptides was able to form a BC with Z-AT, suggesting a stereochemical constraint of the Z-AT–peptide interface [116]. The aforementioned systematic approaches were essential for the library design. These small libraries of alanine scanning, truncation, and D-amino acid scanning peptides defined the structural requirements and minimal peptide length for the next generation of library, that is, the acetylated 4-mer peptide amide backbone.

13.4.2
The β-Strand-Directed Library

Having defined the optimal scaffold of the library by the aforementioned small-peptide libraries, the selection of building blocks was the next step for the synthesis of a larger library if not all of the 20 amino acids were to be used. The pendulum of library screening is swinging back and forth between rational design and molecular diversity. While molecular diversity represents the beauty of combinatorial chemistry, integrating rational design into library is essential in certain cases. In principle, a combinatorial peptide library can attain the maximum molecular diversity if all the 20 proteinogenic amino acids are incorporated as the building blocks. Constructing all the theoretical peptide sequences diligently is an aesthetic goal, but some potential problems and practical issues may arise that have to be taken into account. In these cases, the information of how to exquisitely design a library is pivotal. On the basis of the previous screening conditions, a conservative estimate indicated a potential solubility problem might occur if the building blocks were not selected properly [116]. As the screening of libraries was designed to perform not on solid supports but in solution, partially soluble libraries would be impractical despite the larger theoretical number of compounds. As a result, a selection criterion of building blocks was necessary not only to control the library size but also to wisely design the library in a guided direction.

Insights of the inhibitory mechanism of serpins provide an excellent criterion for the selection of building blocks of the library. In the light of the inhibitory mechanism, RCLs of serpins are flexible and can adopt a number of diverse conformations, but upon snaring their substrates RCLs insert into their corresponding β-sheets and render them a fully antiparallel β-sheet conformation. Therefore, it is reasonable that the residues in the RCLs of inhibitory serpins must adopt a conformation that is favored by the β-sheet to complete the inhibitory processes. As each amino acid has a relative frequency in the α-helix, β-sheet, and turn, most likely the β-sheet-preferring amino acids are more compatible with the targeted s4A site. In order to confirm the hypothesis, the RCL sequences of human serpins were analyzed in conjunction with the Chou–Fasman helix [120] and sheet propensities in the prediction of secondary structure. The analysis showed that 60.1% (69.0 and 72.8% in the P12–P3′ and P7–P2 segments, respectively) of the RCL residues of inhibitory serpins consist of the top 10 β-sheet-forming amino acids (A, F, H, I, L, M, T, V, W, and Y). On the other hand, only 38.3% of the residues were found in the RCLs of non-inhibitory serpins. The diverse trend of the preference of amino acids is clear and supported the hypothesis. Consequently, the 10 amino acids with higher frequency in β-sheets were selected as the building blocks for library construction.

Solid-phase synthesis and the split-and-mix method were employed to construct the β-strand-directed library [121]. The synthetic strategy of this method can rapidly produce a greater number of compounds than with the parallel synthesis approach. However, the screening of a mixture of compounds rather than individual molecules required a judicious screening strategy to identify the active compounds in a muddle. A representative and simplified example of a $3 \times 3 \times 3$ library is shown in Figure 13.3, which gives all the 27 possible sequences of tripeptides. The order of magnitude could be increased dramatically if more amino acids and reaction cycles were introduced. In the first generation of the β-strand-directed library (designated as Ac-$X_1X_2X_3X_4$-NH_2), 10 amino acids were used for the synthesis of the tetrapeptide libraries and this represents a $10 \times 10 \times 10 \times 10$ library (10 000 theoretical peptides). The synthesis of the peptides was initiated at the X_4 position and terminated at the X_1 position, thus the elongation of peptide chains was from the C-terminal to the N-terminal. Note that, upon completion of the fourth reaction cycle (coupling of the amino acids at the X_1 position), the resulting 10 sublibraries (each containing 1000 peptides) were not mixed and thus the 10 N-terminal amino acids (X_1) were known. As a result, the first generation of the acetylated tetrapeptide library consisting of 10 sublibraries with diversity in the last three residues (X_2, X_3, and X_4) was represented by Ac-A$X_2X_3X_4$-NH_2, Ac-F$X_2X_3X_4$-NH_2, Ac-H$X_2X_3X_4$-NH_2, Ac-I$X_2X_3X_4$-NH_2, Ac-L$X_2X_3X_4$-NH_2, Ac-M$X_2X_3X_4$-NH_2, Ac-T$X_2X_3X_4$-NH_2, Ac-V$X_2X_3X_4$-NH_2, Ac-W$X_2X_3X_4$-NH_2, and Ac-Y$X_2X_3X_4$-NH_2. Each randomized position (X_i) was a composition of the 10 amino acids (A, F, H, I, L, M, T, V, W, and Y). The next three

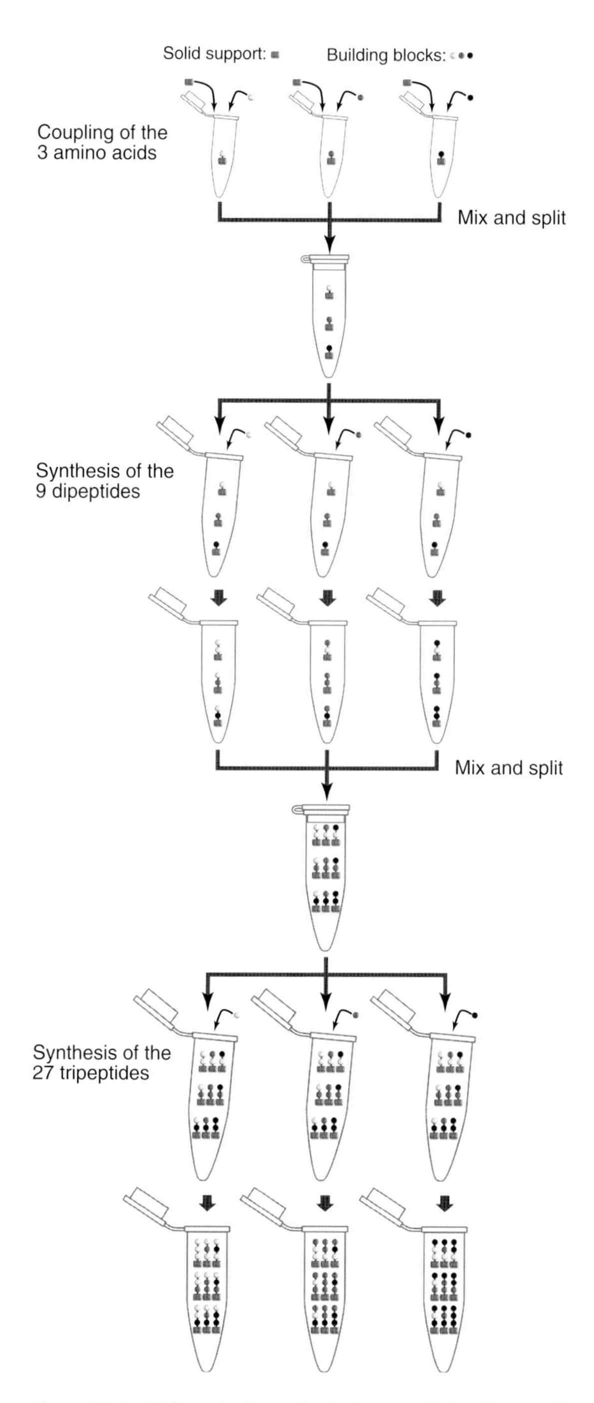

Figure 13.3 Split-and-mix method showing that a $3 \times 3 \times 3$ library with 27 peptides can be achieved in three synthetic cycles.

generations of libraries were synthesized on the basis of the results of iterative deconvolution [122].

Iterative deconvolution was employed to identify the optimal residues stepwise for the bindings to M and Z-AT by cycles of screening and synthesis of libraries. The screen result of the first β-strand-directed library was used to define the next synthesis/screening of library, and so on. Four cycles of the syntheses and screenings of libraries were repeated. In the course of iterative deconvolution, the size of libraries was decreased exponentially until all the peptide sequences were defined in the final library screening (from 10 000 to 1000, then 100, and finally 10 peptides). In other words, except the last cycle of screening, all the previous screenings were performed with the predefined mixtures, and only the last step of screening was against 10 single peptides and hence the optimal sequence could be identified. As shown in Figure 13.4, the first three cycles of screening identified the residues of T, T, and A for the bindings to Z-AT from the libraries of Ac-$X_1X_2X_3X_4$-NH_2, Ac-T$X_2X_3X_4$-NH_2, and Ac-TTX_3X_4-NH_2, respectively. In the last cycle of screening, the optimal Z-AT and M-AT binding peptide was revealed as Ac-TTAI-NH_2 and Ac-TTAF-NH_2, respectively, from the final library of Ac-TTAX_4-NH_2 [123]. This result demonstrated the power of the developed combinatorial approach, as different ligands of M-AT and its point mutant Z-AT were able to be identified. In addition, very recently, the identified peptide was used to support the classical loop–sheet polymerization model. However, the peptide was mistakenly described as the reactive loop tetrapeptide, rather than combinatorially selected ligand [69].

13.4.3 The Positional Scanning Library

A potential problem of the split-and-mix and iterative deconvolution methods is that the most potent compound may not be identified because of synthetic and/or screening artifacts. These effects should be taken into account particularly if the potency of each sublibrary cannot be distinguished unambiguously. On reviewing the four cycles of the screening of β-strand-directed library, the second residue (X_2) was not easily identified without the quantitative densitometric analysis [123]. Although threonine was eventually incorporated for the syntheses and identified potent inhibitors, we suspected whether the selected amino acid was truly the best at the X_2 residue. Positional scanning library is a useful tool for peptide sequence optimization [124]. There are several types of positional scanning library but basically it screens the amino acid(s) of interest at given position(s). We used this technique to confirm the screening results of the β-strand-directed library. A positional scanning library (Ac-TX_2AF-NH_2; X_2 = A, H, I, L, M, F, T, V, W, and Y) was prepared and screened against M-AT. The positional-scanning-retrieved peptide Ac-TTAF-NH_2 was identified, which was identical to that from the result of iterative deconvolution [123]. These data suggested that the incorporation of threonine at the X_2 position was appropriate (unpublished result).

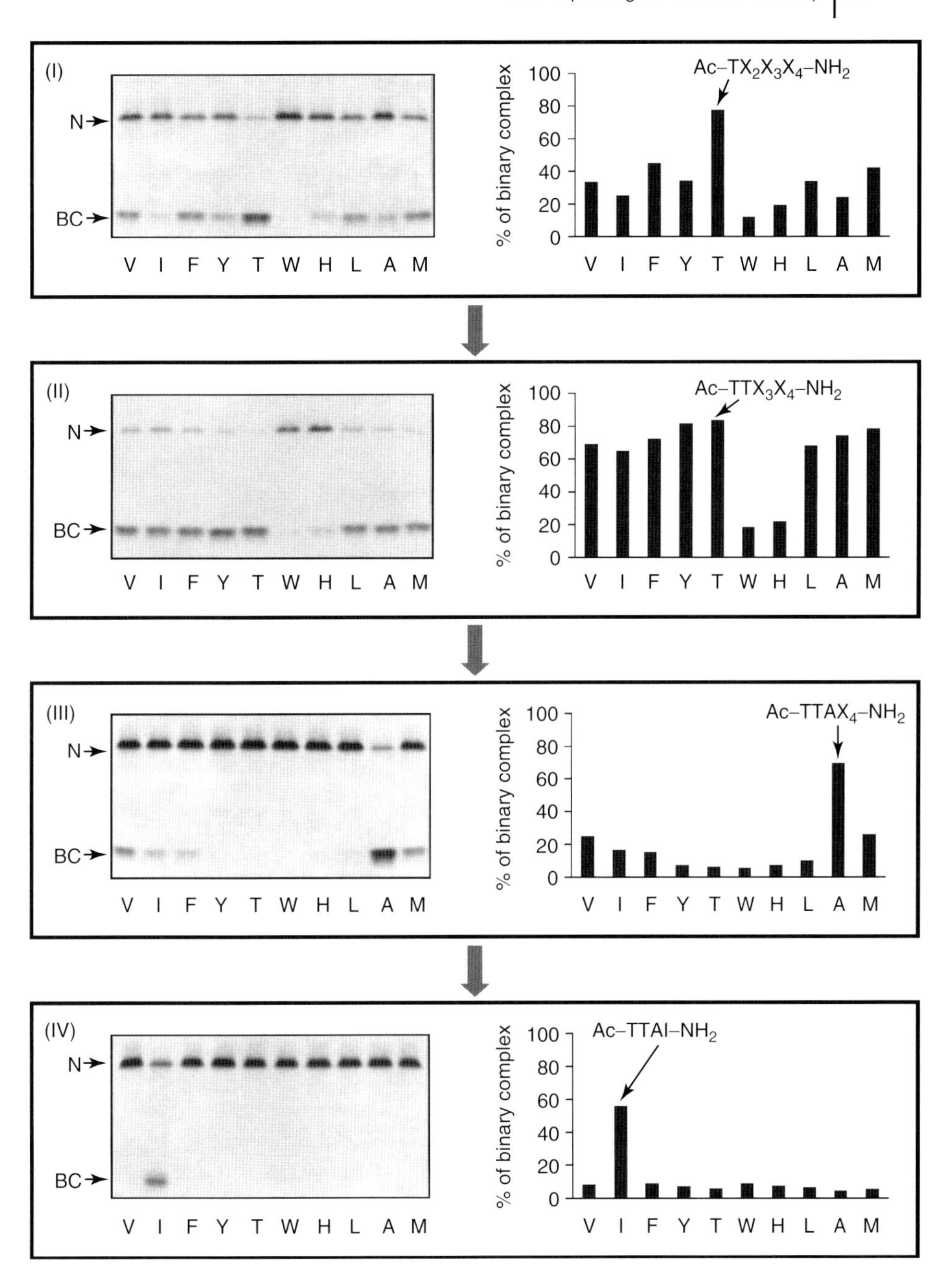

Figure 13.4 Iterative deconvolution of the β-strand-directed tetrapeptide library against Z α_1-antitrypsin. The potency of each sublibrary was determined by 8 M urea gels and further quantitatively measured by densitometric analysis to give the percentage of binary complex formation.

13.5 Characterization of the Combinatorially Selected Peptide

The characterization of combinatorially selected ligands is important for future mimetic design. The combinatorially selected peptide Ac-TTAI-NH_2 was further characterized by biophysical, cellular, and computational methods to determine its intrinsic properties. SPR was employed to validate the binding affinity and specificity of the identified Z-AT binder orthogonally. The cytotoxicity of the peptide was assessed in several types of human cell lines by MTT (3-(4,5-dimethylthiazol-2-yl)-2,5-diphenyltetrazolium bromide) cell proliferation assays. A molecular model has also been suggested to elucidate the noncovalent interactions of the combinatorially selected peptide in detail.

13.5.1 Validation of the Binding by SPR

The binding affinity and specificity of the identified peptide was further confirmed by the SPR technique. This chip-based biosensor is a powerful tool for biomolecular interaction analyses and has served as an excellent platform to validate the gel-based library screening with a different principle. SPR allows the monitoring of protein–ligand interaction in real time and without the need of tags or labels [125]. Two of my previous works have also used SPR for the study of protein–peptide/small molecules interactions. The equilibrium dissociation constants (K_D) were quantitatively measured, which revealed that cyclopeptides bind more tightly (approximately 1000-fold) than their linear counterparts to streptavidin. The advantage of conformational constraint in enhancing protein–ligand affinity was also demonstrated [126]. The binding kinetics (k_{on} and k_{off}) of low molecular weight VanX inhibitors was clearly analyzed. The stronger affinity of these transition-state analogs over its natural substrate represents a step forward to cope with vancomycin resistance [127]. Back to the validation of combinatorially selected Z-AT binding peptide, the macromolecular Z-AT protein was deliberately immobilized onto the sensor chip to avoid any hindrance of protein–peptide interactions that might result from the immobilization chemistry. In addition, the protein chip is reusable and applicable to massive screening of peptides and small molecules if the regeneration protocol can be optimized. The results showed that the binding of the peptide Ac-TTAI-NH_2 to the immobilized Z-AT was specific and dose dependent. The negative control Ac-WWWH-NH_2 was also combinatorially selected (the least active residues from library screening) but hardly recognized by the immobilized Z-AT even at the highest concentration of the performed injections. The SPR results clearly confirmed not only the specificity but also the affinity of the combinatorially selected peptide Ac-TTAI-NH_2 to Z-AT [123].

13.5.2 Cytotoxicity of the Identified Peptide and the Proposed Structure of the Binary Complex

The cytotoxicity of the identified peptide was determined by MTT assays against two normal lung epithelial cell lines (BEAS-2B and NL20), two normal lung fibroblast cell lines (WI-38 and IMR-90), and one cancer cell line (A2058). The viability of cells was not affected even at a high concentration of 10 μM of the peptide Ac-TTAI-NH_2. This result indicated that this small peptide was not cytotoxic, which is excellent for future cell-based assay and development of mimetics [123].

A structure of the BC of the identified peptide and AT was proposed to predict the binding site and explain the tight binding [123]. Briefly, the peptide Ac-TTAI-NH_2 is most likely inserted into the lower part of the s4A site of β-sheet A and lined up with the strands 3 (Ala183, Leu184, Val185, and Asn186) and 5 (Lys331, Ala332,

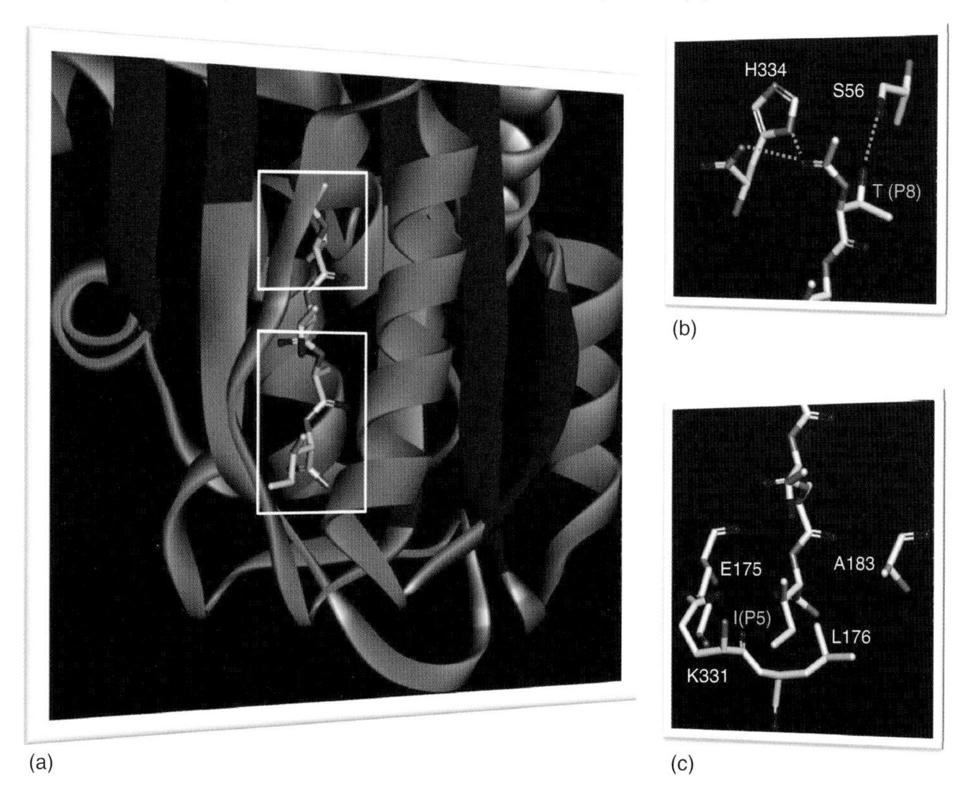

Figure 13.5 (a) Proposed complex of the combinatorially selected peptide Ac-TTAI-NH_2 binding to α_1-antitrypsin. (b) The hydrogen bond (light green dashed line) between the N-terminal Thr (P8) of peptide Ac-TTAI-NH2 and Ser56. Hydrogen bonds are in light green dashed line. (c) The hydrophobic side chain of Ile interacts with a pocket surrounding by the residues of Val173, Glu175, Leu176, Ala183, and Lys331. Carbon, nitrogen, and oxygen atoms are shown in white, light blue, and red, respectively.

Val333, and His334) as the central strand 4. As shown in Figure 13.5, additional hydrogen bonds were found at the N-terminal threonine and its acetyl group with the side chain of His334, the backbone NH of Lys335, and the side chain of Ser56, respectively. The N-terminal threonine may act as the "anchoring residue" to locate the peptide and land at the hot spot. Then the flexible peptide chain may pinpoint and bind to the β-sheet A through those backbone hydrogen bonds. According to the Venn diagram of amino acids [128], all the residues are small amino acids (Ala, Cys, Gly, Ser, and Thr) up to the penultimate residue of the peptide. Furthermore, comparing the P8 and P7 sites of human inhibitory serpin RCL sequences, it was found that 88.9% of both sites were the small amino acids Ala, Gly, Ser, or Thr. The percentage of the small amino acids on the P6 site decreased to 44.0%, but 72.0% were found to be hydrophobic residues and this explained why the small hydrophobic Ala and Val were experimentally identified at the P6 site. Therefore, we reason that these smaller side chains on the peptide are likely to facilitate the incorporation of the β-sheet A by evading the connecting loop of the F-helix, which was reported to undergo a reversible conformational change during the inhibitory mechanism [129]. Finally, the bulky side chain of the C-terminal isoleucine may occupy the cavity surrounded by Val173, Leu176, Ala183, Glu175, and Lys331. Taken together, the tight binding of the BC is stabilized by hydrogen bonds, hydrophobic interactions, and the cavity-filling effect.

13.6 Conclusion and Outlook

Under certain circumstances, an unusual subset of proteins is able to aggregate within or around cells and cause tissue damages. Inappropriate aggregation of proteins is normally prevented by complex cellular quality control processes. However, if the mechanism goes wrong for some reason, the resulting protein tangles are extremely hard to eradicate. Therapeutic strategies for protein aggregation diseases basically include the promotion of the clearance of aggregates and inhibition of polymer formation [130]. The understanding of the mechanism of AT-related diseases provides a potential strategy to block the pathogenic polymerization and opens the opportunity to attenuate the associated diseases and other serpinopathies. The combinatorial approach described herein has facilitated the identification of potent anti-AT polymerization inhibitors. The effectiveness of the "libraries from libraries" concept has been demonstrated by the evolution of the initial alanine scanning library to the truncation, D-amino acid scanning, and eventually the β-strand-directed library and its related positional scanning library. The libraries are complementary to each other, and the optimal peptide sequence may not be identified if one of the libraries were omitted. On the other hand, the peptides identified from different libraries may provide more insights for the design of mimetics. The power of the developed combinatorial approach was also demonstrated, as it could identify the different binding ligands of proteins with only one point mutation. The conformation-sensitive assay 8 M urea PAGE was used throughout the screening of all libraries. The effect of the chaotropic reagent urea immensely improved the readout of gels. The gel-based screening was further validated orthogonally by SPR, which represents a potential alternative of the screening platform, if the screening protocol could be optimized, and renders the protein chip more robust for high-throughput screening. In view of the fact that serpins share the conserved tertiary structure and thus have similar structural basis for the propagation of pathogenic tangles, we reasoned that the developed combinatorial approach should be applicable to other serpinopathies straightforwardly. In addition, the molecular diversity of libraries could be further increased by the incorporation of other building blocks such as unnatural amino acids and small molecules. Taken together, the developed combinatorial strategy provides a rapid and flexible screening platform and paves the way for the discovery of anti-serpin polymerization drugs.

My postgraduate training and postdoctoral work have been at the interface of chemistry and biology. Combinatorial libraries and screening assays have been employed to study a few biological problems such as the conformational disease (AATD), antibiotic resistance (vancomycin resistance), and a lysosomal storage disease. Small peptides and carbohydrate mimetics served as molecular probes in the intervention of these biological systems. To study these protein–ligand interactions, *in vitro* and *in vivo* screening methods have been developed, including cell-free (gel electrophoresis, enzymatic assay, and SPR) and cell-based assays. In addition, molecular and animal models have been used to investigate the effects of molecular probes. Currently, I am working on a cell-based assay and identified several potent inhibitors in the catabolism of protein N-glycosylation pathway. The animal and histological studies are in progress. The enthusiasm of interdisciplinary research and training will continue to prompt me to explore interesting biochemical fields as well as orphaned or rare diseases.

Acknowledgments

I am grateful to my PhD supervisor Prof. Yen-Ho Chu, for his guidance throughout my PhD training. I am thankful to Dr. Wun-Shaing Wayne Chang and Dr. Ravi Mahadeva, for their assistance in the AT project. I also wish to thank Prof. Yen-Ho Chu and Prof. Benjamin G. Davis for giving me the chance to participate the European Young Chemists Award Competition at the third EuCheMS Chemistry Congress (Nürnberg, 2010).

References

1. Carrell, R.W. and Lomas, D.A. (1997) Conformational disease. *Lancet*, **350**, 134–138.
2. Chiti, F. and Dobson, C.M. (2006) Protein misfolding, functional amyloid, and human disease. *Annu. Rev. Biochem.*, **75**, 333–366.
3. Merrifield, R.B. (1963) Solid phase peptide synthesis. I. The synthesis of a tetrapeptide. *J. Am. Chem. Soc.*, **85**, 2149–2154.

4. Leznoff, C.C. (1974) The use of insoluble polymer supports in organic chemical synthesis. *Chem. Soc. Rev.*, **3**, 65–85.
5. Leznoff, C.C. (1978) The use of insoluble polymer supports in generalorganic chemical synthesis. *Acc. Chem. Res.*, **11**, 327–333.
6. Frank, R., Heikens, W., Heisterberg-Moutsis, G., and Blöcker, H. (1983) A new general approach for the simultaneous chemical synthesis of large numbers of oligonucleotides: Segmental solid supports. *Nucleic Acids Res.*, **11**, 4365–4377.
7. Geysen, H.M., Meloen, R.H., and Barteling, S.J. (1984) Use of peptide synthesis to probe viral antigens for epitopes to a resolution of a single amino acid. *Proc. Natl. Acad. Sci. U.S.A.*, **81**, 3998–4002.
8. Geysen, H.M., Barteling, S.J., and Meloen, R.H. (1985) Small peptides induce antibodies with a sequence and structural requirement for binding antigen comparable to antibodies raised against the native protein. *Proc. Natl. Acad. Sci. U.S.A.*, **82**, 178–182.
9. Houghten, R.A. (1985) General method for the rapid solid-phase synthesis of large numbers of peptides: Specificity of antigen-antibody interaction at the level of individual amino acids. *Proc. Natl. Acad. Sci. U.S.A.*, **82**, 5131–5135.
10. Furka, A., Sebestyen, F., Asgedom, M., and Dibo, G. (1991) General method for rapid synthesis of multicomponent peptide mixtures. *Int. J. Pept. Protein Res.*, **37**, 487–493.
11. Lam, K.S., Salmon, S.E., Hersh, E.M., Hruby, V.J., Kazmierski, W.M., and Knapp, R.J. (1991) A new type of synthetic peptide library for identifying ligand-binding activity. *Nature*, **354**, 82–84.
12. Houghten, R.A., Pinilla, C., Blondelle, S.E., Appel, J.R., Dooley, C.T., and Cuervo, J.H. (1991) Generation and use of synthetic peptide combinatorial libraries for basic research and drug discovery. *Nature*, **354**, 84–86.
13. Dömling, A. and Ugi, I. (1993) The seven-component reaction. *Angew. Chem. Int. Ed. Engl.*, **32**, 563–564.
14. Uversky, V.N. (2010) Mysterious oligomerization of the amyloidogenic proteins. *FEBS J.*, **277**, 2940–2953.
15. Lomas, D.A. and Carrell, R.W. (2002) Serpinopathies and the conformational dementias. *Nat. Rev. Genet.*, **3**, 759–768.
16. Lieberman, J., Mittman, C., and Gordon, H.W. (1972) Alpha$_1$ antitrypsin in the livers of patients with emphysema. *Science*, **175**, 63–65.
17. Mahadeva, R. and Lomas, D.A. (1998) Alpha$_1$-antitrypsin deficiency, cirrhosis and emphysema. *Thorax*, **53**, 501–505.
18. Greene, C.M., Miller, S.D.W., Carroll, T., McLean, C., O'Mahony, M., Lawless, M.W., O'Neill, S.J., Taggart, C.C., and McElvaney, N.G. (2008) Alpha-1 antitrypsin deficiency: A conformational disease associated with lung and liver manifestations. *J. Inherit. Metab. Dis.*, **31**, 21–34.
19. Laurell, C.B. and Eriksson, S. (1963) The electrophoretic alpha$_1$-globulin pattern of serum in alpha$_1$-antitrypsin deficiency. *Scand. J. Clin. Lab. Invest.*, **15**, 132–140.
20. Eriksson, S. (1964) Pulmonary emphysema and alpha$_1$-antitrypsin deficiency. *Acta Med. Scand.*, **175**, 197–205.
21. Kubba, A.K. and Young, M. (1998) The long suffering of Frederic Chopin. *Chest*, **113**, 210–216.
22. Stoller, J.K. and Aboussouan, L.S. (2009) Myths and misconceptions about α_1-antitrypsin deficiency. *Arch. Intern. Med.*, **169**, 546–550.
23. Stoller, J.K. and Aboussouan, L.S. (2005) α_1-Antitrypsin deficiency. *Lancet*, **365**, 2225–2236.
24. Kopito, R.R. and Ron, D. (2000) Conformational disease. *Nat. Cell Biol.*, **2**, E207–E209.
25. Carrell, R.W. and Lomas, D.A. (2002) Alpha$_1$-antitrypsin deficiency-A model for conformational diseases. *N. Engl. J. Med.*, **346**, 45–53.
26. Asofsky, R. and Thorbecke, G.J. (1961) Sites of formation of immune globulins and of a component of C'_3: II. Production of immunoelectrophoretically identified serum proteins by human and monkey tissues in vitro. *J. Exp. Med.*, **114**, 471–483.

27. Molmenti, E.P., Perlmutter, D.H., and Rubin, D.C. (1993) Cell-specific expression of alpha1-antitrypsin in human intestinal epithelium. *J. Clin. Invest.*, **92**, 2022–2034.
28. Hu, C. and Perlmutter, D.H. (2002) Cell-specific involvement of HNF-1β in α_1-antitrypsin gene expression in human respiratory epithelial cells. *Am. J. Physiol. Lung Cell Mol. Physiol.*, **282**, L757–L765.
29. Carrell, R.W., Jeppsson, J.O., Laurell, C.B., Brennan, S.O., Owen, M.C., Vaughan, L., and Boswell, D.R. (1982) Structure and variation of human α_1-antitrypsin. *Nature*, **298**, 329–334.
30. Eriksson, S. (1965) Studies in α_1-antitrypsin deficiency. *Acta Med. Scand. Suppl.*, **432**, 1–85.
31. King, M.A., Stone, J.A., Diaz, P.T., Mueller, C.F., Becker, W.J., and Gadek, J.E. (1996) Alpha$_1$-antitrypsin deficiency: evaluation of bronchiectasis with CT. *Radiology*, **199**, 137–141.
32. Griffith, M.E., Lovegrove, J.U., Gaskin, G., Whitehouse, D.B., and Pusey, C.D. (1996) C-antineutrophil cytoplasmic antibody positivity in vasculitis patients is associated with the Z allele of alpha-1-antitrypsin, and P-antineutrophil cytoplasmic antibody positivity with the S allele. *Nephrol. Dial. Transplant.*, **11**, 438–443.
33. Wood, A.M. and Stockley, R.A. (2007) Alpha one antitrypsin deficiency: From gene to treatment. *Respiration*, **74**, 481–492.
34. de Serres, F.J. (2002) Worldwide racial and ethnic distribution of α_1-antitrypsin deficiency. *Chest*, **122**, 1818–1829.
35. Knaupp, A.S. and Bottomley, S.P. (2009) Serpin polymerization and its role in disease – the molecular basis of α_1-antitrypsin deficiency. *IUBMB Life*, **61**, 1–5.
36. Whisstock, J., Skinner, R., and Lesk, A.M. (1998) An atlas of serpin conformations. *Trends Biochem. Sci.*, **23**, 63–67.
37. Johnson, D. and Travis, J. (1978) Structural evidence for methionine at the reactive site of human alpha-1-proteinase inhibitor. *J. Biol. Chem.*, **253**, 7142–7144.
38. Stratikos, E. and Gettins, P.W.G. (1997) Major proteinase movement upon stable serpin-proteinase complex. *Proc. Natl. Acad. Sci. U.S.A.*, **94**, 453–458.
39. Stratikos, E. and Gettins, P.W.G. (1999) Formation of the covalent serpin-proteinase complex involves translocation of the proteinase by more than 70 Å and full insertion of the reactive center loop into β-sheet A. *Proc. Natl. Acad. Sci. U.S.A.*, **96**, 4808–4813.
40. Huntington, J.A., Read, R.J., and Carrell, R.W. (2000) Structure of a serpin-protease complex shows inhibition by deformation. *Nature*, **407**, 923–926.
41. Wilczynska, M., Fa, M., Karolin, J., Ohlsson, P.-I., Johansson, L.B.-Å., and Ny, T. (1997) Structural insights into serpin-protease complexes reveal the inhibitory mechanism of serpins. *Nat. Struct. Biol.*, **4**, 354–357.
42. Gettins, P.G.W. (2002) Serpin structure, mechanism, and function. *Chem. Rev.*, **102**, 4751–4803.
43. Wright, H.T. and Scarsdale, J.N. (1995) Structural basis for serpin inhibitor activity. *Proteins*, **2**, 210–225.
44. Wilczynska, M., Fa, M., Ohlsson, P.-I., and Ny, T. (1995) The inhibition mechanism of serpins. Evidence that the mobile reactive center loop is cleaved in the native protease-inhibitor complex. *J. Biol. Chem.*, **270**, 29652–29655.
45. Shore, J.D., Day, D.E., Francis-Chmura, A.M., Verhamme, I., Kvassman, J., Lawrence, D.A., and Ginsburg, D. (1995) A fluorescent probe study of plasminogen activator inhibitor-1: evidence for reactive center loop insertion and its role in the inhibitory mechanism. *J. Biol. Chem.*, **270**, 5395–5398.
46. Engh, R., Huber, R., Bode, W., and Schulze, A.J. (1995) Divining the serpin inhibition mechanism: a shicide substrate 'springe'. *Trends Biotechnol.*, **13**, 503–510.
47. Sivasothy, P., Dafforn, T.R., Gettins, P.G.W., and Lomas, D.A. (2000) Pathogenic α_1-antitrypsin polymers are formed by reactive loop-β-Sheet

A Linkage. *J. Biol. Chem.*, **275**, 33663–33668.

48. Jeppsson, J.O. (1976) Amino acid substitution Glu → Lys in α_1-antitrypsin PiZ. *FEBS Lett.*, **65**, 195–197.
49. Huber, R. and Carrell, R.W. (1989) Implications of the three-dimensional structure of α_1-antitrypsin for structure and function of serpins. *Biochemistry*, **28**, 8951–8966.
50. Lomas, D.A., Evans, D.L., Finch, J.T., and Carrell, R.W. (1992) The mechanism of Z α_1-antitrypsin accumulation in the liver. *Nature*, **357**, 605–607.
51. Lomas, D.A., Finch, J.T., Seyama, K., Nukiwa, T., and Carrell, R.W. (1993) α_1-Antitrypsin Siiyama (Ser^{53} → Phe): further evidence for intracellular loop-sheet polymerization. *J. Biol. Chem.*, **268**, 15333–15335.
52. Lomas, D.A., Elliott, P.R., Sidhar, S.K., Foreman, R.C., Finch, J.T., Cox, D.W., Whisstock, J.C., and Carrell, R.W. (1995) Alpha$_1$-antitrypsin Mmalton (Phe52-deleted) forms loop-sheet polymers in vivo. Evidence for the C sheet mechanism of polymerization. *J. Biol. Chem.*, **270**, 16864–16870.
53. Tomashefski, J.F., Crystal, R.G., Wiedemann, H.P., Mascha, E., and Stoller, J.K. Jr. (2004) The bronchopulmonary pathology of alpha-1 antitrypsin (AAT) deficiency: findings of the death review committee of the national registry for individuals with severe deficiency of alpha-1 antitrypsin. *Hum. Pathol.*, **35**, 1452–1461.
54. Brantly, M.L., Paul, L.D., Miller, B.H., Falk, R.T., Wu, M., and Crystal, R.G. (1988) Clinical features and history of the destructive lung disease associated with alpha-1-antitrypsin deficiency of adults with pulmonary symptoms. *Am. Rev. Respir. Dis.*, **138**, 327–336.
55. Gishen, P., Saunders, A.J., Tobin, M.J., and Hutchison, D.C. (1982) Alpha 1-antitrypsin deficiency: the radiological features of pulmonary emphysema in subjects of Pi type Z and Pi type SZ: a survey by the British Thoracic Association. *Clin. Radiol.*, **33**, 371–377.
56. Gooptu, B., Hazes, B., Chang, W.S., Dafforn, T.R., Carrell, R.W., Read, R.J., and Lomas, D.A. (2000) Inactive conformation of the serpin alpha$_1$-antichymotrypsin indicates two-stage insertion of the reactive loop: implications for inhibitory function and conformational disease. *Proc. Natl. Acad. Sci. U.S.A.*, **97**, 67–72.
57. Bruce, D., Perry, D.J., Borg, J.-Y., Carrell, R.W., and Wardell, M.R. (1994) Thromboembolic disease due to thermolabile conformational changes of antithrombin Rouen-VI (187 Asn → Asp). *J. Clin. Invest.*, **94**, 2265–2274.
58. Aulak, K.S., Eldering, E., Hack, C.E., Lubbers, Y.P.T., Harrison, R.A., Mast, A., Cicardi, M., and Davis, A.E. III (1993) A hinge region mutation in C1-inhibitor (Ala436 → Thr) results in nonsubstrate-like behavior and in polymerization of the molecule. *J. Biol. Chem.*, **268**, 18088–18094.
59. Davis, R.L., Shrimpton, A.E., Holohan, P.D., Bradshaw, C., Feiglin, D., Collins, G.H., Sonderegger, P., Kinter, J., Becker, L.M., Lacbawan, F., Krasnewich, D., Muenke, M., Lawrence, D.A., Yerby, M.S., Shaw, C.M., Gooptu, B., Elliott, P.R., Finch, J.T., Carrell, R.W., and Lomas, D.A. (1999) Familial dementia caused by polymerization of mutant neuroserpin. *Nature*, **401**, 376–379.
60. Carrell, R.W., Stein, P.E., Fermi, G., and Wardell, M.R. (1994) Biological implication of a 3Å structure of dimeric antithrombin. *Structure*, **2**, 257–270.
61. Mottonen, J., Strand, A., Symersky, J., Sweet, R.M., Danley, D.E., Geoghegan, K.F., Gerard, R.D., and Goldsmith, E.J. (1992) Structural basis of latency in plasminogen activator inhibitor-1. *Nature*, **355**, 270–273.
62. Stein, P.E., Leslie, A.G.W., Finch, J.T., Turnell, W.G., McLaughlin, P.J., and Carrell, R.W. (1990) Crystal structure of ovalbumin as a model for the reactive centre of serpins. *Nature*, **347**, 99–102.
63. Wei, A., Rubin, H., and Cooperman, B.S. (1994) Crystal structure of an uncleaved serpin reveals the conformation of an inhibitory loop. *Nat. Struct. Biol.*, **1**, 251–258.

64. Elliott, P.R., Lomas, D.A., Carrell, R.W., and Abrahams, J.P. (1996) Inhibitory conformation of the reactive loop of α_1-antitrypsin. *Nat. Struct. Biol.*, **3**, 676–681.
65. Huntington, J.A., Pannu, N.S., Hazes, B., Read, R.J., Lomas, D.A., and Carrell, R.W. (1993) A 2.6 Å structure of a serpin polymer and implications for conformational disease. *J. Mol. Biol.*, **293**, 449–455.
66. Yamasaki, M., Li, W., Johnson, D.J.D., and Huntington, J.A. (2008) Crystal structure of a stable dimer reveals the molecular basis of serpin polymerization. *Nature,* **455**, 1255–1259.
67. Whisstock, J.C. and Bottomley, S.P. (2008) Serpins' mystery solved. *Nature,* **455**, 1189–1190.
68. Yamasaki, M., Timothy, J.S., Harris, L.E., Lewis, G.M.W., and Huntington, J.A. (2010) Loop-sheet mechanism of serpin polymerization tested by reactive center loop mutations. *J. Biol. Chem.*, **285**, 30752–30758.
69. Ekeowa, U.I., Freekeb, J., Miranda, E., Gooptu, B., Bush, M.F., Pérez, J., Teckman, J., Robinson, C.V., and Lomas, D.A. (2010) Defining the mechanism of polymerization in the serpinopathies. *Proc. Natl. Acad. Sci. U.S.A.*, **107**, 17146–17151.
70. Krishnan, B. and Gierasch, L.M. (2011) Dynamic local unfolding in the serpin α-1 antitrypsin provides a mechanism for loop insertion and polymerization. *Nat. Struct. Mol. Biol.*, **18**, 222–227.
71. Flotte, T.R., Brantly, M.L., Spencer, L.T., Byrne, B.J., Spencer, C.T., Baker, D.J., and Humphries, M. (2004) Phase I trial of intramuscular injection of a recombinant adeno-associated virus alpha-1 antitrypsin (rAAV2-CB-hAAT) gene vector to AAT-deficient adults. *Hum. Genet. Ther.*, **15**, 93–128.
72. McLean, C., Greene, C.M., and McElvaney, N.G. (2009) Gene targeted therapeutics for liver disease in alpha-1 antitrypsin deficiency. *Biologics: Targets Ther.*, **3**, 63–75.
73. Marcus, N.Y. and Perlmutter, D.H. (2000) Glucosidase and mannosidase inhibitors mediate increased secretion of mutant α1 antitrypsin Z. *J. Biol. Chem.*, **275**, 1987–1992.
74. Burrows, J.A., Willis, L.K., and Perlmutter, D.H. (2000) Chemical chaperones mediate increased secretion of mutant α_1-antitrypsin (α_1-AT) Z: a potential pharmacological strategy for prevention of liver injury and emphysema in α_1-AT deficiency. *Proc. Natl. Acad. Sci. U.S.A.*, **97**, 1796–1801.
75. Cowan, K.N., Heilbut, A., Humpl, T., Lam, C., Ito, S., and Rabinovitch, M. (2000) Complete reversal of fatal pulmonary hypertension in rats by a serine elastase inhibitor. *Nat. Med.*, **6**, 698–702.
76. Powers, J.C., Asgian, J.L., Ekici, Ö.D., and James, K.E. (2002) Irreversible inhibitors of serine, cysteine, and threonine proteases. *Chem. Rev.*, **102**, 4639–4750.
77. McBride, J.D., Freeman, H.N.M., and Leatherbarrow, R.J. (1999) Selection of human elastase inhibitors from a conformationally constrained combinatorial peptide library. *Eur. J. Biochem.*, **266**, 403–412.
78. Tsang, W.Y., Ahmed, N., Harding, L.P., Hemming, K., Laws, A.P., and Page, M.I. (2005) Acylation versus sulfonylation in the inhibition of elastase by 3-Oxo-β-sultams. *J. Am. Chem. Soc.*, **127**, 8946–8947.
79. Cantin, A.M., Woods, D.E., Cloutier, D., Dufour, E.K., and Leduc, R. (2002) Polyethylene glycol conjugation at Cys232 prolongs the half-life of α1 proteinase inhibitor. *Am. J. Respir. Cell Mol. Biol.*, **27**, 659–665.
80. Parfrey, H., Mahadeva, R., Ravenhill, N., Zhou, A., Dafforn, T.R., Foreman, R.C., and Lomas, D.A. (2003) Targeting a surface cavity of α_1-antitrypsin to prevent conformational disease. *J. Biol. Chem.*, **278**, 33060–33066.
81. Elliott, P.R., Abrahams, J.P., Dafforn, T.R., and Lomas, D.A. (1998) Wild-type α_1-antitrypsin is in the canonical inhibitory conformation. *J. Mol. Biol.*, **275**, 419–425.
82. Elliott, P.R., Pei, X.Y., Dafforn, T.R., and Lomas, D.A. (2000) Topography of a 2.0 Å structure of α_1-antitrypsin reveals targets for rational drug design to

prevent conformational disease. *Protein Sci.*, **9**, 1274–1281.

83. Kim, S.-J., Woo, J.-R., Seo, E.J., Yu, M.-H., and Ryu, S.-E. (2001) A 2.1 Å resolution structure of an uncleaved α_1-antitrypsin shows variability of the reactive center and other loops. *J. Mol. Biol.*, **306**, 109–119.
84. Gooptu, B., Miranda, E., Nobeli, I., Mallya, M., Purkiss, A., Brown, S.C.L., Summers, C., Phillips, R.L., Lomas, D.A., and Barrett, T.E. (2009) Crystallographic and cellular characterization of two mechanisms stabilising the native fold of α_1-antitrypsin: Implications for disease and drug design. *J. Mol. Biol.*, **387**, 857–868.
85. Chow, M.K., Devlin, G.L., and Bottomley, S.P. (2001) Osmolytes as modulators of conformational changes in serpins. *Biol. Chem.*, **382**, 1593–1599.
86. Devlin, G.L., Parfrey, H., Tew, D.J., Lomas, D.A., and Bottomley, S.P. (2001) Prevention of polymerization of M and Z alpha1-Antitrypsin(alpha1-AT) with trimethylamine *N*-oxide. Implications for the treatment of alpha1-AT deficiency. *Am. J. Respir. Cell Mol. Biol.*, **24**, 727–732.
87. Zhou, A., Stein, P.E., Huntington, J.A., and Carrell, R.W. (2003) Serpin polymerization is prevented by a hydrogen bond network that is centered on His-334 and stabilized by glycerol. *J. Biol. Chem.*, **278**, 15116–15122.
88. Sharp, L.K., Mallya, M., Kinghorn, K.J., Wang, Z., Crowther, D.C., Huntington, J.A., Belorgey, D., and Lomas, D.A. (2006) Sugar and alcohol molecules provide a therapeutic strategy for the serpinopathies that cause dementia and cirrhosis. *FEBS J.*, **273**, 2540–2552.
89. Teckman, J.H. (2004) Lack of effect of oral 4-phenylbutyrate on serum alpha-1-antitrypsin in patients with α-1-antitrypsin deficiency: a preliminary study. *J. Pediatr. Gastroenterol. Nutr.*, **39**, 34–37.
90. Mallya, M., Phillips, R.L., Saldanha, S.A., Gooptu, B., Brown, S.C.L., Termine, D.J., Shirvani, A.M., Wu, Y., Sifers, R.N., Abagyan, R., and Lomas, D.A. (2007) Small molecules block the polymerization of Z α_1-antitrypsin and increase the clearance of intracellular aggregates. *J. Mol. Biol.*, **50**, 5357–5363.
91. Stoller, J.K., Fallat, R., Schluchter, M.D., O'Brien, R.G., Connor, J.T., Gross, N., O'Neil, K., Sandhaus, R., and Crystal, R.G. (2003) Augmentation therapy with alpha1-antitrypsin: patterns of use and adverse events. *Chest*, **123**, 1425–1434.
92. Gildea, T.R., Shermock, K.M., Singer, M.E., and Stoller, J.K. (2003) Cost-effectiveness analysis of augmentation therapy for severe alpha1-antitrypsin deficiency. *Am. J. Respir. Crit. Care Med.*, **167**, 1387–1392.
93. Ioachimescu, O.C. and Stoller, J.K. (2005) A review of alpha-1 antitrypsin deficiency. *COPD*, **2**, 263–275.
94. Chang, Y.P., Mahadeva, R., Patschull, A.O.M., Nobeli, I., Ekeowa, U.I., McKay, A.R., Thalassinos, K., Irving, J.A., Haque, I.U., Nyon, M.P., Christodoulou, J., Gonzalez, A., Miranda, E., and Gooptu, B. (2011) Targeting serpins in high-throughput and structure-based drug design. *Method. Enzymol.*, **501**, 139–175.
95. Schulze, A.J., Baumann, U., Knof, S., Jaeger, E., Huber, R., and Laurell, C.-B. (1990) Structural transition of α_1-antitrypsin by a peptide sequentially similar to β-strand s4A. *Eur. J. Biochem.*, **194**, 51–56.
96. Schulze, A.J., Huber, R., Degryse, E., Speck, D., and Bischoff, R. (1991) Inhibitory activity and conformational transition of α_1-proteinase inhibitor variants. *Eur. J. Biochem.*, **202**, 1147–1155.
97. Schulze, A.J., Frohnert, P.W., Engh, R.A., and Huber, R. (1992) Evidence for the extent of insertion of the active site loop of intact α_1-proteinase inhibitor in β-sheet A. *Biochemistry*, **31**, 7560–7565.
98. Björk, I., Ylinenjärvi, K., Olson, S.T., and Bock, P.E. (1992) Conversion of antithrombin from an inhibitor of thrombin to a substrate with reduced heparin affinity and enhanced conformational stability by binding of a

tetradecapeptide corresponding to the P_1 to P_{14} region of the putative reactive bond loop of the inhibitor. *J. Biol. Chem.*, **267**, 1976–1982.

99. Carrell, R.W., Evans, D.L., and Stein, P.E. (1991) Mobile reactive centre of serpins and the control of thrombosis. *Nature*, **353**, 576–578.
100. Lomas, D.A., Evans, D.L., Finch, J.T., and Carrell, R.W. (1992) The mechanism of Z α_1-antitrypsin accumulation in the liver. *Nature*, **357**, 605–607.
101. Lomas, D.A., Evans, D.L., Stone, S.R., Chang, W.-S.W., and Carrell, R.W. (1993) Effect of the Z mutation on the physical and inhibitory properties of α_1-antitrypsin. *Biochemistry*, **32**, 500–508.
102. Fitton, H.L., Pike, R.N., Carrell, R.W., and Chang, W.-S.W. (1997) Mechanisms of antithrombin polymerization and heparin activation probed by insertion of synthetic reactive loop peptides. *Biol. Chem.*, **378**, 1059–1063.
103. Crowther, D.C., Serpell, L.C., Dafforn, T.R., Gooptu, B., and Lomas, D.A. (1992) Conformation of the reactive site loop of α_1-proteinase inhibitor probed by limited proteolysis. *Biochemistry*, **31**, 2720–2728.
104. Eitzman, D.T., Fay, W.P., Lawrence, D.A., Francis-Chmura, A.M., Shore, J.D., Olson, S.T., and Ginsbur, D. (1995) Peptide-mediated inactivation of recombinant and platelet plasminogen activator inhibitor-1 in vitro. *J. Clin. Invest.*, **95**, 2416–2420.
105. Kvassman, J.-O., Lawrence, D.A., and Shore, J.D. (1995) The acid stabilization of plasminogen activator inhibitor-1 depends on protonation of a single group that affects loop insertion into β-Sheet A. *J. Biol. Chem.*, **270**, 27942–27947.
106. Chang, W.-S.W., Wardell, M.R., Lomas, D.A., and Carrell, R.W. (1996) Probing serpin reactive loop conformations by proteolytic cleavage. *Biochem. J.*, **314**, 647–653.
107. Xue, Y., Björquist, P., Inghardt, T., Linschoten, M., Musil, D., Sjölin, L., and Deinum, J. (1998) Interfering with the inhibitory mechanism of serpins: crystal structure of a complex formed between cleaved plasminogen activator inhibitor type 1 and a reactive-centre loop peptide. *Structure*, **6**, 627–636.
108. Skinner, R., Chang, W.-S.W., Jin, L., Pei, X., Huntington, J.A., Abrahams, J.P., Carrell, R.W., and Lomas, D.A. (1998) Implications for function and therapy of a 2.9 Å structure of binary-complexed antithrombin. *J. Mol. Biol.*, **283**, 9–14.
109. Crowther, D.C., Serpell, L.C., Dafforn, T.R., Gooptu, B., and Lomas, D.A. (2003) Nucleation of α_1-antichymotrypsin polymerization. *Biochemistry*, **42**, 2355–2363.
110. Mahadeva, R., Dafforn, T.R., Carrell, R.W., and Lomas, D.A. (2002) 6-mer peptide selectively anneals to a pathogenic serpin conformation and blocks polymerization. Implications for the prevention of Z α_1-antitrypsin-related cirrhosis. *J. Biol. Chem.*, **277**, 6771–6774.
111. Zhou, A., Stein, P.E., Huntington, J.A., and Carrell, R.W. (2003) Serpin polymerization is prevented by a hydrogen bond network that is centered on His-334 and stabilized by glycerol. *J. Biol. Chem.*, **278**, 15116–15122.
112. Zhou, A., Stein, P.E., Huntington, J.A., Sivasothy, P., Carrell, R.W., and Lomas, D.A. (2004) How small peptides block and reverse serpin polymerisation. *J. Mol. Biol.*, **342**, 931–941.
113. Jones, J.H. (2007) R.B. Merrifield: European footnotes to his life and work. *J. Pept. Sci.*, **13**, 363–367.
114. Gettins, P. (2002) Serpin structure, mechanism, and function. *Chem. Rev.*, **383**, 1677–1682.
115. Morrison, K.L. and Weiss, G.A. (2001) Combinatorial alanine-scanning. *Curr. Opin. Chem. Biol.*, **5**, 302–307.
116. Chang, Y.-P., Mahadeva, R., Chang, W.-S.W., Shukla, A., Dafforn, T., and Chu, Y.-H. (2006) Identification of a 4-mer peptide inhibitor that effectively blocks the polymerization of pathogenic Z α_1-antitrypsin. *Am. J. Respir. Cell Mol. Biol.*, **35**, 540–548.
117. Ostermeier, M., Nixon, A.E., Shim, J.H., and Benkovic, S.J. (1999) Combinatorial protein engineering by incremental truncation. *Proc. Natl. Acad. Sci. U.S.A.*, **96**, 3562–3567.

118. Svenson, J., Stensen, W., Brandsdal, B.O., Haug, B.E., Monrad, J., and Svendsen, J.S. (2008) Antimicrobial peptides with stability toward tryptic degradation. *Biochemistry*, **47**, 3777–3788.

119. Tugyi, R., Uray, K., Ivan, D., Fellinger, E., Perkins, A., and Hudecz, F. (2005) Partial D-amino acid substitution: improved enzymatic stability and preserved Ab recognition of a MUC2 epitope peptide. *Proc. Natl. Acad. Sci. U.S.A.*, **102**, 413–418.

120. Chou, P.Y. and Fasman, G.D. (1978) Empirical predictions of protein conformation. *Annu. Rev. Biochem.*, **47**, 251–276.

121. Furka, Á. (2002) Combinatorial chemistry: 20 years on . . . . *Drug Discovery Today*, **7**, 1–7.

122. Dooley, C.T., Chung, N.N., Wilkes, B.C., Schiller, P.W., Bidlack, J.M., Pasternak, G.W., and Houghten, R.A. (1994) An all D-amino acid opioid peptide with central analgesic activity from a combinatorial library. *Science*, **266**, 2019–2022.

123. Chang, Y.-P., Mahadeva, R., Chang, W.-S.W., Lin, S.-C., and Chu, Y.-H. (2009) Small-molecule peptides inhibit Z α_1-antitrypsin polymerization. *J. Cell Mol. Med.*, **13**, 2304–2316.

124. Pinilla, C., Appel, J.R., Blanc, P., and Houghten, R.A. (1992) Rapid identification of high affinity peptide ligands using positional scanning synthetic peptide combinatorial libraries. *Biotechniques*, **13**, 901–905.

125. Rich, R.L. and Myszka, D.G. (2006) Survey of the year 2005 commercial optical biosensor literature. *J. Mol. Recognit.*, **19**, 478–534.

126. Chang, Y.P. and Chu, Y.H. (2005) Using surface plasmon resonance to directly determine binding affinities of combinatorially selected cyclopeptides and their linear analogs to a streptavidin chip. *Anal. Biochem.*, **340**, 74–79.

127. Chang, Y.P., Tseng, M.J., and Chu, Y.H. (2006) Using surface plasmon resonance to directly measure slow binding of low-molecular mass inhibitors to a VanX chip. *Anal. Biochem.*, **359**, 63–71.

128. Taylor, W.R. (1986) The classification of amino acid conservation. *J. Theor. Biol.*, **119**, 205–218.

129. Cabrita, L.D., Dai, W., and Bottomley, S.P. (2004) Different conformational changes within the F-Helix occur during serpin folding, polymerization, and proteinase inhibition. *Biochemistry*, **43**, 9834–9839.

130. Aguzzi, A. and O'Connor, T. (2010) Protein aggregation diseases: pathogenicity and therapeutic perspectives. *Nat. Rev. Drug Discovery*, **9**, 237–248.

14
Synthesis and Application of Macrocycles Using Dynamic Combinatorial Chemistry

Vittorio Saggiomo

14.1
Supramolecular Chemistry

"Atoms are letters, molecules are the words, supramolecular entities are the sentences and the chapters" [1]. This brief sentence contains in itself the importance and the great fascination of supramolecular chemistry. The year was 1987. Jean-Marie Lehn, together with Donald J. Cram and Charles J. Pedersen, won the Nobel Prize *"for their development and use of molecules with structure-specific interaction of high selectivity."* Born by genial intuition and lucky accidental synthesis [2], supramolecular chemistry is nowadays of great importance in various and different fields in chemistry and industry.

Supramolecular chemistry was born as a branch of organic chemistry and slowly grew up as a highly interdisciplinary subject. It attracts not only chemists but also biochemists, biologists, engineers, physicists, and theoreticians [3]. Moreover, it plays a crucial role in the development of new nanotechnologies [4]. In a more poetic way we can say, without any doubt, that supramolecular chemistry "is life."

In biology, molecular recognition plays a key role in almost all processes with an amazing selectivity. This high selectivity is difficult to reproduce in laboratories by synthetic hosts. Host–guest complexes have great importance also in signaling and transport, moving in and out guests in different part of cells.

In everyday life, supramolecular chemistry is even more apparent. Plastic materials have completely changed our life, and now this branch is evolving into polymer and materials science. High-tech materials such as thermoplastics, solar panels, adhesives, or new and resistant materials for artificial organs are based on supramolecular and polymer chemistry. These new materials can drastically improve our daily lives. Also, in pharmaceutics, supramolecular chemistry is widely exploited. Cyclodextrines, for example, thanks to their shapes and to the apolar binding site, are used for more than 10 years in many pharmaceutical formulations in order to solubilize synthetic drugs [5]. Moreover, "smart materials," molecular shuttles, motors, and rotors are, in this first part of the twenty-first century, the new era of supramolecules thanks to the work of J. Fraser Stoddart [6], David Leigh

New Strategies in Chemical Synthesis and Catalysis, First Edition. Edited by Bruno Pignataro.

[7], and Ben L. Feringa [8]. Those are only a few applications of supramolecular chemistry.

Since the birth of supramolecular chemistry, many host molecules have been synthesized: from crown and lariat ethers, podands, and cryptands to calixarenes, cyclodextrines, and then to even more complex structures such as catenanes, rotaxanes, foldamers, or helicates [3].

The synthesis of host molecules for specific guests is usually long and difficult. Often, the final product is obtained only by a multistep synthesis, requiring various purification steps and, many a time, also protection/deprotection steps. Once the host is synthesized, it must be tested for the host–guest complexation. And this, unfortunately, is not always successful. In fact, it is not rare that the hard-won host does not bind the desired guest or binds it with low affinity.

An alternative strategy to obtain a host for a specific guest, with less synthetic effort, was developed in the last two decades in an emerging field: dynamic combinatorial chemistry (DCC) [9].

14.2
Dynamic Combinatorial Chemistry

DCC is, by definition, combinatorial chemistry under thermodynamic control. In nonreversible combinatorial chemistry, a large number of different related products are synthesized from a mixture of different starting building blocks. The obtained products will not react with one another, and the final composition of the library will not change. On the contrary, all members of a dynamic combinatorial library (DCL) are in equilibrium and can be interconverted into one another changing the library composition. In order to achieve the interconversion, the link between the components of the dynamic library must be reversible. Pioneers with two different approaches were Jean-Marie Lehn [10] and Jeremy K. M. Sanders [11].

Lehn's group in Strasbourg used DCC to cast a substrate (guest) for an enzyme (host). They describe it as "*casting consist in the receptor-induced assembly of a substrate that fits the receptor*" [10]. The use of DCC allows the guests in the library to interconvert into one another. The spontaneous guest diversity generation is then shifted mainly to one guest that will fit inside the enzyme pocket in what they call "*sort of (supra)molecular Darwinism*" (Figure 14.1a). In short, they discovered an easy and efficient methodology to synthesize a perfectly fitting guest for a specific host.

Using the opposite approach, Sanders' group in Cambridge used DCC in order to obtain a host for a specific guest [11]. In this case, the reaction of different building blocks gave various linear and macrocyclic products. To this resulting library was then added a guest. Perturbing the equilibrium by addition of a guest resulted in a different library composition. Thus, the macrocycles that better bound the given guest were "amplified" to the detriment of other library members (Figure 14.1b).

Since its birth, DCC has evolved discovering and using different reversible covalent reactions (mainly used are imines [12], hydrazones [13], and disulfides [14]

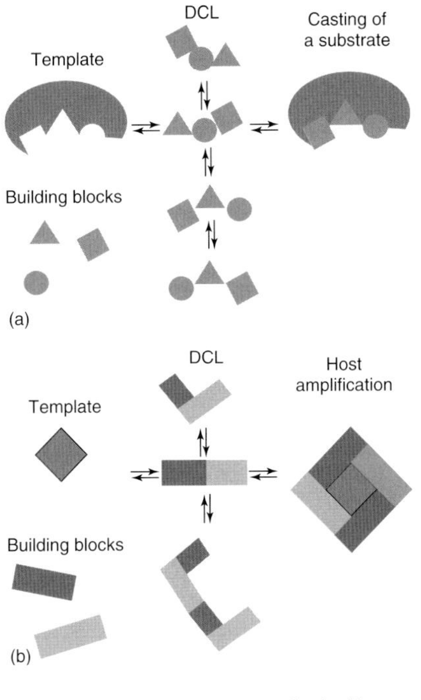

Figure 14.1 Schematic drawing of first experiments on dynamical combinatorial chemistry. (a) Casting of a substrate (guest) [10]. (b) Host amplification [11].

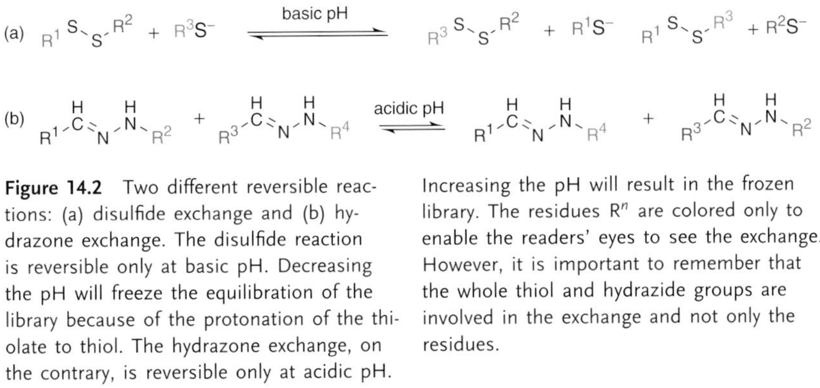

(a) $$R^1{-}S{-}S{-}R^2 + R^3S^- \underset{}{\overset{\text{basic pH}}{\rightleftharpoons}} R^3{-}S{-}S{-}R^2 + R^1S^- \quad R^1{-}S{-}S{-}R^3 + R^2S^-$$

(b) $$R^1{-}CH{=}N{-}NH{-}R^2 + R^3{-}CH{=}N{-}NH{-}R^4 \overset{\text{acidic pH}}{\rightleftharpoons} R^1{-}CH{=}N{-}NH{-}R^4 + R^3{-}CH{=}N{-}NH{-}R^2$$

Figure 14.2 Two different reversible reactions: (a) disulfide exchange and (b) hydrazone exchange. The disulfide reaction is reversible only at basic pH. Decreasing the pH will freeze the equilibration of the library because of the protonation of the thiolate to thiol. The hydrazone exchange, on the contrary, is reversible only at acidic pH. Increasing the pH will result in the frozen library. The residues R^n are colored only to enable the readers' eyes to see the exchange. However, it is important to remember that the whole thiol and hydrazide groups are involved in the exchange and not only the residues.

(Figure 14.2), but in our group there are also examples of boronic ester exchange [15, 16]), as well as noncovalent bonds (e.g., metal–ligand exchange [17] or hydrogen bonding [18]).

Moreover, it was proven that two different reversible reactions can be carried out at the same time, giving an "orthogonal" screening [19]. Without any doubt, the idea to synthesize perfectly fitting guests for a specific host in one step (or the other way around: to synthesize a host for a specific guest) is intriguing and fascinating. Naturally in a novel field like this, many problems arise. The DCL is in thermodynamic equilibrium, and small physical or chemical changes can affect the library distribution. In addition to this, it should be noted that, the more

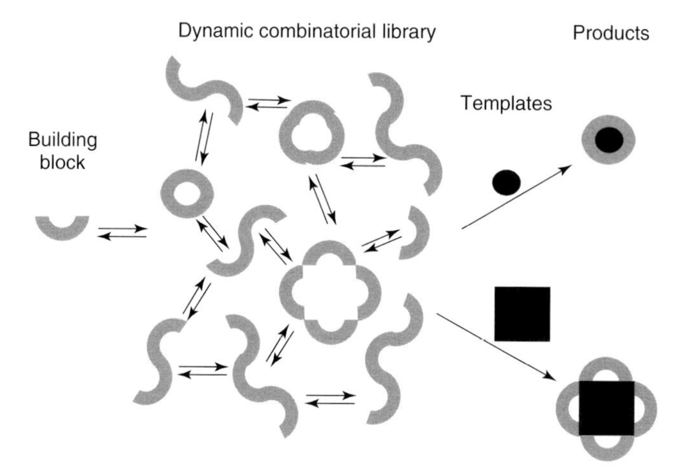

Figure 14.3 In dynamic combinatorial chemistry, only one building block can react to give a dynamic combinatorial library. All members of the dynamic library are in equilibrium one with another. The addition of a guest (solid black) shifts this equilibrium to the products that best bind it [20].[1)] Note that the building block must be a bifunctional molecule (for example, dithiols).

complex the library, the more difficult its screening, and small changes in the library composition can sometimes be overlooked. However, thanks to DCC, many novel hosts have been discovered in the last years.

It is interesting to note that also only one building block can be used for setting up a DCL. Various products can be amplified from this dynamic library using different templates. Starting from only one building block, it is possible to amplify various products (Figure 14.3).

Imine exchange, with disulfide and hydrazone exchange, is one of the most used reversible reactions in DCC. The condensation between aldehydes and amines is a reversible reaction. In order to obtain imines in good yield, it is usually important to remove water. In fact, the imine formation (from amine and aldehydes or ketones) is accompanied by the release of one molecule of water for each imine formed. In the specific case of DCC, hydrolysis plays an important role for the reversibility of the reaction. The freshly formed imines can be hydrolyzed, and the starting material will be free to react again forming a different imine. Imines can also be interconverted by transimination [21].

Beyond DCC, imines were and still are widely used for several reasons. Since their discovery by Hugo Schiff in 1864 (imines are also known as *Schiff's bases*) [22], they have found broad applications in organic and inorganic chemistry. The reason for this interest by organic chemists lies in the fact that condensation between aldehydes and imines is considered an easy reaction and that the generated imines

1) To be more precise: not the Gibb's free enthalpy of each single compound is determining the product distribution but the total Gibb's free enthalpy of the whole system. In most cases, the most stable products are formed but there are exceptions.

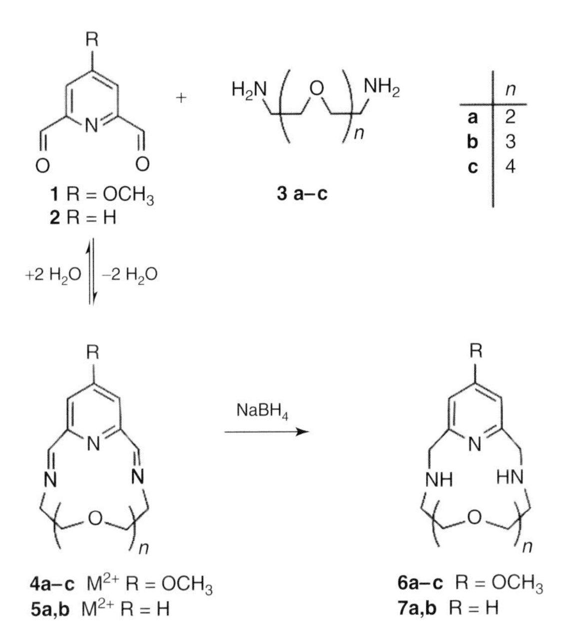

Figure 14.4 Dialdehyde **1** or **2** reacts with different diamino chains **3a–c** forming different products. Among those, macrocycles of different sizes (**4a–c** or **5a,b**) can be amplified by the addition of a proper cation (**4a**, **5a** Mg^{2+}, **4b**, **5b** Ca^{2+}, and **4c** Ba^{2+}). The last step is the reduction to amino macrocycles to stop the equilibration.

can then be used as versatile key intermediates in various synthetic routes: from reductive amination [23, 24] to multicomponent reactions such as the Mannich [25] or the Ugi reaction [26]. The latter was also one of the first reactions used in (nonreversible) combinatorial chemistry [27].

Inorganic imino complexes with different transition-metal ions have been used for a long time as catalysts. Salen, salen-like complexes [28], and bis(imino)pyridines [29] have taken the lion's share within the large plethora of related catalysts. Moreover, imine formation is extensively used in macrocycle syntheses [30, 31].

Even though this reaction was exploited in a large number of syntheses, it has only recently become an essential part of DCC. In 2002, Ole Storm and Ulrich Lüning elegantly demonstrated that imino macrocycles can be efficiently synthesized from a DCL (Figure 14.4) [12]. The reaction of a dialdehyde with diamino chains of different lengths produces various imino macrocycles, oligomers, and polymers.

Then, using different alkaline-earth metal ions, it is possible to shift the equilibria to mainly one of the various products. Cations act as templates for the formation of the macrocycle, amplifying the best binder and shifting all the other products of the library to it. Several products can be selectively obtained starting from a single DCL.

At the end of equilibration, imines are reduced to amines by the addition of $NaBH_4$ in order to "freeze" the dynamic library and to screen it. This step is necessary because imines usually are not stable enough to the purification steps and some of them could easily be hydrolyzed during this process. Unfortunately, once the imino macrocycle is reduced to an amino macrocycle, it will be not exactly the amplified binder anymore, as the imine is reduced to an amine.

Since Lüning's paper first appeared in the literature, imine formation has continued to grow in importance as a reversible reaction in DCC. In the last decade, many publications appeared regarding DCLs made of imines [32–34]. Related work was done by Vicente Gotor's group in Oviedo. Starting from a library composed of pyridine-2,6-dicarbaldehyde and *trans*-cyclohexane-1,2-diamine, they amplified different macrocycles (dimer and trimer) using various cations [35].

A few years later, they discovered a diastereoselective amplification from a DCL of macrocyclic oligoimines. Using cadmium ions, they were able to shift the equilibria to one major product as a single diastereomer from a mixture of stereochemically different species [36].

Although many different DCLs were exploited and many receptors were discovered and synthesized using this methodology, the "*road to fulfilling the promise*" of using DCC for new areas of application is still long [37].

14.2.1
The Next Step: Applications

Since the birth of DCC, DCLs were mainly applied to synthesize the "best binder" for a specific guest. On the contrary, in recent years, a few publications have appeared describing the use of a DCL directly for a function instead of single purified products [38–41]. This perspective is growing in importance day after day [42]. We were the first [43], in parallel to Jeremy Sanders' group in Cambridge [44], to use directly a DCL for the screening of a transport carrier. Transport of molecules from a source phase to a receiver phase passing through a membrane is a remarkable but usually difficult task to achieve. In nature, this task is done in various ways, from channels to carriers and by various mechanisms such as passive or active transport.[2)]

Those channels and carriers are relevant in cells in which the concentrations of metabolites, salts, and pH make the difference between life and death. Chemists have, for a long time, tried to mimic and understand transport in nature by studying synthetic membranes and carriers or channels.

2) It is impossible to give a comprehensive overview of transport in all of its facets without writing a book. However, it will be helpful to remember that passive transport does not require energy. The movement of molecules or ions across the membrane is due to a concentration or electrochemical gradient. The flux is achieved by simple diffusion, facilitated diffusion by the help of carriers or channels, and by osmosis. Active transport, on the contrary, is the movement of molecules or ions against a concentration or electrochemical gradient and requires energy.

Various biological membrane mimics were used to understand how transport works. Nowadays, a number of membranes are being used and studied, for example, bulk membranes, supported liquid membranes (SLMs), and liposomes. Each of these membranes has its weak and strong points. For example, whereas it is possible to use a bulk membrane for months, it will be difficult with the more fragile liposomes.

Parallel to studying different membranes, focus is also placed on synthesizing artificial channels and carriers.

Channels and carriers have great importance in many diverse fields. Ion sensors are mainly based on carriers immobilized in a polymeric film. When in contact with a fitting ion, it will be transported across the polymeric film to the detector, signaling the presence of the ion. Using a specific carrier for a given ion is also helpful in the cleaning of wastewater (for example, to remove toxic transition-metal ions or radioactive particles) or for blood dialysis. Moreover, drug delivery is one of the major research lines in the pharmaceutical industry with a special focus on targeting. Even though we do have effective drugs against cancer, this is still one of the worst killer diseases nowadays. The main problem of those anticancer medications is that they are not target specific. Those drugs will simply "kill" nonpreferentially all cells, both diseased and normal ones.

For in vivo imaging, small molecules are easily accessible but it is important to deliver them to the right cell that needs to be screened; otherwise they become useless. For these reasons, it is essential to focus on the synthesis of water-soluble carriers for a specific guest. It is also important to modify them in order to recognize or be recognized by a targeted cell. These two tasks are usually translated into hard benchwork to synthesize the carrier step by step and to optimize it. To overcome and simplify this task, this work explored the possibility of using directly a DCL to screen for a specific carrier. Synthesizing a performing carrier is not trivial, as it must have well-defined properties: it must complex the guest to transport; it must be lipophilic enough for passing through the membrane; and it must release the guest into the receiver water phase. If the carrier binds the guest very tightly, it will not release the guest and will be not a suitable carrier. A fine balance between binding properties and solubility is thus required for a performing carrier. Therefore, it is difficult to design a carrier with those properties a priori, while it may be possible to screen directly the carrier activity using DCC.

14.3 Ion Transport across Membranes Mediated by a Dynamic Combinatorial Library

The advantage of using the DCC instead of a step-by-step synthesis is clear. In a schematic summary, the differences between the two approaches are even more evident (Figure 14.5).

In a step-by-step synthesis (Figure 14.5a), every single intermediate of the synthetic route must be synthesized and purified. Usually, the first reaction tried does not work with 90% yield immediately. It is then necessary to carry out some

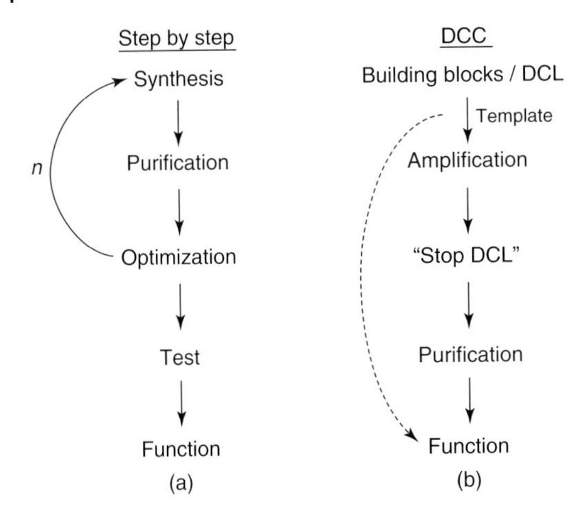

Figure 14.5 Schematic representation of (a) a step-by-step synthesis for *n* synthetic steps and (b) dynamic combinatorial chemistry. It is possible to screen for a function directly by the use of a dynamic library (b, dashed arrow).

optimization work to find out the best reaction conditions. All these steps must be done *n* times in a multistep synthesis. At the end of hard benchwork, it is finally possible to test the product for an application.

On the contrary, in the DCC approach (Figure 14.5b), the building blocks are mixed, the template is added, and the mixture is left to equilibrate. If and when an amplification occurs, it is necessary to stop the equilibration and to purify the final amplified product. At this point, the product is ready to be tested for an application.

Our main goal was to bring DCC one step further: to use a DCL directly to screen for a function. In this way, it should be possible to skip several steps in the DCC flowchart (Figure 14.5b, dashed arrow).

Using the dynamic library directly instead of a purified product should result not only in the amplification of a product but also of a product capable of performing a task. The benefit of synthesizing a product for an application in only one step is, without any doubt, attractive.

Many functions had not yet been screened by a DCL, and we focused on transport as the main function. DCC was still young and there were no proofs of using a library in order to amplify a carrier for transport. The starting point was a well-known DCL. It is based on imines and was used in our laboratory [12]. As a first guest to be transported, calcium ion was chosen. This ion was used in a previous investigation and its ability to shift a specific imine DCL was already proven [12]. In general, the transport is possible for many ions and guest molecules.

For a good and complete study of using a DCL to amplify a carrier, various tests were done. First of all, in fact, it was necessary to study the stability of imines in water and how the ionic guest influences this stability [45].

Water plays an important role in the DCC of imines. Hydrolysis of imines is one of the possible pathways for an imine exchange: the reaction of the imine with water to give back the aldehyde, which then can react with a different amine. Several experiments were set to test the stability of imines at different percentages of water. Three different compounds were synthesized (imino macrocycles **4b** and **9** and the linear imine **11**) and left in methanol to equilibrate using calcium ions as template for the formation of the two macrocycles. After equilibration, various amounts of water were added. The NMR spectra were recorded after each addition of water to check the stability against hydrolysis. For the formation of the imines, three different amines were chosen. One had the proper length to form a good host for complexing calcium ions (Figure 14.6, **4b**). One was a longer diamine but with the same number of donor atoms (Figure 14.6, **9**). The last one formed a product that had almost no affinity to the template (Figure 14.6, **11**). All NMR experiments showed that the (bis)imino pyridine **4b** was extremely stable to hydrolisis at room temperature, up to one week. In pure water, this macrocycle can be obtained in good yield when two equivalents of template were used. This

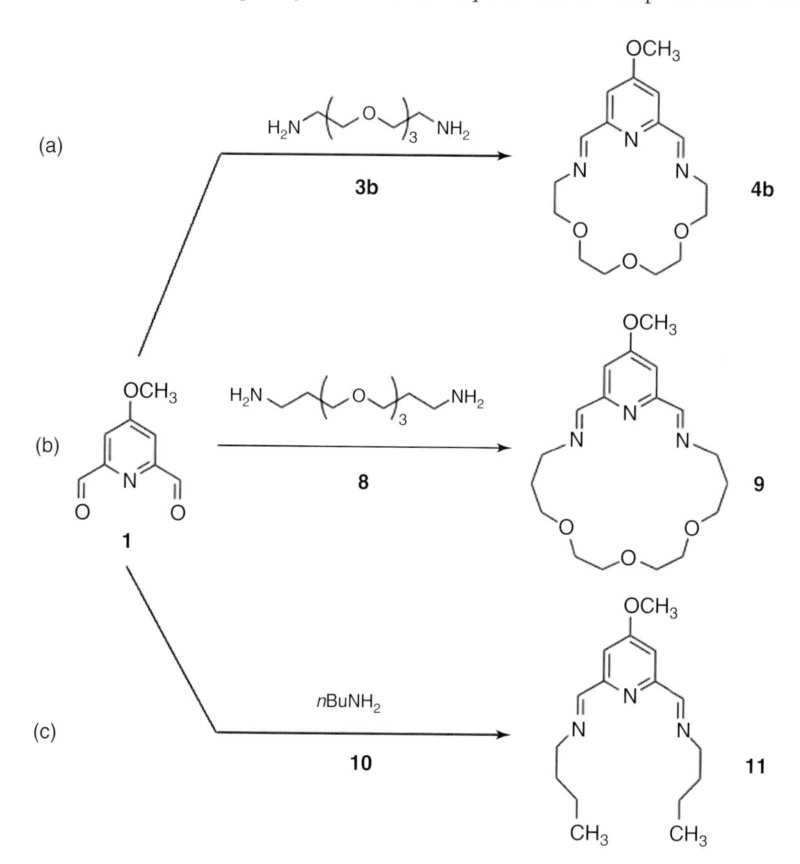

Figure 14.6 (a–c) The experiments were run in CH_3OD and then various aliquots of D_2O were added to test the stability of the three imino products (on the right). Various amounts of $CaCl_2$ as template were used.

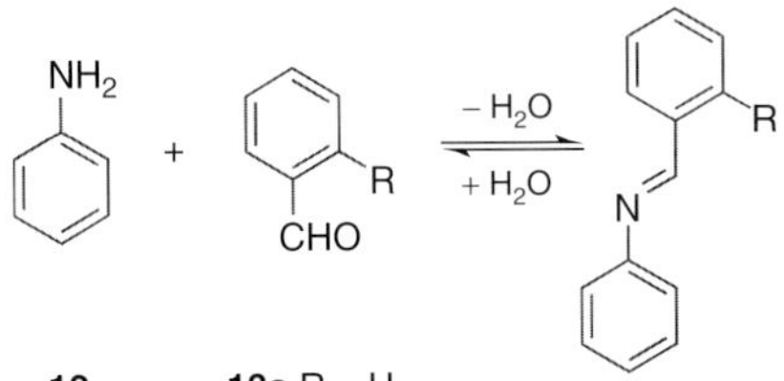

Figure 14.7 The reaction of aniline (**12**) and benzaldehyde (**13a**) or salicylaldehyde (**13b**) produces the respective imine **14a** or **14b**. The reaction is reversible, and water can drive it back to the starting materials.

simple experiment shows the applicability of DCC of imines also using water as solvent.

However, during these experiments, a discrepancy was found in the literature. In two different publications, the same imine synthesis in water had been reported with completely different yields [46, 47]. Starting from the same starting materials (benzaldehyde or salicylaldehyde and aniline, Figure 14.7) and using the same solvent (water), the difference of yields was more than 80% between the results of the two groups. Therefore, both experiments were repeated and the results were carefully analyzed in order to understand and explain the difference [48].

When both experiments were repeated, the same diverging results were obtained. The main problem was that the working groups could not differentiate between experiment and workup. In fact, it became clear that the starting materials and the final products were simply not soluble in water, giving a biphasic system. When the reaction was screened by ^{1}H NMR, the spectra showed that only a small amount of imine dissolved in deuterated water. On the contrary, when the two-layer reaction was extracted with chloroform and then concentrated to dryness, it gave a high yield of imine.

An additional experiment was carried out to reinforce the proof that water was not playing any role in that synthesis. The starting materials were mixed without the addition of any solvent and left under vacuum. Using no solvent, the imines were obtained in high yield. This was probably due to the fact that water from the reaction could be easily removed by reduced pressure driving the reaction to the product.

After proving that the dynamic library composed of dialdehyde **1** and diamine **3b** could be used in water, the same library was studied in a biphasic system (Figure 14.8) [43].

In the biphasic system, the building blocks were dissolved in the two different layers (chloroform/water) and calcium chloride was dissolved in water. Using the deuterated solvent, it was possible to monitor the composition of the two layers by NMR over time. When calcium was used as template, after one week, the imino macrocycle **4b** was found in quite good yield in the water layer. Even though the two building blocks were separated into two different phases, it was possible to set

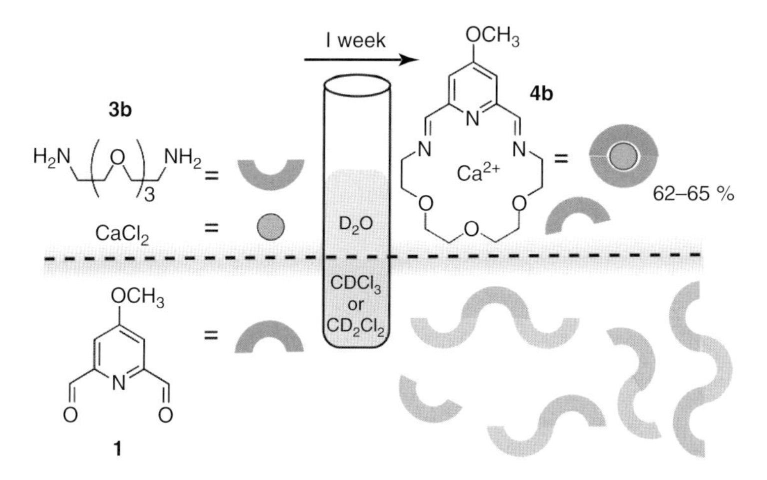

Figure 14.8 Biphasic experimental setup and its cartoon representation.

a DCL and macrocycle **4b** was amplified by the presence of calcium ions. Blank experiments with no calcium added to the water phase showed almost no formation of macrocycles at all.

After this proof, the carrier experiment in a bulk membrane system was setup to study whether a DCL could be used to screen directly for ion transport [43].

In a bulk membrane, the ion source phase is separated from a water receiver phase by a layer of organic solvent in which the carrier is dissolved. In this experiment, diamine **3b** and calcium ions were dissolved in the aqueous source phase (Figure 14.9a), dialdehyde **1** was dissolved in the organic phase (Figure 14.9b), and the receiver phase was pure water (Figure 14.9c). After one week, the imino macrocycle **4b** and calcium ions were found in the receiver phase. This phase was screened by NMR, electrospray ionization mass spectrometry (ESI-MS), and a calcium titration test. ^{1}H NMR of the receiver phase showed the presence of the macrocycle **4b**, while the ESI-MS of the same solution showed the masses of both **4b** and its calcium complex **4b**·Ca^{2+}.

Moreover, with a calcium titration test, it was also possible to detect the presence of calcium ions in the receiver phase. The blank experiment with calcium ions and only one of the two building blocks (diamine or dialdehyde) revealed that none of the two alone transported calcium ions into the receiver phase. Using the two building blocks without calcium ions in the source phase revealed that macrocycle **4b** was almost not synthesized at all.

This is the first proof of a macrocyclic carrier amplified directly by a DCL. These experiments prove that a dynamic library can be directly used for a function. Neither a presynthesized carrier nor an isolated one has been used for transport, but the carrier was amplified directly from the library which then acted as carrier through a bulk liquid membrane. In parallel to our work, a conceptually similar set of observations (using disulfide exchange as reversible reaction) were also reported by Sander's group in Cambridge [44].

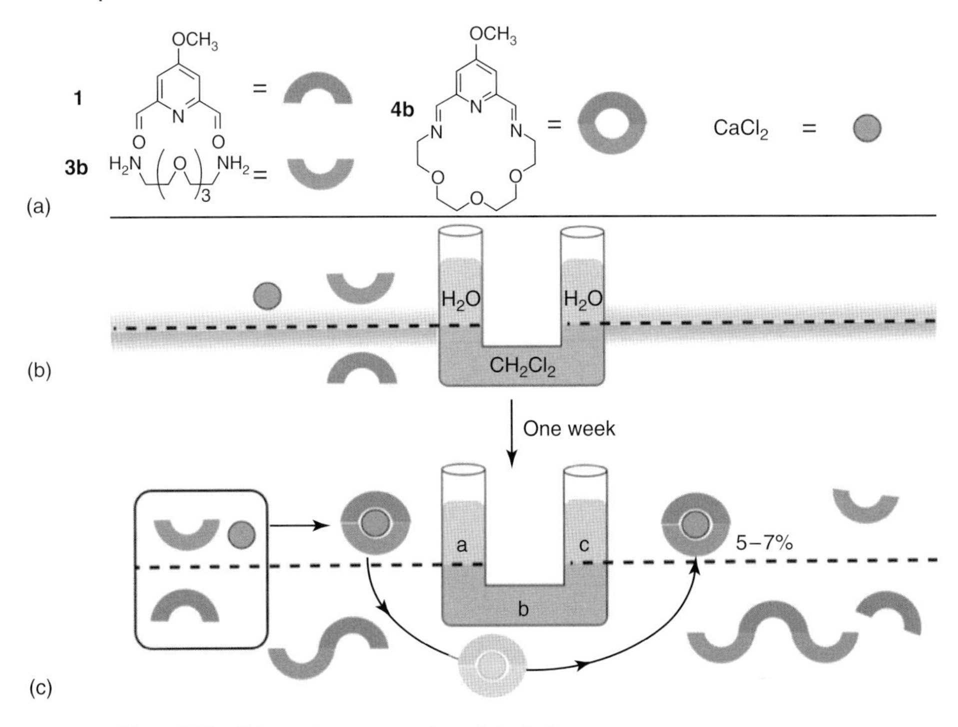

Figure 14.9 Schematic representation of the bulk membrane/DCL experiment. Starting material and the detected final macrocycle are on top. Transport of calcium ions from (a) water source phase to (c) water receiver phase passing through (b) organic bulk membrane.

These exciting and promising results motivated us to test other membranes and to increase the number of building blocks in the dynamic library. One of the drawbacks of liquid bulk membrane is the thickness of the organic membrane itself, which usually is in the range of centimeters. This means a slower transport with respect to thinner membranes.

For this reason, an SLM was chosen as a second membrane to be tested [49]. The thickness of this type of membrane was in the range of millimeters and was composed of a polymeric support soaked with an organic layer. This membrane separated the ion source phase from the receiver phase.

For this transport experiment, a new dialdehyde **15** was synthesized. The low transport rate of calcium ions through the bulk membrane seemed to be caused by the good water solubility of the formed imino macrocycle (in the biphasic experiment, the imino macrocycle **4b** was found only in the water layer, Figure 14.8). The new dialdehyde **15** bears a pentyl group to increase its lipophilicity.

Preliminary experiments of the DCL composed of the lipophilic dialdehyde **15** and the diamine **3b** using the SLM showed six times better calcium transport with respect to the previous DCL composed of dialdehyde **1**.

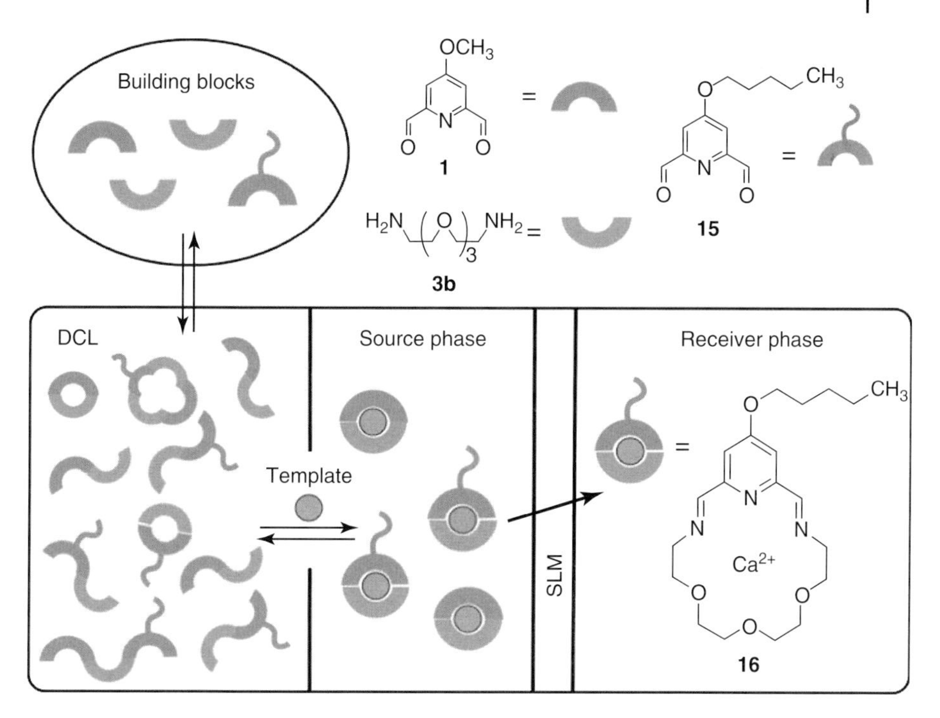

Figure 14.10 Schematic representation of the experiment. Starting from a DCL consisting of calcium ions, one diamine, and two dialdehydes, two macrocycles templated by calcium ions are formed, and one of them transports the ions through a supported liquid membrane (SLM).

After the preliminary experiments, a DCL composed of three building blocks (diamine **3b** and the two dialdehydes **1** and **15**, Figure 14.10) and calcium as the template ion was set. Screening was achieved with the help of a conductivity meter in the receiver phase and by ^{1}H NMR in the source phase. The conductivity meter in the receiver phase detected the passage of charged species (calcium ions) over time. By means of ^{1}H NMR, it was possible to determine the imino species (aromatic imine peaks lie in the region of 8–9 ppm) present in the source phase.

After one day, two macrocycles were formed in the source phase because of their calcium complexing abilities and mainly one of them passed to the receiver phase transporting calcium ions because of its better transport activity with respect to the other. The NMR screening of the source phase showed two peaks in the aromatic region of the spectra (Figure 14.11). Those two species have been assigned to the calcium ion complexes of two different macrocycles (**4b** and lipophilic macrocycle **16**). As the two peaks were well separated, it was also possible to screen the difference in the integration of these two peaks over time. The integration of the two peaks in the NMR spectra revealed a change in composition and in the ratio of the two macrocycles in the source phase over time. With this screening, it was clear that the lipophilic macrocycle was decreasing in the source phase and passing

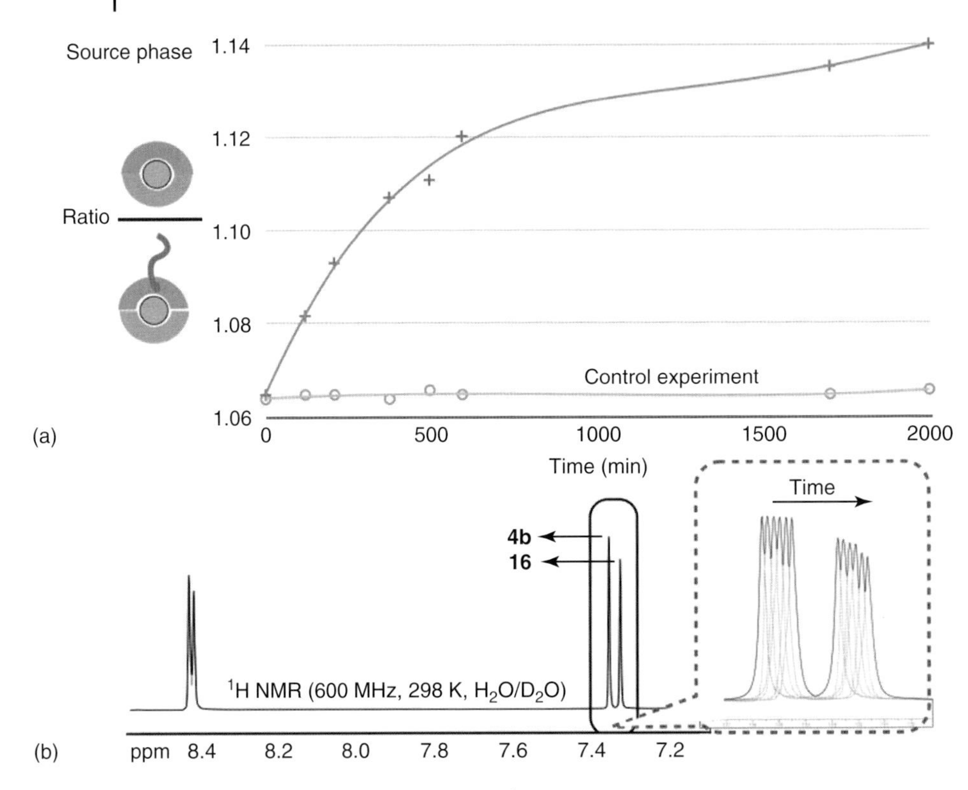

Figure 14.11 Analysis of the experiment shown in Figure 14.10. (a) Ratio of the two macrocycles over time. (b) ^{1}H NMR expansion of the source phase. For the cartoon representation, see Figures 14.9 and 14.10.

into the receiver phase. Moreover, the screening of the receiver phase by the use of the conductivity meter showed an increasing amount of calcium ions over time, proving that the lipophilic macrocycle **16** was transporting the calcium ions. The two curves (ratio of the two macrocycles in the source phase and conductivity in the receiver phase) showed the same shape.

With respect to the bulk membrane experiment, this methodology has a double screening activity: in the first screening, two different macrocycles are amplified from a DCL; during the second screening, the membrane itself chooses the best carrier between the two macrocycles.

This means that in only one step, not only a carrier was amplified, but the better one (the one that has both calcium complexing properties and greater transport properties) of the two was obtained. Moreover the "fittest" species, in terms of best binder and best transporter, will cross the membrane, passing from the source phase to the receiver phase. Here, it can be easily detected and separated from all the other components of the library, reducing to a minimum the effort in the purification step.

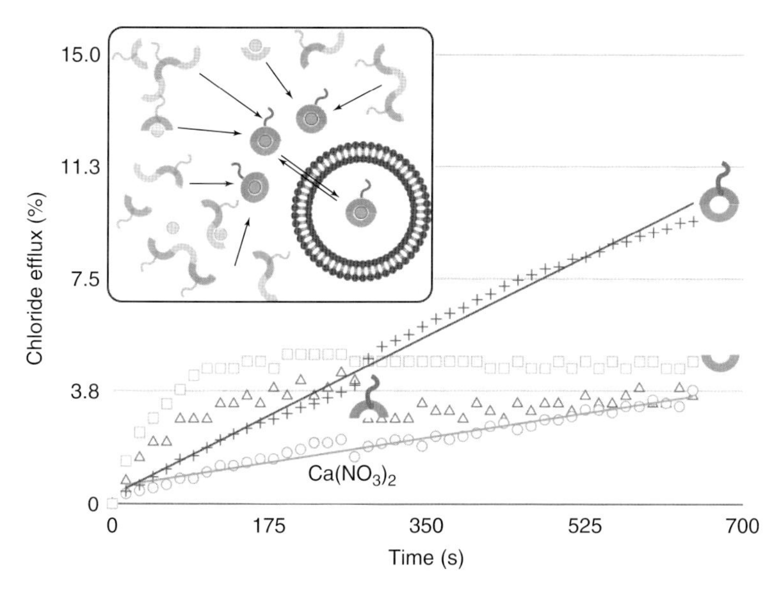

Figure 14.12 Schematic representation of transport of calcium across liposomes using a dynamic combinatorial library (top). At the bottom are chloride efflux over time on addition of the library composed of **15** and **3b** (+) and blank experiments with only one of the building blocks or only calcium nitrate.

Some preliminary experiments with liposomes were also carried out [49]. Liposomes are usually used in laboratories as cell membrane mimics. Also in this case, the dynamic library was used in place of a purified carrier. It was also possible to demonstrate with this kind of membrane that the lipophilic macrocycle was formed and helped in the transport of calcium chloride and calcium nitrate in and out of the liposome. For this experiment, a calcium chloride-loaded liposome was suspended in a calcium nitrate solution and the efflux of chloride was screened over time on addition of the dynamic library composed of dialdehyde **15** and diamine **3b** (Figure 14.12).

The efflux of chloride was compared when the whole library was added and when only one building block was added. Although the two building blocks alone showed an initial efflux of chloride, after some time the efflux stopped and it remained constant over time. This was probably due to the detergent activity of the building blocks. On the contrary, when the whole library was used, the efflux of chloride was constantly increasing over time, showing transport activity.

Using a DCL directly for a function is a remarkable step forward. This methodology allows screening in one step not only for a product but for a product useful for an application.

In this work, the direct use of a DCL for screening a given function was demonstrated. The transport of ions mediated by a carrier amplified from a DCL was tested and successfully achieved. A carrier macrocycle was formed directly

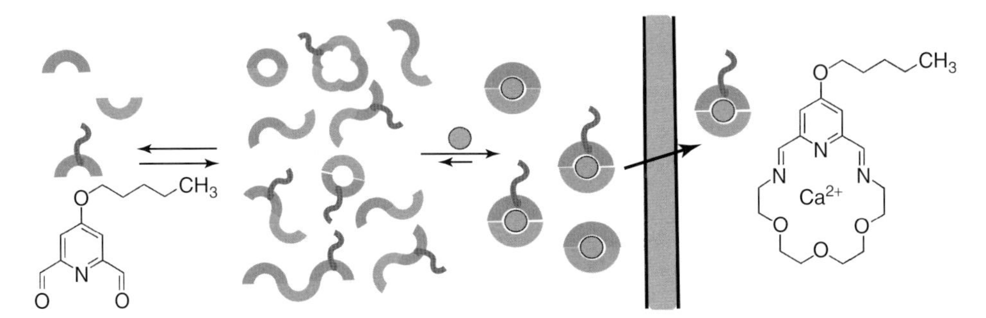

Figure 14.13 Double screening methodology. From three building blocks (left), a dynamic combinatorial library is setup. In a first screening, the addition of a guest template amplifies two related hosts (center). In a second screening, the membrane itself (pear color) selects the host with better carrier activity (right). Equilibrium arrows between the components of the library are omitted for clarity. For a cartoon representation, see Figures 14.9 and 14.10.

from a DCL composed of imines using calcium ions as a template. Then the macrocycle transported calcium ions from a water source phase to a water receiver phase passing through different organic synthetic membranes.

There is great interest and has a wide application of carriers in various fields. This work is a step forward in the application of DCC.

This one-step methodology is extremely flexible and can be used with different synthetic membranes. In fact, its application was successfully tested using membranes of thickness on the order of centimeters (bulk membrane) to liposomes of nanometer thickness. With the double screening methodology developed, it is also possible to screen for the best carrier from a number of possible carriers in an easy and fast way (Figure 14.13). Moreover, the presented methodology is, in principle, adaptable to all reversible reactions. This latter point is indeed interesting for the synthesis of novel and diverse carriers in a one-step manner.

However, the route for this methodology to achieve better performance is still long. The use of larger libraries composed of a larger number of building blocks with different lipophilicities will improve drastically the potential of the screening. Naturally, when more products are screened, the chances of finding a carrier with high performance are higher.

The results of this work now opens also new interesting tasks: testing the methodology with cell membranes.

In fact, the logical continuation of the experiments presented in this work should be to use various cells. Using different cell lines, it should be possible, in principle, to screen and find the best carrier for each type of cell membrane. Moreover, the building blocks can be modified using fluorescent groups. With those groups, it should be easy to follow the movement and the position of the carrier in different cells. Using various drugs and a library composed of fluorescent building blocks, the formation of the carrier and the cellular uptake of the drug could also be screened.

References

1. Lehn, J.-M. (1995) *Supramolecular Chemistry*, Wiley-VCH Verlag GmbH, Weinheim.
2. Pedersen, C.J. (1967) *J. Am. Chem. Soc.*, **89**, 7017–7036.
3. Steed, J.W. and Atwood, J.L. (2009) *Supramolecular Chemistry*, 2nd edn, John Wiley & Sons, Inc., New York.
4. Steed, J.W., Turner, D.R., and Wallace, K.J. (2007) *Core Concepts in Supramolecular Chemistry and Nanochemistry*, Wiley-VCH Verlag GmbH, Weinheim.
5. Frömming, K.-H. and Szejtli, J. (1994) *Cyclodextrins in Pharmacy*, Kluwer Academic Publisher, Dordrecht.
6. Stoddart, J.F. (2009) *Chem. Soc. Rev.*, **38**, 1802–1820.
7. Kay, E.R. and Leigh, D.A. (2005) in *Functional Artificial Receptors* (eds T. Schrader and A.D. Hamilton), Wiley-VCH Verlag GmbH, Weinheim, pp. 333–406.
8. Feringa, B.L. (2001) *Acc. Chem. Res.*, **34**, 504–513.
9. Corbett, P.T., Leclaire, J., Vial, L., West, K.R., Wietor, J.-L., Sanders, J.K.M., and Otto, S. (2006) *Chem. Rev.*, **206**, 3652–3711.
10. Huc, I. and Lehn, J.-M. (1997) *Proc. Natl. Acad. Sci. U.S.A.*, **94**, 2106–2110.
11. Brady, P.A. and Sanders, J.K.M. (1997) *J. Chem. Soc., Perkin Trans. 1*, **21**, 3237–3252.
12. One example of a Dynamic Combinatorial Library based on imines studied in our laboratory: Storm, O. and Lüning, U. (2002) *Chem. Eur. J.*, **8**, 793–798.
13. A recent example of a Dynamic Combinatorial Library of hydazone macrocycles: Klein, J.M., Saggiomo, V., Reck, L., McPartlin, M., Dan Pantos, G., Lüning, U., and Sanders, J.K.M. (2011) *Chem. Commun.*, 3371–3373.
14. Report on selection and amplification of hosts from a DCL of macrocyclic disulfides: Otto, S., Furlan, E.L.R., and Sanders, J.K.M. (2002) *Science*, **297**, 590–593.
15. Lüthje, S., Bornholt, C., and Lüning, U. (2006) *Eur. J. Org. Chem.*, 909–915.
16. Stoltenberg, D., Lüthje, S., Winkelmann, O., Näther, C., and Lüning, U. (2011) *Eur. J. Org. Chem.*, 5845–5859.
17. Saur, I. and Severin, K. (2005) *Chem. Commun.*, 1471–1473.
18. Calama, M.C., Hulst, R., Fokkens, R., Nibbering, N.M.M., Timmerman, P., and Reinhoudt, D.N. (1998) *Chem. Commun.*, 1021–1022.
19. Goral, V., Nelen, M.I., Eliseev, A.V., and Lehn, J.-M. (2001) *Proc. Natl. Acad. Sci. U.S.A.*, **98**, 1347–1352.
20. See for instance: Reek, J.N.H. and Otto, S. (eds) (2010) *Dynamic Combinatorial Chemistry*, Wiley-VCH Verlag GmbH, p. 36ff.
21. Meyer, C.D., Joiner, C.S., and Stoddart, J.F. (2007) *Chem. Soc. Rev.*, **36**, 1705–1723.
22. Schiff, H. (1864) *Ann. Chem.*, **131**, 118–119.
23. Borch reductive amination by $NaBH_3CN$: Borch, R.F., Bernstein, M.D., and Durst, H.D. (1971) *J. Am. Chem. Soc.*, **93**, 2897–2904.
24. Reductive amination by $NaBH(OAc)_3$: Abdel-Magid, A.F., Carson, K.G., Harris, B.D., Maryanoff, C.A., and Shah, R.D. (1996) *J. Org. Chem.*, **61**, 3849–3862.
25. Mannich, C. and Krösche, W. (1912) *Arch. Pharm.*, **250**, 647–667.
26. Ugi, I. (1962) *Angew. Chem.*, **74**, 9–22; (1962) *Angew. Chem. Int. Ed.*, **1**, 8–21.
27. Terret, N.K. (1998) *Combinatorial Chemistry*, Oxford University Press, Oxford.
28. Gupta, K.C. and Sutar, A.K. (2008) *Coord. Chem. Rev.*, **252**, 1420–1450.
29. Gibson, V.C., Redshaw, C., and Solan, G.A. (2007) *Chem. Rev.*, **107**, 1745–1776.
30. Vigato, P.A., Tamburini, S., and Bertolo, L. (2007) *Coord. Chem. Rev.*, **251**, 1311–1492.
31. Borisova, N.E., Reshetova, M.D., and Ustynyuk, Y.A. (2007) *Chem. Rev.*, **107**, 46–79.
32. Epstein, D.M., Choudhary, S., Churchill, M.R., Keil, K.M., Eliseev, A.V., and Morrow, J.R. (2001) *Inorg. Chem.*, **40**, 1591–1596.

33. Lin, J.-B., Xu, X.-N., Jiang, X.-K., and Li, Z.-T. (2008) *J. Org. Chem.*, **73**, 9403–9410.
34. Herrmann, A., Giuseppone, N., and Lehn, J.-M. (2009) *Chem. Eur. J.*, **15**, 117–124.
35. González-Álvarez, A., Alfonso, I., López-Ortiz, F., Aguirre, A., García-Granda, S., and Gotor, V. (2004) *Eur. J. Org. Chem.*, 1117–1127.
36. González-álvarez, A., Alfonso, I., and Gotor, V. (2006) *Chem. Commun.*, 2224–2226.
37. Ladame, S. (2008) *Org. Biomol. Chem.*, **6**, 219–226.
38. Catalysis used as self-screen for substrates from a Dynamic Combinatorial Library: Larsson, R., Pei, Z., and Ramström, O. (2004) *Angew. Chem.*, **116**, 3802–3804; (2004) *Angew. Chem. Int. Ed.*, **43**, 3716–3718.
39. "Dynamic Combinatorial Libraries of Dye Complexes as Sensors": Buryak, A. and Severin, K. (2005) *Angew. Chem.*, **117**, 8149–8152; (2005) *Angew. Chem. Int. Ed.*, **44**, 7935–7938.
40. "Dynamic Combinatorial Libraries of Metal-Dye Complexes as Flexible Sensors for Tripeptides": Buryak, A. and Severin, K. (2006) *J. Comb. Chem.*, **8**, 540–543.
41. Nowadays the use of a chemical library instead of a single compound is growing in importance (system chemistry): Ludlow, R.F. and Otto, S. (2008) *Chem. Soc. Rev.*, **37**, 101–108.
42. Herrmann, A. (2009) *Org. Biomol. Chem.*, **7**, 3195–3204.
43. Saggiomo, V. and Lüning, U. (2009) *Chem. Commun.*, 3711–3713.
44. Pérez-Fernandez, R., Pittelkow, M., Belenguer, A.M., Lane, L.A., Robinson, C.V., and Sanders, J.K.M. (2009) *Chem. Commun.*, 3708–3710.
45. Saggiomo, V. and Lüning, U. (2008) *Eur. J. Org. Chem.*, 4329–4333.
46. Godoy-Alcantar, C., Yatsimirsky, A.K., and Lehn, J.-M. (2005) *J. Phys. Org. Chem.*, **18**, 979–985.
47. Simion, A., Simion, C., Kanda, T., Nagashima, S., Mitoma, Y., Yamada, T., Mimura, K., and Tashiro, M. (2001) *J. Chem. Soc., Perkin Trans. 1*, 2071–2078.
48. Saggiomo, V. and Lüning, U. (2009) *Tetrahedron Lett.*, **50**, 4663–4665.
49. Saggiomo, V., Goeschen, C., Herges, R., Quesada, R., and Lüning, U. (2010) *Eur. J. Org. Chem.*, 2337–2343.

15
Toward Tomorrow's Drugs: the Synthesis of Compound Libraries by Solid-Phase Chemistry

Dagmar C. Kapeller and Stefan Bräse

Abbreviations

DBU	1,8-diazabicyclo[5.4.0]undec-7-ene
DCC	dicyclohexylcarbodiimide
DCE	1,2-dichloroethane
DDQ	2,3-dichloro-5,6-dicyanobenzoquinone
DEAD	diethyl azodicarboxylate
DHP	dihydropyranyl
DIAD	diisopropyl azodicarboxylate
DIC	diisopropylcarbodiimide
DIPEA	diisopropylethylamine
DMA	dimethylamine
DMAP	4-dimethylaminopyridine
DMF	*N,N*-dimethylformamide
DMFDMA	DMF-dimethylacetale
Fmoc	fluroenylmethoxycarbonyl
GABA	γ-aminobutyric acid
HATU	2-(1*H*-7-azabenzotriazol-1-yl)-1,1,3,3-tetramethyl uronium hexafluorophosphate
HMDS	hexamethyldisilazide
HMPA	hexamethylphosphoramide
HOAt	1-hydroxy-7-azabenzotriazole
HOBt	1-hydroxybenzotriazole
MW	microwave
NMP	*N*-methylpyrrolidone
PMP	*p*-methoxyphenyl
PPTS	pyridinium *p*-toluol sulfonatepy
BOP	benzotriazol-1-yl-oxytripyrrolidinophosphonium hexafluorophosphate
TFA	trifluoroacetic acid
TfOH	trifluoromethanesulfonic acid

New Strategies in Chemical Synthesis and Catalysis, First Edition. Edited by Bruno Pignataro.

THC tetrahydrocannabinol
THP tetrahydropyranyl
TMS trimethylsilyl

15.1 Introduction

The term "*drug*" has many different meanings depending on the context it is used in. According to the World Health Organization (WHO), it refers to "*any substance with the potential to prevent or cure disease or enhance physical or mental welfare* in the medical sense, while, in pharmacology, it amounts to *any chemical agent that alters the biochemical physiological processes of tissues or organisms*" [1].

15.1.1 The History of Drug Discovery

Drug research, as an industrial and interdisciplinary endeavor, is not much older than a century [2, 3]. In its early days, the discovery of new drugs was a slow process, largely based on empirical observations. Little was known about their mode of action. The phase between their administration and the appearance of physical symptoms was more or less treated as a black box. Leads came from natural products, clinical observations of the drugs' side effects, conference presentation, scientific publications, or patents.

All of this changed in the 1990s with the emergence of combinatorial chemistry [4–7] and high-throughput screening (HTS) [8–12]. By employing robotics and automation, it became technically feasible to synthesize and screen hundreds of thousands up to a million compounds within a year. The formation of large generalized libraries of diverse compounds as pipelines for new drug candidates soon became a common strategy within many pharmaceutical industries. However, the expected increase in hit productivity failed to deliver on its promises [13]. Often it took months to synthesize and characterize specific libraries. During that time, the lead could have already moved on or problems arose concerning the elucidation of the active species requiring complex deconvolution steps and concerns over synergistic efficacies.

Entering the 2000s, a shift in paradigm occurred toward directed, iterative libraries. Chemical lead optimization ranging from evaluation of screening hits to preclinical candidate identification became a seamless process, with parallel approaches and feedback cycles accelerating the procedure [14]. Furthermore, our steadily growing knowledge of disciplines such as genomics and proteomics multiplied the number of potential biological targets to be exploited. It also increased our understanding of the mechanisms behind their interaction with specific drugs. Nowadays, the synthesis and screening of small compound collections directed against specific targets prevails.

15.1.2
Characteristics of Druglike Molecules

Druglike molecules have to satisfy two different general principles: they must show activity toward a biological target, and they must display properties related to *drugability* [15]. The latter refers to the disposition of a pharmaceutical compound within an organism, usually described by the acronym ADMET (absorption, distribution, metabolism, elimination, toxicity) [16]. That means, the drug has to be soluble for ready absorption and reach its target within the body before losing its activity due to being metabolized. In the end, the drug and its metabolites have to be eliminated by excretion and all of it of course without being toxic. Almost 40% of all drugs in clinical development fail to make it to the market because of ADMET deficiencies [17].

The current strategy in industry is to first screen for receptor activity, closely followed by ADMET studies [16]. Furthermore, it is possible to predict certain properties, such as toxicity, absorption, or biological activity, by employing quantitative structure–activity relationship (QSAR) calculations [18].

According to empirical observations and computational calculations, Lipinski introduced the rule of five [19], which states that poor absorption and permeation are more likely when the following situations are encountered:

1) There are more than five H-bond donors (sums of OHs and NHs).
2) The molecular weight is over 500.
3) The log$P$ is over 5 (P = octanol/water partition coefficient).
4) There are more than 10 H-bond acceptors (sums of O and N).
5) Substrates for biological transporters are exceptions to the rules.

It is estimated that, of all existing small-molecule drugs, over 70% adhere to these rules. Furthermore, polar surface area was found to be another property associated with positive ADMET effects. Lipinski also observed trends within the pharmaceutical industry [20]. As such, he found that new drugs tended either toward a higher molecular weight and an increased number of H-bonds with concomitant poorer permeability or higher molecular weight paired with higher lipophilicity causing poorer solubility.

15.1.3
Drug Targets

Nowadays, drug discovery usually employs an approach called *rational drug design* [21]. It starts by first identifying and selecting molecular targets (biological target space), which are usually receptors or enzymes [22]. If their structure is unknown, ligand-based lead finding is the method of choice. Therefore, natural ligands and their analogs are used for structure–activity relationship (SAR) studies and to determine the direction of the library. On the other hand, if the structure of a target has been elucidated by X-ray crystallography or NMR analysis, it is possible to predict the binding affinity of different drug candidates by computational

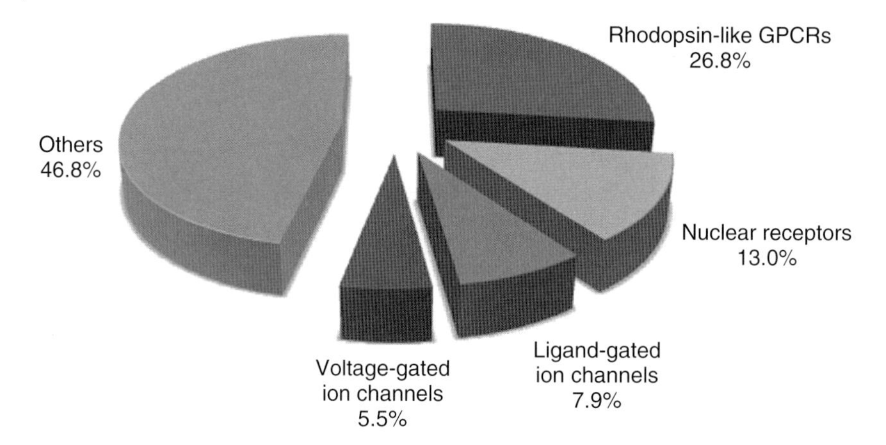

Figure 15.1 Classes of drug targets.

calculations, for example, molecular docking (structure-based drug design) [23]. In many cases, this can also be done for proteins of which only the primary sequence has been determined, for example, by cloning. Then a homology approach with a template protein of known structure with a similar amino acid sequence can be applied. The most promising structures are then used as starting points for the synthesis of a specific library (chemical ligand space).

In 2006, the number of drugs was estimated to be 21 000, which could be reduced to a mere 1357 by eliminating duplicates or salts, for example [15]. These pharmaceuticals act on 324 different biological targets, which can be grouped into only a few key gene families (Figure 15.1). It is interesting to note that over 50% of all existing drugs target various receptors or ion channels.

By far, the largest class of biological targets (>25%) are rhodopsin-like G-protein coupled receptors (GPCRs), which will be the focus of this chapter. They constitute the main part of the GPCR family (the other subclasses are secretin-like and glutamate-like receptors) and are estimated to encompass a total number of 750 receptors [24]. About 350 of these are olfactory for sensing odors, pheromones, or tastes. The remaining receptors, which act as possible drug targets, regulate numerous physiological processes, such as neuronal excitability, metabolism, reproduction, development, hormonal homeostasis, and behavior, on activation.

All GPCRs have a common structural framework composed of seven transmembrane helices. They are located at the cell surface and are responsible for the transduction of an endogenous signal into an intracellular response, the activation of guanine nucleotide-binding (G) proteins. Rhodopsin-like GPCRs share similar amino acid sequences but are activated by ligands of varying structures and characters. This makes them attractive targets for drug discovery, despite the lack of crystallographic data, which is the basis for a good structure-based approach. So far, X-ray data at atomic resolution could be obtained only for the mammalian bovine rhodopsin because of the membrane-bound nature of GPCRs [25, 26]. These data can be used as a starting point for homology modeling [27]. It is estimated that almost half of all newly launched drugs are agonists or antagonists of GPCRs [28].

15.1.4 Privileged Structures

Although it is now principally possible to synthesize and test millions of different compounds, it is not very effective to generate huge amounts of unrelated new structures. Experience and empirical observations, assisted by computational data mining, have shown that the vast world of bioactive compounds can be traced back to only a few structural motifs [29]. For those, Evans introduced the name *privileged structures*, describing them as "*a single molecular framework able to provide ligands for diverse receptors*" [30]. By modifying these scaffolds, it is possible to generate drug candidates for a broad range of biological targets on a reasonable timescale. Accordingly, their exploration is a rapidly emerging theme in medicinal chemistry [31]. It is not quite clear yet why certain scaffolds are privileged. They may display favorable physicochemical characteristics or topological features, or perhaps they mimic secondary structural elements, such as β-turns, as has been proposed for benzodiazepines [32]. The general consensus is that they contain at least two ring systems connected either by fusion (e.g., indoles) or by single bonds (e.g., biaryls) [33]. Furthermore, they are rigid systems that are able to project multiple side-chain substituents in a well-defined three-dimensional space.

In this chapter, several of such common substructures encountered among the ligands of various types of GPCRs are discussed, the main focus being on their synthesis on solid supports.

15.2 Solid-Phase Synthesis of Selected Privileged Structures

15.2.1 Introduction to Solid-Phase Synthesis

The invention of solid-phase synthesis goes back to Robert B. Merrifield [34]. He showed that it was possible to synthesize biopolymers (e.g., peptides) by prior attachment of a starting unit to a polymeric backbone, successive elongation with monomeric building blocks (e.g., amino acids), and cleavage (Scheme 15.1). Immobilization can be done directly on the resin or via prior modification with a chemical handle, a so-called linker, which facilitates cleavage under defined conditions. The main advantages of solid-phase synthesis are easy purification (filtration), high efficiency due to the employment of excess reagents, and facile automation. About 20 years after Merrifield's seminal paper in 1963, he was awarded the Nobel Prize.

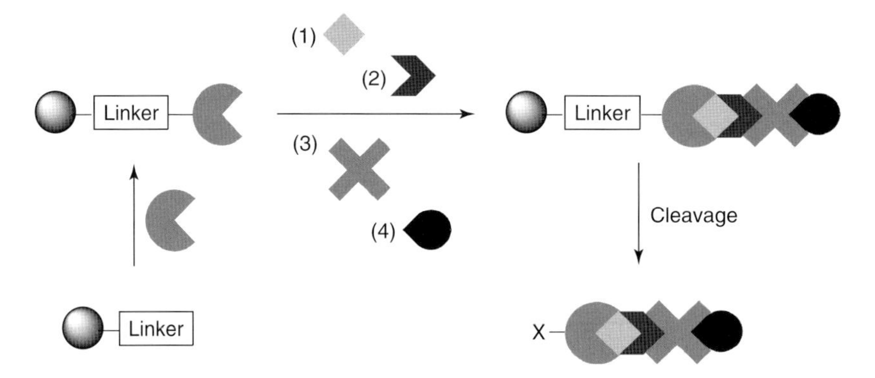

Scheme 15.1 Schematic drawing of a solid-phase synthesis employing a linker.

With the advent of combinatorial chemistry in the 1990s, the original use of solid-phase peptide synthesis was expanded [35, 36]. Its focus now shifted from linear biopolymers to small molecular entities, which are rapidly assembled from various building blocks/reagents. This new concept soon entered the literature as solid-phase organic synthesis (SPOS) [37]. The latter facilitates the generation of large compound libraries for drug discovery by employing novel combinatorial methodologies [36, 38, 39]. One of the main techniques used is the so-called split-and-mix strategy [40]. It involves the division of starting material immobilized on solid supports into a number of equal portions (x). Each of these is individually reacted with a single different reagent. Afterward, the individual portions are recombined, mixed, and divided again. Reaction with a further set of reagents gives all possible dimeric units. The number of compounds grows exponentially with n reaction steps (x to the power of n), leading to libraries of up to a million compounds in a short amount of time. These can be handled by automated technology platforms and modern encoding systems [41, 42], such as the IRORI™ Kan one [43, 44]. The latter is termed a *tea-bag method*, which means that aliquots of the resin are encapsulated in small reactor vessels with unique nonchemical tags for identification purposes. Another possibility is chemical encoding, where at each step, a molecule of a defined mass is added, which can be cleaved separately for deconvolution purposes at the end.

With this background information in hand, one has the basic knowledge to understand the next sections concerning the solid-phase synthesis of selected privileged structures.

15.2.2
Benzodiazepines

Benzodiazepines belong to the earliest recognized classes of bioavailable therapeutic agents. They show widespread activities (e.g., sedative, anxiolytic, and anticonvulsive) for various biological targets [45], and a few of these are discussed here. Research into benzodiazepines started in earnest with Sternbach's patenting

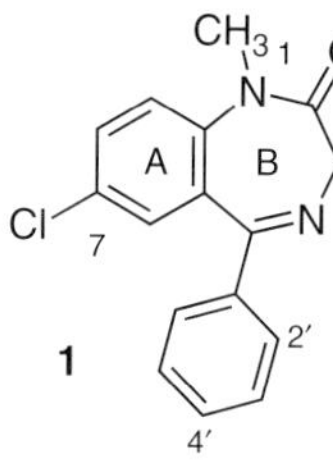

Figure 15.2 Diazepam (1, Valium®).

of the drug Librium® in 1959 and its marketing by Hoffmann-La Roche in the following year [46]. Soon, other derivatives followed, the most well known being diazepam (**1**), which became commercially available as Valium® in 1963 and is still in use (Figure 15.2).

Furthermore, Sternbach performed SAR studies concerning the γ-aminobutyric acid ($GABA_A$) receptor (for neurotransmitter regulation) on well over 3000 benzodiazepine compounds. This was done by drawing correlations between the substitution pattern of the ring system and their pharmacological performance. As such, an electron-withdrawing group at position 7 or a halogen at 2′ was found to increase activity, the same as methylation of $N-1$. Larger groups on *N* were not well tolerated, nor was the substitution at 4′.

The first mention the author came across on the solid-phase synthesis of 1,4-benzodiazepine derivatives stemmed from a Spanish patent in 1977 [47]. The synthesis involved the immobilization of *N*-protected glycine on poly(4-hydroxy-ethylstyrene), followed by deblocking and treatment with 2′-aminobenzophenone.

This method only impacted on the general chemical population in the 1990s. It was strongly promoted by Ellman, who was one of the pioneers in the extension of solid-phase synthesis to nonpolymeric organic compounds [48]. His novel approach started by the construction of a 192-membered library of structurally diverse 1,4-benzodiazepine derivatives containing a variety of chemical functionalities including amides, carboxylic acids, amines, phenols, and indoles [49]. They were then tested concerning their affinity toward the cholecystokinin receptor (CCK_A), which is primarily expressed in the gastrointestinal tract and regulates pancreatic enzyme secretion [50]. The synthesis started by attaching benzophenones, previously modified with a linker functionality, to amine-derivatized polyethylene pins (**2**) on microtiter plates (Scheme 15.2). This was followed by Fmoc deprotection with piperidine and coupling with Fmoc-protected amino acid fluorides. After deprotection of compounds **4**, cyclization was initiated by treatment with acetic acid. The resulting benzodiazepines (**5**) were then alkylated and cleaved under acidic conditions to give free **6**. Thus, the generated compounds were then screened in competitive radioligand-binding assays for their affinity to CCK_A. It was found that, generally, attachment to the resin on the benzophenone ring opposite to the one depicted in Scheme 15.2, led to reduced activity, while incorporating tryptophan (R^C part) led to a higher activity. Furthermore, the nature of the alkyl groups at R^C had a modest effect on binding affinity. The compound with the best

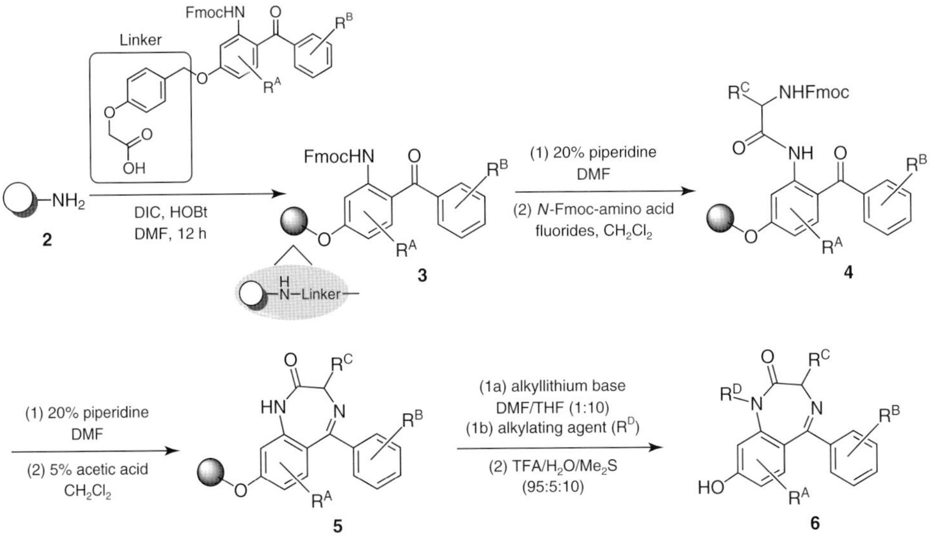

Scheme 15.2 Representative procedure toward benzodiazepines starting from selected benzophenones [49].

result (IC_{50} = 120 nM) was the one with $R^A = R^B$ = H, R^C = 3-indolylmethyl and R^D = ethyl.

Lattmann *et al.* also synthesized a library of 168 1,4-benzodiazepines, which were tested for activity against the second cholecystokinin receptor CCK_B [51]. Contrary to type A, it is expressed not only in the gastrointestinal tract but also in the central nervous system (CNS). It influences neurotransmission in the brain and is thus a relevant target for treating anxiety and panic. Lattmann's efforts were mainly focused on variations concerning position 3 of the benzodiazepine. He used Synphase® crown microreactors, modified with a *p*-hydroxymethyl-phenoxyacetamido linker (**7** in Scheme 15.3) similar to Ellman's route above, but attached to the resin beforehand.

This linker was coupled with 14 different Fmoc-protected amino acids. All compounds **8** were deprotected and each treated with 12 ketones of varying nature. Compounds **9** were then treated either with trifluoroacetic acid (TFA), affording cleavage and subsequent cyclization, or with pyridine and a catalytic amount of DMAP (4-dimethylaminopyridine) in the case of more sensitive amino acids. Consecutive receptor-binding assays were performed to establish the SARs and to identify selective ligands with high affinity for the CCK_B receptor. Similar to the $GABA_A$ receptor, a halogen in position 7 (R^B) increased affinity, while the opposite was found for C-2′. A phenyl group at C-5 (R^D) was essential for binding, and various modifications of substituents at C-3 (R^A) led to the conclusion that an unbranched propyl side chain with (*S*) configuration showed optimal binding. Considering these empirical rules, compound **11** was identified as the best ligand with an IC_{50} value of 180 nM.

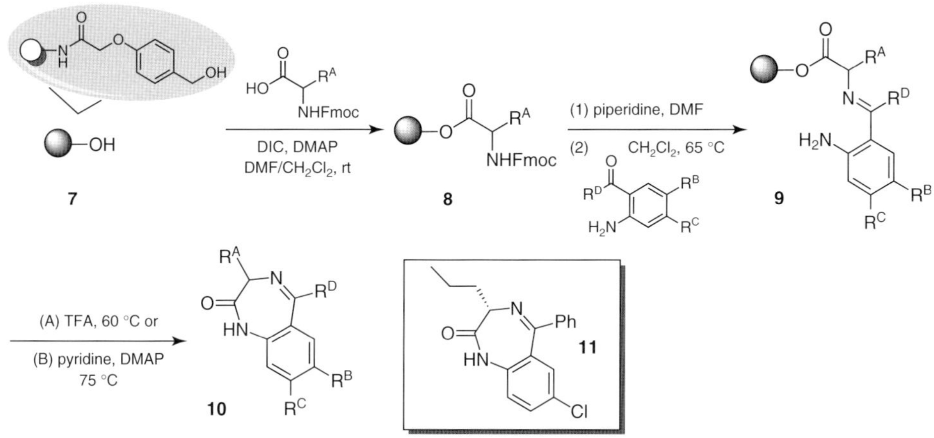

Scheme 15.3 Synthesis of a CCK_B activity-based benzodiazepine library [51].

Another biological target for benzodiazepines is the oxytocin receptor. It plays an important role in the onset of labor and has potential for the therapy for preterm labor. Evans *et al.* constructed a fully encoded differential release library consisting of 1296 benzodiazepine diamides [52]. They employed different chemical codes at each step of the reaction sequence, allowing unique identification of the products. Furthermore, the use of two different linkers, one acid and the other photo cleavable, facilitated orthogonal release strategies. The most potent compound identified from this library was GW405212X (**12**, Scheme 15.4) displaying an IC_{50} value of 5 nM. However, it turned out to be unsuitable as a drug candidate because of its poor pharmacokinetic profile. As such, Wyatt *et al.* set out to conduct SAR studies trying to improve its bioavailability [53]. They employed, among others, a solid-phase approach chemically identical to the sequence by Evans (Scheme 15.4).

It started by attaching an Fmoc-protected amino acid to an amino resin (**13**), followed by deprotection. Afterward, the common benzodiazepine core was coupled onto **14** and its Fmoc group removed. This was followed by another coupling/acylation step [C(O)R^B] toward amides, sulfonamides, and ureas. Treatment with TFA finally released products **16**, which were then tested for their affinity to the oxytocin receptor. Unfortunately, none of them showed any improvement over the original **12**. Nevertheless, important data could be gathered. As such, it was found that the chlorinated aromatic ring was essential for binding. A derivative with much improved pharmacokinetic properties was identified in a solution-phase approach by modifying the amide moiety used as the attachment point for the resin with a *m*-chlorophenylethyl substituent.

A group from Acadia Pharmaceuticals investigated agonists of Mas-related GPCRs (Mrg receptors), which are also called *sensor-neuron specific receptors* [54]. This family consists of over 50 subtypes of orphan receptors, expressed almost exclusively in small diameter sensory neurons. Their physiological function is still unknown, mainly because there are no rodent analogs of these receptors. But their expression pattern suggests a role in nociception, making them attractive potential

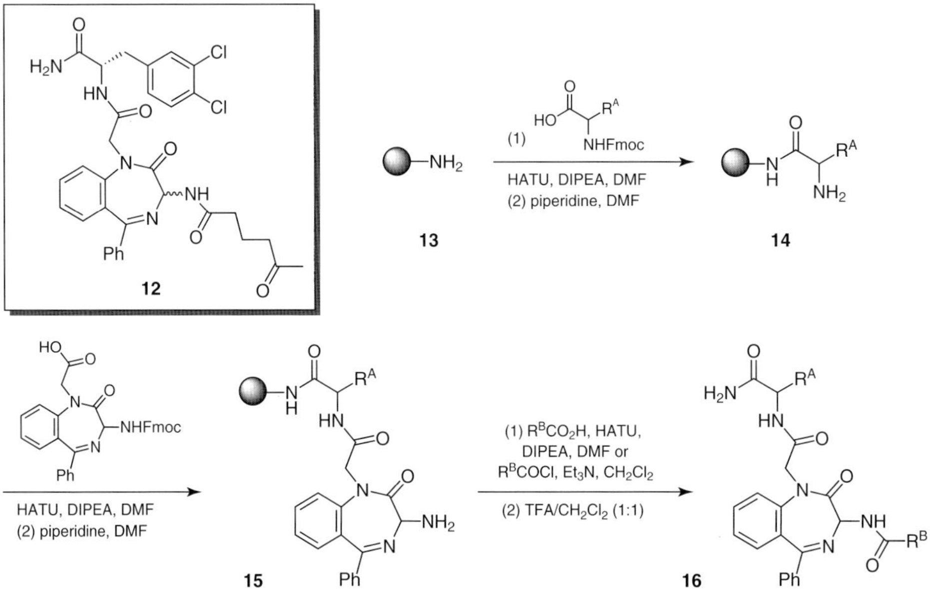

Scheme 15.4 The formation of benzodiazepines with affinity for the oxytocin receptor [53].

drug targets. An initial screen of 250 000 small molecules using recombinant human MrgX1 receptors resulted in the identification of a number of agonist hits from different classes of structures. Subsequent pharmacological profiling revealed several compounds from one chemical class which also showed affinity toward the human MrgX2 receptors. These were used as starting points for the construction of a focused library by solid-phase chemistry. The first step in the synthesis was coupling of N-protected amino acids (R^A) to benzhydrylamine resin (**17**, Scheme 15.5), followed by deprotection.

Afterward, the resin was reacted with 4-fluoro-3-nitrobenzoic acid and the free acid amidated with several amines (R^BNH_2) giving **19**. The nitro group was then reduced to the amine and the following treatment with 5-fluoro-2-nitrobenzaldehyde resulted in benzimidazoles (**20**). A third diversity point was introduced by displacing the fluoride with an amine (R^CNH_2). Next, the nitro group was again reduced and the resulting **21** was cleaved from the resin with TFA/thioanisole/TfOH, (trifluoromethanesulfonic acid). The final ring closure was then facilitated by heating in concentrated HCl under microwave (MW) irradiation, giving benzodiazepines (**22**) in 5–25% overall yield (10 steps) in up to gram quantities of material. The authors also attempted an approach in solution phase, which turned out to be inferior to the solid-phase protocol, due to the extensive purification steps needed. The testing and analysis of 500 thus generated compounds resulted in 30 confirmed hits with **23** being MrgX1 and **24** MrgX2 selective, both with higher efficiency than their respective reference substances. The best results were achieved with the amino acids valine and isoleucine (R^A), while R^B = aryl or R^C = non-aryl provided

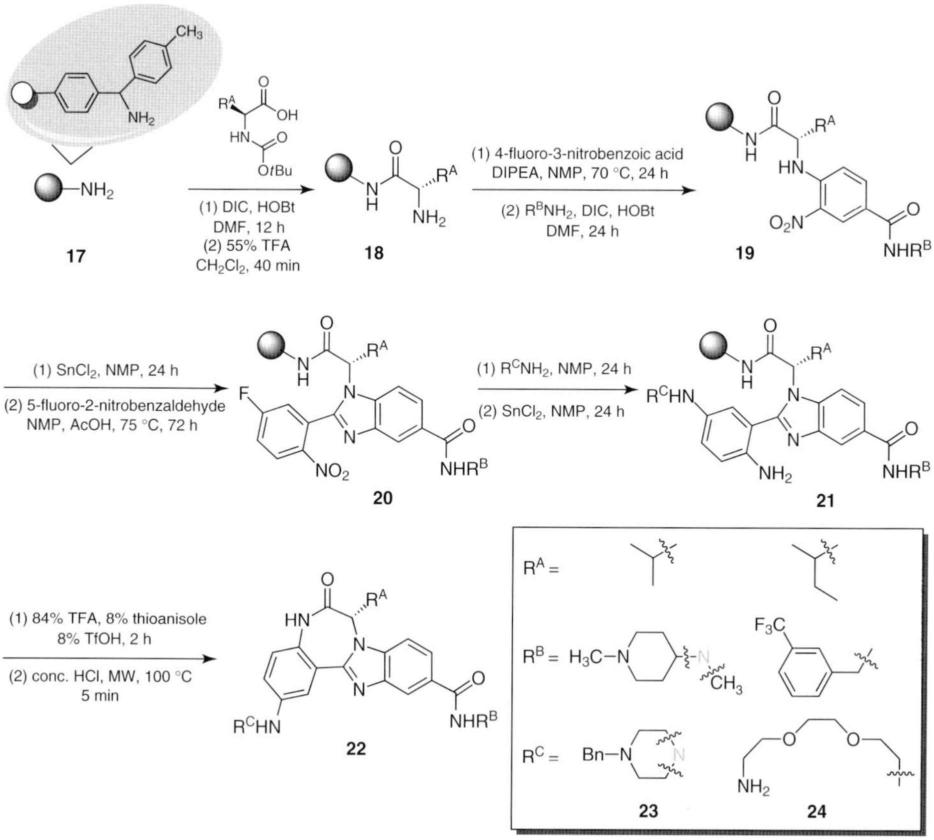

Scheme 15.5 MrgX1 and MrgX2 receptor-selective benzodiazepine synthesis [54].

compounds with little or no affinity for MrgX1. Also, a piperazine moiety at R^B showed preferred MrgX1 agonistic activities.

The next example outlines the synthesis of a 10 000-membered library of different 1,5-benzodiazepine-2-ones. Contrary to the 1,4-analogs so far discussed, this scaffold has stayed largely unexplored, although several compounds exist that are active against a variety of target types (e.g., protease inhibitors, 7-TM-, CCK-, or opioate k-receptors) [55]. Herpin and coworkers used the IRORI technology employing radiofrequency-labeled microreactors loaded with a dimethoxybenzaldehyde-modified backbone amide linker (BAL) resin. They started by attaching primary amines onto the resin via reductive amination, followed by coupling with bromoacetic acid (Scheme 15.6).

Compound **26** was then reacted with a phthalimide-protected benzodiazepine scaffold to **27**, which was in turn alkylated, acylated, sulfonylated, or converted into a urea, shortly depicted as the rest R^B. This was followed by cleavage of the phthalimide with hydrazine. The free amine (**28**) was then coupled with a carboxylic acid or sulfonylated to introduce R^C. Afterward, **29** was released from the resin by

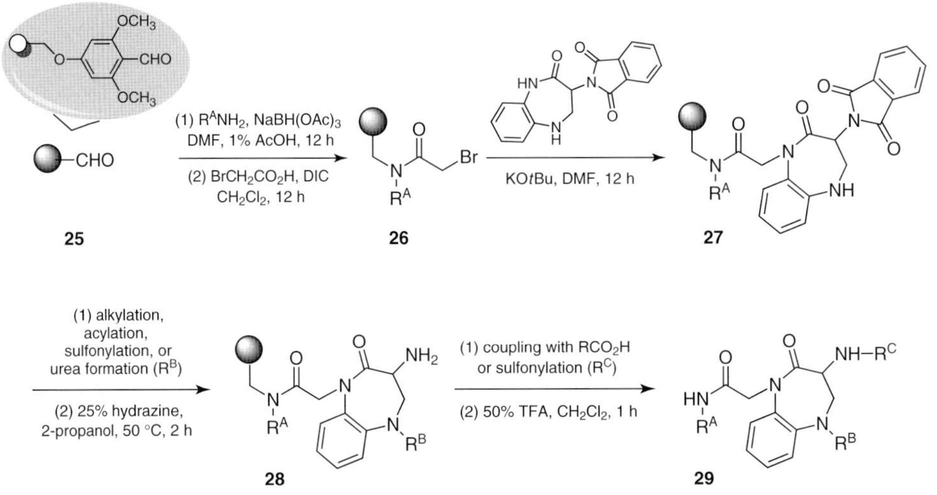

Scheme 15.6 Benzodiazepine library by Herpin *et al.* employing IRORI technology [55].

TFA-mediated cleavage. This gave an array of $18 \times 13 \times 45 = 10\,530$ compounds in high yields with purity exceeding 75%.

Su *et al.* used the benzodiazepine-containing drug clozapine (**30**) as starting point for their parallel synthesis [56]. Clozapine is an antipsychotic medication used to treat the positive and negative symptoms of schizophrenia since 1982. Its mode of action is mainly based on the blockage of the dopamine D_2 receptor, which inhibits cAMP production on activation. But it also shows affinity toward serotonergic, adrenergic, histaminergic, or muscarinic receptors. Su set out to find derivatives with selectivity for the dopamine D_1 receptor, which stimulates the production of cAMP. Therefore, he nitrated the amino group of clozapine, followed by reduction to the hydrazine (**31**) (Scheme 15.7).

The latter was then attached to the Agropore-CHO resin via reductive amination. Compound **32** was treated with 60 different aryl chlorides to give the set of compounds **33** after TFA-mediated cleavage. Screening of their affinities toward the D_1 and D_2 receptors allowed the identification of **34** as the most promising candidate with a K_i value of 1.6 nM for D_1 and 210-fold selectivity compared with D_2. Generally, it was found that the substitution of the phenyl ring improved D_1 affinity regardless of the electronic/steric nature of the substituent with the positioning playing a major role (lower K_i value for ortho substitution).

15.2.3 Benzopyrans

Benzopyrans are found in a wide range of biologically and pharmacologically active compounds, including natural products [57]. For example, HLC-2 (**35**) shows antitumor activity [58] and **36** has an inhibitory effect on a glucose transport assay (Figure 15.3) [59]. (+)-Alboatrin (**37**) exhibits potent anti-HIV and anti-inflammatory

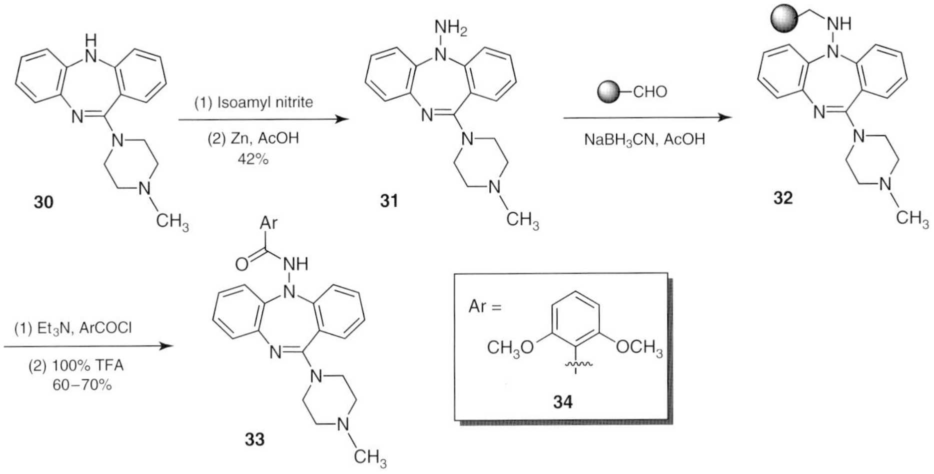

Scheme 15.7 Synthesis of clozapine-like benzodiazepines for various drug targets [56].

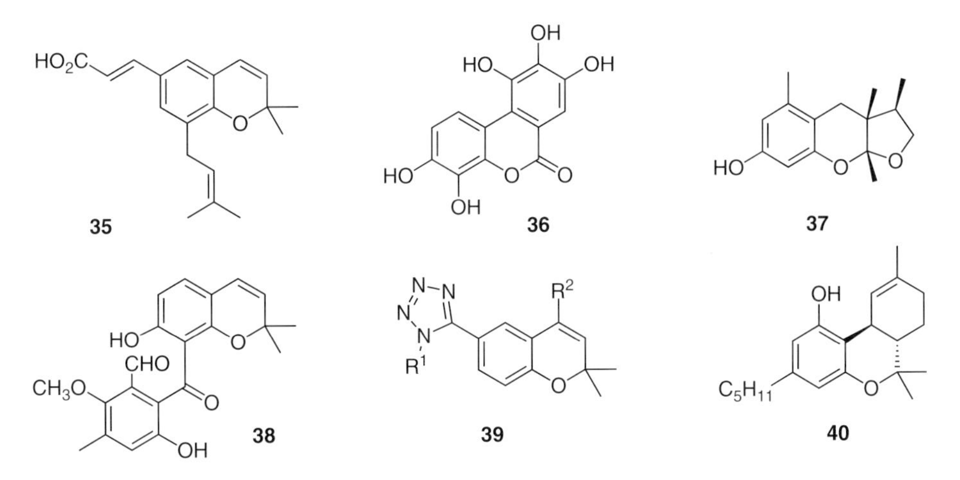

Figure 15.3 Examples of biologically and pharmacologically active benzopyrans.

activities [60]. Compound **38** displays antibacterial properties, having been extracted from *Bacillus subtilis* [61]. Furthermore, **39** is a potassium channel activator [62], and $(-)$-Δ^9-THC (tetrahydrocannabinol) (**40**) derived from the plant *Cannabis sativa* shows analgesic, anticonvulsive, and antiemetic effects because of its interaction with the human cannabinoid receptors CB_1 and CB_2 [63].

One can see that, especially, the 2,2-dimethylbenzopyran moiety is of great interest from a medicinal point of view, having inspired many chemists to develop different solid-phase approaches toward their synthesis. The group of Nicolaou has invested a great deal of research effort into this product class, resulting in the synthesis of over 10 000 different compounds [64–67]. Their approach is based

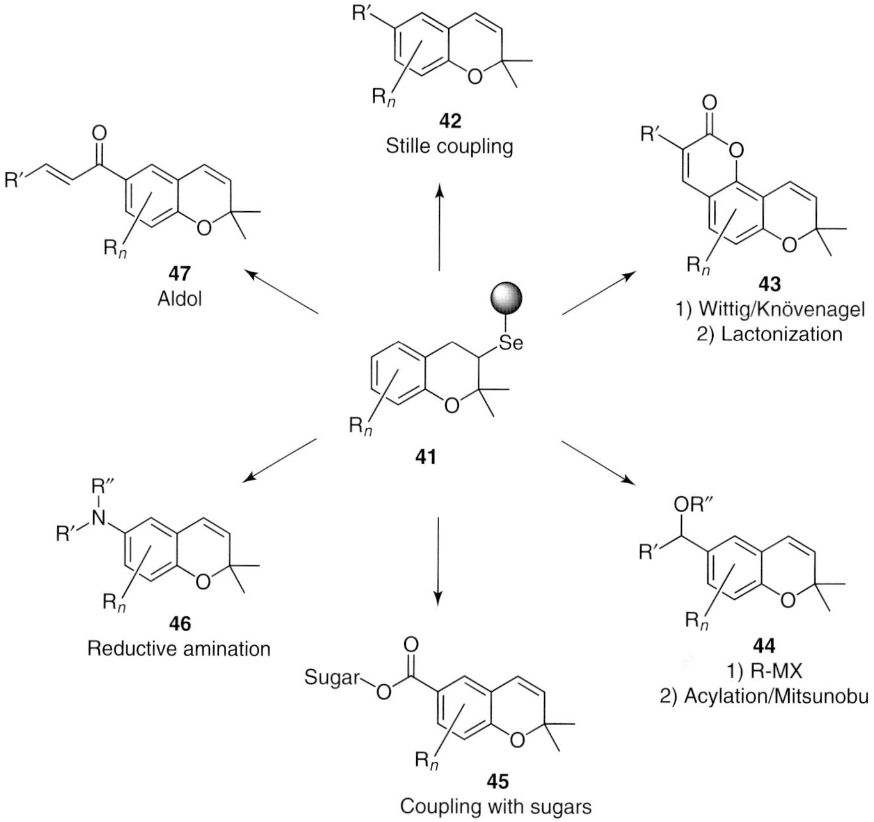

Figure 15.4 Access to various 2,2-dimethylbenzopyran derivatives via selenyl-linked precursor (**41**) [64–67].

on a selenyl-linked 2,2-dimethylbenzopyran core (**41**) that allows access to various derivatives (**42–47**) by different chemical transformations. A few of them are shown in Figure 15.4.

Nicolaou and coworkers employed the IRORI methodology for their larger libraries. They used either macro- or microreactors (capacities of 300 and 30 mg resin, respectively) containing unique radiofrequency tags or so-called NanoKans. The latter can hold 8 mg of resin and are capped by a laser-edged ceramic grid that can be read optically. They have the advantage of allowing complete automation of filling, sorting, washing, and cleaving. One particular approach was made to prepare inhibitors of the NADH:ubiquinone oxidoreductase [68]. This enzyme, which serves to carry electrons from NADH to molecular oxygen during oxidative phosphorylation, plays an essential role in cellular physiology. Its deficiencies are implicated in the pathogenesis of diseases such as Parkinson's or Leber's hereditary optic neuropathy. Furthermore, there is evidence that its inhibition has chemotherapeutic and chemopreventive effects. Nicolaou *et al.* synthesized and

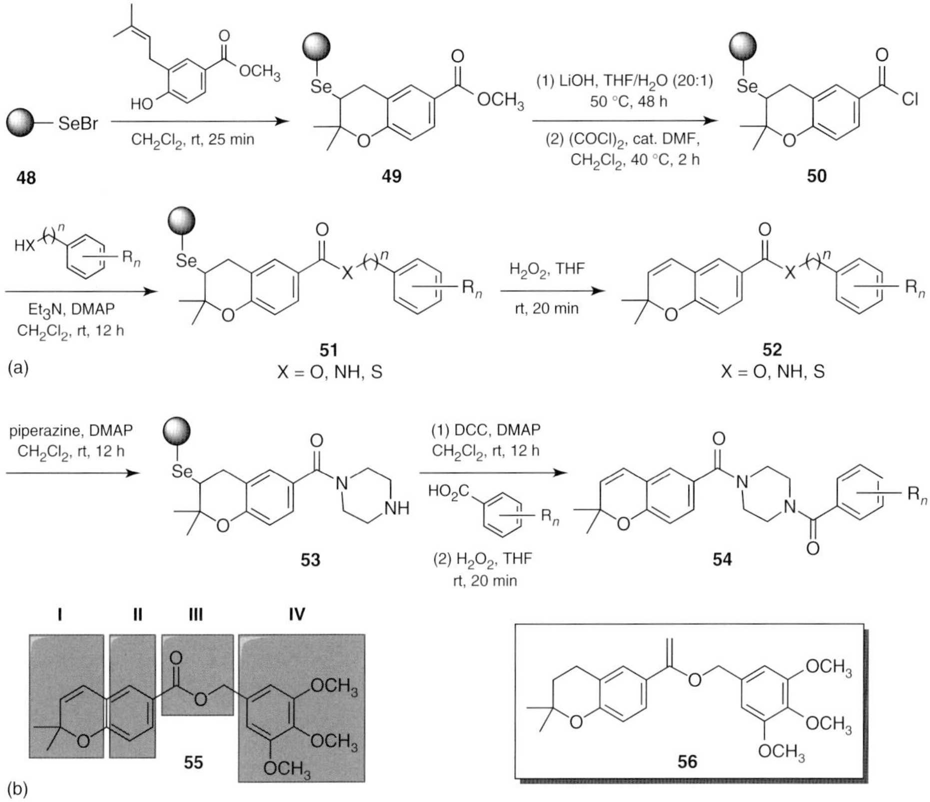

Scheme 15.8 Synthesis of 2,2-dimethylbenzopyrans as NADH:ubiquinone oxidoreductase inhibitors [68].

screened an initial library of 2,2-dimethylbenzopyrans bridged with a terminal aromatic ring in analogy to naturally occurring inhibitors isolated from Cubé resin [69]. Thus, pyran (**49**) was formed by treating selenylbromide resin (**48**) with an ortho prenylated phenol (Scheme 15.8).

This was followed by saponification and transformation into the acid chloride (**50**). From there on, the resin was either coupled with alcohols, thiols, or amines to **51** (Path A) or with piperazine (**53**), followed by attachment of an acid (Path B). Cleavage from the resin was facilitated by H_2O_2-mediated oxidation of selenium and β-hydride elimination. The resulting compounds **52** and **54** were then tested for their biological activity, with **55** being the most active ($IC_{50} = 55$ nM). Accordingly, three further focused libraries were prepared mainly by solution phase, concentrating on variations around the four regions of the molecule (gray areas in Scheme 15.8). The first one focused on the aromatic parts **II** and **IV**. It was found that derivatization of the aromatic benzopyran ring had only a moderate influence on the activity, while the substitution pattern of **IV** turned out to be very important. Apparently, the 3,4,5-trimethoxyphenyl group was at or near the optimum. The second library focused on the pyran ring **I** with the result that lipophilicity was crucial to retain activity. Only hydration of the double bond was tolerated. In a third attempt, the bridging unit **III** was modified. The best results gave benzyl esters, ketones, thioesters, or styrenes. Other groups, such as amides, thioamides, thioesters, oximes, or sulfonates, were not tolerated. The best IC_{50} value (18 nM) was found for compound **56**.

Kapeller and Bräse developed a versatile solid-phase approach toward classical cannabinoids based on the structure of (−)-Δ^9-THC (**40**) [70]. The latter are ligands of CB_1, CB_2 [71, 72], or the newly discovered GPR55 receptor [73]. Cannabinoids are used in medicine as antiemetics and analgesics and show promising results against Alzheimer's and Parkinson's diseases or osteoporosis. But severe side effects (e.g., addiction, psychotic episodes) due to nonselectivity concerning the receptors limit their prescription so far to the terminally ill. In a search for alternatives, Kapeller and Bräse used an open solid-phase synthesis to prepare a wide range of THC derivatives. The first step was the immobilization of hydroxysalicylaldehydes on DHP (dihydropyranyl) resin (**57**) (Scheme 15.9) [74].

This is followed by a domino oxa-Michael-aldol condensation and consecutive methylenation of the aldehyde via Wittig reaction. Diels–Alder cylcoaddition of **59** with various dienophiles (e.g., alkenes and azodicarboxylates) and cleavage gave products **60** and **61** in 20–60% overall yield. In all cases, diastereomeric mixtures were formed, which could be separated by column chromatography. Currently, these compounds are being tested for their affinity to the GPR55 receptor by a β-arrestin PathHunter™ assay [75] by the group of Christa Müller at the University of Bonn, Germany.

Park *et al.* constructed a polyheterocyclic benzopyran library with five diverse core skeletons. They employed a parallel, diversity-oriented synthesis in order to aid in the search for specific small molecular modulators of diverse protein receptors [76]. Eight different benzopyran moieties were immobilized on a novel silyl-modified resin (**62**) (Scheme 15.10).

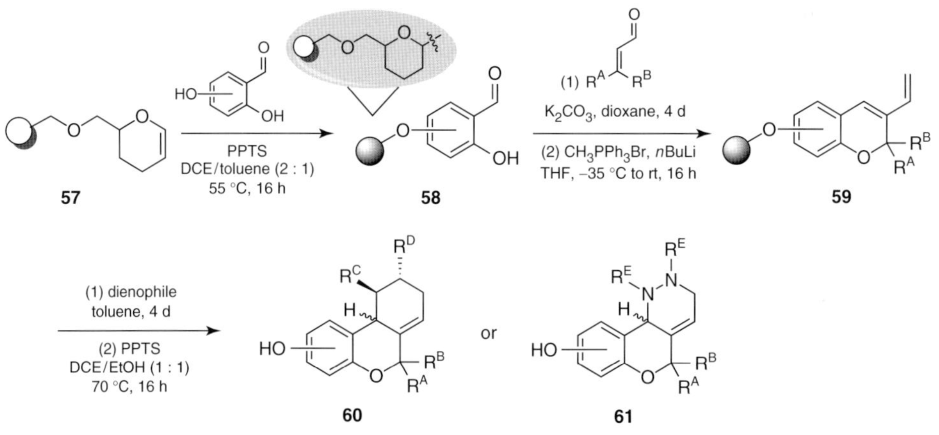

Scheme 15.9 Solid-phase approach toward classical cannabinoids [70].

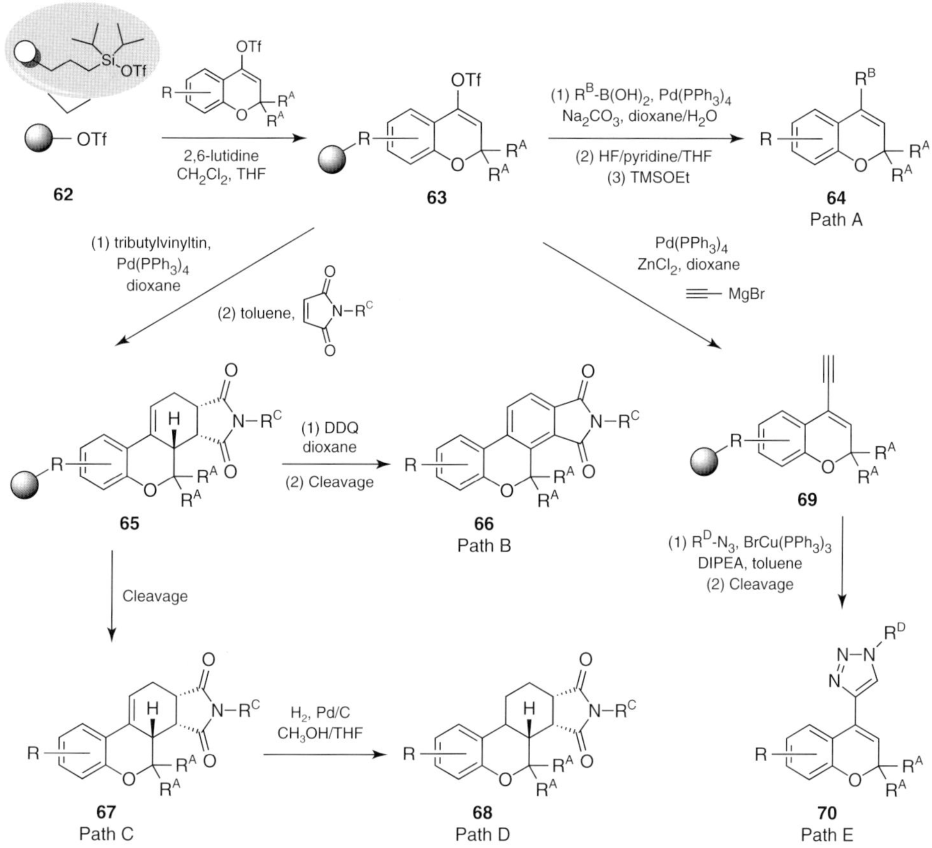

Scheme 15.10 Solid-phase synthesis of benzopyrans displaying five different core structures [76].

The resulting triflate (**63**) formed the basis for the five core structures (Paths A–E). Path A encompassed a Suzuki coupling of 18 different aryl boronic acids, followed by cleavage in HF/pyridine and subsequent quenching with TMSOEt. Paths B–D started with a palladium-mediated Stille vinylation and further Diels–Alder reaction with 17 substituted maleimides. The resulting tetracycles (**65**) were then either aromatized employing 2,3-dichloro-5,6-dicyanobenzoquinone (DDQ) to give **66** after cleavage (Path B), or directly cleaved to diastereochemically enriched **67** (Path C). Further solution-phase asymmetric hydrogenation yielded polyheterocycles (**68**) of Path D. Path E started with a Negishi-type alkynylation of **63**. The resulting product **69** was further elaborated by a Huisgen 1,3-dipolar [3 + 2] cycloaddition (click chemistry) with eight different azides, giving **70** after cleavage. In summary, Park and coworkers generated a library of 434 drug-like molecules with an average purity of 85%.

Gong and coworkers prepared a 2 000-member library of 2,6-disubstituted-2-methyl-2*H*-1-benzopyrans [77], which have been found to act as moderately active

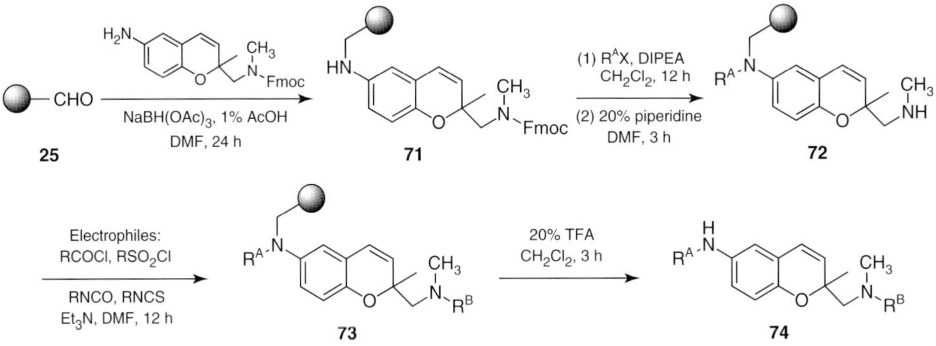

Scheme 15.11 Synthesis of a library of 2,6-disubstituted-2-methyl-2*H*-1-benzopyrans (**74**) [77].

5-lipoxygenase (5-LO) inhibitors [78]. 5-LO is the initial enzyme in the biosynthesis of leukotrienes and implicated in the treatment of asthma and lung diseases [79]. They started by modifying BAL resin (**25**) with an in solution-phase synthesized benzopyran core employing reductive amination (Scheme 15.11).

This was followed by alkylation of the secondary amine (**71**) and cleavage of the Fmoc group. Further conversion with several different electrophiles gave amides, sulfonamides, ureas, and thioureas of type **73**. Cleavage afforded products **74** in yields of 50–80% and excellent purity (>90%). Furthermore, in order to judge the bioavailability of the library members, the authors calculated their physical properties according to general rules, as predicted by Lipinski *et al.* [19]. It was found that all key parameters, such as molecular weight, lipophilicity, number of hydrogen donors, and acceptors, as well as the number of rotatable bonds and the polar surface area, were on a reasonable level with regard to drugability.

15.2.4 Indoles

Indoles present one of the most important structural classes in drug discovery, which can be seen by the large number of pharmacologically relevant molecules based on this heterocyclic framework [80–82]. It is, for example, included in the essential amino acid tryptophan (**75**), the neurotransmitter serotonin (**76**), the anti-inflammatory drug indomethacin (**77**) [83], antitumoral nortopsentins (**78**) [84], protein kinase C activator (−)-indololactam V (**79**) [85], or the selective and potent COX_2 inhibitor L-761,066 (**80**) [86] (Figure 15.5). Furthermore, a variety of GPCRs, such as the neurokinin-, chemokine-, serotonin-, and melanocortin-receptor families have been found to be activated by ligands with an indole skeleton [29]. The latter show widespread biological activities, including dopaminergic, antineoplastic, anticonvulsant, analgesic, sedative, muscle relaxant, antiviral, or antimicrobial effects. Accordingly, considerable attention has been directed toward the synthesis of indoles, many routes having also been translated to solid supports. In this

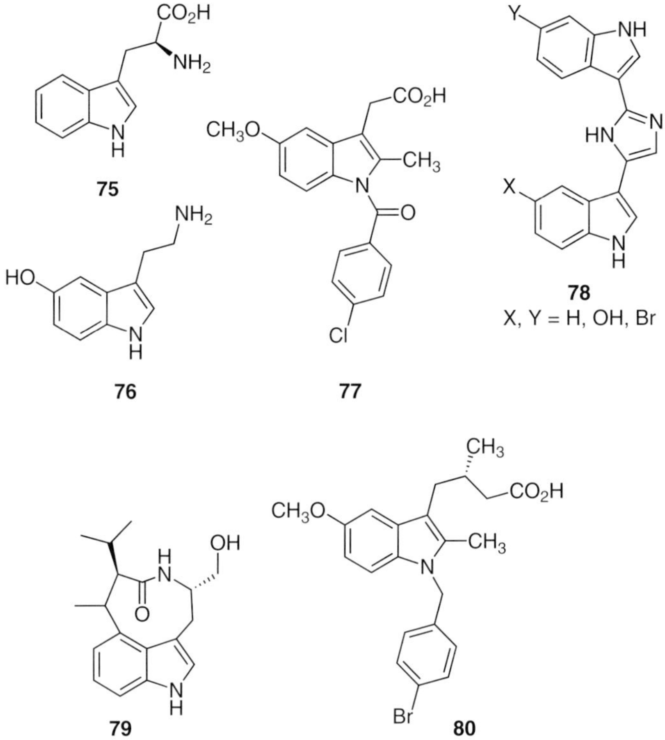

Figure 15.5 Examples of pharmacologically important indoles (**75**–**80**).

section, a few selected examples of indole libraries and their biological evaluation are discussed.

The group of Gmeiner is researching on ligands with high affinity for the dopamine D_4 receptor inspired by the drug clozapine (**30**) [87]. Selective antagonists for this biological target can be used to treat neuropsychiatric disorders, such as attention-deficient hyperactivity, mood disorders, or Parkinson's disease. Gmeiner *et al.* synthesized a library of 13 indole-based derivatives on solid supports that were tested for their dopamine receptor affinity by radioligand-binding studies [88]. Therefore, they applied a traceless linker strategy, in which diethoxymethyl-protected indoles were immobilized on polymer-bound 3-benzyloxypropane-1,2-diol (**81**) by transacetalization (Scheme 15.12) [89].

Compounds **82** and **85** were then modified with various secondary amines by Mannich (**83**) or substitution reactions (**86**), respectively. Acidic cleavage from the resin followed by the addition of NaOH gave the free indoles (**84**) and (**87**) in good yields and purity (80–100%). It was found that substitution of the indole in position 2 instead of 3 gave substrates with higher affinity for the D_4 receptor, the most promising ligands being **88** and **89**. Their K_i values were extremely low (0.52 and 1.0 nM), and they showed very good selectivities (>600 and >8 600 compared to other dopamine receptors).

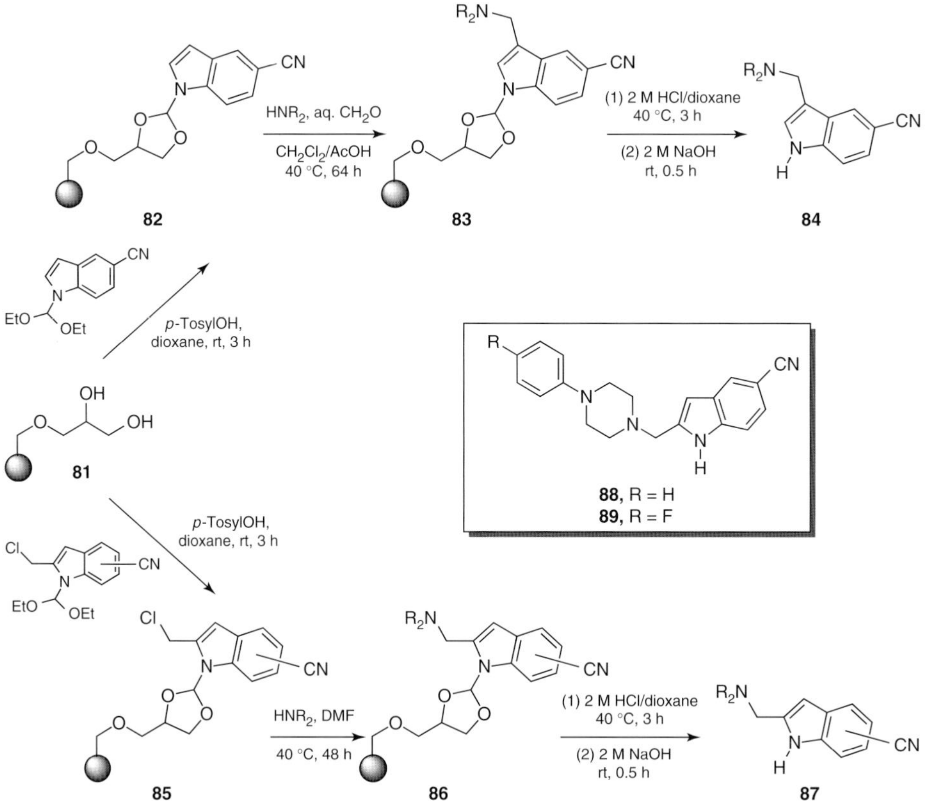

Scheme 15.12 Solid-phase synthesis of dopamine D_4 receptor-selective indoles [89].

Cooper *et al.* used a solid-phase approach to optimize the substitution pattern of 2-aryl indoles acting as neurokinin-1 (NK_1) receptor antagonists [90]. Their inhibition of the mammalian tachykinin substance P is implicated in several medical conditions, such as migraine, cystitis, asthma, depression, and cytotoxin-induced emesis. The synthesis started by immobilizing 4-(4-chlorobenzoyl)butyric acid on Kenner's safety-catch linker (**90**) (Scheme 15.13) [91].

The resulting **91** was then treated with differently substituted phenylhydrazines in a Fischer indol reaction. Cleavage with bromoacetonitrile followed by standard coupling gave compounds of type **93**. *In vitro* radioligand-binding studies employing cloned hNK_1 receptors showed that a substitution at position 5 was necessary for a good binding affinity, while it was detrimental at position 4, 5, or 7. This result led to the synthesis of **94** with an IC_{50} value of 0.16 nM, which was a significant improvement over the original lead structure.

A similar approach was undertaken by research laboratories at Merck, which used a mixture synthesis to form two libraries of 64 000 2-aryl indoles in only a few synthetic steps [92]. They also started from Kenner's linker (**90**), reacting it with two sets of 20 different keto acids (X) according to the conditions shown in

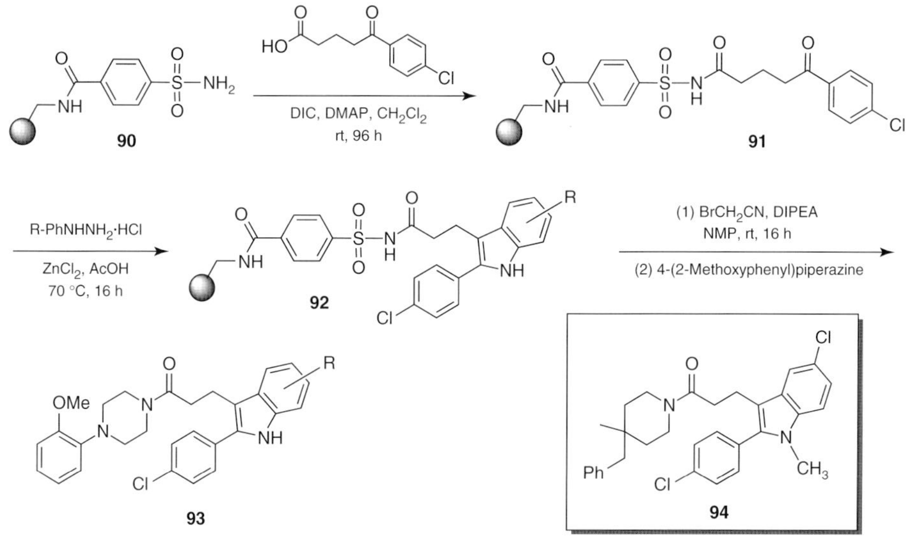

Scheme 15.13 Solid-phase approach toward NK_1 receptor antagonists (**93**) [91].

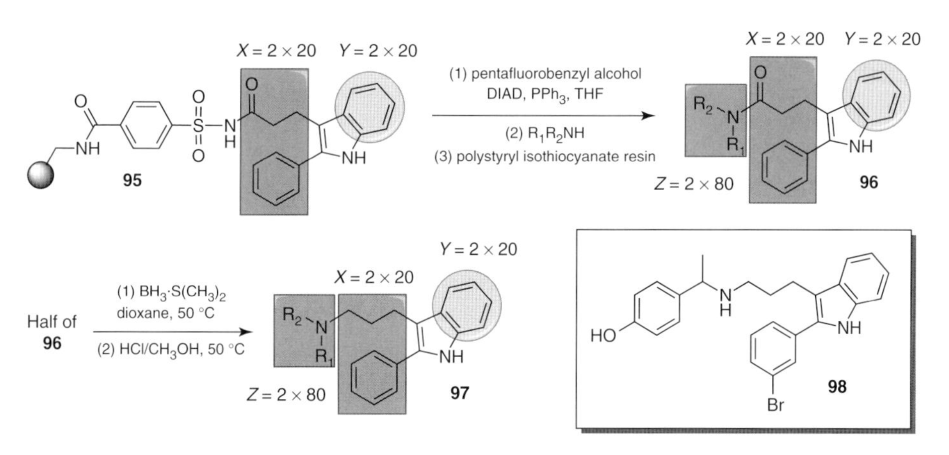

Scheme 15.14 Solid-phase synthesis of over 100 000 indoles for screening in GPCR-binding assays [92].

Scheme 15.13. Each set was mixed and separated in 20 equal portions, which were, in turn, treated with 20 different hydrazine derivatives (Y) to give $2 \times (20 \times 20)$ indoles (**95**) (Scheme 15.14).

After mixing and splitting both libraries, the sulfonamide moiety was alkylated under Mitsunobu conditions and cleaved from the resin via a displacement reaction with 80 amine "Z subunits" each. As such, a total of $2 \times (20 \times 20 \times 80)$ compounds **96** were formed. This number was again doubled by reducing the amide functionality of half of each library to give **97**. Thus, all the generated 2-aryl indoles were then screened in a wide variety of GPCR-binding assays (e.g., neurokinin, chemokine, and serotonin receptors). One promising example is compound **98**, which is potent and selective in respect to the neuropeptide Y receptor NPY_5. It displays an affinity of 0.8 nM while having no activity versus NPY_1. NPY receptors are involved in the control of a diverse set of behavioral processes including appetite, circadian rhythm, and anxiety.

Another research group within Merck investigated 2,3-disubstituted indoles as candidates for the treatment of schizophrenia [93]. Their aim was to find compounds with selectivity for the 5-hydroxytryptamine ($h5\text{-}HT_{2A}$) receptor, which is known to regulate neurotransmission via the natural ligand serotonin. They started from the *p*-nitrophenylcarbonate derivative of Wang resin (**99**) that was loaded with the respective 2-bromoindole (Scheme 15.15).

In the next step, aryl groups were introduced at position 2 by either a Suzuki [$Ar\text{-}B(OH)_2$] or Stille [$Ar\text{-}Sn(CH_3)_3$] reaction. The tetrahydropyranyl (THP) group of compound **101** was then removed and the resulting OH functionality transformed into the triflate (**102**). The latter was reactive enough to allow smooth conversion with secondary amines. Cleavage finally gave the products **103** in 35–65% yield and 89–94% purity. Compound **104** was identified as the most active ($K_i = 2.7$ nM), determined by displacement experiments of tritium-labeled ketanserin at the $h5\text{-}HT_{2A}$ receptor.

Torisu *et al.* explored substances with affinity for the prostaglandin D_2 receptor (DP) [94]. This specific prostanoid is considered to have potential clinical uses in various allergic diseases, including allergic rhinitis, atopic asthma, allergic conjunctivitis, and atopic dermatitis. After identification of a new chemical lead structure (**108** for R = butyl), which is structurally similar to the anti-inflammatory drug indomethacin (**77**) [95], they set out to optimize the substituent R (Scheme 15.16).

Therefore, modified trityl resin (**105**) [96] was derivatized with an indole core structure to give **106**. Deprotection of the acetyl-protected phenol gave **107**, which was then etherified with 19 different alcohols (R = dihydrobenzofuranes, methylenedioxybenzenes, dihydrobenzopyrans, and benzodioxanes) in a Mitsunobu reaction. Cleavage with acetic acid gave compounds **108** in 10–48% yield. These were tested for their affinity to various mouse prostanoid receptors and the human hDP receptor. The most promising structures concerning hDP were **109** and **110** with IC_{50} values of 8 and 21 nM, respectively. They were further evaluated for their pharmacokinetics and tested concerning their inhibitory effects on prostaglandin D_2 (PGD_2)-induced vascular permeability in an animal model. In both cases, promising profiles were found.

15.2.5 Pyrazoles

Another common framework with a wide range of biological activities is the pyrazole one [97]. Depending on its substitution pattern, these can range from analgesic, anti-inflammatory, antipyretic, anti-arrhythmic, tranquilizing, muscle relaxing, and anticonvulsant, to antidiabetic properties. One of the first drugs prepared in a laboratory was the pyrazole derivative phenazone (**111**, Figure 15.6).

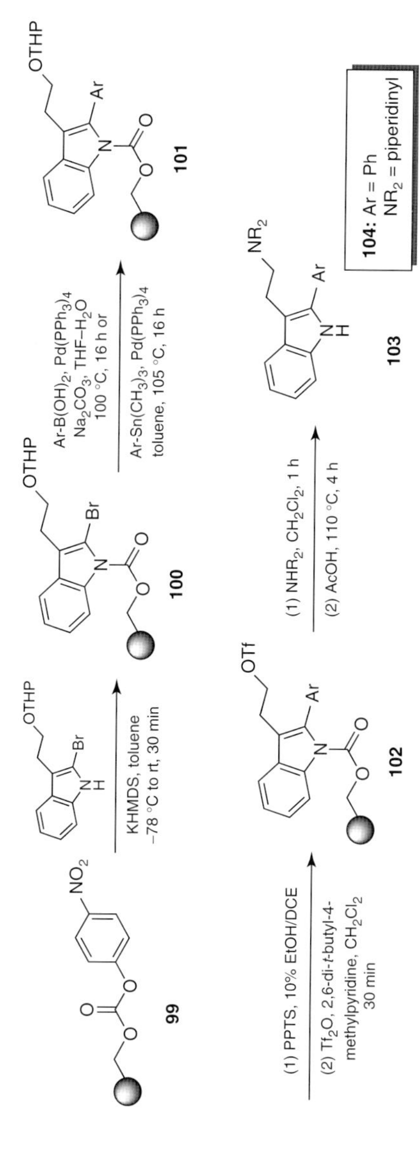

Scheme 15.15 Solid-phase synthesis of 2,3-disubstituted indoles for the treatment of schizophrenia [93].

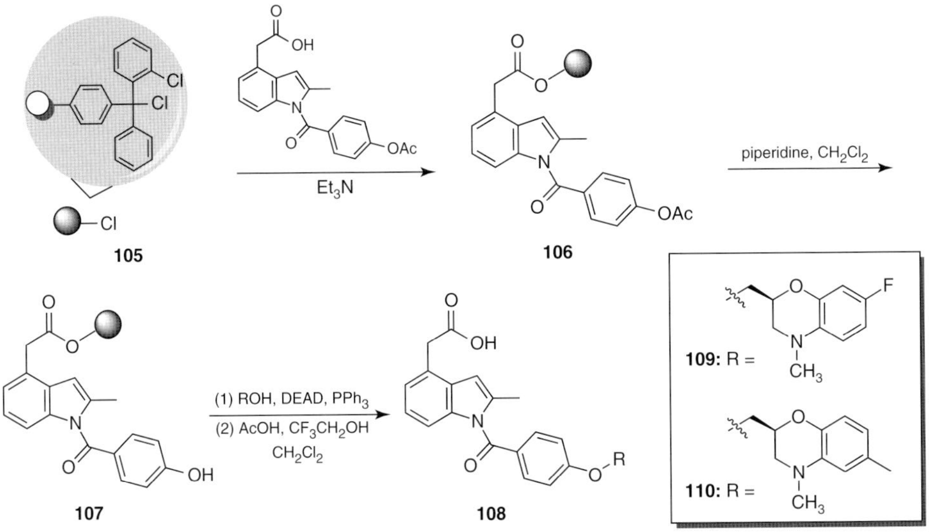

Scheme 15.16 Solid-phase organic synthesis of DP receptor-specific indoles [91].

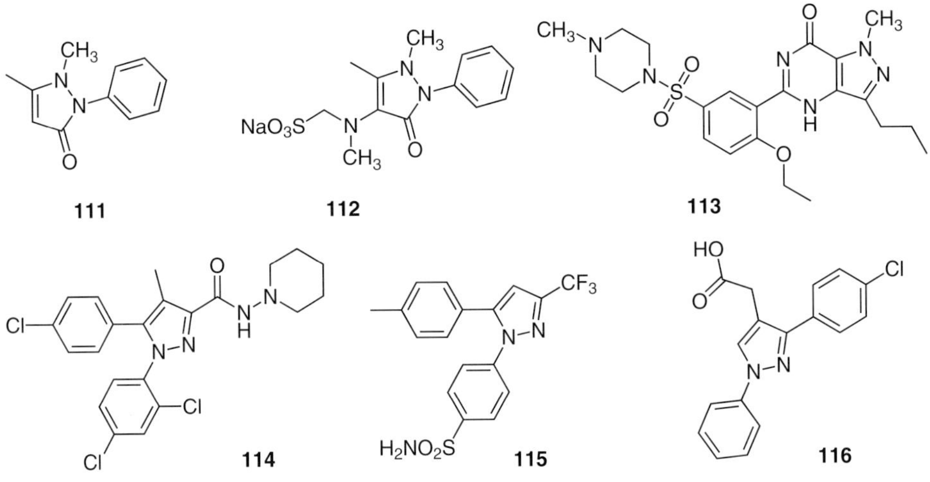

Figure 15.6 Examples of pyrazole-containing drugs (**111–116**).

It was synthesized by Ludwig Knorr in 1883 and marketed by Hoechst as the painkiller Antipyrin [98]. It was later on replaced by the analgesic metamizole (**112**), which is available under the brand name Novalgin®. Sildenafil (**113**) was synthesized and studied at Pfizer for its use in hypertension [99]. But it became commercially available as Viagra® under a very different aspect, namely, the treatment of erectile dysfunction. Rimonabant (**114**), an inverse agonist of the CB_1 receptor and anorectic anti-obesity drug, was withdrawn two years after its launch

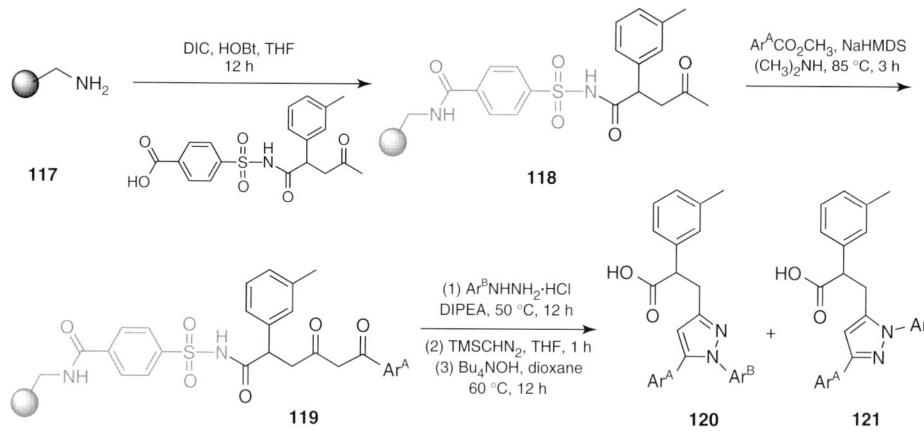

Scheme 15.17 Solid-phase synthesis of CCK_A receptor-active pyrazoles (**120**) [103].

because of psychiatric disorders [100]. The last two compounds are Celecoxib (**115**) [101] and Lonazolac (**116**) [102], drugs against arthritis/pain and inflammation.

The first pyrazole library to be described here has been tested toward activity against the previously mentioned CCK_A receptor [103]. Its antagonists have been clinically evaluated for pancreatic disorders, irritated bowel syndrome (IBS), and biaryl colic. Breitenbucher and coworkers started out by coupling a substituted sulfonamide to aminomethyl polystyrene resin (**117**) (Scheme 15.17).

The former's 3-methylphenyl group was chosen because of its high CCK_A affinity [104]. Initially, Kenner's safety-catch resin (**90**, gray area) had been used to couple the keto acid, but in that case the combined yields of pyrazoles were below 10%. The next step was a Claisen condensation with five different arylcarboxylates to give diketones (**119**). These were then reacted to the pyrazoles employing six aryl hydrazines (Ar^BNHNH_2), followed by cleavage from the resin (combined yield = 20–35%). In all cases, product mixtures of **120** and **121** were formed, favoring the desired 1,5-pyrazoles (**120**). The unwanted 1,3-regioisomer could be removed by column chromatography. Afterward, the authors determined the binding affinity of all 30 compounds **120** (pK_i). They compared these data to the calculated pK_i values (using a modified version of the Fujita–Ban equation) in order to determine the additivity of the system. The latter is important for simultaneously optimizing multiple properties in a series because additivity means that, by altering one portion of a series to optimize a second property, the SAR trends for the primary target will very likely stay the same. In that particular series, no additivity could be determined, but in solution-phase approaches, where R^A and the *m*-methylphenyl moieties were varied, the additive model could be applied [104].

Spivey *et al.* synthesized iodinated analogs of rimonabant (**114**) via a parallel solid-phase approach [105]. As mentioned earlier, rimonabant is an inverse agonist of the CB_1 receptor that belongs to the group of rhodopsin-like GPCRs and is widely distributed in the CNS. Antagonists or inverse agonists of these receptors show

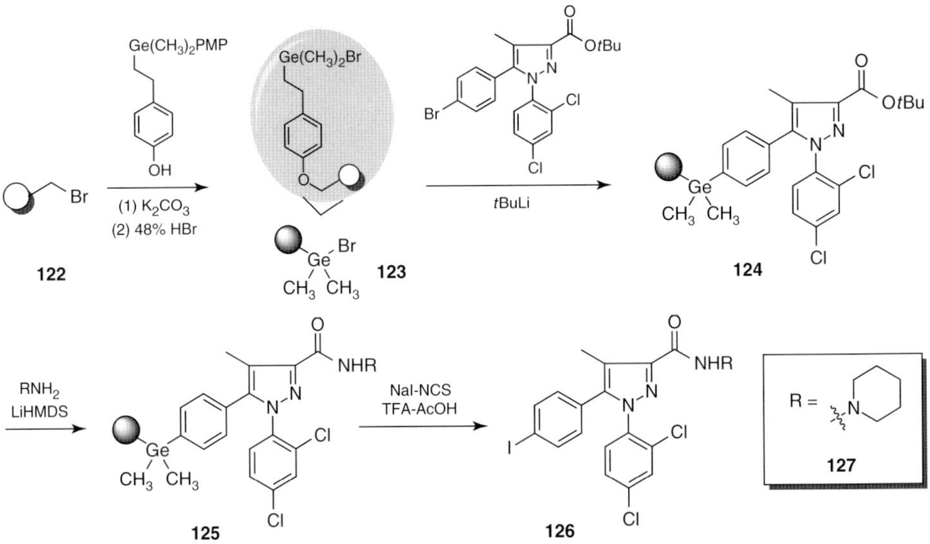

Scheme 15.18 Synthesis of iodinated analogs of rimonabant (**114**) for molecular imagining techniques [105].

promise not only for the treatment of obesity but also against metabolic disorders or various addictions including smoking. In their approach, Spivey and coworkers envisioned to compose a route toward iodinated ligands for high-resolution "molecular" imaging techniques (e.g., positron emission tomography, PET) with variation at the hydrazide moiety, which has a profound effect on affinity and biodistribution. Usually, radio-iodinated aryl iodides ^{123}I and ^{124}I are introduced by *ipso*-iodostannylation, followed by high performance liquid chromatography (HPLC) purification. This solid-phase protocol involved the use of a trialkylgermyl linker that could be cleaved as an iodide, thus avoiding toxic tin and the necessity for purification. Starting from the PEG–polystyrene copolymer, **122** and **123** was generated by etherification with a germyl-containing phenol and exchange of the *p*-methoxyphenyl (PMP) group by a bromide (Scheme 15.18). This was followed by transmetalation and treatment with the pyrazole precursor toward **124**. Parallel amidation of the latter gave five different compounds **125**, which were then cleaved from the resin with concomitant *ipso*-iodination in 42–63% yield and up to 96% purity. Among these products was also **127**, which has already been employed in PET scans because of its high selectivity and affinity for the CB_1 receptor (7.5 nM).

Stauffer and Katzenellenbogen investigated novel pyrazole ligands for estrogen receptors (ERs) [106]. The latter are endogenously activated by the hormone 17β-estradiol. Most ERs belong to the nuclear hormone receptor superfamily [107] with one exception, GPR30, which is a member of the GPCR family [108]. The former class can be divided into ERα and ERβ, which show a high degree of similarity at the amino acid level and are expressed in, for example, prostate, ovary, breast, and certain areas of the brain. Ligands that modulate those ERs

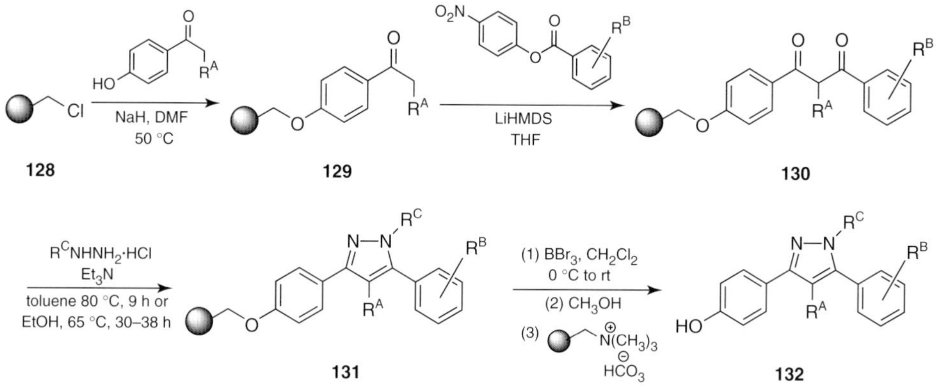

Scheme 15.19 Solid-phase synthesis of estrogen receptor-active pyrazoles (**132**) [106].

display tissue-selective action suitable for menopausal hormone replacement or the treatment and prevention of breast cancer.

Stauffer and Katzenellenbogen found a novel ER ligand consisting of a pyrazole core structure with very high binding affinity and selectivity for ERα. Accordingly, they based their libraries of first 12 and later on 96 members (96-well plate) on this compound. The synthesis started by attaching an acylated phenol (R^A = propyl or isobutyl) on Merrifield resin (**128**), followed by Claisen condensation with six differently substituted phenylesters to **130** (Scheme 15.19). Afterward, the pyrazoles were formed with eight hydrazine derivatives either with toluene (12 membered-library) or with ethanol for the 96-well plate. Cleavage was affected with BBr_3, followed by addition of methanol and removal of the resulting HBr with an ion exchange resin in carbonate form. The average purity of the library members was 50%. The major side product was identified as unreacted β-diketone (**130**), which did not disturb the consecutive biological assay. Overall, the compounds with the highest affinity for ER receptors were those with R^B being *p*-OH and R^C a *p*-hydroxy phenyl group, showing 14% (R^A = propyl) and 23% (R^A = isobutyl) activity based on estradiol.

The next solid-phase synthesis features pyrazoles with chemical decorations to preferentially target tyrosine kinases [109]. These belong to the family of protein kinases, which have become the second most exploited group of drug targets after GPCRs, accounting for 30% of drug discovery projects at many pharmaceutical companies [110]. Tyrosine kinases transfer a phosphate group from adenosine triphosphate (ATP) to a tyrosine residue of a cellular protein, thus regulating many cellular functions. Their inhibitors can, for example, be used in the treatment of certain types of cancer, as shown by the drug imatinib [111]. Leonetti and coworkers initially immobilized 3-hydroxybenzoic acid on Rink-amide resin (**133**) as solid support in over 90% yield (**134**, Scheme 15.20).

This, in turn, was reacted with bromomethyl ketones to yield six different compounds **135** in 67–95% yield. Enaminoketone formation to **136** proceeded quantitatively with *N*,*N*-dimethylformamide-dimethylacetal (DMFDMA), followed

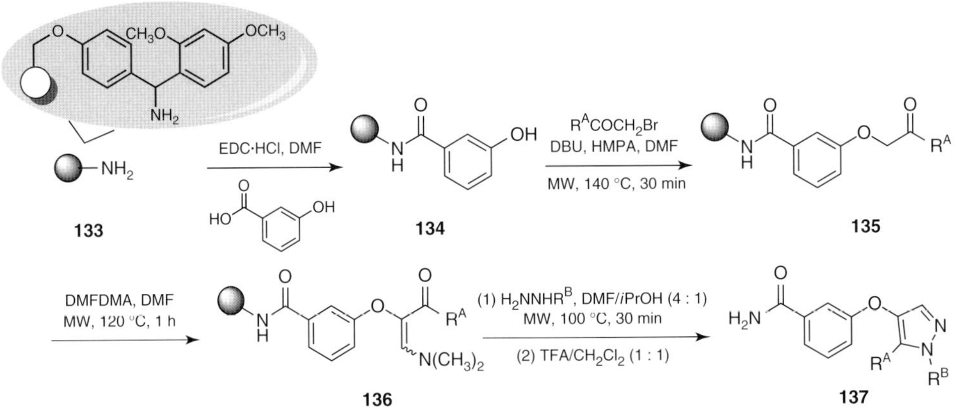

Scheme 15.20 Synthesis of a pyrazole library targeting tyrosine kinases [109].

by cyclization with two aryl-substituted hydrazines. TFA-mediated cleavage gave **137** in 40–81% yield over the last two steps. So far, no biological data for the synthesized pyrazoles are available.

The last example in this section will demonstrate that drugs do not necessarily have to be aimed at human biological targets. Haque and coworkers investigated inhibitors of the dihydroorotate dehydrogenase (DHODase) in *Heliobacter pylori* [112]. These Gram-negative bacteria infect up to 50% of the world's human population causing numerous gastrointestinal disorders, such as gastric ulcers, gastritis, or gastric cancer. DHODases are enzymes involved in the *de novo* pyrimidine synthesis of prokaryotic and eukaryotic cells alike. Pyrimidines are crucial for the survival of bacteria, and *H. pylori* does not have a salvage pathway for an alternative synthesis. As such, DHODases pose attractive targets for an antibacterial therapy, provided of course that they are selective and do not inhibit the corresponding human enzymes, which has an immunosuppressive effect. Haque's strategy was to synthesize several small, focused libraries of compounds with the general structure **140** (Scheme 15.21), to assay them, and then to incorporate the respective results into the planning of the subsequent library. The first library investigated the influence of the rests R^A and R^B, with R^C staying the same. Ten different amines were attached to a polystyrene resin via the BAL linker **25** employing reductive amination (Scheme 15.21).

Afterward, a single pyrazole prepared by solution-phase synthesis was coupled to the amines (**138**). The advantage of not generating the pyrazole directly on the solid support is that no problems due to regioisomers arise, as these can be separated by chromatography beforehand. Pyrazoles (**139**) were then saponified and the second amine side chain (R^B) coupled with the resulting acid, followed by cleavage from the resin. Seventy-seven different pyrazoles thus generated were tested for their activity against *H. pylori* DHODase, none of them showing improvement compared to the lead structures. In the second library, the rest R^B was varied with only two diversity points at R^A and R^C. This gave 44 different compounds **140** in 70–95%

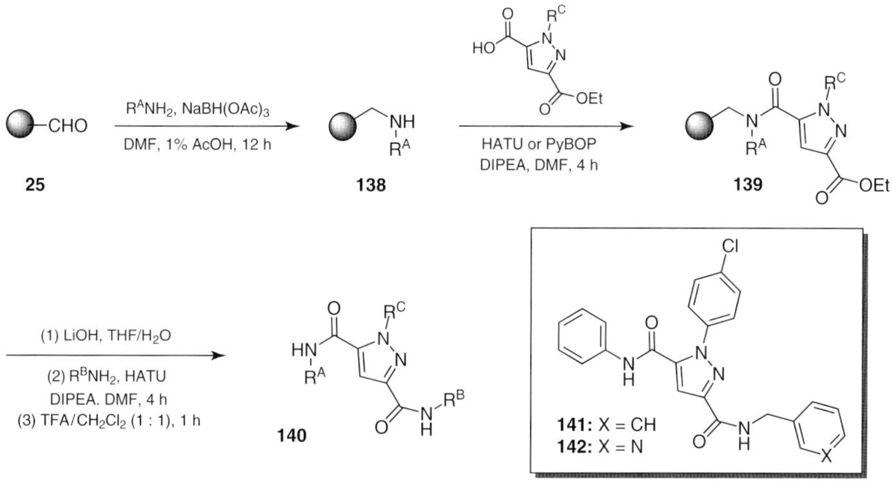

Scheme 15.21 Synthesis of pyrazoles acting as inhibitors of the DHODase in *H. pylori* [112].

yield in over 80% purity. Generally, it was found that compounds with R^B = aryl were more potent inhibitors than those with an alkyl side chain, the same being true for R^A. In the third and last library, 160 different compounds were generated, this time varying all three positions with an emphasis on R^A. This resulted in the identification of two compounds (**141** and **142**) with a K_i value of 4 nM and over 10 000-fold selectivity concerning human DHODase enzymes. Furthermore, in whole-cell antimicrobial assay, minimal inhibition concentrations (MICs) of 0.125 and 0.250 μg/ml were detected, respectively, proving these compounds to be superior to the initial leads.

15.3 Conclusions and Outlook

Combinatorial solid-phase chemistry and high-throughput screening as tools in drug discovery have come a long way since their introduction 20 years ago. In the beginning, they were brutish methods, used to generate large, unrelated collections of compounds (>100 000). But over the years, they became more refined and focused, directed by the increasing knowledge concerning biological targets, as well as computational methods. Now, combinatorial solid-phase chemistry is an established technique with the capacity of being very powerful, provided that the appropriate molecules are selected for the synthesis. Privileged structures present ideal sources for lead-finding libraries, as they produce active compounds for a variety of different targets. In case of an already established hit, SPOS poses an invaluable tool for structure optimization.

In many aspects, such as easy automation without the need to purify after each step, solid-phase synthesis exceeds its solution-phase counterpart. Its only real disadvantage still lies in reaction control and on-bead analysis. Lately, novel techniques such as Raman spectroscopy or magic angle spinning NMR have been applied to polymeric resins, but neither has established itself to date. If suitable fast analytical methods can be introduced that allow continuous monitoring of solid-phase reactions, traditional solution-phase chemistry will have found a very strong competitor in the area of drug discovery.

This review illustrates representative examples of solid-phase-generated libraries including their evaluation against various biological targets. The latter are mainly members of the GPCR family, to which about 50% of all newly launched drugs belong. The selected syntheses are organized by their respective privileged substructures and also give an overview of different reaction conditions and linkers used in solid-phase chemistry.

Acknowledgment

This work was supported by the Alexander von Humboldt-Foundation.

References

1. *www.who.int/substance_abuse/terminology/who_lexicon/en* (accessed January 26 2011).
2. Bleicher, K.H., Böhm, H.-J., Müller, K., and Alanine, A. (2003) *Nat. Rev. Drug Discovery*, **2**, 369–378.
3. Drews, J. (2000) *Science*, **287**, 1960–1964.
4. Zajdel, P., Pawlowski, M., Martinez, J., and Subra, G. (2009) *Comb. Chem. High Troughput Screen.*, **12**, 723–739.
5. Lebl, M. (1999) *J. Comb. Chem.*, **1**, 3–24.
6. Watson, C. (1999) *Angew. Chem. Int. Ed.*, **38**, 1903–1908.
7. Balkenkohl, F., von dem Bussche-Hühnefeld, C., Lansky, A., and Zechel, C. (1996) *Angew. Chem. Int. Ed.*, **35**, 2288–2337.
8. Diller, D.J. (2008) *Curr. Opin. Drug Discovery Dev.*, **11**, 346–355.
9. Macarron, R. (2006) *Drug. Discovery Today*, **11**, 277–279.
10. Davis, A.M., Keeling, D.J., Steele, J., Tomkinson, N.P., and Tinker, A.C. (2005) *Curr. Top. Med. Chem.*, **5**, 421–439.
11. Armstrong, J.W. (1999) *Am. Biotechnol. Lab.*, **17**, 26–27.
12. Hamilton, S.D., Armstrong, J.W., Gerren, R.A., Janssen, A.M., Peterson, J.V., and Stanton, R.A. (1996) *Lab. Robotics Automat.*, **8**, 287–294.
13. Kennedy, J.P., Williams, L., Bridges, T.M., Daniels, R.N., Weaver, D., and Lindsley, C.W. (2008) *J. Comb. Chem.*, **10**, 345–354.
14. Alanine, A., Nettekoven, M., Roberts, E., and Thomas, A.W. (2003) *Comb. Chem. High Troughput Screen.*, **6**, 51–66.
15. Overington, J.P., Al-Lazikani, B., and Hopkins, A.L. (2006) *Nat. Rev. Drug Discovery*, **5**, 993–996.
16. Hodgson, J. (2001) *Nature Biotechnol.*, **19**, 722–726.
17. Kennedy, T. (1997) *Drug Discovery Today*, **2**, 436–444.
18. Oprea, T.I. and Gottfries, J. (1999) *J. Mol. Graph. Model.*, **17**, 261–274.
19. Lipinski, C.A., Lombardo, F., Dominy, B.W., and Feeney, P.J. (1997) *Adv. Drug Delivery Rev.*, **23**, 3–25.

20. Lipinski, C.A. (2000) *J. Pharmacol. Toxicol. Methods*, **44**, 235–249.
21. Hunter, W.N. (1995) *Mol. Med. Today*, **1**, 31–34.
22. Schnur, D., Beno, B.R., Good, A., and Tebben, A. (2004) *Methods Mol. Biol.*, **275**, 355–377.
23. Abad-Zapatero, C. (2007) *Expert Opin. Drug Discovery*, **2**, 469–488.
24. Vassilatis, D.K. *et al.* (2003) *Proc. Natl. Acad. Sci.*, **100**, 4903–4908.
25. Teller, D.C., Okada, T., Behnke, C.A., Palczewski, K., and Stenkamp, R.E. (2001) *Biochemistry*, **40**, 7761–7772.
26. Palczewski, K. *et al.* (2000) *Science*, **289**, 739–745.
27. Simms, J., Hall, N.E., Lam, P.H.C., Miller, L.J., Christopoulos, A., Abagyan, R., and Sexton, P.M. (2009) *Methods Mol. Biol.*, **552**, 97–113.
28. Klabunde, T. and Hessler, G. (2002) *ChemBioChem*, **3**, 928–944.
29. DeSimone, R.W., Currie, K.S., Mitchell, S.A., Darrow, J.W., and Pippin, D.A. (2004) *Comb. Chem. High Throughput Screening*, **7**, 473–493.
30. Evans, B.E. *et al.* (1988) *J. Med. Chem.*, **31**, 2235–2246.
31. Horton, D.A., Bourne, G.T., and Smythe, M.L. (2003) *Chem. Rev.*, **103**, 893–930.
32. Ripka, W.C., De Lucca, G.C., Bach, A.C., Pottori, R.A., and Blaney, J.M. II (1993) *Tetrahedron*, **49**, 3593–3608.
33. Guo, T. and Hobbs, D.W. (2003) *Assay Drug Dev. Technol.*, **1**, 579–592.
34. Merrifield, R.B. (1963) *J. Am. Chem. Soc.*, **85**, 2149–2154.
35. Lee, A. and Breitenbucher, G.J. (2003) *Curr. Opin. Drug Discovery Dev.*, **6**, 494–508.
36. Früchtel, J.S. and Jung, G. (1996) *Angew. Chem. Int. Ed. Engl.*, **35**, 17–42.
37. Schreiber, S.L. (2000) *Science*, **287**, 1964–1969.
38. Knepper, K., Gil, C., and Bräse, S. (2003) *Comb. Chem. High Throughput Screening*, **6**, 673–679.
39. Patel, D.V. and Gordon, E.M. (1996) *Drug Discovery Today*, **1**, 134–144.
40. Furka, A., Sebestyen, F., Asgedorn, M., and Dibo, G. (1991) *Int. J. Pept. Protein Res.*, **37**, 487–493.
41. Seneci, P. (2001) *J. Recept. Signal. Transduction Res.*, **21**, 409–445.
42. Chabala, J.C. (1995) *Curr. Opin. Biotechnol.*, **6**, 632–639.
43. Moran, E.J., Sarshar, S., Cargill, J.F., Shahbaz, M.M., Lio, A., Mjalli, A.M.M., and Armstrong, R.W. (1995) *J. Am. Chem. Soc.*, **117**, 10787–10788.
44. Nicolaou, K.C., Xiao, X.-Y., Parandoosh, Z., Senyei, A., and Nova, M.P. (1995) *Angew. Chem. Int. Ed. Engl.*, **34**, 2289–2291.
45. Gil, C. and Bräse, S. (2005) *Chem. Eur. J.*, **11**, 2680–2688.
46. Sternach, L.H. (1979) *J. Med. Chem.*, **22**, 1–7.
47. Camps Diez, F., Castells Guardiola, J., and Pi Sallent, J. (1977) Patent ES 445831 A1 19770616.
48. Bunin, B.A. and Ellman, J.A. (1992) *J. Am. Chem. Soc.*, **114**, 10997–10998.
49. Bunin, B.A., Plunkett, M.A., and Ellman, J.A. (1994) *Proc. Natl. Acad. Sci. U.S.A.*, **91**, 4708–4712.
50. Dufresne, M., Seva, C., and Fourmy, S. (2006) *Physiol. Rev.*, **86**, 805–847.
51. Lattmann, E., Billington, D.C., Poyner, D.R., Arayarat, P., Howitt, S.B., Lawrence, S., and Offel, M. (2002) *Drug Des. Discovery*, **18**, 9–21.
52. Evans, B., Pipe, A., Clark, L., and Banks, M. (2001) *Bioorg. Med. Chem. Lett.*, **11**, 1297–1300.
53. Wyatt, P.G., Allen, M.J., Chilcott, J., Hickin, G., Miller, N.D., and Woollard, P.M. (2001) *Bioorg. Med. Chem. Lett.*, **11**, 1301–1305.
54. Malik, L., Kelly, N.M., Ma, J.-N., Currier, E.A., Burstein, E.S., and Olsson, R. (2009) *Bioorg. Med. Chem. Lett.*, **19**, 1729–1732.
55. Herpin, T.F., Van Kirk, K.G., Salvino, J.M., Yu, S.T., and Laboudinière, R.F. (2000) *J. Comb. Chem.*, **2**, 513–521.
56. Su, J., Tang, H., McKittrick, B.A., Burnett, D.A., Zhang, H., Smith-Torhan, A., Fawzi, A., and Lachowicz, J. (2006) *Bioorg. Med. Chem. Lett.*, **11**, 4548–4553.
57. Ziegert, R.E., Toräng, J., Knepper, K., and Bräse, S. (2005) *J. Comb. Chem.*, **7**, 147–169.
58. Banskota, A.H., Tezuka, Y., Prasain, J.K., Matsushige, K., Saiki, I., and

Kodota, S. (1998) *J. Nat. Prod.*, **61**, 896–900.
59. Bai, N., He, K., Roller, M., Zheng, B., Chen, X., Shao, Z., Peng, T., and Zheng, Q. (2008) *J. Agric. Food. Chem.*, **56**, 11668–11674.
60. Rodriguez, R., Moses, J.E., Adlington, R.M., and Baldwin, J.E. (2005) *Org. Biomol. Chem.*, **3**, 3488–3495.
61. Fukushima, T., Tanaka, M., and Gohara, M. (2000) Jpn. Kokai Tokkyo Koho, JP 000063376.
62. Mackenzie, A.R. and Monaghan, S.M. (1994) PTC Int. Appl., WO 9420491.
63. Gaoni, Y. and Mechoulam, R. (1964) *J. Am. Chem. Soc.*, **86**, 1646–1647.
64. Nicolaou, K.C., Pfefferkorn, J.A., and Cao, G.-Q. (2000) *Angew. Chem. Int. Ed.*, **39**, 734–739.
65. Nicolaou, K.C., Cao, G.-Q., and Pfefferkorn, J.A. (2000) *Angew. Chem. Int. Ed.*, **39**, 739–743.
66. Nicolaou, K.C., Pfefferkorn, J.A., Roecker, A.J., Cao, G.-Q., Barluenga, S., and Mitchell, H.J. (2000) *J. Am. Chem. Soc.*, **122**, 9939–9953.
67. Nicolaou, K.C., Pfefferkorn, J.A., Mitchell, H.J., Roecker, A.J., Barluenga, S., Cao, G.-Q., Affleck, R.L., and Lilling, J.E. (2000) *J. Am. Chem. Soc.*, **122**, 9954–9967.
68. Nicolaou, K.C., Pfefferkorn, J.A., Schuler, F., Roecker, A.J., Cao, G.-Q., and Casida, J.E. (2000) *Chem. Biol.*, **7**, 979–992.
69. Fang, N. and Casida, J.E. (1999) *J. Nat. Prod.*, **62**, 205–210.
70. Kapeller, D. and Bräse, S. (2011) *Synlett*, 161–164.
71. Poso, A. and Huffman, J.W. (2008) *Br. J. Pharmacol.*, **153**, 335–346.
72. Howlett, A.C., Barth, F., Bonner, T.I., Cabral, G., Casellas, P., Devane, W.A., Felder, C.C., Herkenham, M., Mackie, K., Martin, B.R., Mechoulam, R., and Pertwee, R.G. (2002) *Pharmacol. Rev.*, **54**, 161–202.
73. Sharir, H. and Abood, M.E.G. (2010) *Pharmacol. Ther.*, **126**, 301–313.
74. Thompson, L.A., and Ellman, J.A. (1994) *Tetrahedron Lett.*, **35**, 9333–9336.
75. Yin, H., Chu, A., Li, W., Wang, B., Shelton, F., Otero, F., Nguyen, D.G., Caldwell, J.S., and Chen, Y.A. (2009) *J. Biol. Chem.*, **284**, 12328–12338.
76. Oh, S., Jang, H.J., Ko, S.K., Ko, Y., and Park, S.B. (2010) *J. Comb. Chem.*, **12**, 548–558.
77. Hwang, J.Y., Choi, H.-S., Seo, J.-Y., La, H.-J., Yoo, S.-E., and Gong, Y.-D. (2006) *J. Comb. Chem.*, **8**, 897–906.
78. Gong, Y.-D., Cheon, H.-G., Cho, Y.-S., Seo, J.-S., Hwang, J.-Y., Park, J.-Y., and Yoo, S.-E. (2005) U.S. Pat. Appl. Publ., US 20050203145 A1 20050915.
79. Dahlén, S.-E. (2001) in *New Drugs for Asthma, Allergy and COPD*, Progress in Respiratory Research, Vol. **31** (eds T.T. Hansel and P.J. Barnes), Karger, Basel, pp. 115–120.
80. Patil, S.A., Patil, R., and Miller, D.D. (2009) *Curr. Med. Chem.*, **16**, 2531–2565.
81. Bräse, S., Gil, C., and Knepper, K. (2002) *Bioorg. Med. Chem.*, **10**, 2415–2437.
82. Gil, C. and Bräse, S. (2009) *J. Comb. Chem.*, **11**, 175–197.
83. Hart, F.D. and Boardman, P.L. (1963) *Br. Med. J.*, **2**, 965–970.
84. Gu, X.-H., Wan, X.-Z., and Jiang, B. (1999) *Bioorg. Med. Chem. Lett.*, **9**, 569–572.
85. Meseguer, B., Alonso-Diaz, D., Griebenow, N., Herget, T., and Waldmann, H. (1999) *Angew. Chem. Int. Ed.*, **38**, 2902–2906.
86. Leblanc, Y. *et al.* (1996) *Bioorg. Med. Chem. Lett.*, **6**, 731–736.
87. Haubmann, C., Hübner, H., and Gmeiner, P. (1999) *Bioorg. Med. Chem. Lett.*, **9**, 1969–1972.
88. Hübner, H., Kraxner, J., and Gmeiner, P. (2000) *J. Med. Chem.*, **43**, 4563–4569.
89. Kraxner, J., Arlt, M., and Gmeiner, P. (2000) *Synlett*, 125–127.
90. Cooper, L.C. *et al.* (2001) *Bioorg. Med. Chem. Lett.*, **11**, 1233–1236.
91. Kenner, G.W., McDermott, J.R., and Sheppard, R.C. (1971) *J. Chem. Soc. Chem. Commun.*, **12**, 636–637.
92. Willoughby, C.A. (2002) *Bioorg. Med. Chem. Lett.*, **12**, 93–96.
93. Smith, A.L., Stevenson, G.I., Lewis, S., Patel, S., and Castro, J.L. (2000) *Bioorg. Med. Chem. Lett.*, **10**, 2693–2696.

94. Torisu, K. *et al.* (2004) *Bioorg. Med. Chem.*, **12**, 5361–5378.
95. Torisu, K. *et al.* (2004) *Bioorg. Med. Chem. Lett.*, **14**, 4557–4562.
96. Barlos, K., Gatos, D., Kallitsis, J., Papaphotiu, G., Sotiriu, P., Wenging, Y., and Schäfer, W. (1989) *Tetrahedron Lett.*, **30**, 3943–3946.
97. Tambe, S.K., Dighe, N.S., Pattan, S.R., Kedar, M.S., and Musmade, D.S. (2010) *Pharmacol. online Newsl.*, **2**, 5–16.
98. Brune, K. (1997) *Acute Pain*, **1**, 33–40.
99. Kling, J. (1998) *Mod. Drug. Discovery*, **1**, 31–38.
100. Fong, T.M. and Heymsfield, S.B. (2009) *Int. J. Obesity*, **33**, 947–955.
101. Luong, B.T., Chong, B.S., and Lowder, D.M. (2000) *Ann. Pharmacother.*, **34**, 743–760.
102. Janssen, M., Van Leeuwen, M.H., Albrecht, H., and Dijkmans, B.A. (1988) *Clin. Rheumatol.*, **7**, 545–547.
103. Sehon, C., McClure, K., Hack, M., Morton, M., Gomez, L., Li, L., Barrett, T.D., Shankley, N., and Breitenbucher, J.G. (2006) *Bioorg. Med. Chem. Lett.*, **16**, 77–80.
104. McClure, K., Hack, M., Huang, L., Sehon, C., Morton, M., Li, L., Barrett, T.D., Shankley, N., and Breitenbucher, J.G. (2006) *Bioorg. Med. Chem. Lett.*, **16**, 72–76.
105. Spivey, A.C., Tseng, C.-C., Jones, T.C., Kohler, A.D., and Ellames, G.J. (2009) *Org. Lett.*, **11**, 4760–4763.
106. Staufferm, S.R. and Katzenellenbogen, J.A. (2000) *J. Comb. Chem.*, **2**, 318–329.
107. Dahlman-Wright, K. *et al.* (2006) *Pharmacol. Rev.*, **58**, 773–781.
108. Filardo, E.J. (2002) *J. Steroid Biochem. Mol.*, **80**, 231–238.
109. Pellegrino, G., Leonetti, F., Carotti, A., Nicolotti, O., Pisani, L., Stefanachi, A., and Catto, M. (2010) *Tetrahedron Lett.*, **51**, 1702–1705.
110. Morphy, R. (2010) *J. Med. Chem.*, **53**, 1413–1437.
111. Gambacorti-Passerini, C. (2008) *Lancet Oncol.*, **9**, 600.
112. Haque, T.S. (2002) *J. Med. Chem.*, **45**, 4669–6478.

Index

New Strategies in Chemical Synthesis and Catalysis, First Edition. Edited by Bruno Pignataro.

d